Plasmonics-Based Optical Sensors and Detectors

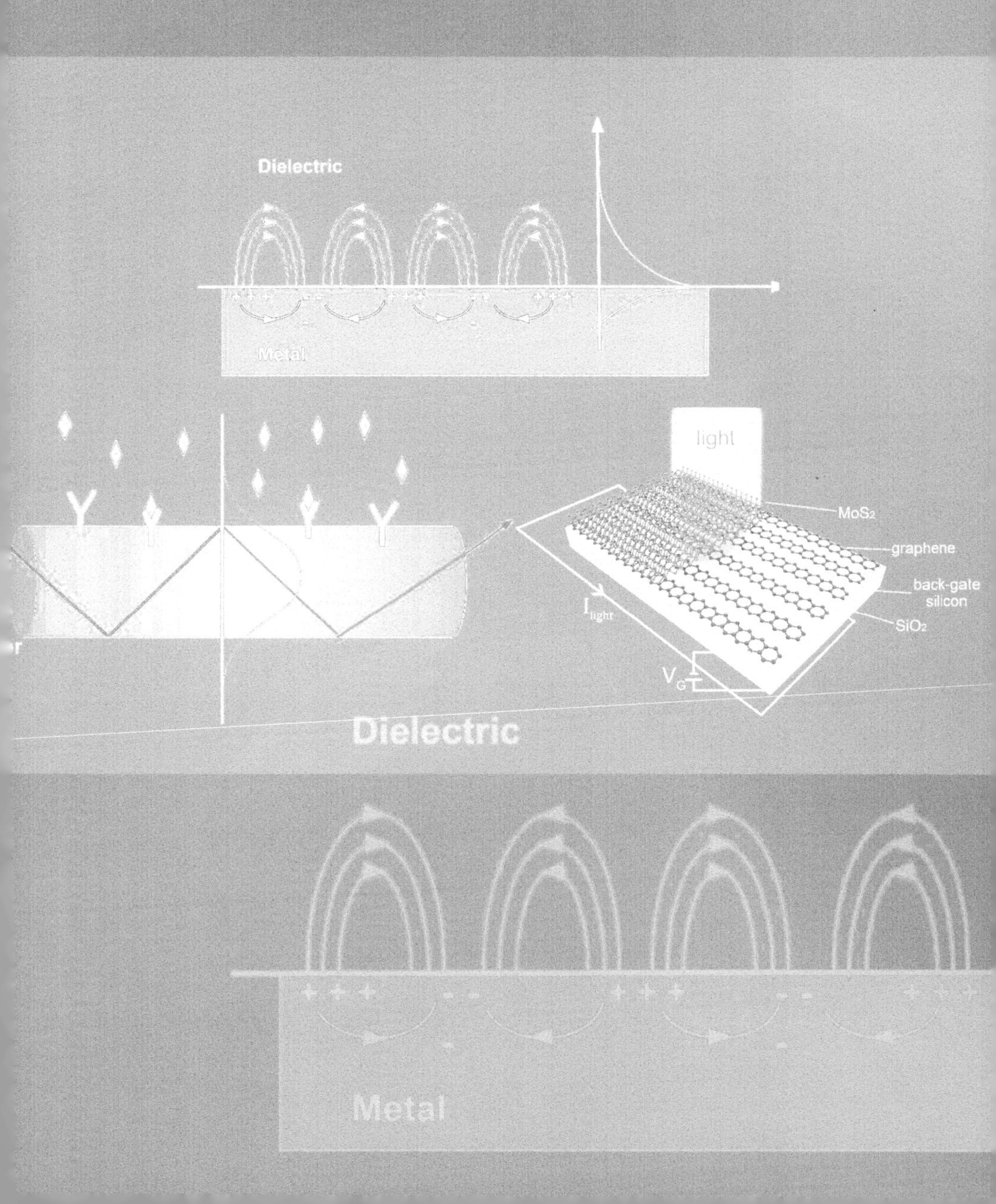

Plasmonics-Based Optical Sensors and Detectors

edited by

Banshi D. Gupta

Anuj K. Sharma

Jin Li

Published by

Jenny Stanford Publishing Pte. Ltd.
101 Thomson Road
#06-01, United Square
Singapore 307591

Email: editorial@jennystanford.com
Web: www.jennystanford.com

British Library Cataloguing-in-Publication Data
A catalogue record for this book is available from the British Library.

Plasmonics-Based Optical Sensors and Detectors

ISBN 978-981-4968-85-0 (Hardcover)
ISBN 978-1-003-43830-4 (eBook)

Contents

Preface

Plasmonics stems from the coherent oscillations of surface-charge density at metal–dielectric interface, leading to extremely strong light–matter interaction. In the past few decades, plasmonics has become one of the most sought-after field/technique to realize high-performance photonic devices. For this purpose, different new concepts, for example, exploration of different radiation frequency regions (visible, IR, THz, etc.), new materials including two-dimensional materials/heterostructures, and different types of substrates for the excitation of plasmons, have been introduced in plasmonics-based devices such as sensors and detectors.

This book is a collection of invited chapters from prominent researchers working on plasmonics-based devices in different parts of the world. The chapters focus on recent and advanced works on optical sensors and detectors based on various plasmonic techniques for their application in optoelectronic devices. Some of the chapters are comprehensive reviews of the related state of the art and will be useful for undergraduate and graduate students in addition to researchers working in various related fields. The book is special in the sense that it describes both plasmonics-based sensors as well as detectors for practical applications, a feature not available in any book so far. The chapters are organized in such a fashion that they are as much connected to one another as possible.

The book is divided into two sections: Section I discusses plasmonics-based optical sensors and Section II presents plasmonics-based optical detectors. Section I begins with the basic framework of plasmonics, starting from the historical context to the latest understanding of the underlying physics (Chapter 1). The chapter aims to develop a basic understanding of plasmonics so that it can be further utilized to translate the concept to the technological level. Although the chapter is quite comprehensive, there have been many other developments in the field, particularly in biological sensing, microscopy, subwavelength device manufacturing, and surface-enhanced Raman spectroscopy (SERS). Chapter 2 describes the concepts behind surface plasmon resonance (SPR) sensing with different performance indicators of the sensors in addition to some of the emerging configurations in SPR-based sensing. In Chapter 3, optical-fiber and integrated plasmonic sensors based on metal–insulator–metal waveguides are discussed. The device configurations discussed provide extraordinary sensing characteristics and the preference for utilizing plasmonic device configuration based on intended applications. Chapter 4 introduces the localized surface plasmon resonance (LSPR) phenomenon predominant in plasmonic metal nanoparticles, followed by its exploitation in realizing various sensing strategies including label-free and labeled biosensors. In addition, single-particle spectroscopy and multiplexed sensor applications of the plasmonic nanostructures are briefly covered. Among the different structures that enable the interaction of light with the surroundings, the metal-coated fiber gratings imprinted on the fiber core provide modal features that allow the generation of surface plasmons at any wavelength. Chapter 5 discusses several methods for fiber-grating fabrication and the different configurations used for plasmonic sensors. In Chapters 6 and 7, microstructured, non-microstructured, and nanostructured fiber-based plasmonic sensors are discussed. Chapter 8 presents the role of plasmonic-nanostructure optimization to maximize SERS signals through the example of plasmonic-nanosculptured thin films. The chapters present various modalities of SERS sensing and the utility of machine-learning techniques for SERS analysis and sensing of species in complex analyte matrix. Chapter 9 is devoted to spectroscopic detection and near-field optical imaging *via* plasmon-enhanced fluorescence and Raman scattering. Selected examples of single-molecule detection are presented to illustrate the capability of each technique. Chapter 10 discusses several sensing approaches utilizing polymer optical fibers (POFs) in different schemes to realize plasmonic platforms. More specifically, the POFs' capabilities are especially used to monitor or realize plasmonic sensors, which are useful in several bio/chemical application fields, and to exploit specific receptor layers. Characterization and optoelectronic properties of various 2D materials such as MXene, graphene, transition metal dichalcogenides, hexagonal boron nitride, and black phosphorus are presented in Chapter 11. The chapter covers the usage of heterostructures and metasurfaces for plasmonic sensing and biosensing applications. Chapter 12 covers the design considerations and instrumental optics of a typical SPR device along with the interrogation schemes, various analytical methods to find the suitable SPR dip position, several parameters affecting the signal noise, and other basic units of the SPR system covering the flow channels and their properties along with sensogram. The last chapter of Section I, Chapter 13, presents the bottom-up fabrication of plasmonic sensors and biosensors.

As far as the second section of the book is concerned, it is important to mention that light-absorption enhancement in various plasmonic devices is greatly achieved by engineered metallic nanostructures that play a crucial role in improving the responsivity of conventional photodetectors. In this context, Chapter 14 provides an insight into the application of grating nanostructures for photodetection by using the plasmonic phenomenon. Apart from metallic nanostructures, graphene plasmons can resonantly enhance incident-light absorption and have the potential of tunable spectral selectivity for mid-infrared detection. In the last chapter of this book, Chapter 15, graphene plasmonic mid-infrared photodetectors and tunable graphene plasmonic mechanism are discussed, which can be very useful for next-generation photodetectors.

In the end, we would like to thank all the contributors for sharing detailed insights and latest research results related to their corresponding topics. We would also like to thank the editorial team of Jenny Stanford Publishing for their support in making this project a reality.

Banshi Gupta
Anuj Sharma
Jin Li

March 2023

Section I

Plasmonics-Based Optical Sensors

Chapter 1

Fundamental Framework of Plasmonics

Rajneesh K. Verma[a,b] **and Banshi D. Gupta**[c]
[a]*Department of Physics, Central University of Rajasthan, Bandarsindri, Ajmer 305817, India*
[b]*Department of Physics, University of Allahabad, Prayagaraj 211002, India*
[c]*Physics Department, Indian Institute of Technology Delhi, New Delhi 110016, India*
rkverma@curaj.ac.in

Plasmonics is primarily about the use of light waves to manipulate the movement of electrons at the metal-dielectric interface. Its foundation is built up on the resonant interaction of light with the free electrons of noble metals. In this chapter, we present a basic framework of plasmonics, starting from the historical context to the latest understanding of the underlying physics. Though, there are three kinds of plasmons, namely, bulk, surface, and localized, but we shall mainly confine our discussion to the surface plasmons, briefly the localized surface plasmons. The surface plasmons are excited at the interface of the metal and dielectric using evanescent

Plasmonics-Based Optical Sensors and Detectors
Edited by Banshi D. Gupta, Anuj K. Sharma, and Jin Li

ISBN 978-981-4968-85-0 (Hardcover), 978-1-003-43830-4 (eBook)
www.jennystanford.com

wave and hence we shall discuss, evanescent wave in detail. After this, the condition of excitation of surface plasmons and how it is achieved shall be discussed. In the end, we shall present its potential applications including the sensing of various physical, chemical, and biological analytes. Besides, a brief discussion is also presented pertaining to the applications in other areas such as microscopy, subwavelength device manufacturing, surface-enhanced Raman scattering (SERS), data storage, and solar cells. Its application in sensing and photodetection will be dealt with in subsequent chapters in greater detail. The present chapter will pave the way for novice readers in understanding the chapters that follow.

1.1 Introduction

More than 1700 years ago, in the Roman Empire, plasmonic materials (metal nanoparticles) have been used by craftsmen to make colored glasses. These craftsmen did certainly not understand the physics behind them but they knew their peculiar applications. One of the oldest plasmonic glass materials, the **Lycurgus Cup** from the fourth century AD is still available for display in the British Museum. It is a Roman glass cage cup made of a dichroic glass that looks red when lit from behind and green when lit from the front side, as shown in Fig. 1.1. The world-acclaimed medieval glass windows in French and German gothic cathedrals also contain plasmonic metal nanoparticles; however, the basic understanding of the physics of nanoparticles dates back to Faraday (1857) and Mie (1908).

Figure 1.1 The Lycurgus cups in British Museum (Wikipedia: https://en.wikipedia.org/wiki/Lycurgus_Cup).

To understand the physics behind them, one must adhere to the interaction of electromagnetic waves with matter. The internal degrees of freedom of a medium get excited when an electromagnetic wave passes through it. The electric field associated with the wave gets coupled with the infrared active phonons of the dielectric medium. This coupling leads the dielectric constant of the medium to differ from its counterpart in free space (i.e., unity). Ipso facto, the phase velocity of the wave in the medium also differs from the velocity of light in free space. Besides, the propagation of the excitation of these degrees of freedom consists of the electric and magnetic fields veiled by the electric and magnetic polarization induced in the medium by the driving fields. Therefore, the electromagnetic wave in a medium becomes a complex entity to deal with. Further, the stored energy of the wave is shared between the driving fields and the excited internal degrees of freedom of the medium. In dielectric as well as in a medium, this composite wave is referred to as ***polariton***.

An electromagnetic wave propagating along the interface between the two media with exponentially decaying amplitude from the interface is termed as a surface electromagnetic wave. Its existence can be easily established by the use of Maxwell's equations with necessary boundary conditions. For a transverse magnetically polarized wave (TM or p-wave), the magnetic vector lies in the plane of the interface and perpendicular to the direction of the propagation. Their normal component of the electric field shows a discontinuity along the interface that leads to the creation of surface charge density. The necessity for the implication of Gauss's law ensures a charge distribution with infinite density wherever there is a discontinuity in the electric field. To understand it further, the field lines will either be starting or ending at the location of the discontinuity. We can think of the fact that the electric field magnitude will be different on one side of the discontinuity than the other. Hence, there will be more field lines on the stronger field side else the lines will be pointing in different directions on different sides if the amplitude is positive on one side and negative on the other. Electric field lines can start and terminate only at the source of the charges or at infinity. To understand physically, we can consider it in terms of the force on a point test charge. The force will be different on both sides of the discontinuity. Since the electric forces are exerted by charges, the only way to produce a jump in the force across the discontinuity is to have some charges there.

An electromagnetic wave coupled to these surface charges is known as the surface electromagnetic mode. The existence of the surface electromagnetic waves at the interface between two non-absorbing media are termed as Fano modes. However, if one of the media is absorbing the surface waves are termed as Zenneck modes. If the decay distances of the surface vibration amplitude and the wavelength are much larger than the interatomic distances between the atoms, then these surface vibrations may be treated as alternating dipole moments as shown in Fig. 1.2 [1]. This allows to a macroscopic electrodynamic treatment of the media described by a simple frequency-dependent dielectric constant, $\varepsilon(\omega)$.

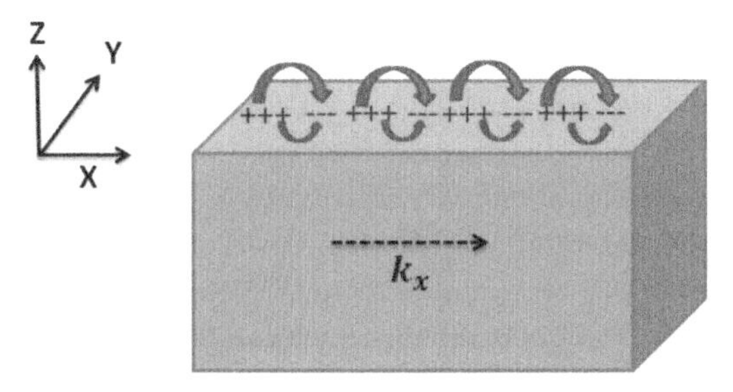

Figure 1.2 Oscillation of free electrons at a metal-dielectric interface in response to the electromagnetic wave.

For the frequencies close to ω_o, where $\varepsilon(\omega_o)$ shows a strong resonance character, the electromagnetic surface wave is described as an admixture of the electromagnetic field and the elementary excitation in the medium gives rise to the resonances in $\varepsilon(\omega_o)$. Such electromagnetic waves are conventionally termed as ***surface polaritons***. One should note that the term plasmonic resonances emerge from an electrodynamically motivated picture, whereas the word plasmon points out the particle nature from a quantum mechanical inclination of the concept. Therefore, to address the complete essence of the whole scenario, the term **plasmon–polariton** should be used to depict the mixed entity made from photon and plasmon. Though in general writing, we either write surface plasmons or surface polaritons to address the surface plasmon polaritons (SPPs). The materials having a negative real part of their dielectric constant ($\varepsilon_{r2} < 0$) are capable of causing a surface polariton at the interface of the second medium with a positive dielectric constant ($\varepsilon_{r1} > 0$), subject to the condition ($|\varepsilon_{r2}| > \varepsilon_{r1}$). The detailed mathematical treatment of the plasma frequency and surface plasmons follow henceforth.

The free electrons inside a metal can be treated as the electron liquid of high density of about 10^{23} cm^{-3} (Fig. 1.3a). From this assumption, we shall estimate the longitudinal density fluctuation and its frequency of oscillation. Suppose n_o is the equilibrium electron density, and owing to the charge neutrality, this will also be the equilibrium ion density. Somewhere the electrons get accumulated (instantaneously some electrons come together in some region), as shown in Fig. 1.3b. This will produce some electric field. This electric field will cause them to move away from there. Since the charge on the electrons is negative, therefore, they have the tendency to move out. When they move out, a larger movement takes place due to inertia. As a result, a positive charge is left over there. This positive charge attracts them back and the electrons start oscillating. The frequency of these oscillations is called plasma frequency.

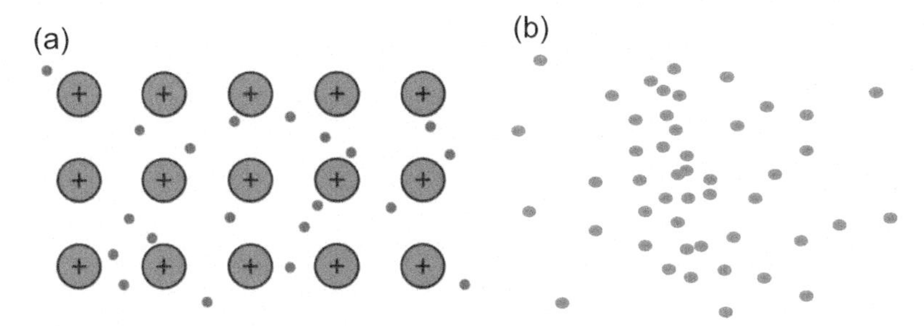

Figure 1.3 (a) Free electrons in a metal, and (b) instantaneous accumulation of free electrons.

Let the density of the electrons get modified as

$$n_e = n_o + n_1, \tag{1.1}$$

where n_1 is the modification in the density of the electrons.

From Poisson's equation, when there is a changing density of electrons, there will be electric field given as

$$\vec{E} = -\vec{\nabla}\phi. \tag{1.2}$$

Also

$$\varepsilon_o \nabla^2 \phi = e(n_e - n_o) \tag{1.3}$$

$$\nabla^2\phi = \frac{en_1}{\varepsilon_o}. \tag{1.4}$$

Hence, the force on an electron,

$$m\frac{d\vec{v}}{dt} = -e\vec{E} = e\,\vec{\nabla}\phi. \tag{1.5}$$

Assuming, $\frac{d\vec{v}}{dt} = \frac{\partial\vec{v}}{\partial t}$

$$\frac{\partial\vec{v}}{\partial t} = \frac{e}{m}\vec{\nabla}\phi. \tag{1.6}$$

Here $\vec{v}$ is the drift velocity of the electron and ϕ is the scalar potential. From the equation of continuity,

$$\frac{\partial n_e}{\partial t} + \vec{\nabla}.(n_e\vec{v}) = 0. \tag{1.7}$$

Since $n_e = n_o + n_1$ (n_1 is very small), we can write to a good approximation

$$\frac{\partial n_1}{\partial t} + \vec{\nabla}.(n_e\vec{v}) = 0. \tag{1.8}$$

Differentiating with respect to time once, we get

$$\frac{\partial^2 n_1}{\partial t^2} + n_o\left(\vec{\nabla}.\frac{\partial\vec{v}}{\partial t}\right) = 0. \tag{1.9}$$

Substituting $\frac{\partial\vec{v}}{\partial t}$ from Eq. (1.6) in Eq. (1.9), we get

$$\frac{\partial^2 n_1}{\partial t^2} + n_o\frac{e}{m}(\vec{\nabla}.\vec{\nabla}\phi) = 0 \tag{1.10}$$

or

$$\frac{\partial^2 n_1}{\partial t^2} + n_o\frac{e}{m}\nabla^2\phi = 0.$$

Putting the value of $\nabla^2\phi$ from Eq. (1.4) in the above equation, we get

$$\frac{\partial^2 n_1}{\partial t^2} + \left(\frac{n_o e^2}{m\varepsilon_o}\right)n_1 = 0. \tag{1.11}$$

The bracketed quantity has the dimension of frequency square, therefore, we define

$$\omega_p{}^2 = \frac{n_o e^2}{m\varepsilon_o}. \tag{1.12}$$

Hence, Eq. (1.11) becomes

$$\frac{\partial^2 n_1}{\partial t^2} + \omega_p{}^2 n_1 = 0. \tag{1.13}$$

The electron density oscillates with time and frequency (ω_p). The solution of Eq. (1.13) will be

$$n_1 = n_o e^{-i\omega_p t}$$

$$\omega_p = \sqrt{\frac{n_o e^2}{m\varepsilon_o}}. \tag{1.14}$$

This implies that the free electrons in the metal support space charge oscillations. These oscillations may carry energy and momentum just like electromagnetic waves do. We call these waves as plasma waves. Further, based on the idea that in a metal, valence electrons are free to move and are not associated with any nucleus, Drude derived a formula to estimate the dielectric constant of the metals given as [2]

$$\varepsilon(\omega) = 1 + \frac{\omega_p^2}{i\gamma\omega - \omega^2}. \tag{1.15}$$

Here, ω is the frequency of the incident radiation, ω_p is as defined in Eq. (1.14), and γ is the collision frequency of the electrons, i.e., the number of collisions suffered by the electrons with ions/electrons in one second. If we assume the electrons to behave completely like a free electron gas (hypothetical situation) then Eq. (1.15) will become

$$\varepsilon(\omega) = 1 - \frac{\omega_p^2}{\omega^2}. \tag{1.16}$$

Consequently, the refractive index N can be written as

$$N = n + ik = \sqrt{\varepsilon(\omega)} = \sqrt{1 - \frac{\omega_p^2}{\omega^2}}. \tag{1.17}$$

Consequences: For, $\omega < \omega_p$, $\varepsilon(\omega)$ will become negative, and hence N will become imaginary, i.e., the whole expression will be dominated by the imaginary part of the refractive index k. Therefore, electrons will follow the incident electric field and may absorb it too. On the other hand, for $\omega > \omega_p$, N will become a real number, i.e., the whole expression will be dominated by the real part of the refractive index, i.e., n. In this regime, the electrons cannot follow the incident wave and behave as mere spectators to it. This behavior of the dielectric constant with frequency can be seen in Fig. 1.4. Metals essentially behave as a dielectric in this regime. Thus, the plasma frequency plays a very crucial role in addressing the behavior of the metals in response to the incident electromagnetic wave.

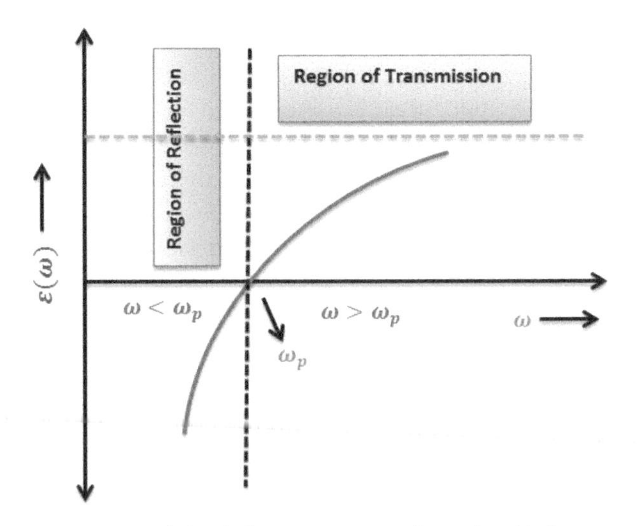

Figure 1.4 Variation of the dielectric constant of metals with frequency.

Also, the high reflectivity of the metals in visible frequency can be understood by the concept of the plasma frequency viz. reflectivity [3],

$$R = \frac{(n-1)^2 + k^2}{(n+1)^2 + k^2}. \tag{1.18}$$

For $\omega < \omega_p$, $n \cong 0$; hence, $R = \frac{1+k^2}{1+k^2} = 1$, i.e., 100% light reflection!!

Whereas for $\omega > \omega_p$, $N = \sqrt{1 - \frac{\omega_p^2}{\omega^2}} \cong 1$ for $(\omega \gg \omega_p)$; hence $k \cong 0$; therefore, $R \cong 0$, i.e., complete non-reflectivity!

As stated earlier, this was an approximate estimation of the metal properties based on the assumption that the electrons behave as free electron gas. In reality, the electrons do suffer collisions with ions and other electrons and hence possess a characteristic collision frequency and relaxation time (average time between two successive collisions ($\tau = \frac{1}{\gamma} \cong 10^{-14}$ s). Therefore, the dielectric constant of the metal will also have a real and imaginary part given as

Real: $$\varepsilon_1(\omega) = 1 - \frac{\tau^2\omega_p^2}{1+\tau^2\omega^2} \tag{1.19}$$

Imaginary: $$\varepsilon_2(\omega) = \frac{\tau\omega_p^2}{\omega(1+\tau^2\omega^2)}. \tag{1.20}$$

Further, the metals like Cu, Au, and Ag possess d-orbitals, which are more tightly bound as compared to the other orbitals. In these metals, energy is absorbed by the transition from the d band to the conduction band. These transitions are called band-to-band transitions, and for copper and gold, the energy required to encounter these transitions falls within the visible range of the electromagnetic spectrum [2]. As a consequence of which the Drude formula has been modified and the Drude-Lorentz model addresses these issues.

The electromagnetic field of an incoming light wave can induce polarization of conduction electrons. As a result, the electrons are displaced with respect to the much heavier ions. Depending upon the dimensionality of the metal body, one can distinguish between different modes of plasmonic oscillations. There exist volume plasmons in a large 3D metallic body (as discussed above), whereas the surface plasmons are of importance at the metal and dielectric interface. Zenneck in 1907 and Somerfield in 1909 had written about these waves in connection with radio broadcasting. They used the interface between the soil and the air. However, Maxwell's equations lead to a special surface-bound mode of plasmons at the interface between metal and dielectric media. We shall deduce the basic framework of the surface plasmon waves and also derive their dispersion relation using the concept of evanescent waves (2D waves), though the same can also be derived by using Maxwell's equations (first principle) with appropriate boundary conditions [3–6].

1.2 Evanescent Wave: A 2D Wave

The dimensionality of an optical wave is defined in terms of the number of real components of the wavevector $\vec{k}$. An optical wave having three real components in $\vec{k}$ is called a 3D wave. The propagation of light in free space is an example of a 3D wave. On the other hand, an optical wave having one or two real components of wave vector $\vec{k}$ is termed as 1D or 2D wave, respectively. The wave with all three components of the wavevector, imaginary is called a zero-dimensional (0D) wave. Now, let us consider the incidence of a plane electromagnetic wave at an angle α from medium 1 to the interface and gets refracted at an angle β in medium 2 as shown in Fig. 1.5. The electric field associated with the wave can be written as

$$\vec{E} = E_o \exp i\left(\vec{k}.\vec{r} - \omega t\right). \tag{1.21}$$

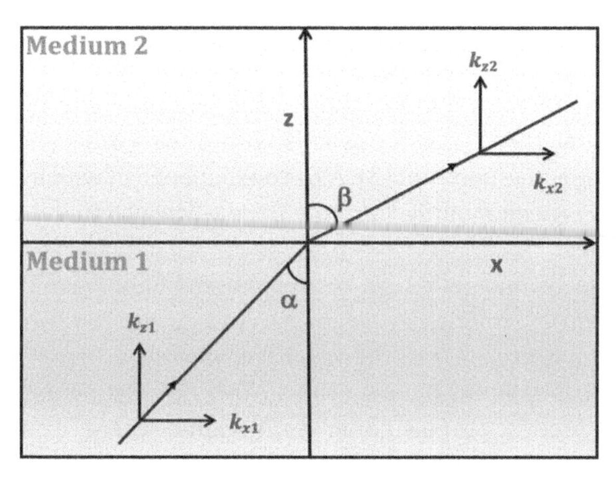

Figure 1.5 Interface between two dielectric media with indent and reflected light waves.

In terms of the components of the wave vector, it is written as

$$\vec{E} = E_o \exp i(k_x x + k_y x + k_z x - \omega t). \tag{1.22}$$

The wavevector magnitude in the medium is written as

$$k^2 = k_x{}^2 + k_y{}^2 + k_z{}^2 = k_o{}^2 n^2 = \frac{\omega^2 n^2}{c^2}. \tag{1.23}$$

Without any loss of generality, we can choose the direction of the light beam in such a way that $k_y = 0$. Now from Snell's law,

$$n_1 \sin \alpha = n_2 \sin \beta, \tag{1.24}$$

where n_1 and n_2 are the refractive indices of medium 1 and medium 2, respectively. Further, the components of the wave vectors in both the media can be written as

$$k_{x1} = k_1 \sin\alpha = k_o n_1 \sin\alpha, \tag{1.25}$$

$$k_{x2} = k_2 \sin\beta = k_o n_2 \sin\beta. \tag{1.26}$$

Therefore, using Eq. (1.23), we can conclude that

$$k_{x1} = k_{x2} = k_x \text{ (say)}. \tag{1.27}$$

To find k_{z2}, we shall make use of Eq. (1.23) and arrive at

$$k_{z2}^2 = \frac{\omega^2 n_2^2}{c^2} - k_{x2}^2 \tag{1.28}$$

$$= \frac{\omega^2 n_2^2}{c^2} - k_{x1}^2 \qquad \because k_{x1} = k_{x2}$$

$$= \frac{\omega^2 n_2^2}{c^2} - \frac{\omega^2 n_1^2}{c^2} \sin^2 \alpha$$

$$= \frac{\omega^2 n_1^2}{c^2} \left[\frac{n_2^2}{n_1^2} - \sin^2 \alpha \right]. \tag{1.29}$$

For a ray to suffer total internal reflection (TIR),

$$\sin \alpha > \sin \alpha_c \tag{1.30}$$

$$\sin\alpha > \frac{n_2}{n_1},$$

Here α_c is the critical angle at the interface of the two media. Therefore, in such a condition, the value of k_{z2} will be an imaginary number. Thus, the only wave that exists in the second medium has real values of k_{x2} and k_{y2} with an imaginary value of k_{z2}. This wave is called the evanescent wave and is essentially a 2D wave, if we say that

$$k_{z2} = i\kappa. \tag{1.31}$$

Then the electric field in the second medium will be

$$\vec{E_2} = E_o \exp i(k_x x + i\kappa z - \omega t) \tag{1.32}$$

$$\vec{E_2} = E_o \, e^{-\kappa z} \exp i\left(k_x x - \omega t\right). \tag{1.33}$$

Thus, the amplitude of this wave is decaying exponentially along the z direction. The field in medium 2 is termed as the evanescent field. It has a penetration depth of about **half a wavelength**. It is primarily responsible for the interface sensitivity toward change in the environment close to the interface.

In a similar line of calculation, we can deduce the complex reflection coefficient for a p-polarized electric field amid the interface between the two media (Fresnel's formula) [7],

$$r_p = \frac{E_r}{E_i} = |r_p| e^{i\varphi} = \left| \frac{\tan(\alpha - \beta)}{\tan(\alpha + \beta)} \right| e^{i\varphi} \tag{1.34}$$

here, φ is the phase of the reflected light relative to the incident light. For reflectance,

$$R_p = |r_p|^2 . \tag{1.35}$$

Now, let us examine the two cases:

Case 1: $\alpha + \beta = \frac{\pi}{2}$. In this case, the denominator will become infinity and hence, R_p will tend toward zero. This is a very familiar case of Brewster's angle that makes the reflectance of a p-polarized light to vanish completely.

Case 2: $\alpha - \beta = \frac{\pi}{2}$. In this case, R_p will tend toward infinite value, which means a finite value of E_r for an almost zero value of E_i. This circumstance is termed as 'resonance'. Using these formulations, we shall deduce the dispersion relation as under.

Let $\alpha - \beta = \frac{\pi}{2}$ or $\alpha = \beta + \frac{\pi}{2}$, it means

$$\cos \alpha = -\sin \beta. \tag{1.36}$$

Further, from Fig. 1.5,

$$\frac{k_{x1}}{k_{z1}} = \tan\alpha = \frac{\sin\alpha}{\cos\alpha} = \frac{\sin\alpha}{-\sin\beta} = -\frac{n_2}{n_1}. \tag{1.37}$$

Since $\quad k_{x1}^2 = k_1^2 - k_{z1}^2 \qquad (\because k_{y1}^2 = 0)$

$$k_x^2 = k_o^2 n_1^2 - k_{z1}^2. \qquad (\because k_{x1} = k_{x2} = k_x)$$

Substituting the value of k_{z1},

$$k_x^2 = k_o^2 n_1^2 - \frac{n_1^2}{n_2^2} k_x^2. \tag{1.38}$$

Since $n = \sqrt{\varepsilon}$, therefore,

$$k_x^{\ 2}\left(1+\frac{\varepsilon_1}{\varepsilon_2}\right)=\frac{\omega^2}{c^2}\varepsilon_1. \tag{1.39}$$

This gives the following dispersion relation

$$k_x = \frac{\omega}{c}\sqrt{\frac{\varepsilon_1\varepsilon_2}{\varepsilon_1+\varepsilon_2}}. \tag{1.40}$$

Alternatively, the dispersion relation can also be derived using Maxwell's equations as described below.

The four Maxwell's equations can be written in two independent sets having six finite field components viz. E_x, E_z, H_y (TM mode) and H_x, H_z, E_y (TE mode) with reference to the Cartesian coordinates as depicted in Fig. 1.5. These field components can be written as [8]

For, z < 0

$$H_y(z) = Ae^{ik_x x}e^{k_{z1}z} \tag{1.41}$$

$$E_x(z) = \frac{-iAk_1}{\omega\varepsilon_o\varepsilon_1}e^{ik_x x}e^{k_{z1}z} \tag{1.42}$$

$$E_z(z) = \frac{-iAk_x}{\omega\varepsilon_o\varepsilon_1}e^{ik_x x}e^{k_{z1}z}. \tag{1.43}$$

For, z > 0

$$H_y(z) = Be^{ik_x x}e^{-k_{z1}z} \tag{1.44}$$

$$E_x(z) = \frac{iBk_2}{\omega\varepsilon_o\varepsilon_2}e^{ik_x x}e^{-k_{z2}z} \tag{1.45}$$

$$E_z(z) = \frac{iBk_x}{\omega\varepsilon_o\varepsilon_2}e^{ik_x x}e^{-k_{z2}z}. \tag{1.46}$$

Here, A and B are constants, and the rest of the symbols have their usual meanings as described in the previous section. Now for transverse magnetic mode, applying the boundary conditions at the interface, i.e., $z = 0$; since the magnetic field has only one component, therefore, its continuity will be ensured only when the y component of the magnetic field is equal on both sides of the interface,

$$H_y\big|\, z>0 = H_y\big|\, z<0 \qquad \text{at } z = 0 \tag{1.47}$$

$$A\exp(ik_x x) = B\exp(ik_x x) \qquad \forall x. \tag{1.48}$$

The tangential component of the electric field must be continuous, i.e.,

$$E_x\big|\, z>0 = E_x\big|z<0 \qquad \text{at } z=0 \tag{1.49}$$

Applying boundary condition in Eqs. (1.42) and (1.43), we get

$$\frac{-Ak_1}{i\omega\varepsilon_o\varepsilon_1}\exp(ik_x x) = \frac{Bk_2}{i\omega\varepsilon_o\varepsilon_2}\exp(ik_x x)$$

or

$$\frac{-k_1}{\varepsilon_1} = \frac{k_2}{\varepsilon_2}. \tag{1.50}$$

Squaring on both sides and substituting, the values we get

$$\frac{k_x^{\ 2} - k_o^{\ 2}\varepsilon_1}{\varepsilon_1^{\ 2}} = \frac{k_x^{\ 2} - k_o^{\ 2}\varepsilon_2}{\varepsilon_2^{\ 2}} \tag{1.51}$$

$$k_x^{\ 2}\left\{\frac{\varepsilon_2^{\ 2} - \varepsilon_1^{\ 2}}{\varepsilon_2^{\ 2}\varepsilon_1^{\ 2}}\right\} = k_o^{\ 2}\left\{\frac{1}{\varepsilon_1} - \frac{1}{\varepsilon_2}\right\} \tag{1.52}$$

$$k_x^{\ 2} = k_o^{\ 2}\frac{\varepsilon_2\varepsilon_1}{\varepsilon_2 + \varepsilon_1}. \tag{1.53}$$

The propagation constant of the surface plasmon wave is

$$\beta = k_x = k_{sp} = \frac{\omega}{c}\left\{\frac{\varepsilon_1\varepsilon_2}{\varepsilon_1 + \varepsilon_2}\right\}^{1/2}. \tag{1.54}$$

This is the same as Eq. (1.40) derived above. Thus, we see that the same dispersion relation can be arrived by using two different approaches.

Also,

$$k_{z1} = \frac{\omega}{c}\sqrt{\frac{\varepsilon_1^{\ 2}}{\varepsilon_1 + \varepsilon_2}} \tag{1.55}$$

$$k_{z2} = \frac{\omega}{c}\sqrt{\frac{\varepsilon_2^{\ 2}}{\varepsilon_1 + \varepsilon_2}}. \tag{1.56}$$

In the case of a metal-dielectric interface, medium 2 will be metal, therefore,

$$\varepsilon_2(\omega) = 1 - \frac{\omega_p^2}{\omega^2}. \tag{1.57}$$

For k_x to be real, $\varepsilon_2(\omega)$ should be negative with $\varepsilon_2 > -\varepsilon_1$. Since ε_2 is negative, to make k_x a real number, the denominator should also be negative and hence ε_2 should be greater than $-\varepsilon_1$. This would make both k_{z1} and k_{z2} imaginary. This means that the wave will be decaying in both the media, i.e., above and below the interface. Thus, at the metal-dielectric interface, a 2D electromagnetic wave (surface plasmon wave) can exist that will propagate along the interface with an evanescent tail extending above and below the interface as portrayed in Fig. 1.6. In this figure, δ_d and δ_m represent the penetration depths of the surface plasmon wave in dielectric and metal, respectively, whereas δ_{SPP} represents the propagation length of the wave. To get an idea about the typical values, suppose we choose the incident light of wavelength $\lambda = 700$ nm falling on the gold-water interface. At this wavelength, $\varepsilon_{\text{gold}} \cong -16$ and $\varepsilon_{\text{water}} \cong 1.77$. With these parameters, the penetration depth in water comes out to be $\delta_{\text{water}} = \frac{1}{k_{z,\,\text{water}}} = 238$ nm and in gold, it is $\delta_{\text{gold}} = \frac{1}{k_{z,\,\text{gold}}} = 26$ nm.

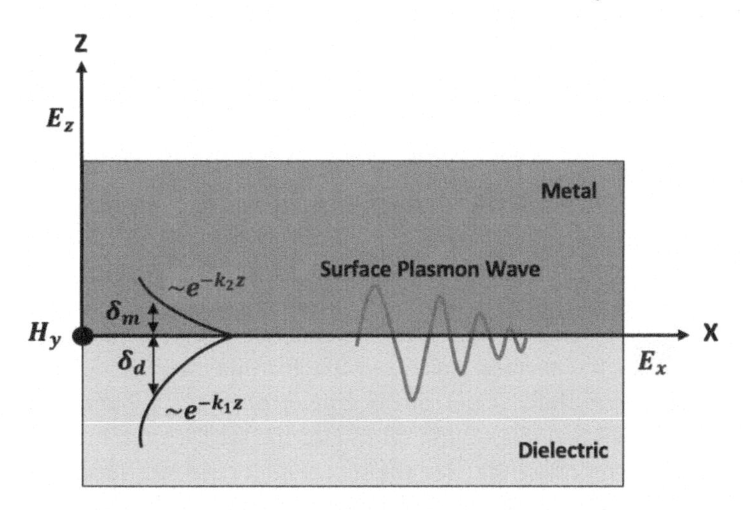

Figure 1.6 Depiction of typical length scales associated with SPPs.

1.3 Excitation of Surface Plasmon Polaritons

If we substitute the value of $\varepsilon_2(\omega)$ from Eq. (1.57) in Eq. (1.54), we will get the frequency-dependent wavevector (dispersion relation) for the surface plasmon wave, and the same has been plotted and shown as SP dispersion curve @ metal-dielectric interface-2 in Fig. 1.7. In the same figure, we have also plotted the dispersion curve for the light propagating in free space (Light line). Since these two curves never intersect each other, this geometry is inappropriate for the excitation of surface plasmon waves because normal light cannot provide the correct wavevector and angular frequency to excite surface plasmons.

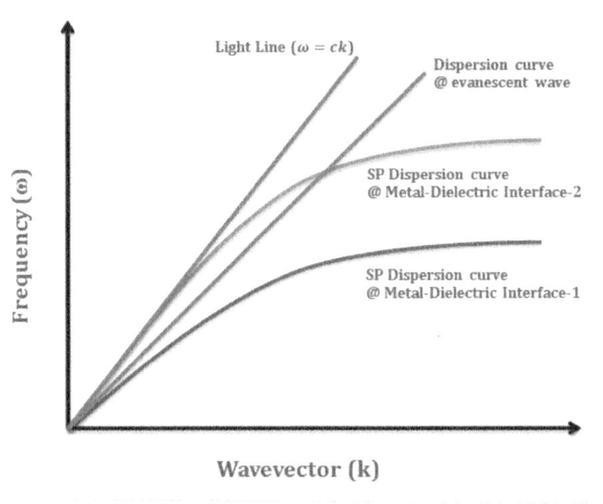

Figure 1.7 Dispersion curves for the surface plasmon waves and direct light.

This issue was resolved by introducing a second dielectric medium (ε_g) of refractive index, n_g, on the other side of the metal layer with a dielectric constant more than that of the other dielectric medium ($\varepsilon_g > \varepsilon_1$). The frequency-dependent wavevector (dispersion relation) for the surface plasmon wave for this interface, known as metal-dielectric interface-1, has been plotted in Fig. 1.7 in red color. Now, we have two dispersion relations; one for the metal-dielectric interface-1 and the other for the metal-dielectric interface-2. Now, the evanescent wave (propagating along the interface-1) dispersion curve may intersect the surface plasmon dispersion curve for the metal-dielectric interface-2 only. This means that the light from the second medium can excite the surface plasmons at a proper value of the incident angle (α), which would lead to an adequate change in the incident wavevector, $k_x = k_o n_g \sin \alpha$, to match the wavevector required for the surface plasmon wave. This entire arrangement of

attenuated TIR for the excitation of surface plasmons is known as the **Kretschmann-Reather configuration**. It is to be noted that by no means, we can excite the surface plasmons at the metal-dielectric interface-1 (that would require a large wavevector value to match with the surface plasmons). Mathematically, the surface plasmon excitation condition can be depicted as

$$k_o n_g \sin\alpha = \frac{\omega}{c}\sqrt{\frac{\varepsilon_{\text{metal}}\varepsilon_{\text{dielectric}}}{\varepsilon_{\text{metal}}+\varepsilon_{\text{dielectric}}}}. \quad (1.58)$$

From this equation, it can be inferred that for a minute change in the dielectric constant of the dielectric medium (medium in the vicinity of the metal layer, also referred to as sensing/surrounding medium), the wavevector matching condition will be satisfied at some other value of either angle or wavelength of the incident light. By suitably measuring the change in the angle/wavelength, we can calibrate the change in the refractive index of the surrounding medium. This is the essence of making a refractive index sensor utilizing resonance conditions.

1.4 SPP Excitation Configurations

Though discussed in brief about the surface plasmon excitations in the previous section, there are several mechanisms that are used to excite these SPPs [9]. Otto and Kretschmann proposed the optical excitations of the SPPs using a prism of high refractive index [1]. In the Kretschmann configuration, a prism of high refractive index is used as a substrate. The base of the prism is coated with a thin layer of metal. The dielectric medium is kept in contact with this metal layer. Light is allowed to fall on one face of the prism at an angle greater than the critical angle. The evanescent wave generated due to the TIR of the light wave may excite the surface plasmon wave at the metal-dielectric interface if the wavevector matching condition (Eq. (1.58)) is satisfied as shown in Fig. 1.8. The condition is satisfied at a particular angle of incidence known as the resonance angle (θ_{res}) and at this angle most of the energy of the incident light is transferred to the surface plasmon wave resulting in the minimum intensity of the reflected light as shown in Fig. 1.8.

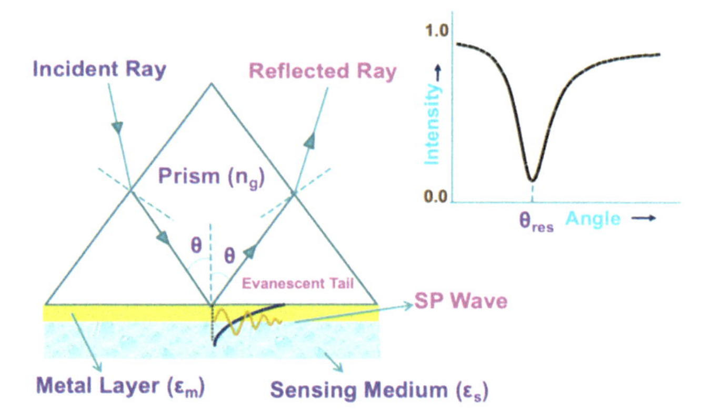

Figure 1.8 SPP excitation using Kretschmann configuration. The reflected light intensity suffers a minimum as a consequence of SPP excitations at the resonance angle.

A similar configuration can also be realized using multimode optical fiber. Here, the core of the fiber serves the purpose of the high refractive index prism. Since the light gets guided through the fiber due to TIR, the evanescent wave at the core-cladding interface may lead to the excitation of the surface plasmon modes at the metal-sensing region interface as shown in Fig. 1.9. The sensing region is a dielectric medium of refractive index lower than that of the core of the fiber.

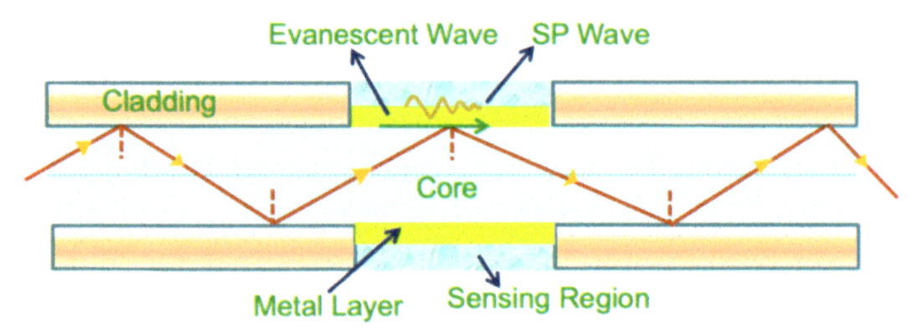

Figure 1.9 SPP excitation using multimode optical fiber.

The other mechanism includes the use of a metallic grating as shown in Fig. 1.10. In this case, when the light ray falls onto the surface of a periodic grating structure, different diffraction orders are produced in the reflected light. The wave vector parallel to the interface can be written as

$$k_{x\|} = k_x + m\frac{2\pi}{\Lambda}, \tag{1.59}$$

where m represents the order of diffraction and Λ is the grating period. Therefore, it is evident that the parallel component of the wave vector can be tuned and made equal to the surface plasmon wavevector by changing the incident angle with an appropriate grating period.

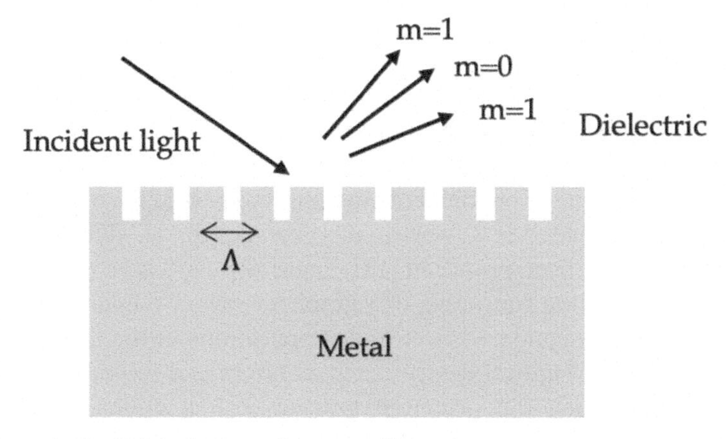

Figure 1.10 SPP excitation using a metallic grating.

1.5 Applications

Till now, we have discussed the basic framework to understand the fundamentals of SPPs and their possible excitation techniques. Now in the subsequent subsections, we shall discuss some of their particular applications.

1.5.1 Surface Plasmon Waveguides

The guidance of a light wave in a waveguide is always limited by the diffraction that leads to constrain on the size of the waveguide. Quinten et al. [10] coined a subwavelength optical waveguide by using a spherical chain of silver nanoparticles. In this chain, when the first particle was irradiated by a light field with polarization parallel and perpendicular to the axis of the nanoparticle chain, the efficient guidance of light was observed for a parallel polarized light field. The intensity distribution along the line was decreased continuously. On the other hand, for perpendicular polarization, the light guidance was damped strongly. This guidance of light was governed by the coupling of parallel light to the SPP oscillation energy. The propagation of light can also be seen on the metal nanowires deposited onto the transparent substrate [11].

Oulton et al. [12] proposed a different approach that integrates the dielectric waveguides with SPPs. The new approach used dielectric nanowires separated from a metal surface by a nanoscale dielectric gap. The capacitor-like energy storage and effective subwavelength transmission in the nonmetallic region were observed due to the coupling between the SPP mode and waveguide mode. In this scheme, a propagation distance as large as ~150 μm with strong mode confinement was observed by SPPs. In a recent study, long-range SPP waveguides composed of double-layer graphene have been proposed [13]. The study deals with the design of a novel grating consisting of a graphene-based cylindrical long-range SPP waveguide array. It finds applications in the fabrication of mid-infrared optical devices such as filters and refractive index sensors. Further, the proposed grating with a double layer of graphene arrays can potentially excite and manipulate the mid-infrared electromagnetic waves in future photonic integrated circuits. Another study presents a remotely excited dual cavity resonance scheme to excite second harmonic generation (SHG) in a CdSe nanobelt on an Au film hybrid waveguide system [14]. Since it is highly desirable to develop an on-chip nanoscale optical platform with SHG for optical sensing, with subwavelength coherent sources and quantum photonic devices, this study paves the way to realize this mechanism in device making. It is further reported that with the effective cooperation of hybrid plasmon modes and Fabry Perot modes, the two orders of magnitude enhancement of the conversion efficiency, is achieved as compared to the off-resonance scheme.

1.5.2 Surface Plasmon Polariton Sources

The miniaturization of laser elements is a crucial demand of the modern, technologically advanced era. Though we have microscopic lasers capable of reaching the diffraction limit based on photonic crystals, metal-clad cavities, nanowires, etc., but they have optical mode size and device dimension larger than half of the wavelength of the light used. Ipso facto, there remains the fundamental quest to achieve ultra-compact lasers capable enough to directly generate coherent optical fields beyond the diffraction limit. The SPP laser at the deep subwavelength scale has been presented by Oulton et al. [15]. This deep subwavelength SPP laser action at visible frequencies suggests new sources that may produce coherent light far below the diffraction limit. These plasmonic lasers are capable of generating strongly confined electromagnetic fields over a narrow range of wavelengths. They may be extremely useful for non-linear effects enhancement, chemical species detection, and on-chip sources of plasmons. The studies have revealed that a plasmonic laser can be fabricated by suitably placing a semiconductor gain layer near a metallic interface with a gap layer in between. Recently a planar metallic laser has been examined and the effect of gain on the lasing behavior has been reported [16]. The development of stable, low-noise laser sources is essential to explore the true potential of optical spectroscopy on the nanoscale. In a study, the development of a low noise super continuum (SC) source has been realized [17]. The propagation of SPP waves has been mapped on monocrystalline gold platelets in the wavelength region of 1.34–1.75 μm in a single measurement, characterizing experimentally the dispersion curve of the SPP in the NIR [17]. In 2019, Zhu et al. proposed a novel double-mode SPP nano-laser consisting of InGaAsP high-index dielectric, Ag metal, and T-shape MgF_2 low-index dielectric. The best performances of the proposed waveguide were obtained in the conditions of $\lambda = 675$ nm, $R = 70$ nm, $g = 3$ nm, and $w = 2$ nm, which correspond to the incident wavelength, radius of nanowires, gap, and width, respectively. It is shown that the proposed device has significant potential in ultrahigh-density plasmonic and photonic integrated circuits [18].

1.5.3 Near-Field Optics

Scanning near-field optical microscope (SNOM) is used to observe and control the light at nanometer-scale resolution. This limit is much smaller than the light resolution of the conventional optical microscope, which is half of the wavelength of the light used. The SNOM consists of a probe that serves as an emitter and the collector of the light. Usually, this probe is in the form of a small hole formed at the end of a metal-coated-tapered optical fiber. Its efficiency is limited by the small amount of light that can be coupled through the hole. An exquisite application of SPPs in SNOM demonstrated by Ketterson et al. [19, 20] used an electrochemically etched tungsten filament as a probe tip with a silver film sputtered onto a microscope cover glass. Kretschmann TIR configuration was used to excite SPPs at the metal film surface leading to the localization of the enhanced fields at the individual surface irregularities by the scattering of SPPs. The scattered SPPs produced conical radiation. Here, a small scatterer served as a probe to avoid the hitches associated with a hole, besides it uses SPPs to enhance the signal leading to an enhanced collection efficiency. If the metal nanostructures smaller than the wavelength of the light can be made then one can scan such nanostructures along the sample surface as a probe and image the sample with high-resolution geometry like other scanning probe microscopes. The strong localization of SPPs in nanostructures can couple to the propagating photons leading to slowing down of the light/reduction in the effective wavelength close to the nanostructure. The microscopy that involves a confined field invariably has a spatial resolution commensurable with the size of the field and the same as the size of the metal nanostructure.

1.5.4 Surface Enhanced Raman Scattering (SERS)

The SERS was first observed by Fleischman in 1974 [21] and later in 1977, it was discovered by Jeanmarie and Van Duyne [22] and Albrecht and Creighton [23]. Kneipp et al. [24] gave a simple theoretical understanding of the SERS phenomenon. It says that if E_o represents the magnitude of the incident field then the average magnitude of the field radiated by the metal particle will be $E_S = gE_o$, with g as a field enhancement averaged over the surface of the metal particle. The Raman scattered light produced by the molecules adsorbed at the surface of the metal particle will have a field strength of $E_R = \alpha_R g E_o$ with α_R as the component of Raman tensor. The metal particles can further scatter this Raman scattered field at Raman shifted wavelength with an enhancement factor, g'. Thus, the amplitude of this surface-enhanced field will be $E_{\text{SERS}} \propto \alpha_R g g' E_o$

and the average SERS intensity will be proportional to the square of the modulus of E_{SERS}, i.e., $I_{SERS} \propto |\alpha_R|^2 |gg'|^2 I_o$, where I_o is the incident intensity. It is obvious that for low-frequency modes when $g \approx g'$, the SERS field intensity will be raised by the fourth power of the field enhancement factor. On the other hand, for higher frequency modes the SERS intensity becomes a more complicated function of the SPP properties of the metal particle. One should understand that it is the scattering through the metal particle, which is responsible for the SERS. Conventionally, SERS experiments allow incident light to be focused on SERS hot spots on metal nanostructures, and the emitted light is detected from the same spot.

1.5.5 Data Storage

The application of LSPR is also being explored in data storage and recordings. The longitudinal LSPR exhibits an extraordinary wavelength and polarization sensitivity with narrow line width. The LSPR-mediated recording and readouts can be governed by photothermal reshaping and two-photon luminescence detection [25]. This technique is highly beneficial for high-density optical data storage with a disc capacity of 7.2 Tb and a recording speed of 1 Gbits^{-1}. Further, it is evident that artificial intelligence gets inspired by the functionalities and structure of the brain to solve complex tasks and allow learning. Despite this, hardware realization that simulates the synaptic activities realized with electrical devices still lags behind computer software implementation though it has improved significantly during the past decade. Recently, the capability to emulate synaptic functionalities by exploiting SPPs has been explored [26]. In this study, photochromic switching molecules (diarylethene) were deposited on a thin film of gold, to demonstrate the possibility of reliably controlling the electronic configuration of the molecules with UV and visible light. These reversible changes led to modulate the dielectric function of the photochromic film and thus enable the effective control of the SPP dispersion relation at the molecule/gold interface. The integration of such plasmonic devices in an artificial neural network can be deployed in plasmonic neuro-inspired circuits for optical computing and data transmission.

1.5.6 Solar Cells

The solid-state devices that are used to convert light energy directly into electricity are called solar cells or photoelectric/photovoltaic cells. Usually, semiconductors with a thickness many times the optical absorption length are used as photovoltaic absorbers for complete light absorption and photocarrier current collection. This conversion efficiency can be increased by decreasing the absorber thickness, which allows the use of low-dimensional structures like quantum dots. However, the efficiencies of nanometer-thickness cells are strongly limited by decreased absorption, carrier excitation, and photocurrent generation. The conventional light trapping schemes use wavelength scale surface texturing on the front panel of the solar cell to enhance light absorption. The size of these structures is large enough to be used with extremely thin films. The nanostructured thin films of metals are capable of confining and guiding the incident light into subwavelength thick absorber layer volumes, owing to the excitation of the SPPs. In this way, the photovoltaic conversion efficiency of the solar cells can be enhanced using SPPs. In a recent study, the potential application of LSPR on organic solar cells has been explored [27]. The researchers have designed a one-pot synthesis of Ag@SiO_2@AuNPs dual plasmons and observed an immense increase in light absorption over a wide range of wavelengths. It was revealed that Ag@SiO_2 plays a crucial role in light absorption enhancement near the ultraviolet band. To further enhance the LSPR effect and prevent recombination on the surface of AgNPs, the silica shell can also be used. Due to LSPR and decreased light reflectance, strong broad visible-light absorption was exhibited by AuNPs on the Ag@ SiO_2 shell. By utilizing Ag@SiO_2@AuNPs, the researchers were able to enhance the light absorption and photoinduced charge generation, thereby increasing the device PCE to 8.57% and J(SC) to 17.67 mA cm^{-2}, which can be attributed to the enhanced optical properties. Whereas, the devices without LSPR nanoparticles and Ag@SiO_2 LSPR, only showed PCEs of 7.36% and 8.18%, respectively [27]. The implementation of nanoparticles with the plasmonic effect is another way for photon and charge carrier management.

1.5.7 Sensors

A sensor is a device capable of transforming a given set of information to an analytically useful signal. In this category, the chemical sensors use the chemical information in the form of the concentration of a specific sample and transform it into other forms of measurable optical signal. Owing to the sensitivity, selectivity, reliability, and prototype user-friendly designs, these sensors are gaining overwhelming popularity in monitoring a number of chemical, biological, and environmental issues. Though a number of optical methods are available to make these sensing probes, nevertheless the probes based on SPPs are gaining special attention. The authors themselves have worked extensively in proposing theoretical models to enhance the performance of SPP-based sensors. Besides, an exhaustive work can be seen in the books published [6, 9]. Nevertheless, a number of pioneers have shown their interest in exploiting the surface plasmons to mold them in sensing applications [28–34]. Further details on different types of sensing configurations are the subject matter of the subsequent chapters.

1.6 Summary

To summarize, we have tried to present a basic framework of plasmonics, starting from the historical context to the latest understanding of the underlying physics. Though the chapter is a bit comprehensive in nature, there are many other developments that took place, thereafter, particularly in the field of biological sensing, microscopy, subwavelength device manufacturing, SERS, and many others. In subsequent chapters, many such applications will be dealt with in greater detail. We believe that after going through this chapter, the readers will develop a basic understanding of the concept to utilize it further in translating it to the proof of the concept/technological level.

Acknowledgments

One of the authors, RKV, gratefully acknowledges DST SERB for providing the Core Research Grant (CRG/2020/005593).

References

1. Reather, H. (1988), *Surface-Plasmons on Smooth and Rough Surfaces and on Gratings* (Springer, Berlin, Heidelberg).
2. Maier, S. A. (2007), *Plasmonics: Fundamentals and Applications* (Springer, New York).
3. Schasfoort, R. B. M. and Tudos, A. J. (2008), *Handbook of Surface Plasmon Resonance*, 1st Ed. (Royal Society of Chemistry Cambridge, United Kingdom).
4. Takahara, J. and Kbayashi, T. (2004), *Subwavelength Optics to Nano-Optics* (Optics and Photonics News, Optical Society of America, USA).
5. Shalaev, V. and Kawata, S. (2006), *Nanophotonics with Surface Plasmons*, 1st Ed. (Elsevier Science, USA).
6. Gupta, B. D., Shrivastav, A. M. and Usha, S. P. (2018), *Optical Sensors for Biomedical Diagnostics and Environmental Monitoring* (CRC Press, London).
7. Griffiths, D. J. (2015), *Introduction to Electrodynamics*, 4th Ed. (Pearson Pearson Education London, UK).
8. Ghatak, A. K. and Thyagarajan, K. (1998), *Introduction to Fiber Optics* (Cambridge University Press, UK).
9. Gupta, B. D., Srivastava, S. K. and Verma, R. (2015), *Fiber Optic Sensors based on Plasmonics* (World Scientific, Singapore).
10. Quinten M., Leitner A., Krenn, J. R. and Ausseneggn, F. R. (1998), Electromagnetic energy transport via linear chains of silver nanoparticles, *Opt. Lett.*, **23**, pp. 1331–1333.
11. Weeber, J. C., Dereux, A., Girard, C., Krenn, J. R. and Goudonnet, J. P. (1999), Plasmon polaritons of metallic nanowires for controlling submicron propagation of light, *Phys. Rev. B*, **60**, pp. 9061–9068.
12. Oulton, F. R., Sorger, V. J., Genov, D. A., Pile, D. F. P. and Zhang, X. (2008), A hybrid plasmonic waveguide for subwavelength confinement and long-range propagation, *Nat. Photonics*, **2**, pp. 496–500.
13. Liu, J. P., Wang, W. L., Xie, F., Zhang, X. M., Zhou, X., Yuan, Y. J., Wang, L. L. (2022), Excitation of surface plasmon polariton modes with double-layer gratings of graphene, *Nanomaterials,* **12** (7), pp. 1144.
14. Shi, J. J., He, X. B., Chen, W., Li, Y., Kang, M., Cai, Y. J. and Xu, H. X. (2022), Remote dual-cavity enhanced second harmonic generation in a hybrid plasmonic waveguide, *Nano Letters,* **22** (2), pp. 688–694.

15. Oultan, F. R., Sorger, V. J., Zentgraf, T. Ma, R. M. Gladden, C. Dai, L. Bartal, G. and Zhang, X. (2009), Plasmon lasers at deep subwavelength scale, *Nature*, **461**, pp. 629–632.

16. Aellen, M., Rossinelli, A. A., Keitel, R.C., Brechbuhler, R., Antolinez, F.V., Rodrigo, S.G., Cui, J. and Norris, D.J. (2022), Role of gain in Fabry Perot surface plasmon polariton lasers, *ACS Photonics,* **9** (2), pp. 630–640.

17. Kaltenecker, K. J., Rao, D. S. S., Rasmussen, M., Lassen, H. B., Kelleher, E. J. R., Krauss, E., Hecht, B., Mortensen, N. A., Gruner-Nielsen, L., Markos, C., Bang, O., Stenger, N. and Jepsen, P. U. (2021), Near-infrared nano spectroscopy using a low-noise supercontinuum source, *APL Photonics*, **6** (6), pp. 066106.

18. Zhu, J. and Xu, Z. J. (2019), Novel SPP nanolaser with two modes of electromagnetic for optoelectronic integration device, *Plasmonics,* **14** (5), pp. 1295–1302.

19. Kim, Y. K., Lundquist, P. M., Helfrich, J. A., Mikrut, J. M., Wong, G. K., Auvil, P. R. and Ketterson, J. B. (1995), Scanning plasmon optical microscope, *Appl. Phys. Lett.*, **66**, pp. 3407–3409.

20. Kryukov, A. E., Kim, Y. K. and Ketterson, J. B. (1997), Surface plasmon scanning near-field optical microscopy, *J. Appl. Phys.*, **82**, pp. 5411–5415. https://aip.scitation.org/doi/10.1063/1.125682

21. Fleischm, M., Hendra, P. J. and Mc Quilla, A. J. (1974), Raman spectra of pyridine adsorbed at a silver electrode, *Chem. Phys. Lett.*, **26**, pp. 163–166.

22. Jeanmaire, D. L. and Van Duyne, R. P. (1977), Surface Raman spectro-electrochemistry: Part I. Heterocyclic, aromatic, and aliphatic amines -adsorbed on the anodized silver electrode, *J. Electroanal. Chem.*, **84** (1), pp. 1–20.

23. Albrecht, M. G. and Creighton, J. A. (1977), Anomalously intense Raman spectra of pyridine at a silver electrode, *J. Am. Chem. Soc.*, **99**, pp. 5215–5217.

24. Kneipp, K., Moskovits, M. and Kneipp, H. (2006), *Surface Enhanced Raman Scattering: Physics and Applications* (Springer, Berlin).

25. Zhang, J. X., Yan, Y. G., Cao, X. L. and Zhang, L. D. (2006), Microarrays of silver nanowires embedded in anodic alumina membrane templates: Size dependence of polarization characteristics, *Appl. Opt.,* **45,** pp. 297–304.

26. Rhim, S. Y., Ligorio, G., Hermerschmidt, F., Hildebrandt, J., Patzel, M., Hecht, S. and List-Kratochvil, E. J. W. (2020), Using active surface plasmons in a multibit optical storage device to emulate long-term synaptic plasticity, *Physica Status Solidi A-Applications and Materials Science,* **217** (20), pp. 2000354.
27. Prasetio, A., Kim, S., Jahandar, M. and Lim, D. C. (2021), Single particle dual plasmonic effect for efficient organic solar cells, *Applied Nanoscience*, **847**. https://doi.org/10.1007/s13204-020-01641-2.
28. Hojjat, J. M., Djaileb, A., Ricard P., Lavallée, E., Cellier-Goetghebeur S., Parker, M. F., Coutu, J., Stuible, M., Gervais, C., Durocher, Y., Desautels, F., Cayer, M.P., De Grandmont, M. J., Rochette, S., Brouard, D., Trottier, S., Boudreau, D., Pelletier, J. N. and Masson, J. F. (2021), Cross-reactivity of antibodies from non-hospitalized COVID-19 positive individuals against the native, B.1.351, B.1.617.2, and P.1 SARS-CoV-2 spike proteins, *Scientific Reports*, **11**, pp. 21601.
29. Djaileb, A., Hojjat, J. M., Coutu, J., Ricard, P., Lamarre, M., Rochet, L. N. C., Cellier-Goetghebeur, S., Macauley, D., Charron, B., Lavallée, E., Thibault, V., Stevenson, K, Forest, S., Live, L.S., Abonnenc, N., Guedon, A., Quessy, P., Lemay, J. F., Farnos, O., Kamen, A., Stuible, M., Gervais, C., Durocher, Y., Cholette, F., Mesa, C., Kim, J., Cayer, M. P., De Grandmont, M. J., Brouard, D., Trottier, S., Boudreau, D., Pelletier, J. N. and Masson, J. F. (2021), Cross-validation of ELISA and a portable surface plasmon resonance instrument for IgG antibodies serology with SARS-CoV-2 positive individuals, *Analyst*, **146**, pp. 4905–4917.
30. Altug, H., Oh, S. H., Maier, S. A. and Homola, J. (2022), Advances and applications of nanophotonic biosensors, *Nat. Nanotechnol.,* **17**, pp. 5–16.
31. Špringer, T., Krejčík, Z. and Homola, J. (2021), Detecting attomolar concentrations of microRNA related to myelodysplastic syndromes in blood plasma using a novel sandwich assay with nanoparticle release, *Biosensors and Bioelectronics,* **194**, pp. 113613.
32. Zhou, H., Suo, L., Peng, Y. P., Yang, F., Ren, S., Chen, N. K. and Grattan, K. T. V. (2022), Micro-tapered fiber few-mode interferometers incorporated by molecule self-assembly fiber grating for temperature sensing applications, *Photonics*, **9** (2). doi:10.3390/photonics9020096.
33. Werner, J., Belz, M., Klein, K.-F., Sun, T. and Grattan, K. T. V. (2022), Design and comprehensive characterization of novel fiber-optic sensor systems using fast-response luminescence-based O_2 probes, *Measurement,* **189**, pp. 110670–110670. doi:10.1016/j.measurement.2021.110670.

34. Ahmad, H., Alias, M. A., Ismail, M. F., Ismail, N. N., Zaini, M. K. A., Lim, K. S. and Rahman, B. M. A. (2022), Strain sensor based on embedded fiber bragg grating in thermoplastic polyurethane using the 3D printing technology for improved sensitivity, *Photonic Sensors*, **12** (3). doi:10.1007/s13.

Chapter 2

Surface Plasmon Resonance–Based Optical Sensors: A Conceptual Framework

Vipul Rastogi
Center for Photonics and Quantum Communication Technology and Department of Physics, Indian Institute of Technology Roorkee, Roorkee-247667, Uttarakhand, India
vipul.rastogi@ph.iitr.ac.in

Surface plasmon resonance (SPR) has redefined sensing technology, particularly biosensing. Enhancement of optical fields at the metal-dielectric interface and their modification even with a minute change in the dielectric medium forms the basis of SPR-based sensing. With continued advancements in the field, the sensitivity values that were earlier considered unachievable are now becoming possible with SPR sensing.

2.1 Introduction

Sensing has become an important tool in various domains such as medical diagnostics, food safety, environmental monitoring, defense,

Plasmonics-Based Optical Sensors and Detectors
Edited by Banshi D. Gupta, Anuj K. Sharma, and Jin Li

ISBN 978-981-4968-85-0 (Hardcover), 978-1-003-43830-4 (eBook)
www.jennystanford.com

security, structural health monitoring, and advanced research [1]. A variety of available sensors work on various principles and among those optics-based sensors have the advantages of no electromagnetic interference, large bandwidth, multiplexing, compactness, and the ability to reach otherwise inaccessible places using fiber optics. Fiber optics-based probes can also help in sensing the samples from a safe distance in case of biologically or chemically unsafe analytes. Among optics-based sensors, SPR-based sensors have offered high sensitivity, high selectivity, and short response time and have become extremely useful, particularly in biosensing [2–16]. A signature of the SPR phenomenon, although it was not termed so initially, was observed in the year 1901 by Wood in the analysis of the diffraction pattern of light caused by metal grating [17]. The theory of such anomalous diffraction of metallic gratings was presented by Fano in 1941 [18] and in 1957, it was linked to plasma oscillation [19]. The phenomenon was experimentally demonstrated in 1968 by two independent works one by Kretschmann and Raether using a thin metal film deposited on the surface of a quartz prism [20] and another by Otto in frustrated total internal reflection (TIR) configuration [21]. Since then, there has been extensive study and tremendous growth in the use of the SPR phenomenon in the realization of various types of sensors in prism and grating configurations in a variety of application areas. There have been various theoretical models to study the phenomenon of SPR and the design of SPR-based sensors [22, 23]. The field has considerably evolved over the time. Various new configurations have emerged and several new performance indicators of SPR sensors have been devised. In this chapter, we would briefly describe the concepts behind SPR sensing, will define different performance indicators of the sensors, and will outline some of the emerging configurations in SPR-based sensing.

2.2 SPR-Based Sensing

SPR-based sensors have applications in wide-ranging areas including medicine, gas sensing, food safety, defense, water pollution, physical parameter sensing, chemical sensing, and biosensing [2–16, 24–28]. SPR sensors primarily work on the effect of change in the refractive index of the surroundings on the SPR response of the structure. The change in the refractive index can be caused by the change in the chemical composition or physical parameters of the surroundings. The design of SPR sensors, thus, requires prior knowledge of the relations between the refractive index and these parameters or corresponding experimental data. Some examples of the parameters that can cause a change in the refractive index of the medium are concentration, temperature, presence of foreign elements in a substance, and pH.

In general, when we shine light through a dielectric medium on a metal surface, we observe strong reflection. In certain circumstances, the reflection of a particular color for a specific angle of incidence is missing. Such a phenomenon is very sensitive to the properties of the dielectric medium and forms the basis of SPR sensors. Metal has a large number of free electrons, which are loosely bound to the metal surface. These electrons oscillate in a certain fashion with a certain frequency called plasma frequency and such oscillations are called plasmons. When monochromatic light is made to incident at the interface of a dielectric and metal, a special kind of mode having exponentially decaying fields in both the media can be excited. These modes travel along the dielectric-metal interface with certain propagation constants, and are called surface plasmon modes or surface plasmon polaritons (SPP). Formation of these modes demands that the polarization of light should be normal to the interface. In planar geometry, p-polarized or TM-polarized light can, therefore, excite SPP. In SPR-based sensors, the analyte works as the dielectric medium immediately next to the metal layer. If ϵ_m is the permittivity of the metal, ϵ_Q is the permittivity of the analyte, and λ_0 is the wavelength of light used to excite SPP then the propagation constant of the SPP, k_{SPP} is given by Eq. (2.1) [20, 22]

$$k_{\text{SPP}} = k_0 \sqrt{\frac{\epsilon_a \epsilon_m}{\epsilon_a + \epsilon_m}}, \tag{2.1}$$

where $k_0 = 2\pi/\lambda_0$ is the free-space propagation constant. Excitation of SPP requires matching the propagation constant of the surface plasmon mode (k_{SPP}) with that of the incident light (k). A typical variation of k and k_{SPP} with the frequency of light is shown in Fig. 2.1 [20]. It can be seen that excitation of SPP by the direct incidence of light on the interface is not possible due to mismatch in k and k_{SPP}. This can be achieved by either incidence of light at a specific angle through a high refractive index prism or by bridging the gap between the two wavevectors using another wavevector obtained by periodic refractive index variation, i.e., a grating.

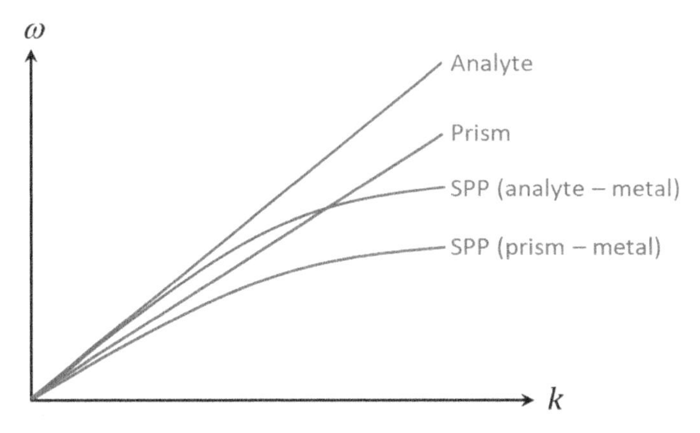

Figure 2.1 Dispersion curves of wavevectors of light in an infinitely extended analyte medium, prism, and SPPs at corresponding interfaces.

2.2.1 Prism-Based Techniques for Excitation of SPP

Two well-known configurations for excitation of SPP using a high-index prism are Kretschmann [20] and Otto [21] configurations as shown in Fig. 2.2. Excitation of SPP requires matching of the horizontal component of the wavevector of light in a prism, k_p with the propagation constant of the SPP, k_{SPP}. This phase matching between k_p and k_{SPP} is obtained for a particular wavelength called resonance wavelength at a specific angle of incidence θ_R called resonance angle and is given by Eq. (2.2). At resonance, the incident light is coupled to the SPP mode and the intensity of reflected light sharply decreases. Such a dip in the reflectance at the resonance angle is the signature of the excitation of SPP mode. Fulfillment of the resonance condition leads to the following relation between k_{SPP} and θ_R

$$k_{SPP} = k_p \sin\left[A - \sin^{-1}\left(\frac{k_0 \sin\theta_R}{k_p}\right)\right] \tag{2.2}$$

where $\theta = \theta_R$ is the resonance angle, A is the angle of prism used, and $k_p = k_0 n_p$ is the propagation constant of light in the prism of refractive index n_p.

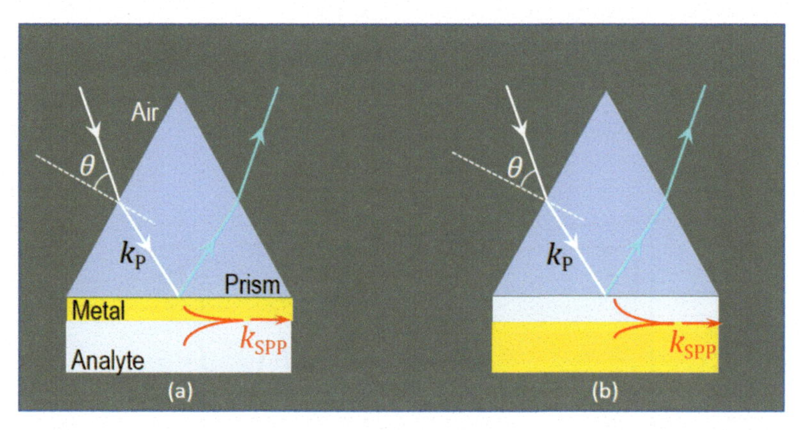

Figure 2.2 (a) Krescthmann configuration [20], (b) Otto configuration [21]. k_p is the propagation constant of light in prism and k_{SPP} is the propagation constant of surface plasmon mode at the metal-analyte interface.

The propagation constant of SPP depends on the refractive index of the dielectric medium in contact with the metal layer. As evident from Eq. (2.1), even a small change in the permittivity ϵ_a and hence in the refractive index of the dielectric medium alters the propagation constant k_{SPP} of the SPP mode. This results in a deviation from phase matching and in turn a shift in the value of the resonance angle in accordance with Eq. (2.2). Measurement of the shift in the resonance angle enables the estimation of change in the refractive index of the dielectric medium. This forms the basis of the SPR-based sensor and is depicted in Fig. 2.3. The figure shows that for a given dielectric medium (analyte) in immediate contact with the metal layer the reflectance drops down at the resonance angle and a rejection band is observed. Such a rejection curve is also termed as the SPR response curve. The strength of the rejection band is defined by the difference between the maximum and minimum values of reflectance and is represented by ΔR. The width of the rejection band is defined in terms of full width at half minimum (FWHM) and is represented by $\Delta\theta$. The two curves shown in the figure correspond to two different analytes having refractive indices n_{A1} and n_{A2}. Measurement of the resonance angles θ_{R1} and θ_{R2} corresponding to Analyte-1 and

Analyte-2, respectively, gives the estimate of refractive indices of the analytes and the shift $\Delta\theta_R = \theta_{R2} - \theta_{R1}$ is used to determine the change in the refractive index of an analyte or difference between the refractive indices of two analytes $\Delta n = n_{A2} - n_{A1}$. The sensitivity of such a refractive index sensor is represented in terms of deg/RIU and is given by Eq. (2.3),

$$S = \frac{\Delta\theta_R}{\Delta n}. \tag{2.3}$$

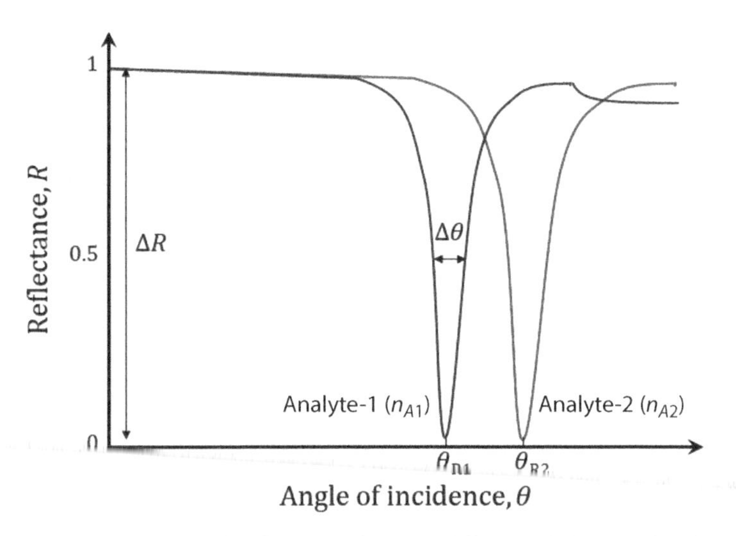

Figure 2.3 Variation of reflectance for two different dielectrics (analytes) in contact with the metal layer. θ_{R1} and θ_{R2} represent the resonance angles corresponding to analytes of refractive indices n_{A1} and n_{A2}, respectively. $\Delta\theta$ represents the FWHM and ΔR represents the rejection strength.

Sensitivity is just one performance indicator for an SPR-based sensor. It only tells how sensitive the shift in the resonance angle is to the change in the refractive index of the analyte. There are circumstances where the phase matching between the incident wave vector and the SPP wave vector is not too critical and the SPR response curve is not very sharp. In such a case, the SPR response curves for two analytes of close refractive indices will overlap and the resolution of the sensor or the detection accuracy would be adversely affected. A better performance indicator of the sensor would be the one that considers both the sensitivity and width of the SPR response curve. Various other parameters of the sensor are defined as detection accuracy (DA) given by Eq. (2.4) and quality factor Q given by Eq. (2.5),

$$DA = \frac{\Delta\theta_R}{\Delta\theta} \tag{2.4}$$

$$Q = \frac{S}{\Delta\theta}. \tag{2.5}$$

Another performance indicator, that considers the strength of the rejection band also, is the figure of merit (FOM) and is defined by Eq. (2.6),

$$\text{FOM} = \frac{\Delta\theta_R}{\Delta\theta}\Delta R. \tag{2.6}$$

As can be seen that in this method, angular interrogation of reflected light is carried out for incident light of a specific wavelength. A monochromatic source of light, usually a laser, is used in this technique. The detection accuracy of the sensor apart from the limit described by Eq. (2.4) also depends on the precision with which the angle of incidence can be measured.

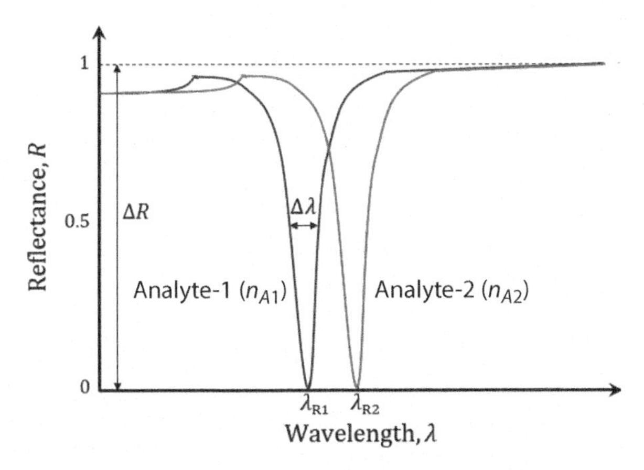

Figure 2.4 Variation of reflectance for two different dielectrics (analytes) in contact with the metal layer. λ_{R1} and λ_{R2} represent the resonance angles corresponding to analytes of refractive indices n_{A1} and n_{A2}, respectively. $\Delta\lambda$ represents the FWHM and ΔR represents the rejection strength.

Another way of using SPP for sensing purposes is to use a broadband light source and carry out wavelength interrogation of reflected light. At a particular angle of incidence, a specific wavelength component of broadband light satisfies the phase-matching condition. This wavelength is termed as resonance wavelength λ_R and is used up in exciting the SPP. The interrogation of the reflection spectrum reveals a rejection band around this wavelength with an FWHM of $\Delta\lambda$ and rejection strength ΔR as shown in Fig. 2.4. A change $\Delta n = n_{A2} - n_{A1}$ in the refractive index of the analyte results in shifting of this SPR response curve by an amount $\Delta\lambda_R = \lambda_{R2} - \lambda_{R1}$. Various performance indicators of the sensors while using the broadband light are defined as sensitivity S_λ, detection accuracy DA_λ, quality factor Q_λ, and FOM_λ. These performance indicators are given by Eqs. (2.7)–(2.10).

$$S_\lambda = \frac{\Delta\lambda_R}{\Delta n} \tag{2.7}$$

$$DA_\lambda = \frac{\Delta\lambda_R}{\Delta\lambda} \tag{2.8}$$

$$Q_\lambda = \frac{S}{\Delta\lambda} \tag{2.9}$$

$$\text{FOM}_\lambda = \frac{\Delta\lambda_R}{\Delta\lambda}\Delta R \tag{2.10}$$

This technique requires a collimated broadband source for input light and a spectrometer for wavelength interrogation of reflected light. Here also the detection accuracy or resolution of the sensor depends on the precision with which the wavelength can be measured, i.e., the resolution of the spectrometer in addition to the limit given by Eq. (2.8).

2.2.2 Grating-Based Techniques for Excitation of SPP

Another method of excitation of SPP is to use a metal grating as shown in Fig. 2.5. In this method, the mismatch between the wavevector of incident light and that of SPP is bridged by the wavevector provided by the grating [23]. If Λ is the periodicity of the grating, then the phase matching condition would be given by Eq. (2.11)

$$k_0 n_A \sin\theta_R + m\frac{2\pi}{\Lambda} = k_{\text{SPP}} \tag{2.11}$$

where n_A is the refractive index of the analyte, θ_R is the resonance angle, and $m = \pm 1, \pm 2, \ldots$ is the diffraction order. This method can also be used with a monochromatic source or broadband source and accordingly angular interrogation or wavelength interrogation technique can be applied. The performance indicators while using angular interrogation with a monochromatic source remain the same as defined by Eqs. (2.3)–(2.6) and those corresponding to wavelength interrogation while using a broadband source are defined by Eqs. (2.7)–(2.10). Grating-based SPR sensors are much more compact compared to prism-based sensors. However, the fabrication of grating-based sensors involves the inscription of gratings, which require special fabrication techniques such as photolithography or e-beam writing.

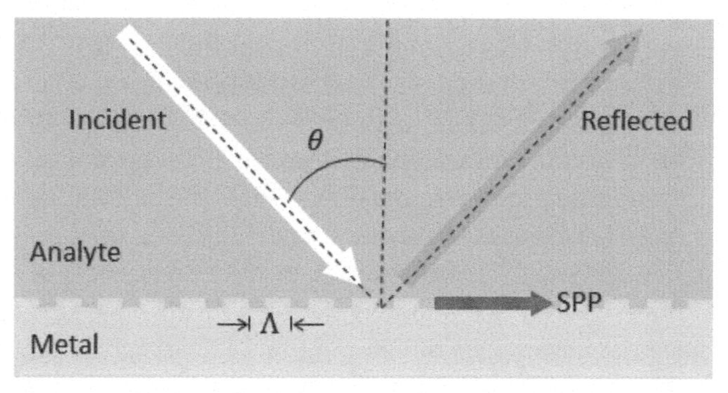

Figure 2.5 Schematic of excitation of SPP at the metal-analyte interface using a metallic grating of periodicity Λ.

2.2.3 Optical Fiber–Based Technique for Excitation of SPP

Optical fibers offer a simple and effective technique for the excitation of SPP. In this technique, a multimode fiber, usually a plastic clad silica (PCS) fiber, is used. Plastic cladding is removed from a short, typically a couple of centimeter length of the fiber and silica core is exposed. A thin layer of noble metal Au or Ag is deposited on the exposed core by the thermal evaporation method. The analyte is brought in contact with the deposited metal layer. SPP at the metal-analyte interface is excited using a broadband light source coupled to one end of the fiber. The spectrum of light from the other end of the fiber is interrogated by a spectrometer or an optical spectrum analyzer (OSA). The schematic of the technique is shown in Fig. 2.6. In this technique, the wavelength that satisfies the phase matching condition is used up in exciting the SPP. This resonance wavelength is missing in the output spectrum and this results in a rejection band around the resonance wavelength in the output spectrum. Any change in the refractive index of the analyte results in the shifting of the rejection band similar to the one shown in Fig. 2.4. The performance indicators of the sensors using the fiber-based technique are again defined by Eqs. (2.7)–(2.10). One of the major advantages of fiber-based SPR sensors is that they can reach otherwise inaccessible places, for example, inside the human body or the interior of an engine. Fiberoptic probes are also useful in diagnosing, for example, a virus sample, from a distance. Other advantages are the possibility of multiplexing several sensors and a large bandwidth. Microstructured optical fibers (MOFs) such as photonic crystal fibers and holey fibers also provide a good platform for SPR-based sensors. The holes running along the length of the MOFs provide channels for the analyte and the incorporation of metal at the appropriate position in the fiber excites the SPP. This modifies the propagation characteristics and transmittance of the fiber in accordance with the refractive index of the analyte [29]. Polymer optical fibers (POFs) are another platform for SPR-based sensors. POFs have the advantages of being more flexible, lightweight, low-cost, and compact. The POFs, however, cannot withstand very high temperatures but they provide an excellent platform for biosensors that operate at and around room temperature [30].

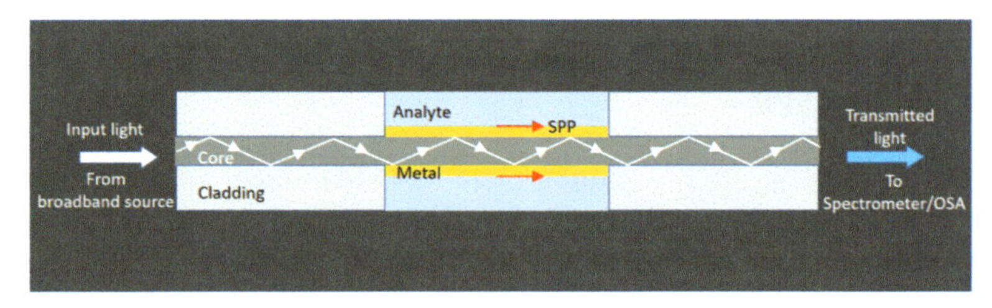

Figure 2.6 Schematic of excitation of SPP at the metal-analyte interface using an etched fiber with a metal coating on the exposed core.

2.3 Configurations to Improve Performance of SPR-Based Sensors

SPR-based sensors have witnessed tremendous growth over the years. There has been a continuous evolution in the designs of sensors for better performance. Various designs with a variety of materials have emerged for applications in specific areas, including biosensing, chemical detection, and gas sensing. Some of the configurations that have come up recently are bi-metallic configuration, palladium grating for hydrogen detection, and configurations involving 2D materials.

2.3.1 Bi-Metallic Configurations

In the configurations described in the previous section, there is a metal layer, and the analyte is in contact with it. Widely used metals in these sensors are noble metals such as gold and silver, owing to their stability and less susceptibility to being oxidized. Oxidation of the metal results in an insulating layer between the metal and the analyte. This degrades the performance of the sensor. It has been observed that in SPR-based sensors, the use of aluminum as a metal layer gives much sharper resonances than

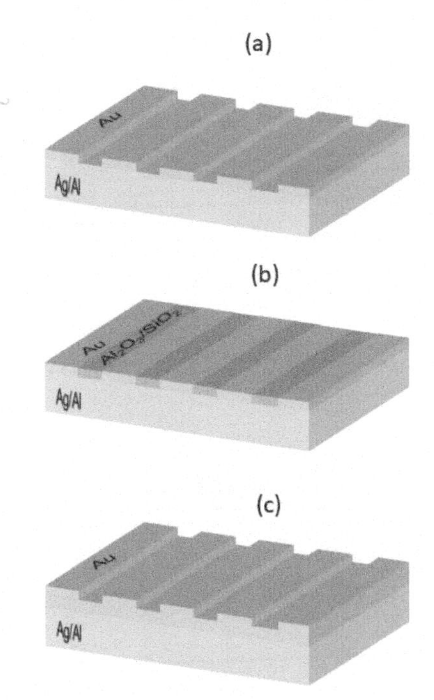

Figure 2.7 (a) Au grating on Ag or Al layer. (b) Air gaps of the grating are shown in (a) filled dielectric Al_2O_3 or SiO_2. (c) Ag or Al layer covered by a thin layer of Au and Au grating is formed on that [38, 39].

gold or silver and hence results in much better performance of the sensor. However, aluminum is oxidized easily and is less stable. In order to take advantage of better performance offered by aluminum while tackling the problem of its oxidation, the bi-metallic configuration of Al with Au or Ag have been studied [31–39]. Various bi-metallic configurations are shown in Fig. 2.7. Figure 2.7a shows a configuration where an Au grating is formed on Ag or Al substrate to improve detection accuracy. The issue of corrosion of exposed regions of Ag or Al can be addressed by either filling the air gaps with dielectrics such as SiO_2 or Al_2O_3 as shown in Fig. 2.7b or by introducing a thin layer of Au underneath the grating as shown in Fig. 2.7c. Such configurations significantly improve the performance of the sensor while preventing the oxidation of Al [37–39]. Analysis of such a configuration can be carried out by a multi-layer analysis model combined with rigorous coupled wave analysis (RCWA) [40–42].

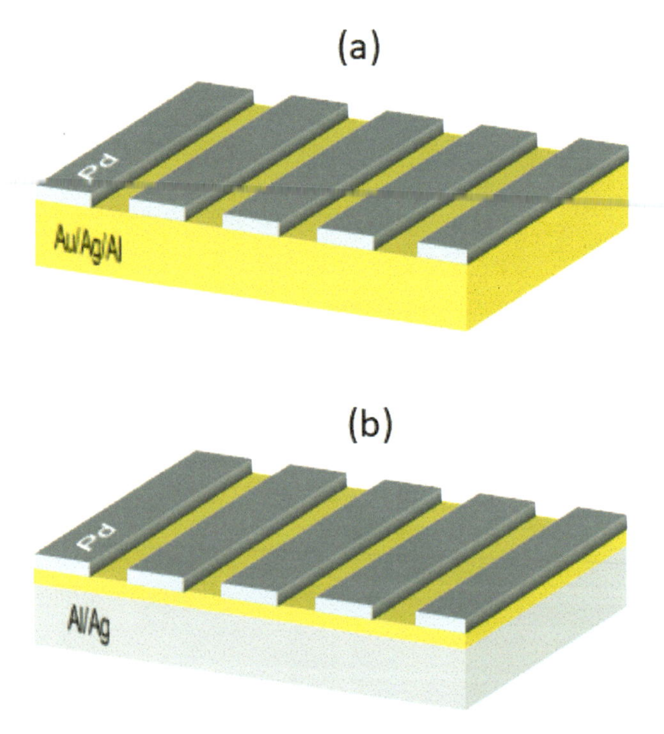

Figure 2.8 (a) Pd grating on Au, Ag, or Al layer. (b) Pd grating on the bi-metallic configuration of Al/Ag and Au in order to prevent oxidation of Al/Ag [45].

2.3.2 Pd in Bi-Metallic Configuration

Pd is a metal whose refractive index changes when it comes in contact with hydrogen. Pd absorbs hydrogen and forms a compound PdHx, where x is the atomic ratio of Pd and H [43]. The complex permittivity of the compound thus formed depends on the concentration of hydrogen [44]. Pd grating can be used in SPR sensors to estimate the concentration of hydrogen in the ambiance. A typical SPR-based hydrogen sensor can be realized by forming Pd grating on Au, Ag, or Al layer as shown in Fig. 2.8a. Performance of the sensor can be improved by using Al and the oxidation of Al can be prevented by introducing a thin Au layer between the Pd grating and Al layer as shown in Fig. 2.8b [45].

2.3.3 2D Materials

2D materials such as graphene or WSe_2 can also be used in SPR-based sensors to enhance sensitivity. A thin film of these materials can be used on a metal layer or on a multilayer structure [46–48]. The configuration of such a sensor with graphene on Al-Au bi-metallic formation is shown in Fig. 2.9a [46]. The performance of the sensor can further be enhanced by using graphene nano-ribbons as shown in Fig. 2.9b. There has been a fairly good amount of study on graphene-based SPR sensors for biosensing applications, the details of which can be found in Ref. [49].

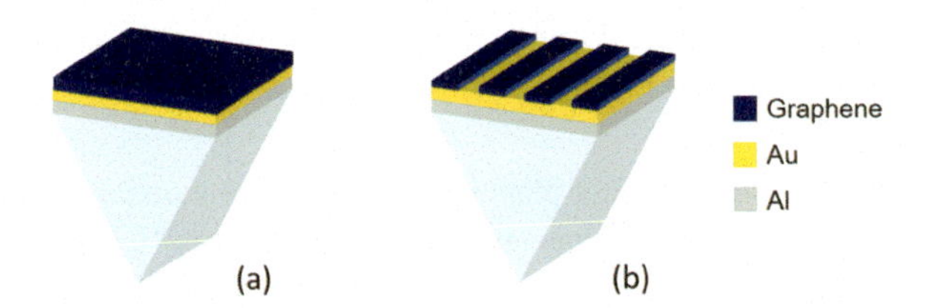

Figure 2.9 (a) Graphene on Al-Au bi-metallic layer. (b) Graphene nano-ribbons on Al-Au bi-metallic layer [48].

2.4 Conclusion

SPR has opened a new route in sensing technology with enhanced performance in terms of sensitivity and resolution. The SPR-based sensors can work in angular interrogation mode while using a monochromatic light source and in wavelength interrogation mode while using a broadband light source. The sensors work on the principle of shifting of SPR response curve by the change in the refractive index of the analyte. The prominent techniques used in the sensors are prism-based, grating-based, and fiber-based. New performance indicators have evolved to assess the performance of the sensors while considering all the parameters of the SPR response curve. These indicators are sensitivity, detection accuracy, quality factor, and FOM. The sensors can be designed by solving the multilayer structure and/or grating structure for the reflectance of light using standard techniques. The design parameters of the sensors can be optimized for obtaining maximum values of performance indicators. The optimization of the sensors has also led to some unconventional configurations such as the use of bimetallic layers and 2D materials. The technology is promising and rapidly evolving.

References

1. Grimes, C.A., Dickey, E.C. and Pishko, M.V. (2006). *Encyclopedia of sensors*, American Scientific Publishers.
2. Liedberg, B., Nylander, C. and lundstorm, I. (1983) Surafce plasmon resonance for gas detection and biosensing, *Sens. Actuators,* **4**, pp. 299–304.
3. Homola, J. (2006). *Surface Plasmon Resonance Based Sensors*, Springer.
4. Sharma, A.K., Jha, R. and Gupta, B.D. (2007). Fiber-optic sensors based on surface plasmon resonance: A comprehensive review, *IEEE Sens. J.*, **7**, pp. 1118–1129.
5. Homola, J. (2008). Surface plasmon resonance sensors for detection of chemical and biological species, *Chem. Rev.,* **108**, pp. 462–493.
6. Fan, X., White, I.M., Shopova, S.I., Zhu, H., Suter, J.D. and Sun, Y. (2008). Sensitive optical biosensors for unlabeled targets: A review, *Anal. Chim. Acta.,* **620**, pp. 8–26.
7. Abdulhalim, I., Zourob, M. and Lakhtakia, A. (2008). Surface plasmon resonance for biosensing: A mini-review, *Electromagnetics,* **28**, pp. 214–242.
8. Gupta B.D. and Verma, R.K. (2009). Surface plasmon resonance-based fiber-optic sensors: Principle, probe design, and some applications, *J. Sensors,* **2009**, pp. 979761: 1–12.

9. Wijaya, E., Lenaerts, C., Maricot, S., Hastanin, J., Habraken, S., Vilcot, J., Boukherroub, R. and Szunerits, S. (2011). Surface plasmon resonance-based biosensors: From the development of different SPR structures to novel surface functionalization strategies, *Curr. Opin. Solid State Mater. Sci.,* **15**, pp. 208–224.
10. Roh, S., Chung, T. and Lee, B. (2011). Overview of the characteristics of micro- and nano-structured surface plasmon resonance sensors, *Sensors,* **11**, pp. 1565–1588.
11. Helmerhorst, E., Chandler, D.J., Nussio, M. and Mamotte, C. D. (2012). Real-time and label-free bio-sensing of molecular interactions by surface plasmon resonance: A laboratory medicine perspective, *Clin. Biochem. Rev.,* **33**, pp. 161–173.
12. Tahmasebpour, M., Bahrami, M. and Asgari, A. (2014). Design study of nanograting based surface plasmon resonance biosensor in the near-infrared wavelength, *Appl. Opt.,* **53**, pp. 1449–1458.
13. Li, M., Cushing, S.K. and Wu, N. (2015). Plasmon-enhanced optical sensors: A review, *Analyst,* **140**, pp. 386–406.
14. Gupta, B.D., Shrivastav, A.M. and Usha, S.P. (2016). Surface plasmon resonance-based fiber optic sensors utilizing molecular imprinting, *Sensors,* **16**, pp. 1381.
15. Teotia, P.K. and Kaler, R.S. (2018). 1D grating based SPR biosensor for the detection of lung cancer biomarkers using Vroman effect, *Opt. Commun.,* **406**, pp. 188–191.
16. Shrivastav, A.M., Cvelbar, U. and Abdulhalim, I. (2021). A comprehensive review on plasmonic-based biosensors used in viral diagnostics, *Commun. Biology,* **4**, pp. 1–12.
17. Wood, W.R. (1902). On a remarkable case of uneven distribution of light in a diffraction grating spectrum, *in Proc. of the Physical Society of London,* **18**, pp. 269–275.
18. Fano, U. (1941). The theory of anomalous diffraction gratings and of quasi-stationary waves on metallic surfaces (Sommerfeld's waves), *J. Opt. Soc. A,* **31**, pp. 213–222.
19. Ritchie, R.H. (1957). Plasma losses by fast electrons in thin films, *Phys. Rev.,* **106**, pp. 874–881.
20. Kretschmann, E. and Raether, H. (1968). Radiative decay of non-radiative surface plasmons excited by light, *Zeitschrift für Naturforschung A,* **23**, pp. 2135–2136.
21. Otto, A. (1968). Excitation of nonradiative surface plasma waves in silver by the method of frustrated total reflection, *Zeitschrift für Physik,* **216**, pp. 398–410.

22. Maier, S.A. (2007). *Plasmonics: Fundamentals and Applications*, Springer.

23. Raether, H. (1988). Surface-plasmons on smooth and rough surfaces and on gratings, *Springer Tracts in Modern Physics*, **111**, pp. 1–133.

24. Homola, J., Dostáleka, J., Chen, S., Rasooly, A., Jiang, S. and Yee, S.S. (2002). Spectral surface plasmon resonance biosensor for detection of staphylococcal enterotoxin B in milk, *Int. J. Food Microbiol.*, **75**, pp. 61–69.

25. Ibrahim, J., Masri, M.A., Verrier, I., Kampfe, T., Veillas, C., Celle, F., Cioulachtjian, S., Lefèvre, F. and Jourlin, Y. (2019). Surface plasmon resonance-based temperature sensors in liquid environment, *Sensors*, **19**, pp. 3354.

26. Perrotton, C., Westerwaal, R.J., Javahiraly, N., Slaman, M., Schreuders, H., Dam, B. and Meyrueis, P. (2013). A reliable, sensitive and fast optical fiber hydrogen sensor based on surface plasmon resonance, *Opt. Express*, **21**, pp. 382–390.

27. Piliarik, M., Parova, L. and Homola, J. (2009). High-throughput SPR sensor for food safety, *Biosens. Bioelectron.*, **24**, pp. 1399–1404.

28. Takeuchi, H., Thongborisute, J., Matsui, Y., Sugihara, H., Yamamoto, H. and Kawashima, Y. (2005). Novel mucoadhesion tests for polymers and polymer-coated particles to 117 design optimal mucoadhesive drug delivery systems, *Adv. Drug Deliv. Rev.*, **57**, pp. 1583–1594.

29. Liu, C., Lü, J., Liu, W., Wang, F. and Chu, P.K. (2021). Overview of refractive index sensors comprising photonic crystal fibers based on the surface plasmon resonance effect, *Chin. Opt. Lett.*, **19**, p. 102202.

30. Liu, L., Deng, S., Zheng, J., Yuan, L., Deng, H. and Teng, C. (2021). An enhanced plastic optical fiber-based surface plasmon resonance sensor with a double-sided polished structure, *Sensors*, **21**, p. 1516.

31. Zynio, S.A., Samoylov, A.V., Surovtseva, E.R., Mirsky, V.M. and Shirshov, Y.M. (2002). Bimetallic layers increase sensitivity of affinity sensors based on surface plasmon resonance, *Sensors*, **2**, pp. 62–70.

32. Yuan, X.C., Ong, B.H., Tan, Y.G., Zhang, D.W., Irawan, R. and Tjin, S.C. (2006). Sensitivity–stability-optimized surface plasmon resonance sensing with double metal layers, *J. Opt. A Pure Appl. Opt.*, **8**, pp. 959–963.

33. Chen, Y., Zheng, R.; Zhang, D., Lu, Y., Wang, P., Ming, H., Luo, Z. and Kan, Q. (2017). Bimetallic chips for a surface plasmon resonance instrument, *Appl. Opt.*, **50**, pp. 387–391.

34. Huang, Y., Xia, L., Wei, W., Chuang, C.J. and Du, C. (2014). Theoretical investigation of voltage sensitivity enhancement for surface plasmon resonance based optical fiber sensor with a bimetallic layer, *Opt. Commun.*, **333**, pp. 146–150.

35. Chen, S. and Lin, C. (2016). High-performance bimetallic film surface plasmon resonance sensor based on film thickness optimization, *Optik*, **127**, pp. 7514–7519.

36. Sharma, A.K. and Gupta, B.D. (2007). On the performance of different bimetallic combinations in surface plasmon resonance based fiber optic sensors, *J. Appl. Phys.*, **101**, p. 093111.

37. Bijalwan, A. and Rastogi, V. (2017). Sensitivity enhancement of a conventional gold grating assisted surface plasmon resonance sensor by using a bimetallic configuration, *Appl. Opt.*, **56**, pp. 9606–9612.

38. Bijalwan, A. and Rastogi, V. (2018). Design analysis of refractive index sensor with high quality factor using Au-Al_2O_3 grating on aluminum, *Plasmonics*, **13**, pp. 1995–2000.

39. Bijalwan, A. and Rastogi, V. (2018). Gold–aluminum-based surface plasmon resonance sensor with a high-quality factor and figure of merit for the detection of hemoglobin, *Appl. Opt.*, **57**, pp. 9230–9237.

40. Moharam, M.G., Grann, E.B., Pommet, D.A. and Gaylord, T.K. (1995). Formulation for stable and efficient implementation of the rigorous coupled- wave analysis of binary gratings, *J. Opt. Soc. Am. A*, **12**, pp. 1068–1076.

41. Moharam, M.G., Pommet, D.A., Grann, E.B. and Gaylord, T.K. (1995). Stable implementation of the rigorous coupled-wave analysis for surface-relief gratings: Enhanced transmittance matrix approach, *J. Opt. Soc. Am. A*, **12**, pp. 1077–1086.

42. Lee, W. and Degertekin, F.L. (2004). Rigorous coupled-wave analysis of multilayered grating structures, *J. Lightwave Technol.*, **22**, pp. 2359–2363.

43. Bevenot, X., Trouillet, A., Veillas, C., Gagnaire, H. and Clement, M. (2002). Surface plasmon resonance hydrogen sensor using an optical fibre, *Meas. Sci. Technol.*, **13**, pp. 118–124.

44. Tobiška, P., Hugon, O., Trouillet, A. and Gagnaire, H. (2011). An integrated optic hydrogen sensor based on SPR on palladium, *Sens. Actuators B: Chem.*, **74**, pp. 168–172.

45. Bijalwan, A. and Rastogi, V. (2019). Design and simulation of a palladium-aluminum nanostructure based hydrogen sensor with improved figure of merit, *IEEE Sensor J.*, **19**, pp. 6112–6118.

46. Zeng, S., Sreekanth, K.V., Shang, J., Yu, T., Chen, C.K., Yin, F., Baillargeat, D., Coquet, P., Ho, H.P., Kabashin, A.V. and Yong K.T. (2015). Graphene–gold metasurface architectures for ultrasensitive plasmonic biosensing, *Adv. Mater,* **27**, pp. 6163–6169.

47. Zhao, X., Huang, T., Ping, P.S., Wu, X., Huang, P., Pan, J., Wu, Y. and Cheng, Z. (2018). Sensitivity enhancement in surface plasmon resonance biochemical sensor based on transition metal dichalcogenides/ graphene heterostructure, *Sensors,* **18**, pp. 2056.

48. Bijalwan, A, Singh, B.K. and Rastogi, V. (2020). Surface plasmon resonance based sensors using nano ribbons of graphene and WSe_2, *Plasmonics*, **15**, pp. 1015–1023.

49. Nurrohman, D.T. and Chiu, N.F. (2021). A review of graphene-based surface plasmon resonance and surface-enhanced Raman scattering biosensors: Current status and future prospects, *Nanomaterials,* **11**, p. 216.

Chapter 3

Plasmonic Sensing Devices

Muhammad Ali Butt, Svetlana Nikolaevna Khonina, and Nikolay Lvovich Kazanskiy
Samara National Research University, Samara 443086, Russia
IPSI RAS-Branch of the FSRC "Crystallography and Photonics" RAS, Samara 443001, Russia
butt.m@ssau.ru

This chapter stipulates the fundamental knowledge related to the recent advances in plasmonic sensing devices. Plasmonics refers to the creation, recognition, and management of light at optical frequencies along with metal-dielectric interfaces on a nanometer scale. Several sensor configurations can be formed based on this phenomenon. However, we aim to discuss Kretschmann configuration, optical fiber plasmonic sensors, and integrated plasmonic sensors based on metal-insulator-metal waveguides. The device configurations discussed in this chapter provide extraordinary sensing characteristics and the preference for utilizing plasmonic device configuration is based on preferred applications.

Plasmonics-Based Optical Sensors and Detectors
Edited by Banshi D. Gupta, Anuj K. Sharma, and Jin Li

ISBN 978-981-4968-85-0 (Hardcover), 978-1-003-43830-4 (eBook)
www.jennystanford.com

3.1 Introduction

The word "plasmonics" defines photonics related to surface plasmon (SP) and has stimulated substantial interest since the 1990s. The antiquity of the SP study goes back as early as the 1970s to the finding of surface plasmon resonance (SPR) on thin metal layers and light dispersion from nanoscale metallic specks. SPs are of two kinds: the localized type (localized surface plasmon polaritons, LSPRs) and the propagating type (propagating surface plasmon polaritons, PSPPs). LSPR occurs when the length of the metallic nanostructure is smaller than the operational wavelength, resulting in mutual yet non-propagating oscillations of free electrons (e^{-}s) in the metallic nanostructure. The LSPR relies on the refractive index (RI) of the ambient medium, leading to the foundation of colorimetric plasmonic sensors. The LSPR-associated electromagnetic field expands into the surrounding medium (typically 30 nm) and decays exponentially. Unlike the LSPR, SPPs are optical waves traveling along with a dielectric-metal boundary, making them attractive for biosensing applications. SPPs are closely linked to metal-dielectric interfaces that penetrate about 10 nm into metal and usually more than 100 nm into dielectrics. SPR has been deliberated as an important tool for numerous sensing applications over the last three decades. SPR sensors are commonly used in food safety [1, 2], colorimetric [3], alcohol sensing [4], fuel adulteration [5], early disease detection [6], biosensing [7, 8], medical diagnostics [9–11], pregnancy sensing [12], bioimaging [13], telemedicine [14], underwater [15], temperature sensing [16, 17], and early disease detection, among others [18, 19]. Some of the most relevant applications of SPR sensors are shown in Fig. 3.1.

Due to advancements in SPR technology, it has also been utilized in optoelectronic devices such as SPR imaging, filters, and modulators. In Japan, an SPP-based research article for the fast detection of COVID-19 was recently released. Antibiotic-coated gold nanoparticles experience a resonance peak shift when the viruses are captured, resulting in a unique color transition. Similar methods are widely used in pregnancy testing. Furthermore, SPR technology enables the combination of nanoelectronic and nanophotonic materials to create ultra-compact optoelectronic devices. Zenneck proposed the SPR foundation for the first time in 1907. When one surface is lossless and the other is a lossy dielectric (or metal), EM surface waves (SWs) can be sensed at the interface of two surfaces. SWs are fast and exponentially attenuated with height over the boundary, as Sommerfeld found in 1909 [20]. Ritchie demonstrated the authenticity of SWs on the metal-dielectric border in 1957 [21], proving the SPR's actual development.

3.2 Working Mechanism of SPR Sensing Devices

A conductor (metal) has many free electrons, and an association of these electrons can be thought of as plasma excitations. At the same time, the amount of positively charged lattice ions is equal, resulting in a charge density of zero in the conductor. The electrons shift to the positive zone when an external field is applied, and the positive ion moves toward the electrons at the same time. SPs are longitudinal oscillations that are injected into the conductor because of such transport. To keep the SPs going, a conductor and a dielectric contact are needed. Surface plasmon waves (SPWs) propagate over the metal-dielectric interface because of these fluctuations and a defined resonance state. Because there is no explanation for the TE-polarized light of Maxwell's equations, these SPs only support TM-polarized EM fields. SPW is degrading exponentially and is described by the propagation constant as follows:

$$\beta = \frac{\omega}{c}\sqrt{\varepsilon_S} \tag{3.1}$$

where ε_S is the dielectric permittivity of the sensing medium. The propagation constant of the SP wave is larger than the propagation constant of light in the dielectric medium, according to the specification. As a result, the SP cannot be stimulated by typical light; it needs light with additional momentum or energy that has the same polarization state as the SPW. In addition, the propagation constant must be in sync with the SPW.

SPR sensors established on optical fibers function based on the directed evanescent field. Most of the EM field propagates through the optical fiber's core, but a tiny amount of it spreads from the cladding as an evanescent field, which interacts with the plasmonic metal surface and stimulates the free electrons. When the frequency of the input photon and the free electrons are balanced, the electrons start to resonate, and SPW is generated at the metal-dielectric boundary. At the resonance state, a strong loss peak arises, which is extremely receptive to the dielectric medium's small RI change. SPR sensors based on optical fibers were initially commercialized using silicon optical fibers (SOFs) and later polymer optical fibers (POFs). These sensors do not require expensive optical equipment and may be utilized to produce a dense miniaturized system with remote sensing capabilities. Temperature, pressure, environmental monitoring, and food safety, to name a few, have all profited from SPR-based optical sensing in recent years. The optical fiber employed in such sensors has been substituted by conventional platforms for SPR sensors, such as prisms and electrodes.

So far, SOFs have been the clear winner in virtually all current applications. SOFs and wireless/copper cables have long outperformed POFs in terms of popularity [22]. POFs are made of inexpensive plastic materials such as polymethyl methacrylate (PMMA), polycarbonate, and polystyrene, and have transmission windows in the visible spectrum of light [23]. POFs are significantly safer than SOFs, which must be handled with care since they are more resistant to bending, stress, and vibration. Technical advancements have made POF networks highly effective in many essential short-distance data connections, such as home setups, car data linkages, and production controls, thanks to all the hard effort put in over the years. Fiber sensing is another important use of POFs [24]. In comparison to SOFs, POFs are significantly more stable at a large core size and have a higher collapse pressure. PMMA-based POF, for example, may recover effectively from strains of more than 10%, whereas SOFs tolerate strains of less than 1%. In general, POFs are useful for detecting stress, fractures, and twists in composite materials [25]. POFs are often more adaptable to these functional polymers or biomedical materials, allowing for the creation of functional POFs using a variety of synthesis methods and materials for a variety of applications, including radiation detection, biomedical and chemical sensing, and structural health control [26].

A comprehensive overview of optical sensors based on POF is presented in Ref. [27]. Several groups have recently demonstrated optical sensors in POF based on SPR and are being employed in a variety of applications in the field of life sciences, electrochemistry, environmental safety, and biomedical diagnostics [28–30]. An overview of wavelength-based optical fiber biosensors is presented in Ref. [31]. An SPR-based optical fiber tip sensor is ideal for chemical or biological sensing since it has a smaller footprint than SPR-based optical fiber sensors [32].

The propagation constant of the SP changes when the RI of the superstrate changes. This fluctuation changes the state of coupling between a light wave and the SP, which is seen as a change in one of the SP-interacting optical wave's properties. Depending on the characteristics of the light wave interacting with the SP are detected, SPR sensing devices can be classified as SPR sensors with angular, wavelength, intensity, phase, or polarization modulation (see Fig. 3.1).

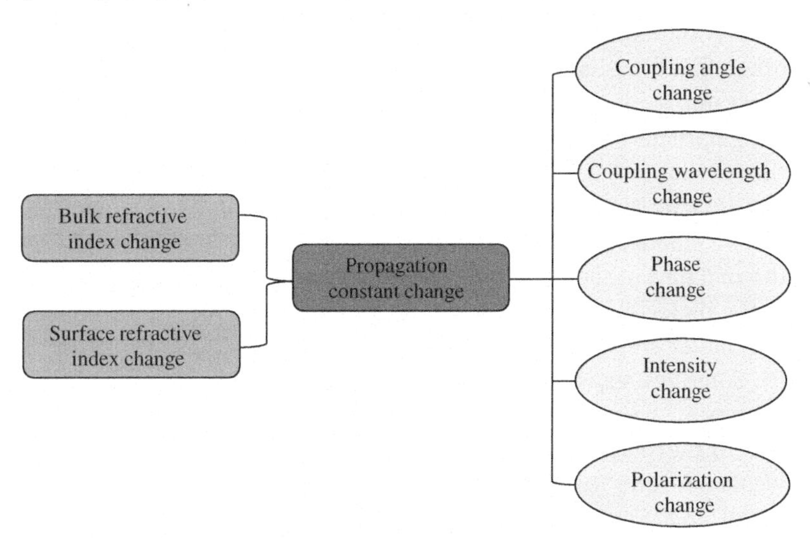

Figure 3.1 The working mechanism of SPR sensing devices.

With angular modulation, a monochromatic light wave activates an SP in SPR sensors. The coupling strength between the incident wave and the SP is evaluated at various angles of incidence light waves, and the angle of incidence giving the robust coupling is determined and utilized as a sensor output. A collimated light pulse that comprises several wavelengths stimulates an SP in SPR sensing devices with wavelength modulation. The intensity of the incident light wave acts as the sensor's output and SPR sensing devices work on intensity modulation work on a single angle of incidence and wavelength measurement of the coupling strength between the incident light wave and the SP. The swing in the phase of the light wave interacting with the SP is detected and used as the sensor's output in SPR sensors with phase modulation at a single angle of incidence and wavelength of the light wave. Variations in the polarization of light waves interacting with an SP are monitored using SPR sensors with polarization modulation.

3.3 Widely Used Plasmonic Materials

The metallic coating on a prism base or fiber core is made of silver (Ag), gold (Au), or copper (Cu) [33]. When the RI of the sensor layer varies, Au displays a greater resonance parameter shift. Ag, on the other hand, has a smaller SPR curve width (FWHM), resulting in a greater signal-to-noise ratio (SNR) and detection accuracy. The imaginary part of the metal's dielectric constant is used to determine the resonance curve's FWHM. Ag with a larger imaginary part value has a narrower FWHM of the SPR curve, resulting in a higher SNR. The resonance curve's shift, on the other hand, is determined by the real part of the metal's dielectric constant. Because the real part of the dielectric constant is greater in Au than in Ag, the resonance parameter shifts more in response to changes in the sensing layer's RI. Some issues with Au, however, include the creation of islands in thin Au layers, band-to-band variations, and surface roughness due to thermal evaporation [34]. The chemical strength of Ag has weakened due to oxidation, making it unable to get a repeatable result and, for practical purposes, the sensor remains inaccurate [35]. The application of the thin and dense coating on the Ag surface is required. A study of Ag, Au, and bimetallic Ag/Au layers as active phases for chemical sensors and biosensors established on SPR found that the bimetallic layer had significant advantages (with Ag as an inner layer) [36]. This configuration, like Au layers, may produce a significant displacement of the resonance angle in response to changes in ambient RI, but the resonance minimum is smaller. The SNR of SPR devices improves because of this. Furthermore, the outer Au layer is chemically inert, preventing oxidation of the inner Ag layer. This new resonance layer may be employed in all sorts of chemical and biological sensing devices established on SPR. Consumers can track the binding of smaller molecules thanks to an increase in the SNR, which might be important in pharmacology, biology, and medicine. Several researchers have recently explored the ability of bimetallic amalgamations to be used in fiber optic sensors using the SPR technique [37, 38].

Copper (Cu) is an alternative plasmonic material with almost comparable interband transition and optical dampening as Au in the wavelength range 600–750 nm [39]. Unfortunately, Cu is also susceptible to oxidation. To overcome the corrosion problem, graphene has recently been used on top of Cu or Ag [40]. Because graphene is chemically inert and physically strong, it should be kept away from fluidics [41]. Cu-graphene [40] and Ag-graphene [42] optical fiber SPR sensors exhibit long-term stability and reliability. Niobium is another one-of-a-kind plasmonic material with high chemical resistance and high mechanical strength [43, 44]. The connection between niobium film and silica glass is robust and no extra bonding layer is required. SPR sensors established on indium tin oxide (ITO) have recently attracted a lot of attention due to their low bulk plasma frequency [45]. Besides, it has the same optical damping as Ag and Au [46].

3.4 Plasmonic Sensors Based on Kretschmann (KR) Configuration

The SPR phenomenon occurs when SP waves, associated with mutual excitations of electrons in a metal, are produced by light at a metal/liquid border, as Otto [47], KR [48], and Raether [49] demonstrated in the late 1960s. The bulk of currently available SPR biosensors still use KR and Raether's suggested approach of focusing TM-polarized light via a prism and then reflecting it from a 50 nm thin Au sheet put on its surface, which was first presented 40 years ago. When the tangential x-component of the incident optical wave vector, $\vec{k}_x$, is harmonized with $\vec{k}_{sp}$, the SP wave vector, the pumping light energy is transferred to SPs. Under total internal reflection (TIR) circumstances, the SPR effect causes a strong resonant dip in the reflected light intensity at a specific angle (θ). Because of the strict need for optical and evanescent wave vector matching, any change in the RI of the ambient medium that associates Au (within 200 nm to 300 nm) would have a significant impact on the specific angle. Figure 3.2a shows a schematic of the typical KR scheme. The sensitivity of the normal KR arrangement with that of the nanograting-on-top-of-metal-layer integration is investigated in Ref. [50] as illustrated in Fig. 3.2b. In [50], the author compared the sensitivity of the standard KR configuration with the one which is realized on the integration of nanograting on top of the metal layer. The existence of absorptive nanograting near the metal film allows for significant field amplification and localization, as well as control of the dispersion relation that was previously defined by a typical SPR structure. The Q-factor improves as a result and is found to be 3–4 times larger than with the traditional KR arrangement. The impact of the incidence angle on the resonance wavelength is also proven computationally and experimentally, where increasing the incidence angle for the plasmonic nanograting arrangement results in just a little change in wavelength [50].

Because of its strong optical and chemical characteristics, and because it promotes SPP transmission in the visible wavelength range, a thin Au sheet is mostly used in SPR sensors. However, analyte adsorption on the Au surface is extremely poor, limiting the efficacy of SPR biosensing devices. Over the last several decades, some innovative approaches for increasing the responsiveness of SPR sensors have been developed, including diverse layouts and materials [51, 52]. Graphene is a single sheet of carbon atoms laid out in a honeycomb pattern that is rapidly becoming the utmost intensively studied topic of the time. Graphene has an extraordinary surface-to-volume ratio and excellent binding/adsorption affinity for analytes due to its carbon ring structure from π–π stacking interactions. Graphene-based electrochemical sensors have also been shown to have high sensitivity for medicinal drug identification [53]. Little research on the application of graphene on a metal sheet to improve sensitivity has been reported in recent years [54, 55]. A computational study of SPR biosensor sensitivity improvement utilizing adjustable wavelengths and graphene layers is presented in Ref. [55]. Figure 3.2c shows a schematic depiction of a graphene-based SPR sensor in a KR configuration. A unique hybrid spectroscopic method based on the combination of SPR and surface-enhanced Raman scattering (SERS) microscopy is presented in Ref.

[56]. With consistent enhancement on flat surfaces, this method opens new vistas for SERS microscopy.

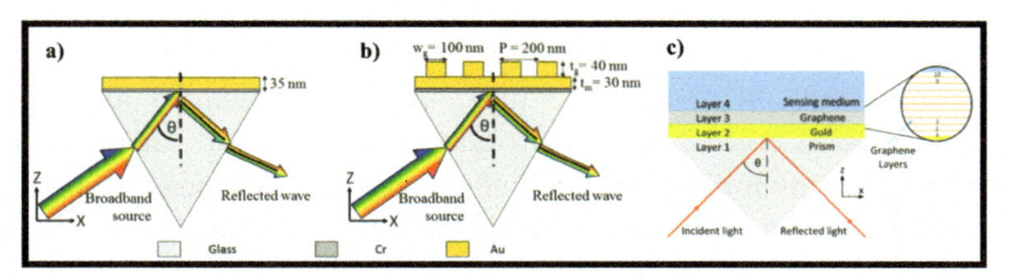

Figure 3.2 Schematic of (a) typical, (b) plasmonic nanograting established KR arrangement [50], and (c) a schematic illustration of graphene-established SPR sensor in KR arrangement [55].

3.5 Optical Fiber–Based Plasmonic Sensor

Traditional SPR sensing devices established on prism-coupling are divided into two types: KR and Otto setups. All these sensor designs function based on attenuated total reflection (ATR). Even though KR-based SPR pattern devices are widely utilized because of their remarkable performance in sensing utilizations, they have several limitations. These sensing devices are usually large and have moving components. Consequently, they are not transportable and cannot be utilized for distant sensing applications. Furthermore, the realistic use of a spectral-based measurement is expensive, and the ability to scale down the sensor size is restricted. To properly handle those future difficulties, optical fiber-based SPRs are utilized. The optical fibers are light and easy to use. TIR is used to transmit light over optical fibers, and these sensors have several advantages over prism-based SPR sensors. Furthermore, because of the optical fiber's small nature, the sensor size that may be employed for distant sensing utilizations can be considerably decreased. Optical fiber-established SPR sensors offer a wide range of sensing and excellent resolution, but they are only viable for limited acceptance angles. Many optical fiber SPR sensors have been described in both theoretical and experimental research [57, 58].

Jorgenson et al. proposed the first optical fiber-based SPP configuration without the bulk prism in 1993 [59]. The interaction of evanescent waves with SPPs was shown in a fiber-based SPP

RI sensor by partly removing the fiber cladding and putting a highly reflective coating at the exposed region. The transmission or reflection properties of the propagating light are generally used to determine the working mechanism of plasmonic sensors implemented on fiber optics [60]. To detect unknown analytes, transmission probe-based fiber optic sensors are made up of noble metal and immobilized ligands. Reflective probe sensors, on the other hand, use a mirror to reflect backlight to the fiber. Single-mode fiber (SMF), multi-mode fiber (MMF), wagon wheel fiber, U-shaped fiber, and Fiber Bragg grating (FBG), among others, have recently been described as plasmonic sensors based on fiber optics, where noble metals are attached on the engraved cladding section.

Optical fiber sensors are small and cost-effective, and provide label-free detection. Furthermore, narrow resonance peaks with great responsivity and stability are achievable. By fine-tuning the geometry of holes in microstructured optical fibers (MOF), it is possible to change the sensor's performance. Figure 3.3 shows a graphical depiction of the optical fiber SPR sensor. In general, the cladding around the core of the optical fiber's sensing area is etched and coated with a thin metal layer that can also be coated with a buffer layer before being covered with the sensing layer. The input light is linked to one end of the optical fiber, and the modulated light is received at the other end, transmitting data about the sensing layer's characteristics. The light wavelength, fiber shape, and metal layer characteristics all influence optical fiber sensing efficiency. SMFs and MMFs, for example, might have distinct light coupling processes. This is because optical mode propagation is dependent on the number of modes that fiber can maintain. A tapered and un-tapered optical fiber, for example, shows distinct light coupling strengths because the penetration depth (PD) of the evanescent field is varied due to various geometric features.

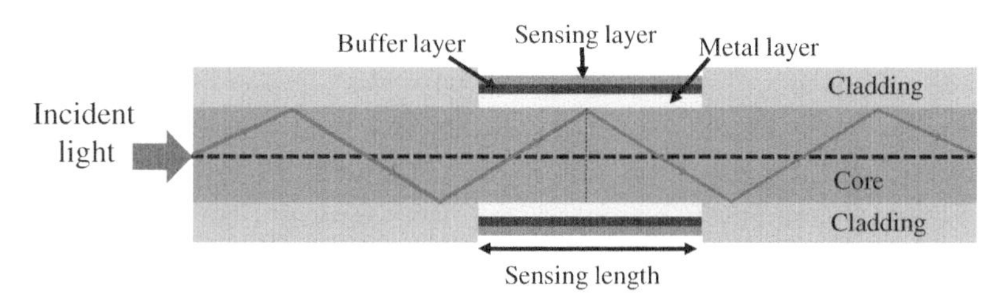

Figure 3.3 Graphical representation of a standard optic fiber SPR sensor.

An un-tapered optical fiber displays a persistent PD of the evanescent field alongside the sensing area, but a tapered optical fiber indicates a significant deviation in evanescent field PD alongside the sensing area. Furthermore, the PD of the evanescent field and the intensity of light coupling with SPs are dependent on a critical fiber parameter known as the numerical aperture (NA), which is linked to the optical fiber's light acceptance limit.

In 1978, the photonic crystal fiber (PCF) was presented for the first time. A fiber core might be covered with BG, which is like 1D-PC [61]. The first PCF was announced in 1996 at the Optical Fiber Conference, and it was made of a 2D-PC with an air core [62]. Figure 3.4 depicts a timeline of PCF development. PCFs have a core and cladding that are like standard optical fibers, however, PCFs have periodic air holes in the cladding area that control light propagation. By altering the number of rings and modifying the geometries of the air holes, it is feasible to manipulate light propagation [63]. PCFs with outstanding abilities have unbolted a window to fix the differences between conventional prisms and optical fiber-based SPR sensors. Furthermore, PCF SPR sensors are compact and can be integrated into microscales [64, 65]. PCF geometry may be improved to create the best evanescent field thanks to design freedom. The core-guided-mode transmission may be regulated utilizing several kinds of PCF assemblies, for example, square, octagonal, hexagonal, and hybrid, and their guiding characteristics can be enhanced by altering their geometric parameters [66].

Kim et al. were the first to demonstrate the fiber-based plasmonic phenomena by replacing graphene for the metal film and employing thermal chemical vapor deposition to create a thin graphene layer [68]. The detecting effectiveness of SPR sensing devices with a graphene layer on top of Au/Ag is also shown to be higher than that of SPR sensors without a graphene layer [69, 70]. In Ref. [71], the utilization of Au as an active plasmonic medium and the selective penetration of the PCF core with a high RI to implement the coexistence of positive and negative RI effects have been studied.

Due to its simplicity of manufacture and deployment, D-shaped PCF has flickered a lot of curiosity in SPR sensing [72, 73]. The top surface of the fiber cladding is adjusted to make it flat, like how the metal layer and sample are placed on top of the flat fragment in a D-shape. The metal layer in D-type PCF can be employed near the core, promoting intimate interaction with the sample, and enhancing device performance. Many D-shaped PCF sensors have been presented for a variety of utilizations to date [74–77].

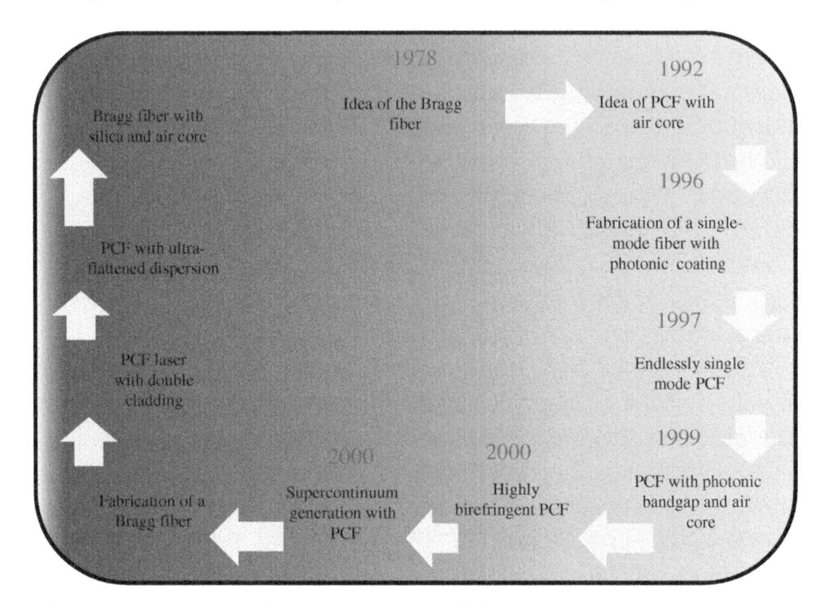

Figure 3.4 Overview of PCF development [67].

The first SPR sensor based on MOF was announced in 2006, in which plasmons on the inner surface of massive, metalized channels holding the analyte may be stimulated by a single-mode MOF's fundamental mode [78]. The metalized holes in the outer layer are full of analytes and are broader than those in the inner layer. As illustrated in Fig. 3.5a, a small air hole was placed in the core to lower the n_{eff} of the core and improve the phase match. Holes in the core and first layer are filled with air, while the analytes are injected into metal-coated holes in the second layer. A PCF based on an SPR sensor using Au as the sensitive material is used to create a sensor model [79]. Because Au film is put on the exterior of the fiber core as a substitute for the holes filled with analytes inside the core, this construction helps reduce the difficulty of Au deposition. This model has a sensitivity of about 4875 nm/RIU, which may be increased to 7500 nm/RIU by adding a thin layer of graphene to the Au layer. Figure 3.5b shows a graphical depiction of the PCF-SPR sensor with an annular-shaped analytes channel.

Using the dual characteristic of PCF packed with variable concentrations of analyte and Ag nanowires to conduct temperature sensing, a temperature sensor based on PCF SPR is experimentally shown in Ref. [80]. The temperature sensitivity of a PCF-based SPR temperature sensor, or any other PCF-based SPR sensing, has been measured to be as high as 2.7 nm/°C in experiments, giving a guideline for the execution and utilization of a PCF-based SPR temperature sensor or any other PCF-based SPR sensing. Figure 3.5c depicts the graphical depiction of the PCF-based SPR temperature sensor.

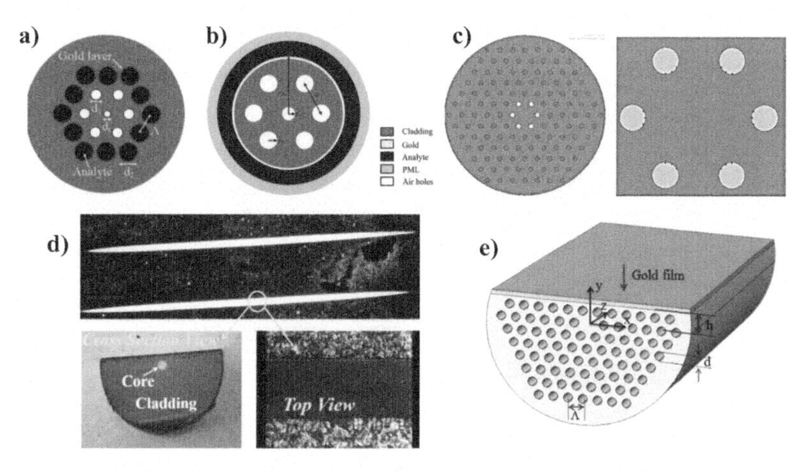

Figure 3.5 (a) The graphical representation of a hexagonal solid-core micro-structured optical fiber-based SPR sensor [78]. (b) The model of a PCF-SPR sensor [79]. (c) Cross-section of the PCF-SPR temperature sensor (left) and cross-section of analyte channel filled with nanowires (right) [80], (d) Photograph of the side-polished single-mode fiber, fixed in the aluminum mount, under light reflection. The bottom images display the cross-section and the top view [81], and (e) a schematic illustration of the SPR sensor based on a D-shaped PCF in a 3D model [82].

A new SPR-based sensor that detects dew formation in smart textiles made of optical fibers is described in Ref. [81]. Moisture condensation and water molecule evaporation in the air may both be tracked using the suggested SPR sensor. This sensor detects the formation of dew in less than a quarter of a second and accurately regulates the temperature of the dew point to within 4%. It has a resolution of 7 nm and may be used to detect variations in the depth of the water layer during dew formation and evaporation in a plasmon depth probe range of 250 nm. Furthermore, by using the plasmon depth probe to calculate the evaporation time, it is possible to detect the relative humidity of an environment throughout a dynamic range of 30–70%. Figure 3.5d shows two parallel side-polished optical fibers that are secured in an aluminum base. The two bottom pictures in Fig. 3.5d show an SEM image of a cross-section of a side-polished fiber and a top view of one of the polished fibers [81].

Wu et al. conducted computational and experimental research on a D-shaped PCF-SPR sensor [82]. On the PCF polished surface, a thin Au film and the analytes are both placed. The sensor has an extraordinarily high sensitivity of 21700 nm/RIU, and the RI of the PCF material is 1.36. The outcome suggests that the model estimation and experiment have a close relationship. The D-shaped fiber SPR sensor, which is based on the PCF material's reduced RI, is especially useful in chemical, biological, and industrial utilizations. Figure 3.5e shows a graphical representation of the SPR sensor implemented on a D-shaped PCF.

3.6 Integrated Plasmonic Sensors Based on Plasmonic W/G

SPP is an EM wave phenomenon that occurs at the dielectric-metal contact. The conventional optical diffraction limit may be overcome, and optical signal transmission and modulation can be achieved in the subwavelength region. Because of their flexibility on-chip incorporation, minimum bent loss, extended propagation length, subwavelength confinement, and relative ease of fabrication, plasmonic metal-insulator-metal (MIM) waveguides (W/Gs) established on SPPs as a possible field of optical W/Gs are widely researched [83]. Owing to the requirement for ultrahigh-sensitivity biological sensors, plasmonic RI sensors constructed on MIM W/Gs have gotten a lot of attention. A dielectric is sandwiched between two plasmonic metal claddings in MIM W/Gs, which are plasmonic structures. The scheme's main features are its simple design and the capacity of these W/G structures to limit light at the subwavelength level. Using MIM W/Gs for the comprehension of highly integrated optical circuits, researchers have thoroughly investigated the concept of a wide range of devices. Filters [84–86], sensors [87–90], splitters [91], couplers [92], BG [93], and lasers [94] are just a few examples.

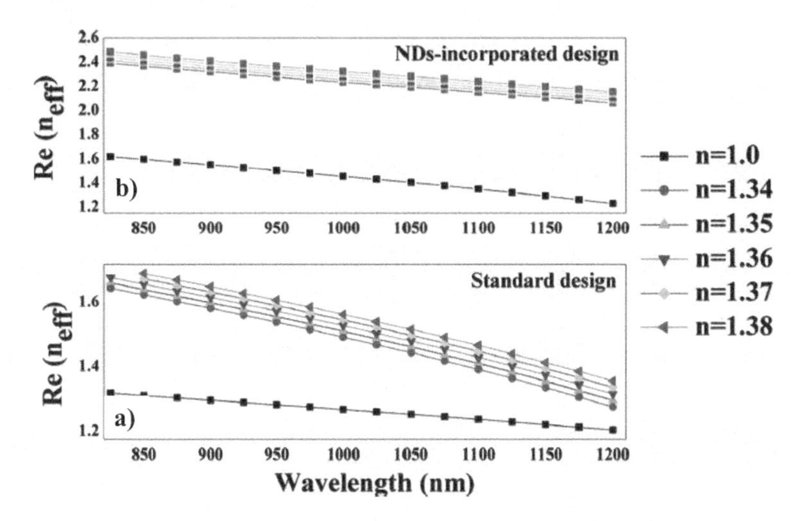

Figure 3.6 The real part of n_{eff} (a) conventional MIM W/G configuration and (b) nanodots integrated MIM W/G [98].

In comparison to traditional approaches such as fluorescence analysis, sensors based on the SPP phenomenon are more compatible with analytes and do not require additional processing processes, such as labeling. SPPs have attracted a lot of interest in optical sensing domains since their initial gas sensing demonstration [95]. Plasmonic sensors made with MIM W/Gs may be used for a variety of utilizations, including temperature, pressure, and RI sensing. In biological disciplines, RI sensors offer a variety of utilizations. For example, any change in RI may be used to determine the solution concentration and pH value. An EM field is generated by stimulating the detecting element with a light source that creates extremely stringent SPs at the metal's surface [96, 97]. When a sample under inspection comes into touch with the sensing device, the MIM W/G's n_{eff} changes, causing λ_{res} to redshift. As shown in Fig. 3.6a,b, we provided a comparative analysis on the variation in n_{eff} of the conventional MIM W/G sensor design and nanodots integrated MIM W/G sensor design in Ref. [98]. The fact that nanodot-filled cavities deliver a significant shift in n_{eff} due to a slight variation in the ambient RI is well-known.

SP wave is exceptionally responsive to variations in the surrounding surface's RI. The variations in one of the attributes of the light related to the SP, such as a shift in the frequency, intensity, or phase, can be used to determine the RI of the ambient medium [99]. A wavelength interrogation approach is used to describe ring resonator sensors in general [100–104]. The measurement of $\Delta\lambda_{res}$ is the most frequent study approach employed in ring resonator constructions. Plasmonic sensors are more sensitive than optical sensors built on a silicon platform [105–108]. Optical platforms that employ the wavelength interrogation approach often use a polychromatic light source, like a halogen lamp, and cover the whole spectrum area where res is anticipated to be seen. Halogen lights have a greater spectrum due to their technology. This light source is needed when the immovable incident angle pattern and wavelength interrogation technique are employed.

The light spectrum linked to an SP wave is acquired using charge-coupled devices, complementary metal-oxide-semiconductor, or silicon photodiode array-based spectrometers, and the spectral position of the plasmonic spectrum is adjusted employing suitable feature-tracing methods. The characteristics of spectrometers used in wavelength inquisition-based setups impact not only the sensor's efficiency but also its degree of miniaturization since they require a specific route length to effectively scatter light to a position-sensitive detector. Miniaturization of spectroscopic optical devices for plasmonic devices has lately received a lot of interest. In recent publications, we have numerically demonstrated the novel designs of plasmonic biosensors comprised of highly sensitive cavities such as Bow-tie cavities, racetrack-integrated circular cavities, nanodot-loaded semi-circular cavities, and BGs. Figure 3.7a–d illustrates the graphical representation of the sensor designs [109–111].

These sensors have exceptional sensing capabilities, including high sensitivity (S), a high-quality factor (Q-factor), and a high figure of merit (FOM). The sensitivity (S) may be written as $S = \Delta\lambda/\Delta n$, where Δn is the variance of the ambient RI and $\Delta\lambda$ is the change in λ_{res}. The capacity to have great wavelength selectivity in a wavelength filter is a significant feature that indicates a high-quality component (Q-factor). Q-factor can be expressed as λ_{res}/FWHM, where FWHM is the full width at half maximum of the transmittance spectrum. FOM is computed as S/FWHM. A compact plasmonic coupled system based on MIM W/G contains a resonance cavity with a simple and broad adjustable spectrum for sensing utilizations. Consequently, enhancing the sensor design is an important challenge in creating a plasmonic MIM W/G RI sensor. The expansion of analytical and numerical simulation tools, such as Lumerical and COMSOL, which are commercially accessible, and offer a greater prospect to grasp the optical characteristics of plasmonic sensor structures. The majority of previous plasmonic sensor research papers used 2D numerical simulations, which assume one dimension as infinite. This simplifies sensor performance analysis by reducing computing time and lowering loss. Nevertheless, the height of the MIM W/G has a considerable impact on the system's loss, which should be taken into consideration for practical processing. The comprehensive examination of 2D and 3D-FEM techniques is discussed in Ref. [111].

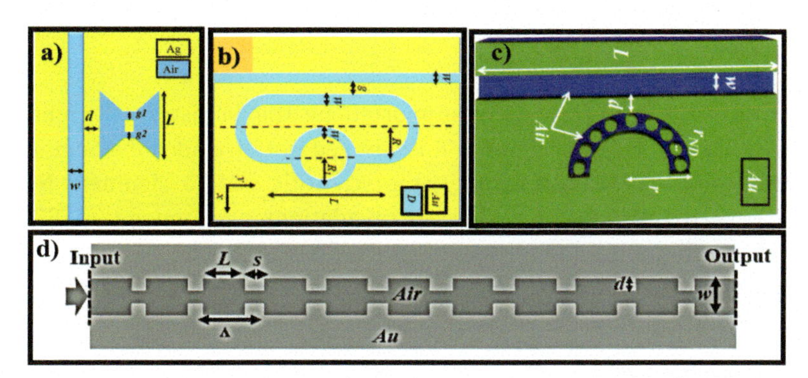

Figure 3.7 Plasmonic sensor based on MIM W/G. (a) Bow-tie configuration [110]. (b) Racetrack-integrated circular cavity [109]. (c) Nanodot-loaded semi-circular cavity [98]. (d) Bragg grating [93].

3.7 Conclusion and Future Prospects

Considering the prospects of SPP sensing, it is conceivable to develop tiny sensing devices that are user-friendly, sophisticated, and efficient for data transfer. Artificial intelligence may be used to aid with signal processing and analysis. Both SPR and SPP may be customized using nanostructured materials and design, providing enormous promise and diversity for the development of plasmon-enhanced sensors. Future plasmonic sensor development is primarily dependent on the efficiency of plasmonic materials/architectures and the use of contemporary production technologies. We concentrated on the most recent breakthroughs in fiber-based plasmonic sensors and integrated plasmonic sensors realized on metal-insulator-metal (MIM) waveguide architectures in this chapter. Even though Kretschmann-based SPR pattern devices are widely employed because of their excellent performance in sensing applications, they face several problems. These sensors are usually large and made with moveable optical components. As a result, they are immobile and cannot be employed for applications such as remote sensing. Furthermore, the actual use of a spectral-based measurement is costly, and the option of scaling down the sensor size is limited. To adequately address these future issues, optical fiber-based SPRs and MIM waveguide-based SPPs are being used.

Acknowledgment

This work was financially supported by the Ministry of Science and Higher Education of the Russian Federation under the Samara National Research University (the scientific state assignment No. 0777-2020-0017) for numerical calculations and under the FSRC "Crystallography and Photonics" of the Russian Academy of Sciences (the state task No. 007-GZ/Ch3363/26) for theoretical research.

References

1. S.-Y. Tseng, S.-Y. Li, S.-Y. Yi, A. Sun, D.-Y. Gao and D. Wan, "Food quality monitor: Paper-based plasmonic sensors prepared through reversal nanoimprinting for rapid detection of biogenic amine odorants," *ACS Appl. Mater. Interfaces,* vol. 9, no. 20, pp. 17306–17316, 2017.
2. H. Thenmozhi, M. Rajan and K. Ahmed, "D-shaped PCF sensor based on SPR for the detection of carcinogenic agents in food and cosmetics," *Optik,* vol. 180, pp. 264–270, 2019.
3. A. Ameen, M. Gartia, A. Hsiao, T.-W. Chang, Z. Xu and G. Liu, "Ultra-sensitive colorimetric plasmonic sensing and microfluidics for biofluid diagnostics using nanohole array," *Journal of Nanomaterials,* vol. 2015, p. 460895, 2015.

4. K. Ahmed, M. Haque, M. A. Jabin, B. K. Paul, I. S. Amiri and P. Yupapin, "Tetra-core surface plasmon resonance based biosensor for alcohol sensing," *Physica B: Condensed Matter,* vol. 570, pp. 48–52, 2019.
5. K. Ahmed, M. A. Jabin and B. K. Paul, "Surface plasmon resonance-based gold-coated biosensor for the detection of fuel adulteration," *Journal of Computational Electronics,* vol. 19, pp. 321–332, 2020.
6. N. Bellassai, R. D'Agata, V. Jungbluth and G. Spoto, "Surface plasmon resonance for biomarker detection: Advances in non-invasive cancer diagnosis," *Frontiers in Chemistry,* vol. 7, p. 570, 2019.
7. B. K. Paul, K. Ahmed, H. J. El-Khozondar, R. F. Pobre, J. Pena, M. Merciales, N. M. Zainuddin, R. Zakaria and V. Dhasarathan, "The design and analysis of a dual-diamond-ring PCF-based sensor," *Journal of Computational Electronics,* vol. 19, pp. 1288–1294, 2020.
8. H. Abdullah, K. Ahmed and S. A. Mitu, "Ultrahigh sensitivity refractive index biosensor based on gold coated nano-film photonic crystal fiber," *Results in Physics,* vol. 17, p. 103151, 2020.
9. J.-F. Masson, "Surface plasmon resonance clinical biosensors for medical disgnostics," *ACS Sens.,* vol. 2, pp. 16–30, 2017.
10. T. Parvin, K. Ahmed, A. M. Alatwi and A. N. Rashed , "Differential optical absorption spectroscopy-based refractive index sensor for cancer cell detection," *Optical Review,* vol. 28, pp. 134–143, 2021.
11. K. Ahmed , B. K. Paul, B. Vasudevan, A. Z. Rashed, R. Maheswar, I. Amiri and P. Yupapin, "Design of D-shaped elliptical core photonic crystal fiber for blood plasma cell sensing application," *Results in Physics,* vol. 12, pp. 2021–2025, 2019.
12. S. A. Mitu, K. Ahmed, F. Zahrani, A. Grover, M. Senthil, M. Rajan and M. A. Moni, "Development and analysis of surface plasmon resonance based refractive index sensor for pregnancy testing," *Optics and Lasers in Engineering,* vol. 140, p. 106551, 2021.
13. T. Vo-Dinh, H.-N. Wang and J. Scaffidi, "Plasmonic nanoprobes for SERS biosensing and bioimaging," *J. Biophotonics,* vol. 3, pp. 89–102, 2010.
14. P. Preechaburana, M. Gonzalez, A. Suska and D. Filippini, "Surface plasmon resonance chemical sensing on cell phones," *Angew. Chem. Int. Ed.,* vol. 51, pp. 11585–11588, 2012.
15. Y. Minagawa, M. Ohashi, Y. Kagawa, A. Urimoto and H. Ishida, "Compact surface plasmon resonance sensor for underwater chemical sensing robot," *Journal of Sensors,* vol. 2017, p. 9846780, 2017.
16. X. Yang, L. Zhu, M. Dong and J. Yao, "Multiplex localized surface plasmon resonance temperature sensor based on grapefruit fiber filled with

a silver nanoshell and liquid," *Optical Engineering,* vol. 58, no. 11, p. 117104, 2019.

17. I. S. Amiri, B. K. Paul, K. Ahmed, A. H. Aly, R. Zakaria, P. Yupapin and D. Vigneswaran, "Tri-core photonic crystal fiber based refractive index dual sensor for salinity and temperature detection," *Microwave and Optical Technology Letters,* vol. 61, no. 3, pp. 847–852, 2019.
18. H. Abdullah, S. A. Mitu and K. Ahmed, "Magnetic fluid-injected ring-core-based micro-structured optical fiber for temperature sensing in broad wavelength spectrum," *Journal of Electronic Materials,* vol. 49, pp. 4969–4976, 2020.
19. S. A. Mitu, K. Ahmed, M. N. Hossain, B. K. Paul, T. K. Nguyen and V. Dhasarathan, "Design of magnetic fluid sensor using elliptically hole assisted photonic crystal fiber (PCF)," *Journal of Superconductivity and Novel Magnetism,* vol. 33, pp. 2189–2198, 2020.
20. A. Sommerfeld, "Uber die Ausbreitung der Wellen in der drahtlosen Telegraphie," *Ann. Phys.,* vol. 333, pp. 665–736, 1909.
21. R. Ritchie, "Plasma losses by fast electrons in thin films," *Phys. Rev.,* vol. 106, p. 874, 1957.
22. P. Polishuk, "Plastic optical fibers branch out," *IEEE Commun. Mag.,* vol. 44, no. 9, pp. 140–148, 2006.
23. C. Koeppen, R. F. Shi, W. D. Chen and A. F. Garito, "Properties of plastic optical fibers," *Journal of the Optical Society of America B,* vol. 15, no. 2, pp. 727–739, 1998.
24. L. Yi and Y. Changyuan, "Highly strechable hybrid silica/polymer optical fiber sensors for large-strain and high-temperature application," *Optics Express,* vol. 27, no. 15, pp. 20107–20116, 2019.
25. S. Kamimura and R. Furukawa, "Strain sensing based on radiative emission-absorption mechanism using dye-doped polymer optical fiber," *Appl. Phys. Lett.,* vol. 111, p. 063301, 2017.
26. P. Stajanca and K. Krebber, "Radiation-induced attenuation of perfluorinated polymer optical fibers for radiation monitoring," *Sensors,* vol. 17, no. 9, p. 1959, 2017.
27. L. Bilro, N. Alberto, J. L. Pinto and R. Nogueira, "Optical sensors based on plastic fibers," *Sensors,* vol. 12, pp. 12184–12207, 2012.
28. N. Cennamo, D. Massarotti, L. Conte and L. Zeni, "Low cost sensors based on SPR in a plastic optical fiber for biosensor implementation," *Sensors,* vol. 11, no. 12, pp. 11752–11761, 2011.
29. N. Cennamo, G. D'Agostino, R. Galatus, L. Bibbo, M. Pesavento and L. Zeni, "Sensors based on surface plasmon resonance in a plastic optical

fiber for the detection of trinitrotoluene," *Sensors and Actuators B:Chemical,* vol. 188, pp. 221–226, 2013.

30. J.-F. Masson, "Portable and field-deployed surface plasmon resonance and plasmonic sensors," *Analyst,* vol. 145, pp. 3776–3800, 2020.
31. A. B. Socorro-Leranoz, D. Santano, I. D. Villar and I. R. Matias, "Trends in the design of wavelength-based optical fibre biosensors (2008–2018)," *Biosensors and Bioelectronics:X,* vol. 1, p. 100015, 2019.
32. Y. Yuan, D. Hu, L. Hua and M. Li, "Theoretical investigations for surface plasmon resonance based optical fiber tip sensor," *Sensors and Actuators B: Chemical,* vol. 188, pp. 757–760, 2013.
33. S. Singh, S. K. Mishra and B. Gupta, "Sensitivity enhancement of a surface plasmon resonance based fibre optic refractive index sensor utilizing an additional layer of oxides," *Sens. Actuators A,* vol. 193, pp. 136–140, 2013.
34. M. Kanso, S. Cuenot and G. Louarn, "Roughness effect on the spr measurements for an optical fibre configuration: Experimental and numerical approaches," *J. Opt. A Pure Appl. Opt.,* vol. 9, p. 586, 2007.
35. M. L. Zheludkevich, A. G. Gusakov, A. G. Voropaev, A. A. Vecher, E. N. Kozyrski and S. A. Raspopov, "Oxidation of silver by atomic oxygen," *Oxidation of metals,* vol. 61, pp. 39–48, 2004.
36. S. A. Zynio, A. V. Samoylov, E. R. Surovtseva, V. M. Mirsky and Y. M. Shirshov, "Bimetallic layers increase sensitivity of affinity sensors based on surface plasmon resonance," *Sensors,* vol. 2, pp. 62–70, 2002.
37. A. K. Sharma and B. D. Gupta, "On the performance of different bimetallic combinations in surface plasmon resonance based fiber optic sensors," *Journal of Applied Physics,* vol. 101, p. 093111, 2007.
38. R. Tabassum and B. D. Gupta, "Performance analysis of bimetallic layer with zinc oxide for SPR-based fiber optic sensor," *Journal of lightwave technology,* vol. 33, no. 22, pp. 4565–4571, 2015.
39. P. R. West, S. Ishii, G. V. Naik, N. K. Emani, V. M. Shalaev and A. Boltasseva, "Searching for better plasmonic materials," *Laser Photonics Rev.,* vol. 4, pp. 795–808, 2010.
40. V. Kravets, et al., "Graphene-protected copper and silver plasmonics," *Sci. Rep.,* vol. 4, p. 5517, 2014.
41. M. M. Huq, C.-T. Hsieh, Z.-W. Lin and C.-Y. Yuan, "One-step electrophoretic fabrication of a graphene and carbon nanotube-based scaffold for manganese-based pseudocapacitors," *RSC Adv.,* vol. 6, pp. 87961–87968, 2016.

42. A. A. Rifat, G. A. Mahdiraji, D. M. Chow, Y. G. Shee, R. Ahmed and F. M. Adikan, "Photonic crystal fiber-based surface plasmon resonance sensor with selective analyte channels and graphene-silver deposited core," *Sensors,* vol. 15, pp. 11499–11510, 2015.

43. T. Wieduwilt, A. Tuniz, S. Linzen, S. Goerke, J. Dellith, U. Hubner and M. A. Schmidt, "Ultrathin niobium nanofilms on fiber optical tapers: A new route towards low-loss hybrid plasmonic modes," *Scientific Reports,* vol. 5, p. 17060, 2015.

44. S. Bagheri, N. Strohfeldt, M. Ubl, A. Berrier, M. Merker, G. Richter, M. Siegel and H. Giessen, "Niobium as alternative material for refractory and active plasmonics," *ACS Photonics,* vol. 5, no. 8, pp. 3298–3304, 2018.

45. K. Shah, N. K. Sharma and V. Sajal, "SPR based fiber optic sensor with bi layers of indium tin oxide and platinum: A theoretical evaluation," *Optik,* vol. 135, pp. 50–56, 2017.

46. S. Franzen, "Surface plasmon polaritons and screened plasma absorption in indium tin oxide compared to silver and gold," *J. Phys. Chem. C,* vol. 112, pp. 6027–6032, 2008.

47. A. Otto, "Excitation of nonradiative surface plasma waves in silver by the method of frustrated total reflection," *Zeitschrift fur Physik A Hadrons and nuclei,* vol. 216, pp. 398–410, 1968.

48. E. Kretschmann, "Die Bestimmung optischer Konstanten von Metallen durch Anregung von Oberflachenplasmaschwingungen," *Zeitschrift fur Physik A Hadrons and nuclei,* vol. 241, pp. 313–324, 1971.

49. E. Kretschmann and H. Raether, "Radiative decay of nonradiative surface plasmons excited by light," *Zeitschrift fur Naturforschung A,* vol. 23, pp. 2135–2136, 1968.

50. P. Arora, E. Talker, N. Mazurski and U. Levy, "Dispersion engineering with plasmonic nano structures for enhanced surface plasmon resonance sensing," *Scientific Reports,* vol. 8, p. 9060, 2018.

51. D. Kim, "Nanostructure-based localized surface plasmon resonance biosensors," *Optical guided-wave chemical and biosensors I,* vol. 7, pp. 181–207, 2010.

52. T. Wang and W. Lin , "Electro-optically modulated localized surface plasmon resonance biosensors with gold nanoparticles," *Applied Physics Letters,* vol. 89, p. 173903, 2006.

53. S. Kruanetr, P. Pollard, C. Fernandez and R. Prabhu, "Electrochemical oxidation of acetyl salicylic acid and its voltammetric sensing in real samples at a sensitive edge plane pyrolytic graphite electrode modified

with graphene," *International journal of electrochemical science,* vol. 9, pp. 5699–5711, 2014.

54. S. Szunerits, N. Maalouli, E. Wijaya, J. Vilcot and R. Boukherroub, "Recent advances in the development of graphene-based surface plasmon resonance (SPR) interfaces," *Analytical and Bioanalytical Chemistry,* vol. 405, pp. 1435–1443, 2013.
55. K. Bhavsar and R. Prabhu, "Investigations on sensitivity enhancement of SPR biosensor using tunable wavelength and graphene layers," *IOP Conf. Ser.: Mater. Sci. Eng.,* vol. 499, p. 012008, 2019.
56. S. A. Meyer, B. Auguie, E. L. Ru and P. G. Etchegoin, "Combined SPR and SERS microscopy in the Kretschmann configuration," *J. Phys. Chem. A,* vol. 116, pp. 1000–1007, 2012.
57. M. Piliarik, J. Homola, Z. Manikova and J. Ctyroky, "Surface plasmon resonance sensor based on a single-mode polarization-maintaining optical fiber," *Sens. Actuators B Chem.,* vol. 90, pp. 236–242, 2003.
58. B. D. Gupta and A. K. Sharma, "Sensitivity evaluation of a multi-layered surface plasmon resonance-based fiber optic sensor: A theoretical study," *Sens. Actuators B Chem.,* vol. 107, pp. 40–46, 2005.
59. R. Jorgenson and S. Yee, "A fiber-optic chemical sensor based on surface plasmon resonance," *Sens. Actuat. B Chem.,* vol. 12, pp. 213–220, 1993.
60. Y. Lin, "Comparison of transmission-type and reflection-type surface plasmon resonance based fiber-optic sensor," in *International Symposium on Computer, Consumer and Control (IS3C),* pp. 436–438, Xi'an, 2016.
61. P. Yeh, A. Yariv and E. Marom, "Theory of Bragg fiber," *Journal of the Optical Society of America,* vol. 68, no. 9, pp. 1196–1201, 1978.
62. J. Knight, T. A. Birks, P. Russell and D. M. Atkin, "All-silica single-mode optical fiber with photonic crystal cladding," *Optics Letters,* vol. 21, no. 19, pp. 1547–1549, 1996.
63. A. M. Pinto and M. Lopez-Amo, "Photonic crystal fibers for sensing applications," *Journal of Sensors,* vol. 2012, pp. 1–21, 2012.
64. Y. Liu and H. M. Salemink, "Photonic crystal-based all-optical on-chip sensor," *Optics Express,* vol. 20, no. 18, pp. 19912–19920, 2012.
65. D. Yong, X. Yu, C. Chan, Y. Zhang and P. Shum, "Photonic crystal fiber integrated microfluidic chip for highly sensitive real-time chemical sensing," in *International Conference on Optical Fiber Sensors, 775343,* Ottawa, Canada, 2011.
66. Y. Zhao, Z.-Q. Deng and J. Li, "Photonic crystal fiber based surface plasmon resonance chemical sensors," *Sensors and Actuators B: Chemical,* vol. 202, pp. 557–567, 2014.

67. R. Buczynski, "Photonic crystal fibers," *Acta Physica Polonica A,* vol. 106, pp. 141–167, 2004.

68. J. A. Kim, T. Hwang, S. R. Dugasani, R. Amin, A. Kulkarni, S. H. Park and T. Kim, "Graphene based fiber optic surface plasmon resonance for bio-chemical sensor applications," *Sens. Actuator B Chem.,* vol. 187, pp. 426–433, 2013.

69. C. Lin and S. Chen, "Design of high-performance Au-Ag-dielectric-graphene based surface plasmon resonance biosensors using genetic algorithm," *Journal of Applied Physics,* vol. 125, p. 113101, 2019.

70. S. Chen and C. Lin, "Sensitivity comparison of graphene based surface plasmon resonance biosensor with Au, Ag and Cu in the visible region," *Materials Research Express,* vol. 6, no. 5, p. 056503, 2019.

71. B. Shuai, L. Xia and D. Liu, "Coexistence of positive and negative refractive index sensitivity in the liquid-core photonic crystal fiber based plasmonic sensor," *Opt. Express,* vol. 20, pp. 25858–25866, 2012.

72. A. A. Melo, M. F. S. Santiago, T. B. Silva, C. S. Moreira, R. M. S. Cruz, "Investigation of a D-shaped optical fiber sensor with graphene overlay," *IFAC-PapersOnLine,* vol. 51, pp. 309–314, 2018.

73. G.An, S. Li, T. Cheng, X. Yan, X. Zhang, X. Zhou, Z. Yuan, "Ultra-stable D-shaped optical fiber refractive index sensor with graphene-gold deposited platform," Plasmonics, vol. 14, pp. 155–163, 2019.

74. M. Mollah, S. R. Islam, M. Yousufali, L. F. Abdulrazak, M. B. Hossain and I. S. Amiri, "Plasmonic temperature sensor using D-shaped photonic crystal fiber," *Results in Physics,* vol. 16, p. 102966, 2020.

75. N. Chen, M. Chang, X. Lu, J. Zhou and X. Zhang, "Numerical analysis of midinfrared D-shaped photonic-crystal-fiber sensor based on surface-plasmon-resonance effect for environmental monitoring," *Applied Science,* vol. 10, p. 3897, 2020.

76. Y. Chen, Q. Xie, X. Li, H. Zhou, X. Hong and Y. Geng, "Experimental realization of D-shaped photonic crystal fiber SPR sensor," *Journal of Physics D: Applied Physics,* vol. 50, no. 2, p. 025101, 2017.

77. A. Yasli, H. Ademgil and S. Haxha, "D-shaped photonic crystal fiber based surface plasmon resonance sensor," in *26th Signal Processing and Communications Applications Conference (SIU),* pp. 1–4, Izmir, 2018.

78. A. Hassani and M. Skorobogatiy, "Design of the microstructured optical fiber-based surface plasmon resonance sensors with enhanced microfluidics," *Optics Express,* vol. 14, no. 24, p. 11616, 2006.

79. C. Liu, L. Yang, W. Su, F. Wang, T. Sun, Q. Liu, H. Mu and P. K. Chu, "Numerical analysis of a photonic crystal fiber based on a surface plasmon resonance sensor with an annular analyte channel," *Optics Communications,* vol. 382, pp. 162–166, 2017.

80. Y. Lu, M. T. Wang, C. J. Hao, Z. Q. Zhao and J. Q. Yao, "Temperature sensing using photonic crystal fiber filled with silver nanowires and liquid," *IEEE Photonics Journal,* vol. 6, no. 3, p. 6801307, 2014.

81. H. Esmaeilzadeh, M. Rivard, E. Arzi, F. Legare and A. Hassani, "Smart textile plasmonic fiber dew sensors," *Optics Express,* vol. 23, no. 11, p. 14981, 2015.

82. T. Wu, Y. Shao, Y. Wang, S. Cao, W. Cao, F. Zhang, C. Liao, J. He, Y. Huang, M. Hou and Y. Wang, "Surface plasmon resonance biosensor based on gold-coated side-polished hexagonal structure photonic crystal fiber," *Optics Express,* vol. 25, no. 17, p. 20313, 2017.

83. M. Singh and S. K. Raghuwanshi, "Metal-insulator-metal waveguide based passive structures analyzed by transmission line model," *Superlattices and Microstructures,* vol. 114, pp. 233–241, 2018.

84. Y.-F. C. Chau, C.-T. C. Chao and H.-P. Chiang, "Ultra-broad bandgap metal-insulator-metal waveguide filter with symmetrical stubs and defects," *Results in Physics,* vol. 17, p. 103116, 2020.

85. M. A. Butt, S. N. Khonina and N. L. Kazanskiy, "A plasmonic colour filter and refractive index sensor applications based on metal-insulator-metal square micro-ring cavities," *Laser Physics,* vol. 30, p. 016205, 2020.

86. D.V. Nesterenko, "Resonance characteristics of transmissive optical filters based on metal/dielectric/metal structures," *Computer Optics,* vol. 44, pp. 219–228, 2020.

87. D.V. Nesterenko, R.A. Pavelkin, S. Hayashi, "Estimation of resonance characteristics of single-layer surface-plasmon sensors in liquid solutions using Fano's approximation in the visible and infrared regions," *Computer Optics,* vol. 43, pp. 596–604, 2019.

88. Y. Chen, Y. Xu and J. Cao, "Fano resonance sensing characteristics of MIM waveguide coupled square convex ring resonator with metallic baffle," *Results in Physics,* vol. 14, p. 102420, 2019.

89. X. Yang, E. Hua, M. Wang, Y. Wang, F. Wen and S. Yan, "Fano resonance in a MIM waveguide with two triangle stubs coupled with a split-ring nanocavity for sensing application," *Sensors,* vol. 19, no. 22, p. 4972, 2019.

90. J. Zhou, H. Chen, Z. Zhang, J. Tang, J. Cui, C. Xue and S. Yan, "Transmission and refractive index sensing based on fano resonance in MIM waveguide-coupled trapezoid cavity," *AIP Advances,* vol. 7, p. 015020, 2017.
91. M. A. Butt, S. N. Khonina and N. L. Kazanskiy, "Ultra-short lossless plasmonic power splitter design based on metal-insulator-metal waveguide," *Laser Physics,* vol. 30, p. 016201, 2020.
92. M. Pu, N. Yao, C. Hu, X. Xin, Z. Zhao, C. Wang and X. Luo, "Directional coupler and nonlinear Mach-Zehnder interferometer based on metal-insulator-metal plasmonic waveguide," *Optics Express,* vol. 18, no. 20, pp. 21030–21037, 2010.
93. M. A. Butt, "Numerical investigation of a small footprint plasmonic Bragg grating structure with a high extinction ratio," *Photonics Letters of Poland,* vol. 12, no. 3, pp. 82–84, 2020.
94. M. T. Hill, et al., "Lasing in metal-insulator-metal sub-wavelength plasmonic waveguides," *Optics Express,* vol. 17, no. 13, pp. 11107–11112, 2009.
95. C. Nylander, B. Liedberg and T. Lind, "Gas detection by means of surface plasmon resonance," *Sens. Actuators,* vol. 3, pp. 79–88, 1982.
96. M. A. Butt, S. N. Khonina and N. L. Kazanskiy, "Plasmonic refractive index sensor based on M-I-M square ring resonator," in *International Conference on Computing, Electronic and Electrical Engineering (ICE Cube), 1–4,* Quetta, Pakistan, 2018.
97. M. A. Butt, S. N. Khonina and N. L. Kazanskiy, "A multichannel metallic dual nano-wall square split-ring resonator: Design analysis and applications," *Laser Physics Letter,* vol. 16, p. 126201, 2019.
98. N. L. Kazanskiy, M. A. Butt and S. N. Khonina, "Nanodots decorated MIM semi-ring resonator cavity for biochemical sensing," *Photonics and Nanostructures-Fundamentals and Applications,* vol. 42, p. 100836, 2020.
99. M. A. Butt, S. N. Khonina and N. L. Kazanskiy, "Metal-insulator-metal nano square ring resonator for gas sensing applications," *Waves in Random and Complex Media,* vol. 31 (1), pp. 146–156, 2021.
100. N. L. Kazanskiy, M. A. Butt, S. A. Degtyarev and S. N. Khonina, "Achievements in the development of plasmonic waveguide sensors for measuring the refractive index," *Computer Optics,* vol. 44, no. 3, pp. 295–318, 2020.
101. M. A. Butt, N. L. Kazanskiy and S. N. Khonina, "Modal characteristics of refractive index engineered hybrid plasmonic waveguide," *IEEE Sensors Journal,* vol. 20, no. 17, pp. 9779–9786, 2020.

102. N. L. Kazanskiy, S. N. Khonina and M. A. Butt, "Subwavelength grating double slot waveguide racetrack ring resonator for refractive index sensing application," *Sensors,* vol. 20, no. 12, p. 3416, 2020.

103. M. A. Butt, S. N. Khonina and N. L. Kazanskiy, "A highly sensitive design of subwavelength grating double-slot waveguide microring resonator," *Laser Physics Letters,* vol. 17, no. 7, p. 076201, 2020.

104. M. A. Butt, S. N. Khonina and N. L. Kazanskiy, "Sensitivity enhancement of silicon strip waveguide ring resonator by incorporating a thin metal film," *IEEE Sensors Journal,* vol. 3, pp. 1355–1362, 2020.

105. N. L. Kazanskiy, M. A. Butt and S. N. Khonina, "Carbon dioxide gas sensor based on polyhexamethylene biguanide polymer deposited on silicon nano-cylinders metasurface," *Sensors,* vol. 21, no. 2, p. 378, 2021.

106. N. L. Kazanskiy, M. Butt and S. N. Khonina, "Silicon photonic devices realized on refractive index engineered subwavelength grating waveguides: A review," *Optics and Laser Technology,* vol. 138, p. 106863, 2021.

107. M. A. Butt, S. N. Khonina and N. L. Kazanskiy, "Ultrashort inverted tapered silicon ridge-to-slot waveguide coupler at 1.55 micrometer and 3.392 micrometer wavelength," *Applied Optics,* vol. 59, no. 26, pp. 7821–7828, 2020.

108. M. A. Butt and S. A. Fomchenkov, "A polarization-independent highly sensitive hybrid plasmonic waveguide structure," in *7th International School and Conference "Saint Petersburg OPEN 2020": Optoelectronics, Photonics, Engineering and Nanostructures, April 27–30,* Saint Petersburg, Russia, 2020.

109. M. Butt, A. Kazmierczak, N. L. Kazanskiy and S. N. Khonina, "Metal-insulator-metal waveguide-based racetrack integrated circular cavity for refractive index sensing application," *Electronics,* vol. 10, p. 1419, 2021.

110. M. A. Butt, N. L. Kazanskiy and S. N. Khonina, "Highly sensitive refractive index sensor based on plasmonic Bow Tie configuration," *Photonic Sensors,* vol. 10, no. 3, pp. 223–232, 2020.

111. M. A. Butt, N. L. Kazanskiy and S. N. Khonina, "Highly integrated plasmonic sensor design for the simultaneous detection of multiple analytes," *Current Applied Physics,* vol. 20, no. 11, pp. 1274–1280, 2020.

Chapter 4

LSPR-Based Optical Sensors and Biosensors

Divagar M.,[a] Athira E. T.,[b] V. V. R. Sai,[c] and Jitendra Satija[d]
[a]*Department of Biotechnology, Indian Institute of Technology Madras, Chennai, Tamil Nadu 600036, India*
[b]*Centre for Nanobiotechnology, Vellore Institute of Technology, Vellore, Tamil Nadu 632014, India*
[c]*Department of Applied Mechanics, Indian Institute of Technology Madras, Chennai, Tamil Nadu 600036, India*
[d]*Centre for Nanobiotechnology, Vellore Institute of Technology, Vellore, Tamil Nadu 632014, India*
bt17d302@smail.iitm.ac.in, athiraet85@gmail.com,
vvrsai.iitm.ac.in, jsatija@vit.ac.in

4.1 Optical Sensors

Optical sensing, based on color or fluorescence, has been the most established and also preferred method for several decades, mainly due to the outcomes that can be appreciated either by the naked eye directly or through optical microscopy [1, 2]. A tremendous

Plasmonics-Based Optical Sensors and Detectors
Edited by Banshi D. Gupta, Anuj K. Sharma, and Jin Li

ISBN 978-981-4968-85-0 (Hardcover), 978-1-003-43830-4 (eBook)
www.jennystanford.com

amount of research has gone into the fields of physical and chemical sciences to realize an umpteen number of sensing schemes for various applications ranging from the identification of chemical elements and compounds to biological molecules and systems [3, 4]. Especially in the field of optical sensors involving biological molecules, significant advances have been made in the later half of the previous century, where complex biochemical systems involving enzymes as biocatalysts and simpler chromophoric molecules were developed to demonstrate analyte detection down to picomolar as well as femtomolar concentrations [5]. These systems are capable of generating an optical output in the form of absorbance, fluorescence, or luminescence. While the first generation of these systems is chromogenic in nature [6], the later generation systems exploit luminescence that offers three-order improvement in sensitivity [7, 8]. However, multiple factors, including rising disease outbreaks, demand for better healthcare diagnostics, and stringent safety standards for water, food, and air to combat increasing levels of environmental pollution or contamination, have created greater challenges for sensors to be more sensitive, robust, reliable, and sometimes field-deployable.

Over the past five decades, nanoscience and technology have been empowered to redefine the frontiers in sensors in a much similar way to the silicon semiconductor and its microfabrication have enabled highly efficient and miniaturized electronics. Among the vast range of nanomaterials developed, the plasmonic nanoparticles stand tall with a list of a variety of applications in research as well as commercial space, owing to their exceptional optical properties including absorption and scattering (e.g., the bright ruby color of colloidal gold). Especially, the strong scattering of light exhibited by individual nanoparticles facilitates optical observation of each individual nanoparticle as small as 30 nm AuNP or 20 nm AgNP under dark-field microscopy. It is interesting to note that the scattering flux of a single 80 nm AgNP under typical white light has been reported to be equivalent to 5×10^6 fluorescein molecules. In addition, these noble metallic nanoparticles neither blink nor bleach, unlike conventional fluorophores. Besides, the plasmonic nanostructures also enable real-time monitoring, short-response time, label-free detection schemes, high reusability, and simple sample processing. In addition, their photostability, high extinction coefficient, and ability to conjugate with proteins make them highly suitable as labels in biomolecular assays [9].

This chapter briefly introduces the LSPR phenomenon of the plasmonic gold and silver nanoparticles, followed by its exploitation in realizing various sensing strategies including label-free and labeled biosensors. The use of gold and silver colloids for colorimetric sensing of biomolecules is discussed in detail. In line with the conventional SPR systems, the development of label-free plasmonic sensors by an assembly of colloidal gold nanoparticles over flat substrates and sensitive fiber optic transducers is presented. When used as labels, the exceptional ability of the plasmonic nanoparticles toward the development of highly simplified and/or sensitive bioassays is elucidated. In addition, single particle spectroscopy and multiplexed sensor applications of the plasmonic nanostructures are briefly covered.

4.1.1 Localized Surface Plasmon Resonance (LSPR)

The LSPR phenomenon found in the noble-metal nanostructures imparts excellent optical properties in the visible and near-infrared regions of the electromagnetic spectrum, thereby making them an invaluable candidate for several colorimetric and fluorometric sensing applications. Oscillations of the conducting electrons confined to the nanoparticle surface upon excitation by the incident light are responsible for this phenomenon. When the size of the metal nanoparticle/nanostructure is less than 1/10 of the interacting wavelength of the electromagnetic field, the conduction band electrons are set into coherent oscillation, which is termed as a localized surface plasmon [10]. The localized surface plasmon resonates with the interacting electromagnetic field at specific frequencies, which, in turn, depends on the dielectric constant, size, and shape of the nanoparticle. This resonating LSPR band is highly sensitive to the surrounding medium. For instance, any change in the refractive index of the surrounding medium or adsorption of molecules onto the nanoparticle surface could alter the LSPR band [11].

The intense red color of the colloidal AuNPs is the manifestation of the LSPR phenomenon. In the early 20th century, Gustav Mie solved Maxwell's equation and derived equations to explain the

LSPR band of small spheres ($d<<\lambda$). According to Mie's theory, for a spherical particle of radius R, which is much smaller than the wavelength of the light, the magnitude of the scattering cross-section is proportional to R^6, while the absorption cross-section is proportional to R^3 [12]. The sum of scattering and absorption cross-section gives rise to extinction cross-section. Thus, the LSPR extinction is dominated by absorption for smaller particles, and as the particle size increases, the scattering takes over. This implies that the LSPR extinction of the noble metal nanoparticles can be tuned. For instance, the LSPR band of gold nanoparticles (AuNPs) can be tuned over 60 nm by varying the size of the particle between 10 nm and 100 nm. In addition to the size of the particle, the aspect ratio also plays a major role in determining the LSPR extinction peak. For instance, the LSPR spectrum of AuNPs is different from that of the gold nanorods (AuNRs), due to the varying distribution of electron cloud [13]. It has been shown that the particle shape plays a major role in determining the sensitivity to its surrounding medium. In particular, it has been reported the nanostructures with sharp tips such as nanotriangles, nanostars (AuNSs), and bipyramids exhibit high refractive index sensitivity [14]. The improved sensitivities can either be attributed to the varying plasmon resonance wavelengths or the intrinsic effects of particle shape. Miller and Lazarides predicted a linear relationship between the particle sensitivity and the plasmon resonant wavelength [15]. However, it is not correlating well with particles of unique shapes such as AuNSs.

In addition, it was also reported that the sensitivity varies with material composition. The majority of the LSPR experiments have been carried out using gold or silver nanoparticles. AuNPs are often considered due to their resistance to oxidation and chemical stability. However, silver nanoparticles (AgNPs) have sharper resonance and high RI sensitivity. For example, AuNPs of 50–60 nm size have an RI sensitivity of 60 nm/RIU at 530 nm [16], while AgNPs of similar size have a RI sensitivity of about 160 nm/RIU at 435 nm [17]. These results oppose Miller and Lazarides' prediction and deviate from the linear relationship between the resonant wavelength and RI sensitivity. Instead, the difference in the RI sensitivity between AgNPs and AuNPs lies in the dielectric functions of these metals. The real dielectric function of bulk silver varies with the wavelength more than that of the bulk gold within 400–800 nm, where the plasmon resonance lies for a reasonable value of the dielectric constant of the medium [18]. Also, the imaginary part of the dielectric function of silver is less than that of gold across the visible region, less plasmon damping occurs, thus resulting in higher scattering efficiency and narrow plasmon line widths. Link et al. studied gold–silver alloys and demonstrated material composition has a greater effect on sensitivity compared to particle size (Link et al., 1999).

4.2 Solution Phase–Based LSPR Sensors

These sensors typically rely on a chemical reaction-triggered mechanism, which either induces the aggregation of the plasmonic nanoparticles or transforms their geometry. The sensing modality mainly includes aggregation, anti-aggregation, etching, and metallization of plasmonic nanoparticles [19–24]. The sensing is usually carried out in small vials, cuvettes, or 96-well plates, which makes this process very simple, easy, and fast [25]. To date, a wide range of sensing schemes has been developed by utilizing various types of plasmonic nanoparticles and specific chemical and biochemical reactions (Table 4.1).

Table 4.1 Overview of the solution phase–based LSPR sensors

Type of nanoparticle	Analyte	Mechanism of sensing	LOD	Ref.
AuNPs	AFM1	Aggregation	0.002 ng/mL	[27]
AuNPs	MCF-7 cells	Aggregation	10 cfu/mL	[31]
AgNPs	Creatinine	Aggregation	53.4 nM	[33]
AuNPs	17β-estradiol	Aggregation	3 pg/mL	[32]
AuNPs	Chromium (III) Chromium (VI)	Aggregation	2 ng/mL 3 ng/mL	[43]
AuNRs	AFM1	Etching	0.11 ng/mL	[37]
AuNRs	ALP	Etching	100 pg/mL 3 ng/mL (Visual)	[34]
Au@Ag nanorod	BPO	Etching	0.75 μmol/L	[35]
AuNRs	Listeria monocytogenes	Etching	10 cfu/mL	[36]

(*Continued*)

Table 4.1 (*Continued*)

Type of nanoparticle	Analyte	Mechanism of sensing	LOD	Ref.
AuNRs	AFB1	Etching	12.5 pg/mL	[44]
AgNPs@ AuNCs	Vc	Growth	24 nM	[42]
AgNPs@ AuNRs	β-gal	Growth	128 pM	[39]
AgNPs@AuNS	ALP DNA	Growth	0.5 pM 2.6 fM	[45]
AgNPs@Au NBPs	Influenza Virus	Growth	1 pg/mL	[40]
AuNRs	PSA	Growth	3 fg/mL	[41]

AuNPs: Gold nanoparticles, AuNRs: Gold nanorods, AgNPs: Silver nanoparticles, AuNCs: Gold nanocages, AuNs: Gold nanostars, AFM1: Aflatoxin M1, ALP: Alkaline phosphatase, BPO: Benzoyl peroxide, AFB1: Aflatoxin B1, VC: Vitamin C, β-gal: Beta-galactosidase

4.2.1 Aggregation-Based LSPR Biosensor

Label-free aggregation-based biosensors have become increasingly popular in recent years, due to their ease of use, visible color change, and cost-effectiveness [26, 27]. The aggregation of noble metal nanoparticles causes a defined color change and LSPR peak shift [28, 29]. Since the optical characteristics of nanostructures of noble metals are strongly distance-dependent, when these come into proximity, plasmonic overlap occurs that eventually modulates the absorption spectra and scattering profile, resulting in changes in the color of the nanoparticle solution [30, 31]. Attaining the controlled nanoparticle orientation in the solution, especially for anisotropic nanoparticles, is extremely difficult to establish. As a result, spherical or quasi-spherical nanoparticles are being used in most aggregation-based plasmon coupling-based biosensors [19]. Further, the strength of the observed signals is influenced by the size of aggregated particles used for the detection. It has been shown that even at relatively large interparticle distances, a quantifiable signal can be obtained if the aggregate size is large enough. Therefore, the sensitivity of plasmonic nanoparticle aggregation-based sensors can be improved by increasing the aggregate size.

Aggregation of AuNPs has been realized to develop a colorimetric sensor for quick quantitative visual detection of aflatoxin M1 (AFM1) in the milk sample [27]. In this approach, an aptamer (72-Mers ssDNA) was used as a selective toxin-recognizing moiety, and the AuNPs were employed as a colorimetric probe that produced a predictable color change from red to blue upon triggering the aggregation (Fig. 4.1). The sensing is based on the fact that positively charged bases of aptamers interact with citrate stabilized negatively charged AuNPs through the coordination bond, and thus enhance their stability even in the presence of NaCl (i.e., aggregation inducing agent). In the presence of specific target molecules, i.e., AFM1, the randomly structured aptamer transforms into secondary structures, such as hairpins, hydrocarbon loops, and beacons, which could not bind to AuNPs and lead to destabilization and aggregation of AuNPs occurs. A detectable change in color from red to blue in the colloidal solution of AuNPs solution indicated a quantitative determination of the analyte in the sample. The plasmonic sensor could achieve the LOD value of 0.002 ng/mL with 80–110% of the recovery.

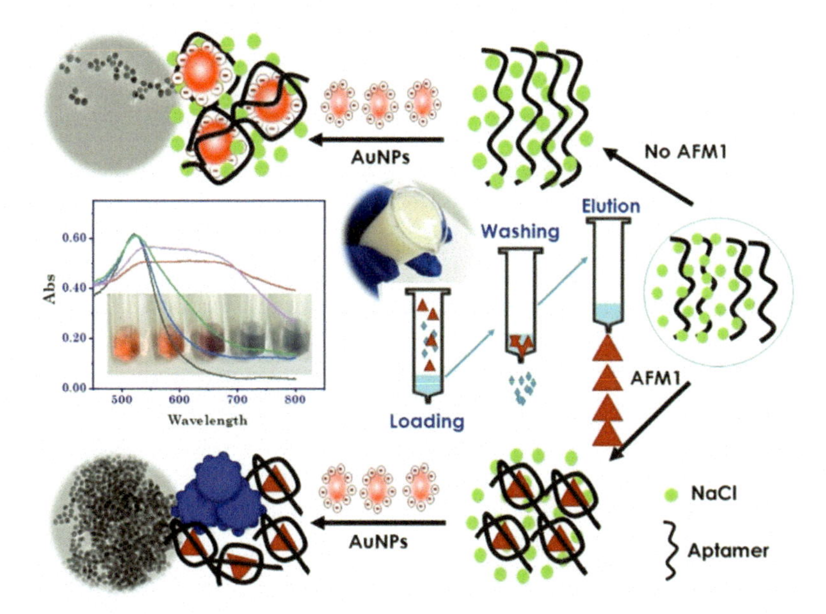

Figure 4.1 Schematic representation of AuNPs aggregation-based sensing of aflatoxin M1 (AFM1). Reproduced from Ref. [27], Copyright (2021), with permission from Elsevier.

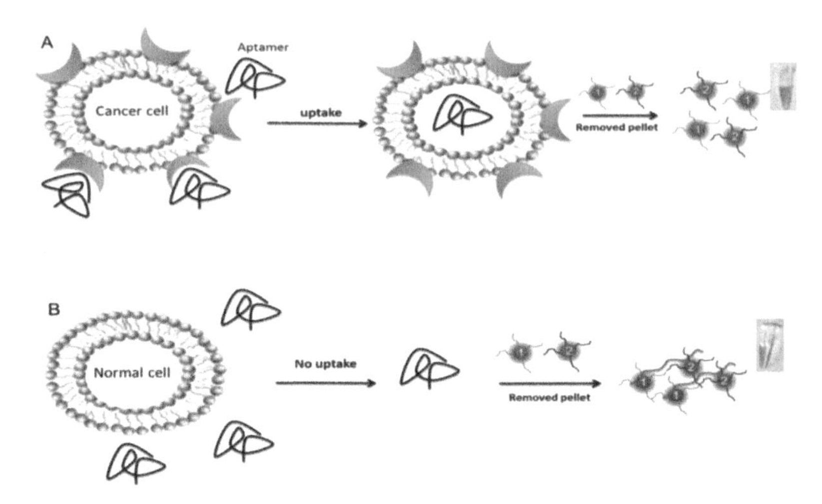

Figure 4.2 Schematic representation of colorimetric detection of cancer cells via AuNPs aggregation. Reproduced from Ref. [31], Copyright (2016), with permission from Elsevier.

In a similar approach, AuNP aggregation-based strategy has been utilized to detect circulating MCF-7 cells (a human breast cancer cell) via colorimetric [31]. Since phosphoprotein nucleolin is overexpressed on the cancer cells, nucleolin-specific aptamer AS 1411 was used as a selective recognizing moiety (Fig. 4.2). In order to develop the sensor, two different DNA functionalized AuNPs were used as probes where both the DNA had the specific sequence that is partially complementary to the aptamer. In the absence of cancer cells, the aptamer binds with the complementary DNA strands, and a change in the AuNPs solution color from pink to purple color is observed, which is associated with the aggregation of AuNP. In contrast, in the presence of cancer cells, the aptamer binds strongly with cancer cells due to their high affinity toward nucleolin receptors. As a result, the aptamer does not interact with AuNPs and therefore, no change in the AuNP solution and the plasmonic peak is observed. The developed sensors exhibited a detection limit of 10 cells with a linearity range of $10–10^5$ cells. In a slightly different approach, an LSPR-based multicolorimetric biosensor has been developed for the detection of 17β-estradiol (E2) in tap water [32]. In this approach, E2 antigen-specific antibodies are immobilized on the surface AuNPs (Fig. 4.3). These functionalized AuNPs are then treated with samples containing E2 antigen, resulting in their binding on the free antigen binding F_c region of the antibody. The E2 antigen acts as a linker due to multiple binding sites inducing the aggregation of the nanoparticles. The aggregation of AuNPs produced a variety of colors, which could be easily observed with the naked eye at a glance. The detection limit of the developed assay was found to be 3 pg/mL.

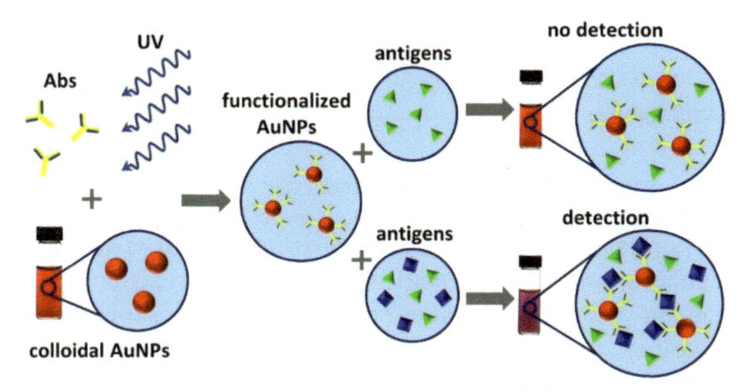

Figure 4.3 Schematic illustration of colorimetric immunosensor based on E2 antigen (blue square) induced aggregation of antibody functionalized AuNPs. Reproduced from Ref. [32], Copyright (2020), with permission from Elsevier.

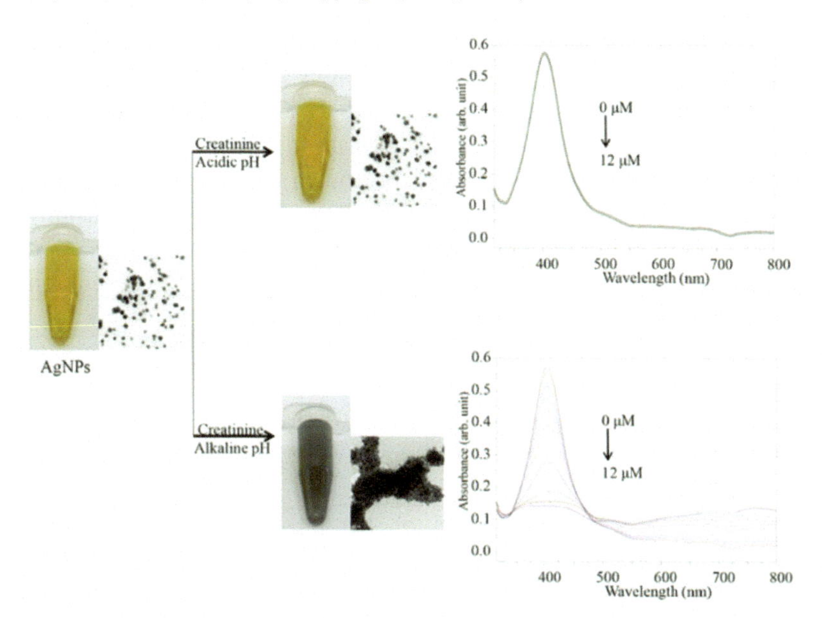

Figure 4.4 Detection of creatinine in the urine based on AgNPs aggregation. Reproduced from Ref. [33], Copyright (2018), with permission from Elsevier.

In a different approach, a citrate-capped AgNPs aggregation-based sensor was developed for the quantification of creatinine in the urine sample, as shown in Fig. 4.4 [33]. The sensing was based on the fact that under alkaline conditions creatinine is converted from its imino tautomer to amino tautomer. The amino tautomer forms a hydrogen bond with the negatively charged citrate-capped AgNPs, causing their aggregation and turning the solution color yellow to dark blue/grey along with a redshift in the LSPR peak. This method could quantify the creatinine in the urine sample with a LOD value of 53.4 pg/mL.

4.2.2 Etching-Based LSPR Biosensor

In this strategy, the etching of the nanoparticle is induced by means of any enzymatic or non-enzymatic reaction and the reaction between plasmonic nanoparticles and biocatalytic products or oxidizing agent causes the etching of the particles in an analyte concentration-dependent manner [21]. This eventually results in a change in the size and shape of nanoparticles that can be easily observed with the naked eye or via measuring the change in characteristic LSPR peak absorbance/wavelength [23]. Etching-based strategies can be categorized based on the type of etchants such as alloys, enzymes, intermediates, inhibitors, and direct etching by targets.

One of the most popular strategies is the plasmonic-ELISA in which the color development is based on plasmonic nanoparticles etching by following a standard sandwich ELISA procedure. For example, iodine-triggered AuNRs etching has been successfully explored for developing a plasmonic ELISA strategy for the detection of alkaline phosphatase (ALP) [34]. In this strategy, the human immunoglobulin G (HIgG) was utilized as an analyte protein model and goat anti-HIgG antibody as the capture antibody. Antibody specific to HIgG and tagged with ALP was utilized as detector antibody. After forming a sandwich immunocomplex in the well, ascorbic acid 2-phosphate (AA-P) was added and hydrolyzed by ALP enzyme to ascorbic acid (AA) (Eq. (4.1)) (Fig. 4.5). Then sodium bromide (NaBr), cetyltrimethylammonium bromide (CTAB), potassium iodate (KIO_3), and AuNRs were introduced into the reaction well. The iodate gets reduced to iodine by AA which eventually reacts with

AuNRs and forms AuI_2^--$(CTA^+)_2$ and led to AuNRs etching (Eq. (4.2)). A change in color of the solution from blue to red was observed with the decrease in the aspect ratio of AuNRs or as a function of HIgG concentration. This assay exhibited the detection limit of 100 pg/mL for HIgG in bovine serum and the visual detection limit is found to be 3.0 ng/mL.

$$5C_6H_8O_6 + 2IO_3^- + 2H^+ \rightarrow 5C_6H_6O_6 + I_2 + 6H_2O \quad (4.1)$$

$$2Au + I_2 + 4CTA^+ \rightarrow 2AuI_2^-\text{-}(CTA^+)_2 \quad (4.2)$$

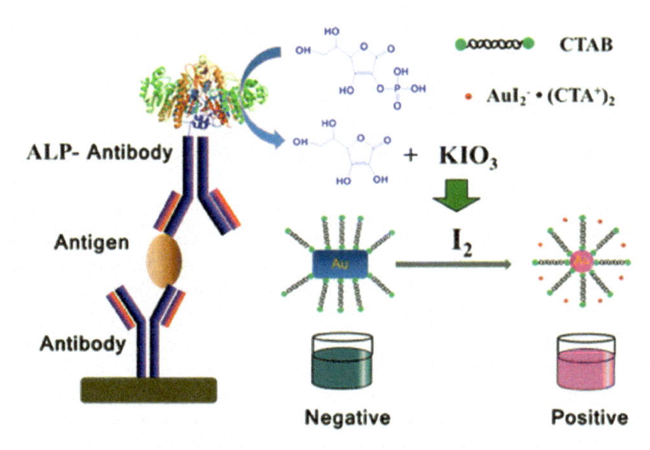

Figure 4.5 Schematic illustration of plasmonic ELISA based on ALP-triggered AuNRs etching. Reproduced with permission from Ref [34], Copyright (2015), American Chemical Society.

In another study, bimetallic nanoparticle etching was exploited for developing a plasmonic sensor for benzoyl peroxide (BPO) detection [35]. The sensing was based on the fact that BPO has the ability to oxidize the Ag^0 into Ag^+, resulting in the removal of the Ag-coating from the Au@Ag nanorod (Fig. 4.6). This portable sensor could visually detect the BPO in the flour sample with a LOD of 0.75 μM. With an increase in the concentration of BPO, the rate of etching of Ag nanoshells was also found to be increased and consequently, a variation in LSPR peak and solution color was observed. However, the presence of any other oxides or peroxides may also cause the etching of nanoparticles, therefore the selectivity of the assay is limited.

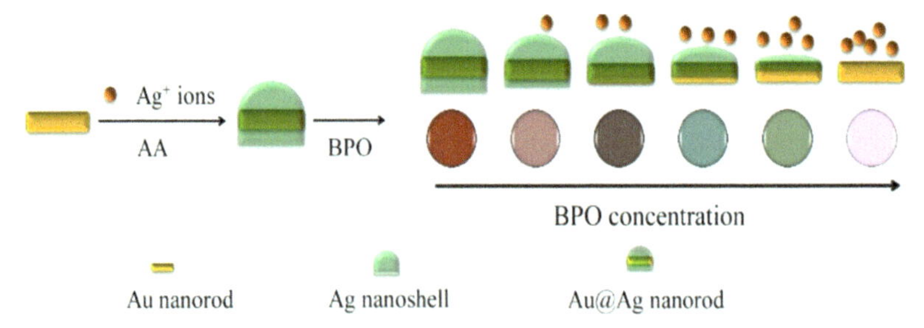

Figure 4.6 Gold-silver nanorods etching mediated colorimetric detection of benzoyl peroxide (BPO). Reproduced from Ref. [35], Copyright (2018), with permission from Elsevier.

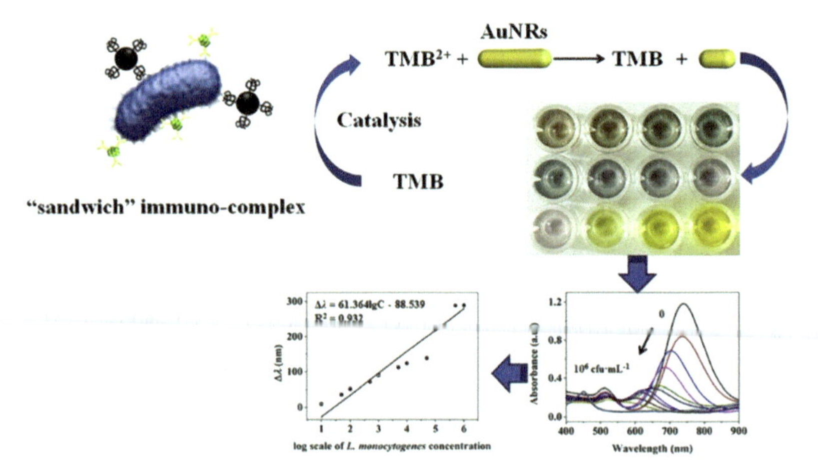

Figure 4.7 Schematic representation of *Listeria monocytogenes* detection with an AuNRs etching-based multicolorimetric assay. Reproduced from Ref. [36], Copyright (2019), with permission from Elsevier.

In another approach, a quantitative multicolorimetric sensor was developed for food poisoning bacteria *Listeria monocytogenes* by employing the 3,3',5,5'-Tetramethylbenzidine (TMB)-triggered etching of gold nanorods [36]. In this study, aptamer-functionalized citric acid-coated iron oxide nanoparticles were used as capture probes, which form a complex with *Listeria monocytogenes* and subsequently added IgY-functionalized manganese oxide (MnO_2) nanoparticles (Fig. 4.7). Thereafter, TMB was added to the reaction solution, which was catalyzed by MnO_2 nanoparticles and yielded TMB^+ (blue color), which was further converted into TMB^{2+} (yellow color) by the addition of hydrochloric acid (HCl). The TMB^{2+} catalyzes the etching of gold nanorods, resulting in a vivid color change proportional to the bacterial concentration. The developed sensor could detect 10 cfu/mL of bacteria with a 97.4–101.3% recovery rate from the spiked food samples. This newly proposed multicolorimetric strategy has the potential to be a reliable and sensitive test for the quick sensing of a wide range of bacteria.

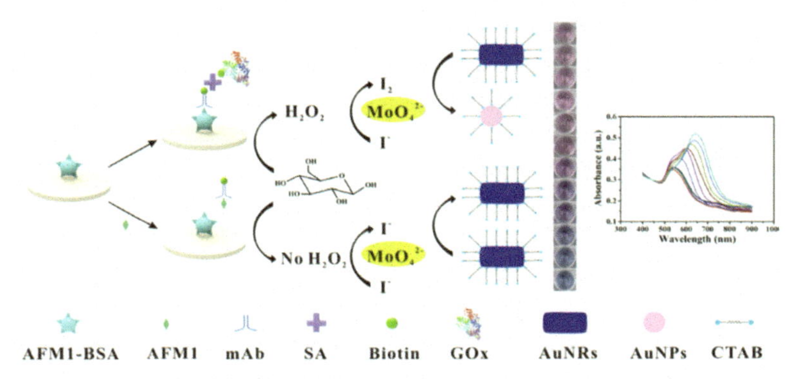

Figure 4.8 Gold nanorods etching-based plasmonic ELISA for the detection of AFM1 in the milk sample. Reproduced from Ref. [37], Copyright (2020), with permission from Elsevier.

In another approach, chemical reaction-induced etching of noble metal nanoparticles resulting in a prominent shift in the LSPR peak and visual color change has been realized for analyte detection. For example, Fang et al. developed a competitive ELISA for AFM1 detection in milk by etching of AuNRs [37]. In this indirect competitive plasmonic ELISA, the biotinylated anti-AFM1 monoclonal antibody treated with a test sample was introduced into a 96-well plate, which is pre-coated with BSA-AFM1 competing for antigen (Fig. 4.8). This is followed by conjugation of biotinylated GOx enzyme to this complex by means of streptavidin as a linker. Only in the absence of an AFM1 analyte, the anti-AFM1 monoclonal antibody binds to the BSA-AFM1 and subsequently, binding of biotinylated GOx occurs via streptavidin linkage. Thereafter, GOx is used to catalyze the glucose oxidation yielding H_2O_2. This was followed by the introduction of iodide (I^-) and AuNRs. The produced H_2O_2 oxidizes the I^- into I_2 in the presence of molybdate $(MoO_4)^{2-}$, which causes the etching of the AuNRs and results in a prominent change in the LSPR peak and solution color. According to the concentration of antigen, the resultant solution showed different shades of colors from pink to blue. In the presence of AFM1, the anti-AFM1 antibody binds with the antigen and thus will not be available to form the immunocomplex (bio-mAb-SA-bio-GOx) with competitor antigens. Therefore, without the GOx enzyme, the etching of AuNRs will not happen, thus LSPR peak and solution color will remain unaltered. This approach demonstrated the visual limit of detection of 0.11 ng/mL toward the AFM1 detection.

In another approach, a robust real-time-multiplexed plasmonic sensor for the simultaneous detection of H_2O_2 and glucose was developed [38]. In this approach, a 96-well plate was modified with polydopamine film functionalized with catechin (CT) coated AuNPs (PDA@Au/CT) (Fig. 4.9). Thereafter, a nanostructured Ag network was created by CT-mediated reduction of Ag^+ ions forming a self-assemble layer of Ag on PDA@Au/CT. In the presence of H_2O_2, the Ag layer gets etched and results in the LSPR peak shift as a function of analyte concentration. This technique was further extended to develop a glucose sensor where the GOx enzyme was used to catalyze the oxidation of glucose and generate H_2O_2 as a by-product. This H_2O_2 etches the Ag layer and a shift in the LSPR peak shift is noted. The LOD of the developed sensing approach was found to be 0.2 μM and 0.4 μM for H_2O_2 and glucose, respectively.

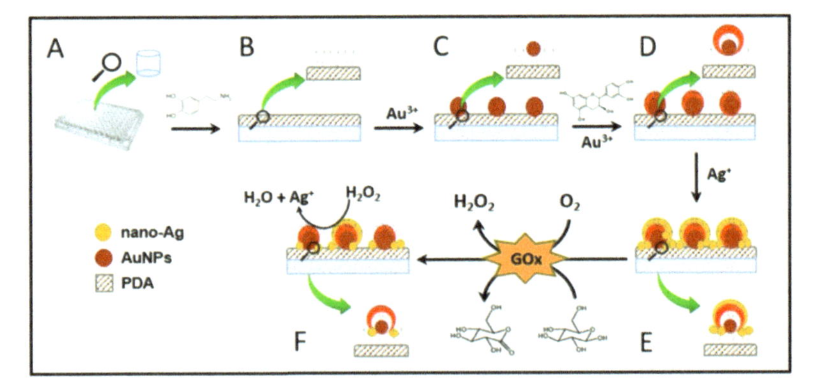

Figure 4.9 PDA@Au-CT@Ag-film formation onto plasmonic ELISA for the detection of H_2O_2 and glucose ELISA microwells and the resulting architecture. Reproduced from Ref. [38], Copyright (2020), with permission from Elsevier.

4.2.3 Growth-Based LSPR Biosensor

The growth or metallization of noble metal nanoparticles is based on the deposition of freshly reduced noble metal ions to form a stable thin film on the template nanoparticles. The change in the composition or the size of the nanoparticles results in a prominent color and LSPR peak shift. Growth of the template nanoparticles can be induced (i) directly by the target analyte, (ii) enzyme-promoted growth, and (iii) enzyme-passivated growth.

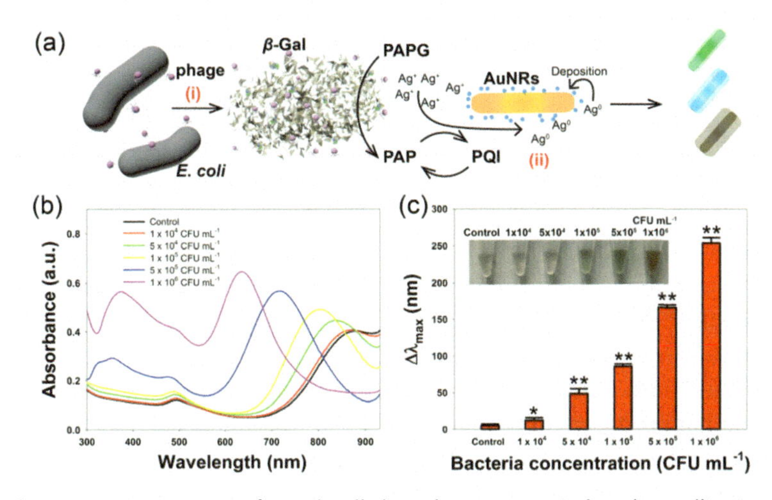

Figure 4.10 Detection of *E. coli* cells based on enzyme-induced metallization of Ag@AuNRs. Reproduced with permission from John Wiley & Sons, Inc. [39].

Silver metallization on the AuNRs surface has been achieved by enzyme-catalyzed reaction and successfully utilized for quantitative colorimetric detection of beta-galactosidase (β-gal) activity [39]. β-gal is an enzyme coded by *lacZ* genes present in *Escherichia coli* (*E. coli*), hence this method can be indirectly used to measure the bacterial cells in a sample (Fig. 4.10). Here, T7 bacteriophages were utilized to induce the β-gal production by *E. coli*, which hydrolyses the substrate p-aminophenyl β-d-galactopyranoside (PAPG) into galactoside and p-aminophenol (PAP). The PAP is a mild reducing agent, and it reduces the Ag^+ into Ag^0, which preferably gets deposited on the surface of AuNRs added during the bioassay and results in the size increment and composition change of the nanoparticles. This is manifested as a change in the color of the solution from light red to green and the LSPR peak. In the absence of *E. coli*, the PAPG remains unchanged and the solution appears pink. This method allows for the effective sensing of the β-gal and indirectly counts the *E. coli* cells with the naked eye. The detection limit was found to be 128 pM and 10^4 cfu/mL toward the β-gal and *E. coli*, respectively. In a similar pattern, a highly sensitive enzyme-assisted colorimetric sensor was developed for quantification determination of H5N1 influenza virus antigen by silver metallization on gold nanobipyramids (AuNBPs) [40]. This sensor could achieve the LOD of 1 pg/mL with good linearity in the range of 0.001–2.5 ng/mL.

In a different approach, AuNRs growth-based plasmonic assay was developed for prostate-specific antigen (PSA) using sandwich ELISA [41]. First, the capture antibody was immobilized on the 96-well plates followed by the sequential addition of antigen and ALP-tagged primary antibody (Fig. 4.11). In the presence of antigen, the immunocomplex is formed; while in the absence of antigen, the ALP-tagged antibody is washed off from the ELISA plate, so no immunocomplex is formed. The ALP converts the AA-P (i.e., enzyme-substrate) into AA, which reduces Au^{3+} ions, present in the growth solution (consisting of gold (III) chloride, CTAB, gold seeds, and silver nitrate), into Au^0, which eventually gets deposited on the surface of gold nanorod and thereby increases the size and aspect ratio of the resultant particle. A change in the refractive index of AuNRs and the color of the solution is correlated with antigen concentration qualitatively and quantitatively. Under the optimized conditions, the sensors could detect 3 fg/mL of PSA in human serum.

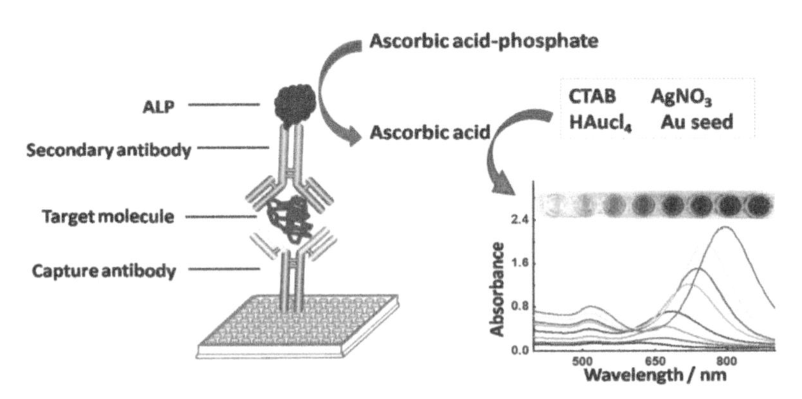

Figure 4.11 AuNRs etching-based plasmonic ELISA for PSA detection. Reproduced from Ref. [41], with permission from The Royal Society of Chemistry.

An easy and robust method for the detection of Vitamin C (VC) and other antioxidants with the metallization of AgNPs with very high sensitivity was reported by Wang et al. [42]. VC acts as a mild reducing agent and converts the ionic silver into the metallic form which preferably gets deposited on template gold nanocages (AuNCs). The refractive index of the resultant bimetallic nanoparticles and their size and morphology differs from the native AuNCs and thereby result in a color change and the LSPR peak shift quantitatively and qualitatively as a function of the concentration of VC. The LOD of the assay was found to be 24 nM.

4.3 Label-Free Plasmonic Chip-Based LSPR Sensors

It is important to realize the fact that the solution phase LSPR sensors monitor the optical properties of nanoparticles under Brownian motion. The number of nanoparticles under observation may have to be significant to pick up a measurable signal. The optical properties of these plasmonic nanoparticles may be efficiently monitored by immobilizing them on a substrate and limiting the binding reactions or refractive index changes to the limited number of nanoparticles. Thus, flat-surface or chip-based LSPR sensors are fabricated as an alternative to solution-based LSPR sensors. Here, gold nanostructures are immobilized onto a substrate made of silica, polymer, or silicon surface to realize portable and reproducible sensing technologies [46]. The sensitive chip-based LSPR sensors have numerous applications for detecting nucleic acids, biomolecules, carbohydrates, and toxins, due to their high reliability, reproducibility, and label-free strategy. However, the sensitivity of the LSPR chip-based sensors is inferior compared to labeled sensing strategies and SPR-based sensors. Despite being less sensitive, LSPR sensors are less affected by external temperature and have a high potential to be miniaturized [47].

The chip or flat substrate-based LSPR sensors could be fabricated through annealing and deposition procedures such as nanosphere lithography, holographic lithography, focused ion beam lithography, soft lithography, and template stripping (Bhalla et al., 2019). Utilizing the aforementioned techniques, nanostructures such as AuNPs, nanoshells, nanoantennas, AuNs, and nanopyramid arrays could be formed on a flat surface and exploited for realizing sensitive LSPR sensors.

The immobilized nanostructures exhibiting LSPR property are sensitive to the changes in the surrounding medium. An increase in the refractive index (RI) of the medium surrounding the LSPR structure induces a redshift, which is measurable [11]. In 2003, Mock and co-workers identified that the addition of oil to silver nanoparticles induced a color change from blue to green, which has also been characterized by a redshift in the LSPR spectra [17]. It has also been found that as the refractive index of oil increases, the LSPR peak gradually red-shifted and exhibited a linear relationship with the peak wavelength and refractive index of the medium. This inherent sensitivity of the LSPR peak to the surrounding medium is utilized to release LSPR-based sensors. In addition, it has been found that the shape of the nanostructures strongly influences the LSPR sensitivity, and nanoparticles with a higher aspect ratio exhibit superior RI sensitivity [25].

In addition, it has also been found that the interaction between the substrate and fabricated nanostructures may change the RI environment of the nanostructure, thereby delocalizing the field into the substrate [48]. It mainly occurs due to the fact that the electric field of the LSPR is highly dependent on the dielectric property of the surrounding environment, and any change in the dielectric property of the substrate affects the LSPR. Here, three different substrates, namely, glass, polymer, and silicon, reported in the literature, are briefly discussed.

4.3.1 Glass Substrate

Most of the chip or flat substrate-based LSPR sensors use glass as a substrate due to its established surface chemistry and adequate availability. The simplest way of making a glass substrate-based LSPR sensor is to dip the glass in amine or thiol-terminated silane and use it to immobilize plasmonic nanostructures, especially gold nanostructures including AuNPs, AuNRs, gold nanoshells, and gold nanopyramids.

Chilkoti and co-workers extensively worked on AuNRs-based LSPR glass chips. The nanorods coated on a glass substrate showed a

RI sensitivity of 252 nm/RIU, and it is at least five-folder higher than that of the AuNPs-based LSPR [46]. The developed sensor chips have also been investigated for studying biomolecular interactions using biotin-streptavidin as models. The results showed a detection limit down to 5 ng/mL in PBS and 1 ug/mL in serum [49]. The improved sensitivity of the AuNRs is also in correlation with the report by Chen et al., showing AuNRs-based glass chips exhibiting 366 nm/RIU, while AuNPs show a sensitivity of 76.4 nm/RIU. Another study in 2014 reported the development of LSPR lab-on-a-chip utilizing AuNRs forming an array on a glass slide. The developed sensor has been shown to detect cancer biomarkers human-alpha fetoprotein and prostate-specific antigen down to 500 pg/mL [50].

Oh and co-workers developed LSPR-based using 20 nm AuNPs on glass and plastic chips and subsequently compared them for the detection of *Salmonella typhimurium* in pork meat samples using aptamers (Fig. 4.12). It has been shown that after conjugation of the aptamers, the detected absorbance signals were in a higher range in the case of LSPR chips based on glass, compared to the plastic substrate [51].

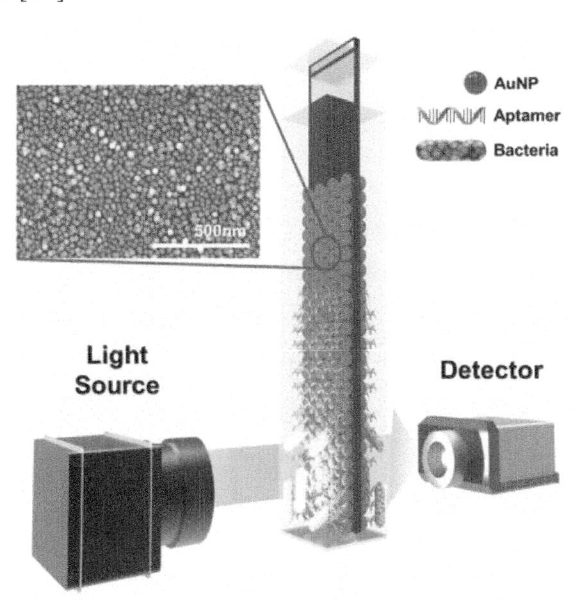

Figure 4.12 Schematic representation of bacterial detection using AuNPs-based LSPR sensor. Reproduced with permission from Springer Nature: Scientific Reports [51], Copyright (2017).

Considering the importance of substrate, the same group in the year 2019 investigated the influence of substrate material by comparing eight different sensor chips such as polycarbonate film, 0.4 mm glass, 0.5 mm glass, 0.4 mm tempered glass, 0.5 mm chemical-strengthened glass, microscopic glass slide, and cove slip (Fig. 4.13). The interesting outcomes from the investigations include that the thin-walled flexible polycarbonate film exhibited the weakest absorption characteristics, which has been attributed to the microscopic bends and the loss of light due to its low planarity compared to glass substrates. It has also been found that the absorbance is inversely proportional to the thickness of the glass. The coverslip that has been thinner than the glass chips showed high absorbance, however, its low durability limits its further usage. Overall, 0.5 mm chemical-strengthened tempered glass was found to be a superior choice for fabricating a sensitive LSPR chip. Later, the fabricated sensor chip was demonstrated to detect C-reactive protein as low as 10 ng/mL [52].

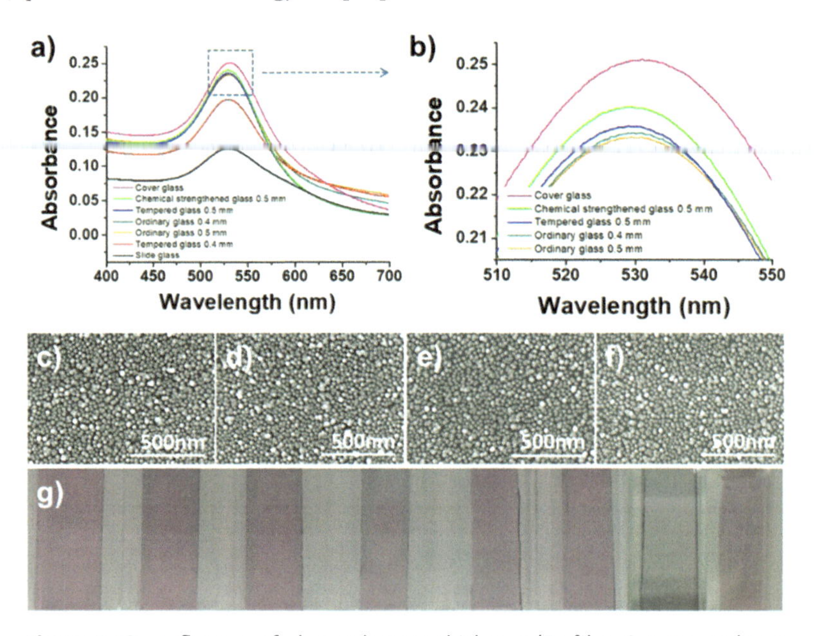

Figure 4.13 Influence of glass substrate thickness (Pc film, 0.4 mm ordinary glass, 0.5 mm ordinary glass, 0.4 mm tempered glass, 0.5 mm tempered glass, 0.5 mm chemical-strengthened glass, glass slide, and coverslip) on LSPR-sensitivity. Reproduced with permission from Ref. [52].

Sensitivity enhancement: It has been well-reported that the AuNRs-based LSPR sensors are sensitive compared to AuNPs, which has been discussed earlier. Further, the sensitivity of the gold nanorods has been reported to be improved using signal enhancer molecules, i.e., berberine. Berberine intercalates itself into the G-quadruplex structures when the capture aptamer molecules bind with the analyte resulting in an enhanced redshift. The berberine-based enhanced LSPR sensor has been demonstrated to detect various molecules including OTA, aflatoxin B1, ATP, and K+, as low as picomolar concentrations. Despite AuNRs showing high sensitivity compared to AuNPs, the tedious synthesis protocol to prepare uniform AuNRs limits its widespread application.

In addition to the use of varying gold nanoparticle shapes, the attempt to improve the sensitivity was also made by using plasmonic labels, i.e., forming a sandwich immunocomplex. Xie et al. proposed an LSPR array by immobilizing 39 nm of AuNPs to a hydrophilic-hydrophobic patterned glass slide and integrated a 13 nm AuNP labels-based enhancement for the sensitive detection of ATP (Fig. 4.14). It was shown that the detection of ATP was amplified by five-fold using AuNP labels [53]. Similarly, a hetero-assembled sandwich structure has been demonstrated for the detection of hepatitis B surface antigen (HBsAg). In the mentioned work, the authors have shown a significant detection of HBsAg as low as 100 fg/mL [54].

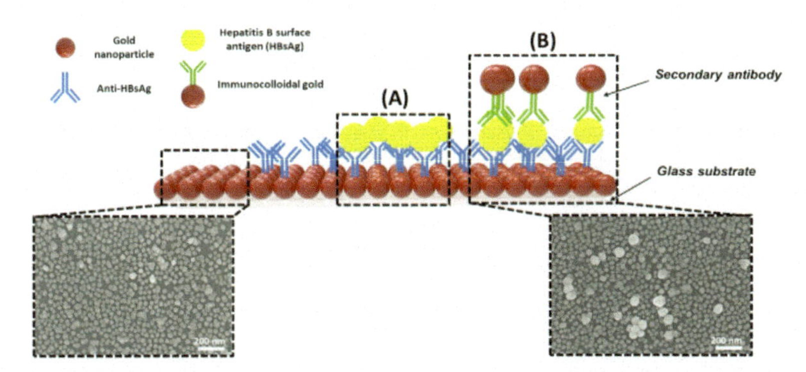

Figure 4.14 Schematic representation of hetero-assembled AuNPs sandwich immunoassay. Reproduced from Ref. [54], Copyright (2018), with permission from Elsevier.

Alternatively, the enhancement in the sensitivity could be achieved using enzymes, especially enzymes such as polymerases, ligases, and nucleases are of high interest due to their specificity and characteristic functional mechanisms. Ki and co-workers demonstrated an interesting miRNA LSPR sensor using a duplex-specific nuclease with signal amplification. The use of duplex-specific nuclease induces a three-level enhancement and a detection limit of 2.45 pM of miRNA [55]. In addition to single analyte sensing, researchers have also demonstrated multiplexed sensing using AuNPs of different shapes. A shape-code biosensor detecting three different biomarkers related to Alzheimer's disease has been demonstrated using spherical AuNPs (diameter 50 nm), short AuNRs (aspect ratio 1.6), and long AuNRs (aspect ratio 3.6), as shown in Fig. 4.15 [56].

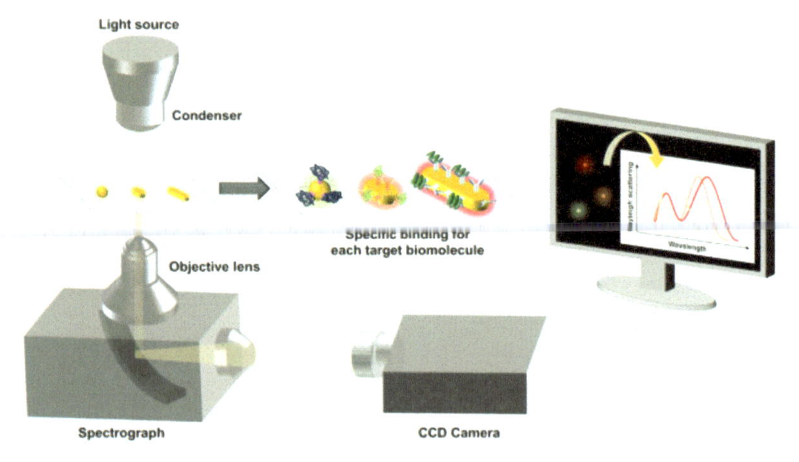

Figure 4.15 Schematic illustration of shape-code LSPR chip-based sensor. Reproduced from Ref. [56], Copyright (2018), with permission from Elsevier.

4.3.2 Polymer Substrate

Glass substrates have been widely used for the fabrication of various sensors, while polymer substrates have gained increasing interest due to their low-cost design and ease of construction. Sensitive LSPR biosensors can be fabricated on polymers such as polydimethylsiloxane (PDMS). PDMS is known to have high acid-alkali resistance and high optical transparency. Chang and co-workers have developed an LSPR sensor with metal-insulator-metal (MIM) nanodisks on PDMS and demonstrated its potential to detect A549 cancer cells. The fabricated flexible LSPR sensor was shown to have a significant sensitivity of 1500 nm/RIU; and in the presence of A549 cancer cells (5×10^5 cells/mL), a redshift of about 75 nm was observed [57].

It has been demonstrated that surface functionality employed to immobilize AuNPs onto the polymeric strips would influence the sensitivity of the LSPR sensor. Jin, Wong, and Granville studied the influence of different surface modification strategies to immobilize AuNPs in PMMA chips (Jin et al., 2016). In the study, the PMMA chips were subjected to thiolation following a sulphuric acid treatment (3 M and 6M) and compared to aminated surface. Firstly, the efficacy of the surface treatments has been characterized using XPS, and subsequently, the treated surfaces were used to immobilize AuNPs. The prepared LSPR sensor chips have been tested against 1-thio-β-D-glucose and subsequent Concanavalin A (ConA) binding. It has been shown that the sensitivity of the LSPR biosensor was substantially affected by the chemistries and the thiolated chips outperformed amine-treated PMMA chips.

Subsequently, in the same year, Jin and co-workers extended the research in PMMA-based LSPR chips for the precise detection of Con A. The detection of Con A, a lectin, using carbohydrates as capture molecules, is shown in Fig. 4.16. However, the detection of lectins through carbohydrates involves several considerations, due to the larger nature of lectins that may induce steric hindrance, carbohydrate-binding sites within the lectins could be inaccessible, and weak binding affinities. In order to overcome such limitations, polymer linkers have been utilized. In the aforementioned study, glycopolymer-bearing pendant sugar, inducing a multivalent "glycocluster effect" has been deployed (Jin et al., 2016). Poly(pentafluorostyrene) has been successfully synthesized via reversible addition-fragmentation chain transfer polymerization. The synthesized poly(pentafluorostyrene) has been converted to glycopolymers through a para-fluoro-thiol 'click' reaction. A detection of about 1.3 nmol/L has been demonstrated by exploiting the developed LSPR sensor.

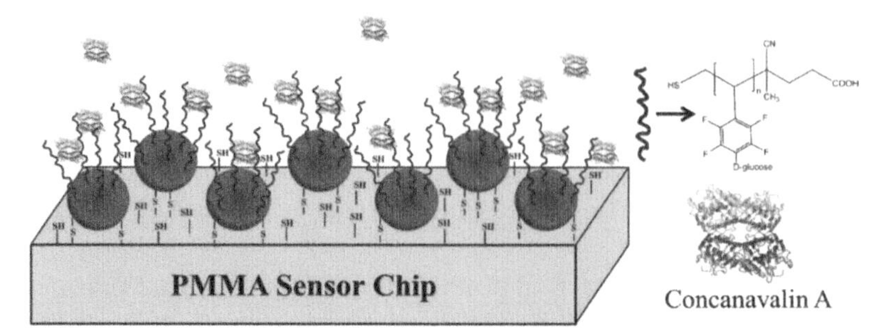

Figure 4.16 Schematic representation of Con A detection using glycopolymer immobilized LSPR sensor chip. Reproduced from Ref. [58], Copyright (2016), with permission from Elsevier.

4.3.3 Silicon and Other Substrates

Silicon substrates enable the large-scale fabrication of single-use sensor units. In 2001, Malinsky and co-workers studied the optical contributions of different substrates such as fused silica, borosilicate optical glass, mica, and SF-10 to the LSPR sensitivity using silver nanoparticles formed via nanosphere lithography (NSL). The Ag nanoparticles were 100 nm in in-plane width and 25 nm in out-of-plane height. It has been shown that the refractive index of the substrate being used is directly proportional to the shift in the extinction peak during the formation of the LSPR substrate. Subsequently, the influence of substrate refractive index has been correlated with the bulk refractive index sensitivity of the LSPR sensor. However, sensitivity with mica substrate was found to be 206 nm/RIU, while for SF-10, it has been 258 nm/RIU. Hence, it has been concluded that no systemic dependence or at most weak dependence of substrate refractive index to the LSPR sensitivity [60].

Bhalla and co-workers studied the influence of four different silicon-based ceramic substrates, namely, Si, SiO_2, Si_3N_4, and SiC (Fig. 4.17). The LSPR sensor has been formed following dewetting protocol, which involves the deposition of a thin metal film, followed by annealing at temperatures below the melting point of the metal. Thus, the metal film breaks and forms nanoislands on the substrate. The Si, SiO_2, and Si_3N_4 substrates with gold nanoislands showed a 50 nm/RIU sensitivity, while SiC with gold nanoislands showed a significant sensitivity of 247.80/RIU [61]. Similarly, Austin Suthanthiraraj and co-workers fabricated a spheroidal silver nanostructure (20–80 nm with spacing ranging from a few tens to a hundred nanometers) based LSPR sensor on a silicon substrate and demonstrated the detection of dengue NS1 antigen in whole blood [62]. A reported sensor consists of a PDMS microfluidic channel to direct the flow of whole blood and a polyethersulfone membrane to filter the excess red blood cells and albumin.

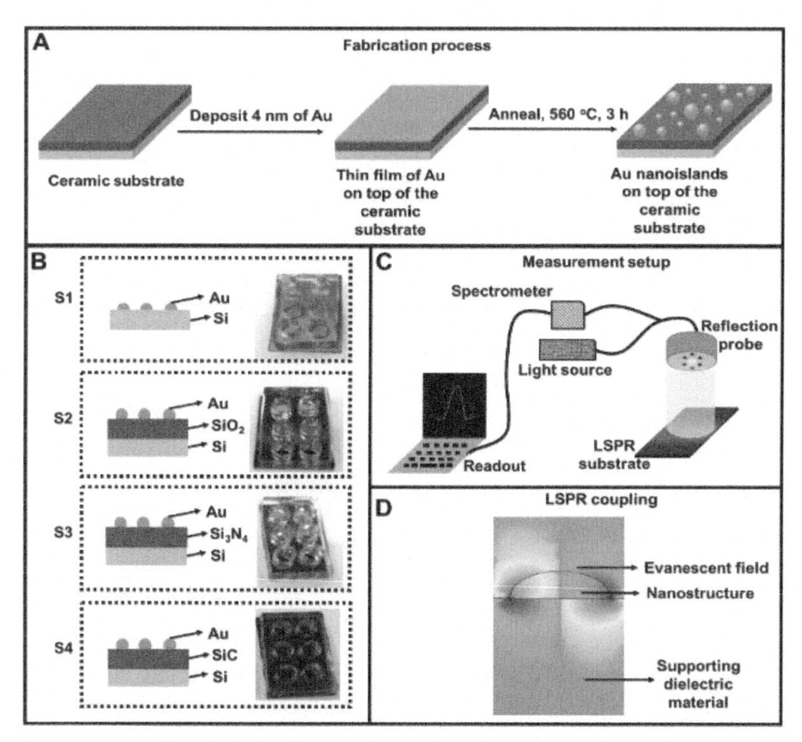

Figure 4.17 Influence of silicon-based substrates with varying dielectric properties on LSPR sensitivity formed via dewetting of Au film to form Au nanoislands. Reproduced with permission from Ref. [61].

4.4 Label Free Plasmonic Fiberoptic Sensors

Optical fibers, owing to their evanescent wave excitation/absorption or evanescent coupled fluorescence at the fiber core-medium interface, can be highly sensitive transducers for monitoring the optical properties of the plasmonic nanoparticles. Optical fibers as LSPR substrates have several advantages, including resistance to electromagnetic interference, chemical passivity affordability, ease of use, and small footprint. In addition, the optical fiber probes are reusable and can be recycled. Plasmonic optical fibers are available in different configurations, including tapered, D-shaped, and U-shaped, showing varying sensitivities [63]. In general, an optical fiber is a cylindrical dielectric waveguide made of low-loss materials that are optically transparent in the desired wavelength of transmission. It is commonly made using glass and polymers such as PMMA and polystyrene. Typically, an optical fiber consists of two components, namely, a central core with a refractive index (n_{co}), and a surrounding cladding or lower refractive index (n_{cl}). Conventional fused silica/glass optical fibers (GOFs) are made with polymer cladding and core of RI 1.45 and 1.42, respectively. The RI of the core is slightly higher than that of the clad to facilitate total internal reflection (TIR). Hence, according to Snell's law, the rays that have an angle of incidence greater than the critical angle (θ_c) will be reflected to the core. In addition, the light transmission property of an optical fiber depends on the numerical aperture (NA), fiber diameter, and wavelength of the light source.

The light that is propagating within the optical fiber has two components, a guided field within the core and the evanescent field in the core-cladding interface. In a cladded optical fiber, the evanescent field decays almost to zero within the clad, hence not interacting with the surrounding medium. One way to enable the interaction of the evanescent field and the surrounding is to reduce or remove the cladding layer. The distance to which the evanescent wave can extend into the surrounding from the core-cladding/medium interface is measured as the depth of penetration, which is the distance where the evanescent field is decreased to 1/e of its initial value at the interface. This evanescent field has widely been exploited to realize several LSPR-based sensors detecting proteins, hormones, and whole cells. The functionalization of glass optical fibers remains similar to that of the flat-glass substrate, i.e., using amine or thiol-terminated silane molecules.

4.4.1 Fiberoptic LSPR-Based Refractometric Sensors

Refractive index sensors are in high demand in the field of biochemistry, analytical chemistry, food preservation, beverage, and medicine. Several geometries such as reflection-based, tapered, and U-shaped optical sensor probes have been explored to realize sensitive LSPR refractive index sensors. A recent study on a U-bent plastic optical fiber (POF) sensor probe coated with graphene and silver nanoparticles hybrid structure over the silver thin film and polyvinyl alcohol (PVA) (Fig. 4.18). The developed sensor showed a significant RI sensitivity of 700.3 nm/RIU [64].

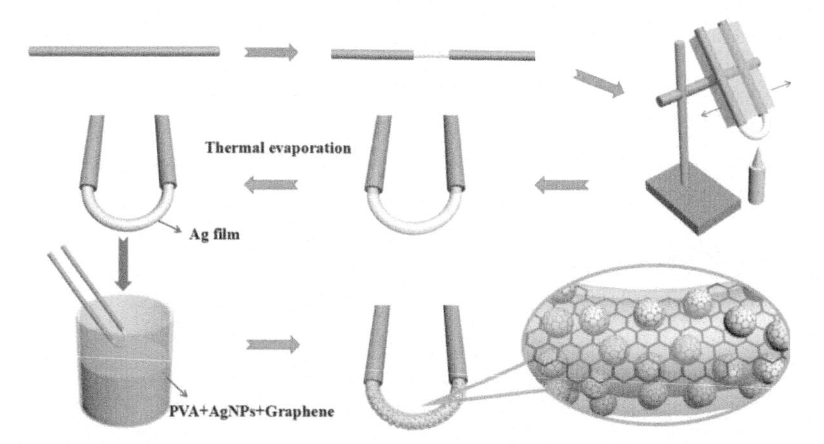

Figure 4.18 The schematic representation of the preparation procedure of the PVA/G/AgNPs@Ag film U-POF. Reproduced with permission from Ref. [64].

Another exciting study by Tu and co-workers involves the use of hollow gold cages and has reported a significantly high sensitivity of 1933 nm/RIU. Other nanostructures reported in the literature are mentioned in Table 4.2.

Table 4.2 Overview of the LSPR-based fiberoptic refractometric sensors

Nanostructure	Sensor configuration	Sensitivity	Ref.
AgNPs	Straight probe, reflection based	387 nm/RIU	[65]
AuNPs	Tapered multimode fiber (MMF)	51 nm/RIU	[66]
AuNPs	MMF-singlemode fiber (SMF)-MMF	765 nm/RIU	[67]
AuNPs	Straight polished POF	84 nm/RIU	[68]
AuNPs	U-bent POF	17.55 ΔAbs/RIU	[69]
AgNPs	Tapered fiber	178 nm/RIU	[70]
Au nanoshell	U-bent probe	18.8 ΔAbs/RIU	[71]
AuNPs	Reflection based	571 nm/RIU	[72]
Hollow AuNPs	Reflection based	1933 nm/RIU	[73]
AgNPs and AuNPs	U-bent probe	15.8 ΔAbs/RIU 18.9 ΔAbs/RIU	[74]
AuNPs	U-bent POF	15 ΔAbs/RIU	[75]

While many plasmonic refractive index sensors mainly report wavelength interrogation with a peak wavelength shift for a given RI change, simpler optical loss-based fiber optic plasmonic sensors have also been reported by L K Chou and co-workers [76, 77] and subsequently S Mukerji and co-workers [78, 79] using straight and U-bent fiber probes, respectively. The U-bent fiber optic sensor probes exhibit an order of improvement in the evanescent wave absorbance (EWA) sensitivity and hence will be able to monitor the minute changes in the spectral properties. A novel U-bent sensor probe-based plasmonic sensor demonstrated a RI sensitivity of 35 Δabs/RIU, with a solution of about 3.8×10^{-5} RIU. Later, Jitendra and co-workers reported the use of hollow gold nanostructures to develop LSPR U-bent sensor probes and showed an improved sensitivity of 1.5 fold compared to gold nanospheres. In 2016 and 2017, POF-based LSPR sensor probes were developed and demonstrated appreciable sensitivity. Interestingly, the study reported on developing LSPR sensor probes through sputtering showed a significant transition from LSPR to SPR with varying deposition times of 30, 60, 90, and

120 s. The results showed an RI sensitivity of 15.5 Δabs/RIU and 1040 nm/RIU for LSPR and SPR, respectively (Fig. 4.19). Optical intensity-based interrogation has the potential to offer better RI sensitivity with simpler and affordable instrumentation, thanks to the low-cost LASERS and sensitive photodetectors.

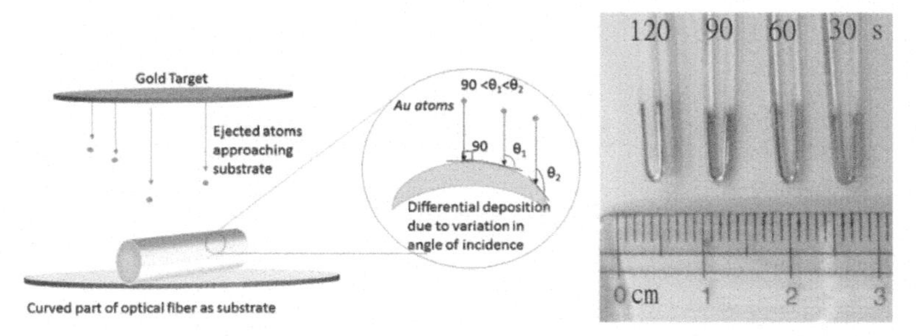

Figure 4.19 Schematic representation of Au sputtering on curved surfaces to form LSPR and SPR optical fiber substrate. Reproduced with permission from Springer Nature: Plasmonics [80], Copyright (2018).

4.4.2 Fiberoptic LSPR Chemical and Biosensors

Numerous biological sensors have been reported utilizing optical fiber-based LSPR sensors (Table 4.3). For example, a tapered optical fiber probe has been coated with AuNPs via 3-mercaptopropyl trimethoxysilane followed by sequential treatment with MUA and EDC/NHS to immobilize cholesterol oxidase (ChOx) [81]. This ChOx immobilized LSPR sensor probe has been utilized to detect varying concentrations of cholesterol.

In another study, Lee and co-workers developed a straight optical fiber LSPR sensor probe coated with aptamer-modified AuNRs to detect a small mycotoxin, ochratoxin A (OTA) (Fig. 4.20) [82]. The binding of OTA to the aptamer leads to a conformational change in the aptamer and also increases the local refractive index around the gold nanorods. The developed sensor has been demonstrated to detect OTA as low as 12 pM with a dynamic range of 10–100 pM. Also, the sensor has been reported to show high selectivity when tested over other mycotoxins such as zearalenone (ZEN) and ochratoxin B (OTB).

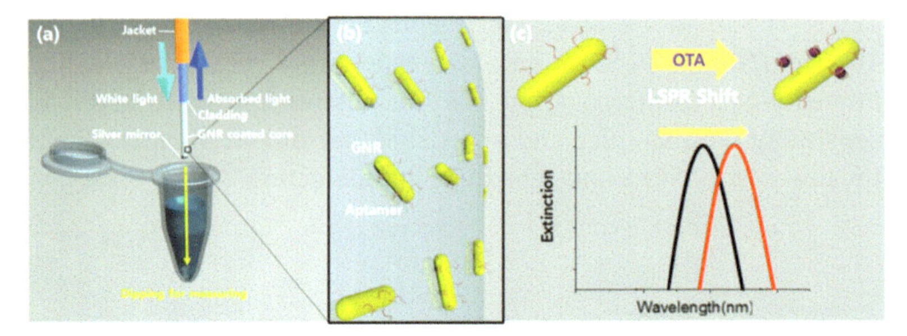

Figure 4.20 Schematic representation of optical fiber-based LSPR aptasensor for simple and rapid in situ detection of ochratoxin A. Reproduced from Ref. [82], Copyright (2018), with permission from Elsevier.

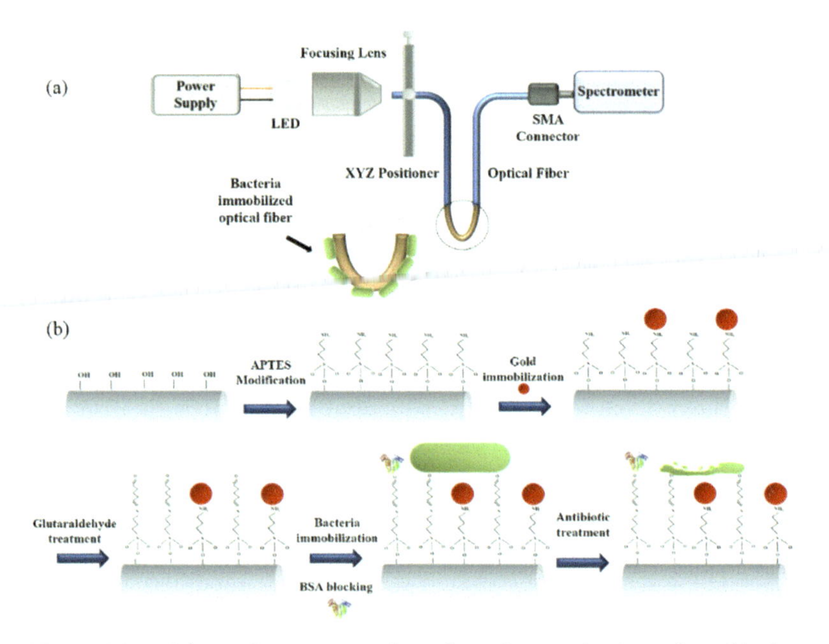

Figure 4.21 Schematic representation of sensing mechanism of antibiotics with bacterial cells on U-bent LSPR sensor probe. Reproduced from Ref. [83], Copyright (2020), with permission from Elsevier.

In 2020, interesting work has been reported for the assessment of antimicrobial resistance and subsequent quantification of antibiotics. In the reported work, an LSPR-bacteriolysis of bacteria on the U-bent fiberoptic probe has been utilized for quick beta-lactam susceptibility, as shown in Fig. 4.21 [83]. A proof-of-concept has been demonstrated using bacterial strains such as *E. coli* and *P. aeruginosa* against a spectrum of third-generation antibiotics including ceftazidime, ceftriaxone, and cefotaxime. *P. aeruginosa* RS1 strain has been found to be highly selective for cephalosporins and subsequently utilized to detect cephalosporins in tap water samples down to 0.01 μg/mL. The reported scheme can be a potential alternative to tedious drug susceptibility testing methods followed in hospitals.

Table 4.3 Overview of the LSPR-based fiberoptic chemical and biosensors

Sensor probe	Fiber-probe design	Analyte	Detection limit	Ref.
Green synthesized AgNPs	Straight fiber	Dopamine	0.2 μM	[84]
L-tyrosine capped AgNPs	U-bent polished POF	Dopamine	6 μM	[85]
AuNPs and cholesterol oxidase	Tapered fiber	Cholesterol	53.1 nM	[81]
AgNPs and PEG	Tapered fiber	Dopamine	0.58 μM	[86]
AuNPs	Hollow-core fiber	Cholesterol	25.5 nM	[87]
AuNPs	U-bent sensor probe	*E. coli*	1000 cfu/mL	[88]

Other than biomolecules, optical fiber-based LSPR sensors have been shown to detect heavy metals such as Hg^{2+}. In 2018, Jia and co-workers developed an LSPR sensor by immobilizing AuNPs to a straight optical fiber probe to detect Hg^{2+} based on thymine-Hg^{2+}-thymine base pair mismatches and coupled plasmonic resonance effect (Fig. 4.22) [89]. Briefly, the LSPR sensitivity of the sensor probe has been investigated using different sizes of gold nanoparticles, including 20 nm, 40 nm, 60 nm, and 80 nm. Among the sizes compared to 80 nm, AuNPs have shown a high RI sensitivity of about 2016 nm/RIU. Subsequently, the LSPR sensor probe with 80 nm gold nanoparticles has been utilized to detect Hg^{2+} using a capture DNA sequence attached to gold nanoparticles on the sensor surface and a detector DNA sequence conjugated to 20 nm gold nanoparticle labels. The combined effect of the high sensitivity of the LSPR probe and plasmonic coupling leads to a detection limit as low as 0.7 nM of Hg^{2+}.

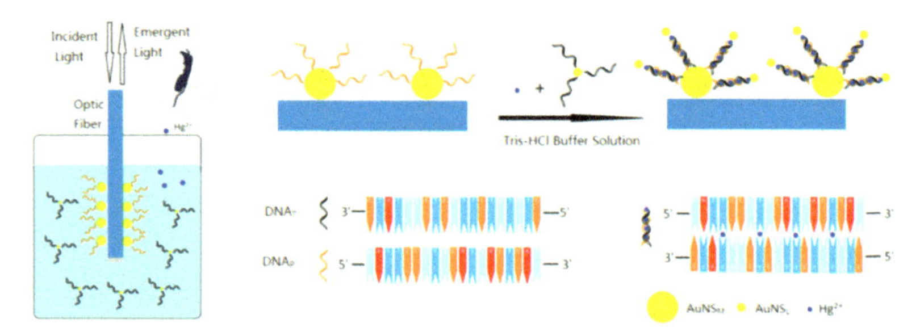

Figure 4.22 Schematic representation of Hg^{2+} detection using T-Hg^{2+}-T structure and plasmon coupling effect. Reproduced from Ref. [89], Copyright (2018), with permission from Elsevier.

In the study reported by Bharadwaj in 2014, the LSPR sensor was made using AuNPs for the vapor phase detection of explosives such as 2,4,6-trinitrotoluene (TNT) and cyclotrimethylenetrinitramine (RDX) [90]. The LSPR sensor probes have been immobilized with receptors such as 4-mercaptobenzoic acid (4-MBA), L-cysteine, and cysteamine to bind nitro-based explosive molecules. It has been found that the use of 4-MBA modified LSPR sensor probes showed a high sensitivity for 2,4-DNT, which is a common decomposition product of TNT and commonly found along with TNT. The reported sensor has a detection limit as low as 10 ppb and enables the direct detection of explosives from the air, thus avoiding solution-based sensing mechanisms.

4.5 Plasmonic Labels for Biosensor Development

Apart from the utilization of the LSPR phenomenon for label-free sensing, where a change in the optical properties of the plasmonic nanoparticles due to the refractive index changes around them is exploited, their significant optical extinction itself due to their presence or absence is utilized for sensing applications. Here, the plasmonic nanoparticles are exploited as labels, where the number of labels present on a sensor surface will be either directly or inversely proportional to the analyte concentration. Lateral flow immunoassay is the most commercially successful technology that utilizes plasmonic gold nanoparticles as labels in a sandwich assay format. The technical origin of lateral flow immunoassay dates to 1956, when Plotz and Singer developed the latex agglutination assay. Several refinements were made in the basic principle to realize a rapid and easy-to-use point-of-care (PoC) lateral flow test [91]. Besides the plasmonic nanoparticles, several other technologies, including antibody generation and tagging, fluid sample handling, nitrocellulose membrane development, and manufacturing, have culminated in the successful development of an efficient lateral flow assay (LFA) platform. Initial success in all these technologies resulted in the introduction of the first lateral flow product to the market in the late 1980s. The main application driving the early development was the pregnancy test. Later, this technology has been widely adapted to diverse areas, including molecular diagnostics, theranostics, agriculture, biowarfare, food safety, and environmental monitoring.

Typically, an LFA strip consists of a variety of materials, and each serves different purposes. Usually, the sample is added to the sample pad, where the samples are treated, making it compatible to run with the rest of the strip. Then, the sample flows to the next part, the conjugate pad, through capillary action. The conjugate pad contains gold nanoparticles or fluorescent or paramagnetic monodisperse latex particles, which are conjugated with specific biological components of the assay, i.e., antibodies specific to the analyte of interest. These bioconjugates are commonly named labels. At the conjugate pad, the sample interacts with the conjugates and migrates further into the next part of the strip, which is the reaction pad. The reaction pad is the nitrocellulose membrane on which the other specific biological component, i.e., antibodies against the analyte of interest is immobilized as a band (test line). In addition to the test line, another band containing antibodies against the conjugates is also present that serves as a control line. Excess samples along with the reagent move past the control line and are entrapped in a wick or absorbent pad. In this section, we mainly focus on bioconjugates or labels, especially colloidal gold nanoparticles.

4.5.1 Colloidal Gold Nanoparticle–Based LFA and Its Evolution

Most LFA strips utilize colloidal gold nanoparticles or monodispersed latex particles loaded with colored dyes and fluorescent molecules as labels. However, the choice of label depends on varying parameters, including the requirement of surface functionality to immobilize antibodies, assay output quantitative or quality analysis, dynamic range and sensitivity requirement, development of multiplexed analysis, and the overall manufacturing cost [92]. Colloidal gold is perhaps the identical choice and is widely used in commercial LFA strips, due to its ease in synthesis and conjugation, intense color without any development process, stability in liquid as well as in solid forms, non-bleaching property after staining onto membranes, and its commercial availability.

The first use of colloidal gold bioconjugates for immunoassay-based diagnostics was reported by Leuvering et al. in 1981. Subsequently, in 1984, Brada and Roth and Moeremans et al. reported the use of immunogold and protein A-gold bioconjugates in western blots and dot blots on nitrocellulose membranes. For use in LFA strips, the macromolecules such as antibodies must be conjugated with colloidal gold. The conjugation of biomolecules to colloidal gold nanoparticles highly depends on the pH used during conjugation. The pH should be adjusted slightly above the isoelectric point of the antibody molecules before conjugation. Below the pKi value, the antibodies induce flocculation/aggregation of colloidal gold. In addition to the pH for conjugation, the minimum concentration of protein/antibody required to stabilize the colloidal gold nanoparticle solution is also required to be determined by adding a constant volume of varying concentrations of protein/antibodies. Consequently, the antibody mixed colloidal gold nanoparticle solution is added with 10% sodium chloride solution, and the aggregation is visualized. The minimum protein/antibody concentration leading to stable gold nanoparticle colloids on the addition of sodium chloride is preferred for subsequent processes.

The overall sensitivity of the LFA strip is significantly influenced by the number of gold bioconjugates or labels accumulated on the test line for a given concentration of the antigen. Several studies have reported that the size of the gold nanoparticle influences the sensitivity of the LFA assay. In 2012, Sha Lou et al. studied the influence of nanoparticle size on LFA assay by comparing AuNPs of size 14 nm, 16 nm, 35 nm, and 38 nm, and it has been found that 16 nm gold nanoparticles have the minimum requirement of antibody concentration for stabilization [93]. However, 38 nm gold nanoparticles have been found highly sensitive. Similarly, in another study reported by Kim and co-workers, AuNPs of size 34 ± 1.2 nm, 42.7 ± 0.8 nm, 64.7 ± 0.2 nm, 106.5 ± 0.1 nm, and 137.8 ± 0.45 nm have been compared to detect Hepatitis B surface antigen [94]. Among the optimized conjugation conditions, the AuNPs of size 42.7 ± 0.8 nm exhibited superior performance compared to other sizes (Fig. 4.23).

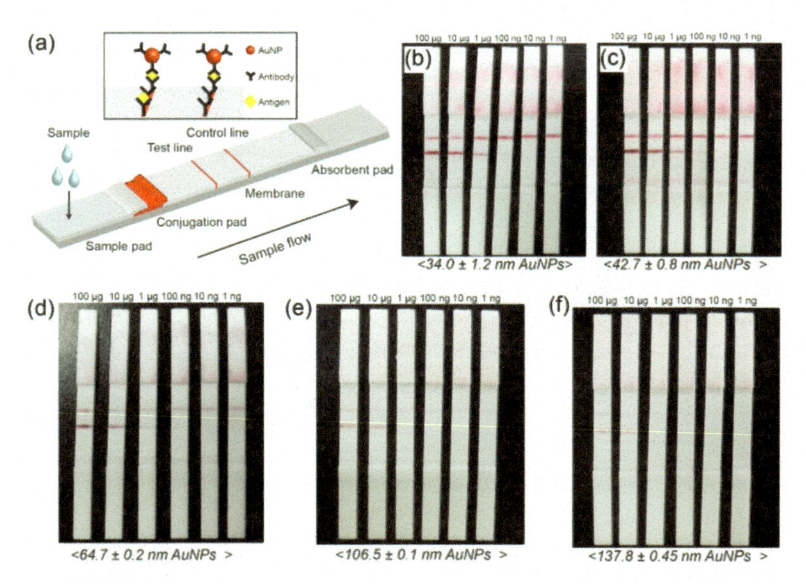

Figure 4.23 Effect of AuNP label size on LFA assay. Reproduced with permission from Ref. [94].

Various studies have found that the AuNPs of size 40 nm are sensitive to LFA-based assay. However, the reported sensitivity of the AuNP-based LFA assay is inferior to other immunoassays like fluorescent-linked immunosorbent assay (FLISA), or radioactive immunoassay (RIA). Thus, additional enhancement is usually preferred in order to improve sensitivity. Several methods have been reported to enhance the sensitivity of the LFA, such as the use of alternative labels or catalytic enhancement.

Catalytic enhancement: Chemical enhancement utilizing the catalytic activity of AuNPs to reduce silver is well-established [92]. Shyu et al. reported the use of silver enhancement to detect ricin through LFA and showed an improved detection limit down to 100 pg/mL using silver enhancement from that of 50 ng/mL using colloidal gold without silver enhancement [95]. Similarly, silver enhancement was shown to improve sensitivity by 15 times and the detection limit from 3 ng/mL to 200 pg/mL for the rapid detection of potato leafroll virus (Fig. 4.24). The silver enhancement approach has an additional step of liquid handling that contradicts the basic claim of the one-step assay.

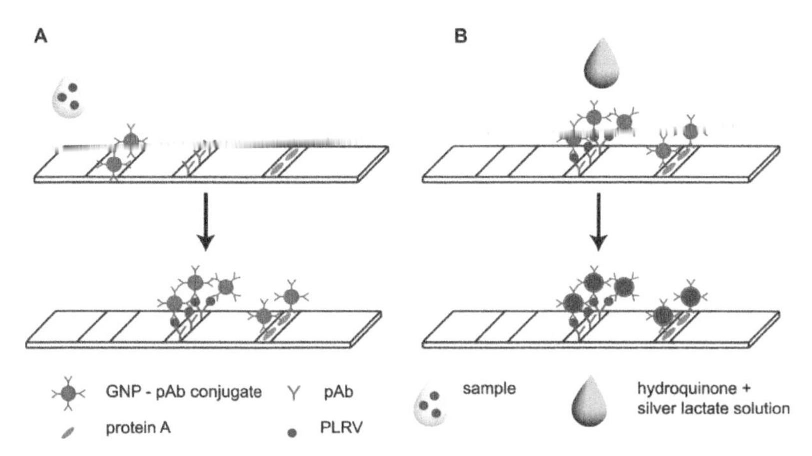

Figure 4.24 LFA with silver enhancement. Reproduced with permission from Ref. [96].

Alternate labels: Other than silver enhancement, AuNPs of different shapes and dual AuNP conjugates. Different AuNP shapes, such as AuNRs, nanoshells, and AuNCs, were utilized to improve the LFA sensitivity. Yang et al. reported an improved sensitivity of 2.5-fold using gold nanocages compared to colloidal AuNPs. However, thermal and optical contract detection methods indicated that gold nanorods and gold nanoshells are highly promising for improving the sensitivity of LFA [97]. Gold nanorods-based LFA was explored for the detection of C-reactive protein, and a detection limit of 1.3 ng/mL was reported. Similarly, another report comparing the gold nanorod and colloidal AuNPs for the detection of single-stranded DNA sequences showed an improved sensitivity of 250 folds with gold nanorods compared to colloidal AuNPs [98]. In addition to different shapes, dual AuNPs have also been used to improve sensitivity. Hwan et al., in 2010, reported a fascinating 100-fold increase in sensitivity by using dual AuNP conjugates for the detection of troponin I (Fig. 4.25). Briefly, the first conjugate

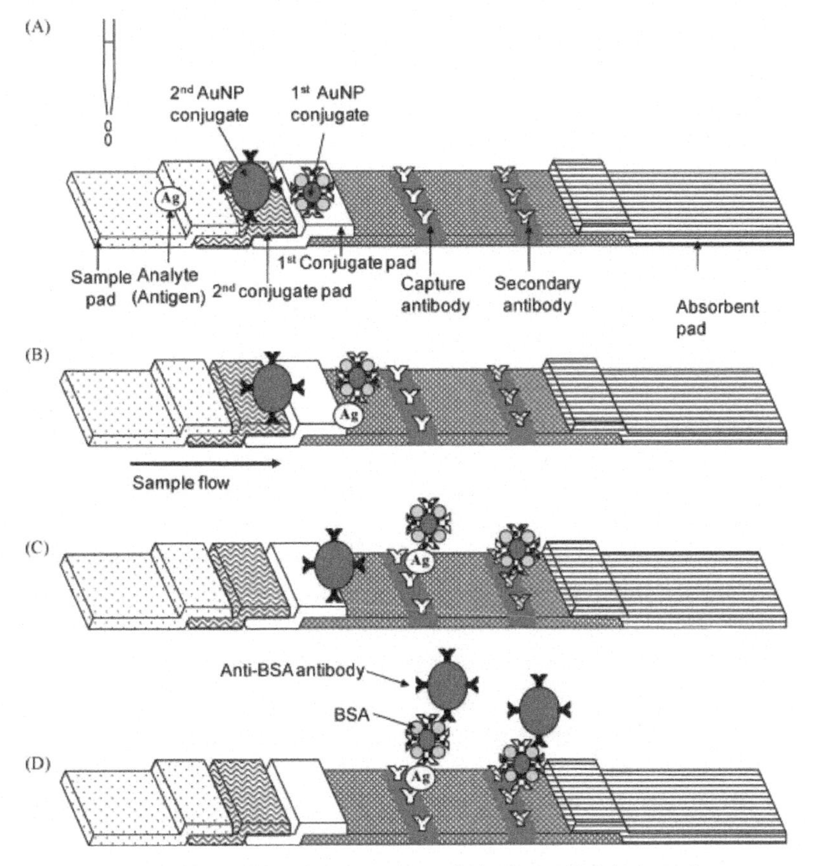

Figure 4.25 Schematic representation of dual AuNP label conjugates to improve the LFA sensitivity. Reproduced from Ref. [99], Copyright (2010), with permission from Elsevier.

has been made of 10 nm AuNP conjugated with anti-troponin I antibody and blocked with bovine serum albumin (BSA), while the second conjugate has been made of 40 nm AuNP conjugated with anti-BSA antibody and blocked with human serum albumin [99]. The developed dual-conjugate method has also been employed for the rapid detection (10 mins) of troponin I in the serum samples of patients with myocardial infarction. In addition to assay sensitivity, the data interpretation has also improved different gold structures for test and control lines. For instance, Petravoka et al. used gold nanoflowers for the test line and colloidal AuNPs for the control line for the detection of T-2 toxin and showed a detection limit down to 30 pg/mL [100]. There is no unique accepted method that is solely recommended for the enhancement of LFA-based assays. Despite the enhanced sensitivity using different methods, these additional steps lead to practical difficulties in realizing a large-scale production.

4.5.2 Plasmonic Fiberoptic Absorbance Biosensor (P-FAB)

While the LFA technology is undoubtedly a versatile biosensor technology, it has several limitations including low sensitivity, limited analyte dynamic range and detection limits, and difficulties in the quantification of analyte concentration. Various sensitive biosensing technology platforms have been developed as an alternative to the existing point-of-care diagnostic technology [101]. Chad Mirkin and co-workers have developed Bio Barcode Assay technology, capable of detecting analytes down to attomolar concentrations [102]. This sensing scheme utilizes gold nanoparticles as plasmonic labels and also exploits their catalytic properties for signal enhancement. This method involves at least two to three amplification stages over five steps: (i) Magnetic sub-microparticle (MNP) based on the separation of analytes from the bulk solution by realizing a sandwich assay with gold AuNP labels conjugated with detector antibodies as well as DNA barcode oligonucleotide (ON) labels, where there could be 100s of barcode ONs for every analyte molecules. (ii) Retrieval of these barcode ONs from the MNP-analyte-AuNP-ON complexes. (iii) Quantification of barcode ONs using AuNP labeled sandwich nucleic acid assay. (iv) An optional amplification step with the silver enhancement of AuNP labels. (v) A microfluidic chip has also been realized to carry out these complex reactions in a sequential manner [103]. It may be noted that the BBA-based analyte detection involves multiple reagents and washing steps, pushing the cost of the assay manifold. An important aspect to realize is that the BBA is a biochemical assay with nanotechnological tools to achieve unprecedented very high sensitivities, but not exactly a biosensor, which involves the use of an efficient optical, electrochemical, or any other kind of transducer that can pick up the presence of gold nanoparticle labels very sensitively and thereby minimize the number of laborious amplification steps as well as expensive reagents.

U-bent optical fiber probes are an excellent candidate due to this established high evanescent wave absorbance sensitivity and low-lost instrumentation facilitating large-scale production. Recently, a plasmonic fiber optic absorbance biosensor has been established and proven to detect and quantify the attomolar concentrations of the analyte. P-FAB relies on the realization of a plasmonic sandwich immunoassay on a U-bent fiber optic sensor probe that can pick up the presence of AuNP labels on its surface with high sensitivity, and give out an absorbance response that is proportional to the analyte concentration, and P-FAB has been initially realized using human immunoglobulin G (HIgG) as model analytes [104]. Briefly, the plasmonic sandwich immunoassay involves two or three simple incubation steps and a wash step for signal enhancement: (i) The addition of a sample containing the analyte to the AuNP reagent vial, where the AuNPs are conjugated with detector antibodies and incubation for 5–15 min. (ii) Dipping of the U-bent fiber optic probe immobilized with capture antibodies on its surface and incubation for another 5–15 min, while connected to the readout device (a reference signal to be taken a priori, by dipping the probe in a buffer). (iii) An optional silver enhancement-based amplification step, where the probe is washed after the second step and dipped in a simple silver enhancer solution over 5 min for absorption signal amplification (Fig. 4.26). The formation of the plasmonic sandwich immunocomplex with AuNP labels and/or the silver island formation over AuNP labels on the U-bent region of the fiber probe could be correlated with the drop in intensity of light collected using a simple LED-photodetector set-up. This assay gave rise to an unprecedented detection limit down to 0.17 zeptomole of HIgG in buffer solution within an assay time of 25 min. Later, using the same technology, Divagar et al., established a sensing scheme for the detection of lipoarabinomannan (LAM), a potential TB biomarker present in the urine at the early stage of the disease [105]. The plasmonic sandwich assay was realized by interacting various concentrations of Mtb LAM between 1 fg/mL and 1 ng/mL, with the capture antibody anti-Mtb LAM immunoglobulin M (IgM) immobilized on U-bent optical fiber, followed by incubation with detector antibody anti-Mtb LAM immunoglobulin G conjugated with AuNP. Further, it has been extended for the sensitive detection of chikungunya non-structural protein, lipopolysaccharides, and SARS-CoV-2 N-protein [74, 106, 107]. P-FAB is a versatile strategy that can be adopted for several applications involving not only single and multiplexed analysis but also high throughput analysis, similar to ELISA/CLIA platforms.

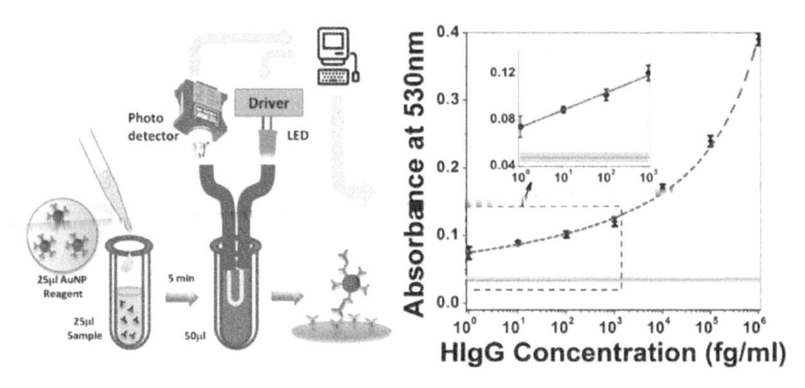

Figure 4.26 Schematic representation of plasmonic fiberoptic absorbance biosensor (P-FAB) using model analytes. Reproduced from Ref. [104], Copyright (2020), with permission from Elsevier.

4.6 Single Nanoparticle LSPR Spectroscopy

Single nanoparticle spectroscopy is usually performed via dark-field microscopy, where a high numerical aperture condenser brings white light to the sample, and a low-numerical aperture microscope objective collects the scattered light at low angles. Later, the scattered light is analyzed by the spectrometer to obtain spectral characteristics. Silver nanoparticles, gold nanostructures such as gold nanorods, gold nanorings, and gold nanoholes have been employed in developing single-nanoparticle-based LSPR spectroscopy. For example, in 2003, McFarland and Duyne utilized a single silver nanoparticle and reported the first measurement of LSPR response to the formation of SAM on single AgNPs and also showed a zeptomole sensitivity (Fig. 4.27) [108].

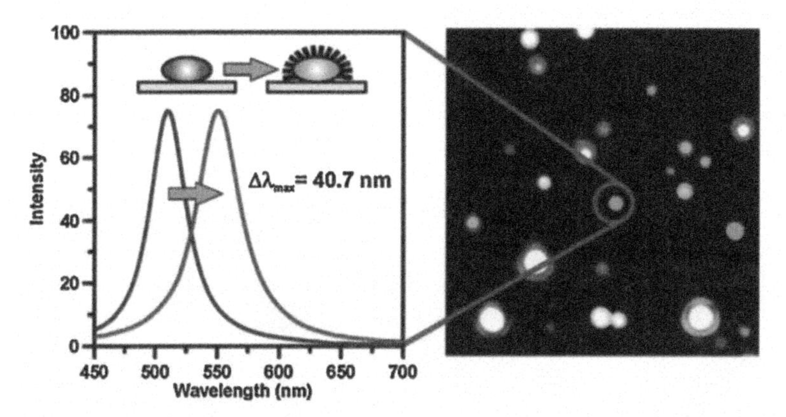

Figure 4.27 Single nanoparticle LSPR spectroscopy. Reproduced with permission from Ref. [108], Copyright (2003), American Chemical Society.

4.7 Multiplexed LSPR Sensors

Multiplexed sensors are gaining attention, due to their ability to simultaneously detect more than one analyte with minimum time and minimum resources in a cost-effective manner. In addition, these sensors are useful for detecting various diseases, which typically involve a plethora of parameters for decision-making purposes [109]. For example, the difference in the LSPR band of AuNPs and AgNPs can be utilized to develop multiplexed sensors. For example, Sciacca and Monro developed a label-free dip-type sensor using AuNPs and AgNPs [110]. Similarly, Ma et al., developed a multiplexed sensor for the detection of two different cancer biomarkers using AuNPs and AgNPs labels (Fig. 4.28) [111]. A sandwich ELISA was employed as a platform for the quantification and imaging of alpha-fetoprotein (AFP) and carcinoembryonic antigen (CEA) at the same time. The anti-AFP and anti-CEA antibodies were immobilized on a glass slide, followed by the introduction of a mixture containing AFP and CEA antigens. Thereafter, the anti-AFP antibodies conjugated with AuNPs and anti-CEA antibodies conjugated with AgNPs were added to the reaction mixture to form the specific immunocomplexes. Under a dark field microscope, the AuNRs scatter green color, while the AgNPs scatter blue color; therefore, the quantity of antigen present in the sample could be easily counted using the color dots and their intensity. This approach could detect the antigen concentration between 0.5 and 10 ng/mL, with no mutual interference between CEA and AFP detection, showing the feasibility of simultaneous detection of two targets.

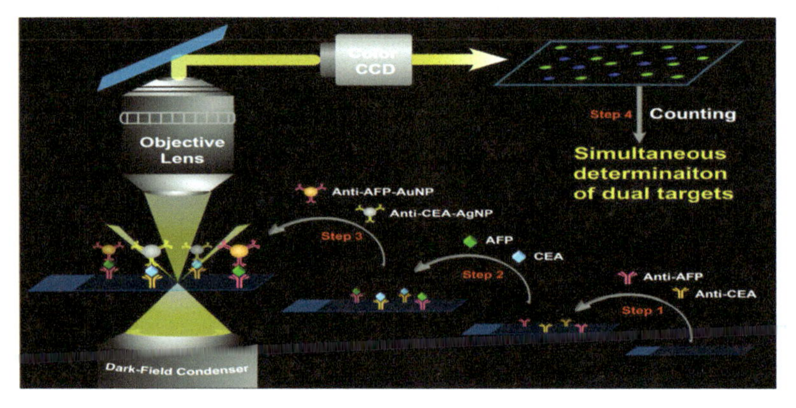

Figure 4.28 Schematic representation of the determination of AFP and CEA using sandwich-type immunoassay. Reproduced with permission from Ref. [111], Copyright (2017), American Chemical Society.

4.8 Summary

In summary, this chapter has elucidated various techniques of utilizing the plasmonic nanoparticles and their properties to develop highly sensitive LSPR-based sensors and biosensors capable of unprecedented analyte detection over a wide range from nM down to aM concentrations, which were unimaginable 20 years ago. The inherent color of the nanoparticles enables detection through the naked eye both in the solution phase and on a wide variety of substrates. Further, the diversity in the color spectrum of the plasmonic nanoparticles, specifically gold nanostructures, also enhances the feasibility of developing multiplexed portable sensing systems. Despite the advantages, there are several challenges involved in realizing LSPR-based sensing technologies.

The predominant challenge is the limited shelf-life or the stability of the plasmonic nanoparticles for their use in solution-based sensing methods. On the other hand, substrate-based LSPR sensors possess additional challenges to controlling the plasmon-coupling due to the aggregation of nanoparticles. Secondly, the LSPR sensors suffer low selectivity in complex sample matrices like blood and serum. The non-specific interactions are usually inhibited to some extent by employing blocking agents like short-chain polymers, saccharide-based molecules, or zwitterionic compounds.

Furthermore, compared to SPR-based sensors, LSPR sensors are characterized by a low depth of evanescent field, leading to less sensitivity toward large-sized analytes such as microbial cells. At the other extreme, the LSPR-based sensors also suffer limitations in detecting small molecules less than 1 kDa, due to negligible change in the plasmonic properties and thus limiting the LSPR sensors to detect only a narrow spectrum of proteins and nucleic acids. However, novel strategies have been reported to amplify the LSPR signal even with small molecules and gases by coating nanohybrid recognition layers such as metal-organic frameworks (MOFs). These MOFs not only improve the sensitivity but also offer size-based selectivity. In addition to surface modifications, sensitive optical transducers such as U-bent fiberoptic probes have also been explored to realize sensitive LSPR-based sensor systems. Such systems have a great potential to meet the challenges in chemical and biochemical analysis such as field-deployable chemical sensors, point-of-care diagnostic devices, and high throughput systems. Lastly, the development of the LSPR-based sensing system also demands an advanced microfluidic platform to enable near-field sensing.

References

1. Hardy, A.C. (1936). *Handbook of Colorimetry*, Cambridge University Press.
2. Markus, S., Johan, H., and Jörg, E. (2011). *Handbook of Fluorescence Spectroscopy and Imaging: From Single Molecules to Ensembles*, Wiley-VCH Verlag GmbH.
3. Fifield, F.W. and Kealey, D. (2000). *Principles and Practice of Analytical Chemistry*, 5th Edition, Wiley-Blackwell.

4. Skoog, D.A., West, D.M., Holler, F.J., and Crouch, S.R. (2013). *Fundamentals of Analytical Chemistry,* 9th Edition, Cengage Learning.
5. Wild, D. (2013). *The Immunoassay Handbook,* Elsevier Science.
6. Lequin, R.M. (2005). *Enzyme Immunoassay (EIA)/ Enzyme-Linked Immunosorbent Assay (ELISA),* **2418**, 2415–2418.
7. Darwish, I.A. (2006). Immunoassay methods and their applications in pharmaceutical analysis: Basic methodology and recent advances, *Int. J. Biomed. Sci.*, **2** (3), 217–235.
8. Wang, C., Wu, J., Zong, C., Xu, J., and Ju, H.X. (2012). Chemiluminescent immunoassay and its applications, *Chinese J. Anal. Chem.*, **40** (1), 3–10.
9. Jans, H. and Huo, Q. (2012). Gold nanoparticle-enabled biological and chemical detection and analysis, *Chem. Soc. Rev.*, **41** (7), 2849–2866.
10. Chung, T., Lee, S.Y., Song, E.Y., Chun, H., and Lee, B. (2011). Plasmonic nanostructures for nano-scale bio-sensing, *Sensors,* **11** (11), 10907–10929.
11. Yi-Tao, L. and Chao, J. (2014). *Localized Surface Plasmon Resonance Based Nanobiosensors*, Springer.
12. Mayer, K.M. and Hafner, J.H. (2011). Localized surface plasmon resonance sensors, *Chem. Rev.*, **111** (6), 3828–3857.
13. Homola, J. (2006). *Surface Plasmon Resonance Based Sensors,* Vol.4, Springer Berlin Heidelberg.
14. Jeon, H. Bin, Tsalu, P.V., and Ha, J.W. (2019). Shape effect on the refractive index sensitivity at localized surface plasmon resonance inflection points of single gold nanocubes with vertices, *Sci. Rep.*, **9** (1), 1–8.
15. Miller, M.M. and Lazarides, A.A. (2006). Sensitivity of metal nanoparticle plasmon resonance band position to the dielectric environment as observed in scattering, *J. Opt. A Pure Appl. Opt.*, **8** (4), 21556–21565.
16. Sun, Y. and Xia, Y. (2002). Increased sensitivity of surface plasmon resonance of gold nanoshells compared to that of gold solid colloids in response to environmental changes, *Anal. Chem.*, **74** (20), 5297–5305.
17. Mock, J.J., Smith, D.R., and Schultz, S. (2003). Local refractive index dependence of plasmon resonance spectra from individual nanoparticles, *Nano Lett.*, **3** (4), 485–491.
18. Johnson, P.B. and Christy, R.W. (1972). Optical constant of the nobel metals, *Phys. Rev. B*, **6** (12), 4370–4379.

19. Guo, L., Jackman, J.A., Yang, H.H., Chen, P., Cho, N.J., and Kim, D.H. (2015). Strategies for enhancing the sensitivity of plasmonic nanosensors, *Nano Today*, **10** (2), 213–239.

20. Satija, J., Punjabi, N., Mishra, D., and Mukherji, S. (2016). Plasmonic-ELISA: Expanding horizons, *RSC Adv.*, **6** (88), 85440–85456.

21. Kermanshahian, K., Yadegar, A., and Ghourchian, H. (2021). Gold nanorods etching as a powerful signaling process for plasmonic multicolorimetric chemo-/biosensors: Strategies and applications, *Coord. Chem. Rev.*, **442**, 213934.

22. Ma, X., He, S., Qiu, B., Luo, F., Guo, L., and Lin, Z. (2019). Noble metal nanoparticle-based multicolor immunoassays: An approach toward visual quantification of the analytes with the naked eye, *ACS Sensors*, **4** (4), 782–791.

23. Zhang, Z., Wang, H., Chen, Z., Wang, X., Choo, J., and Chen, L. (2018). Plasmonic colorimetric sensors based on etching and growth of noble metal nanoparticles: Strategies and applications, *Biosens. Bioelectron.*, **114** (April), 52–65.

24. Rao, H., Xue, X., Wang, H., and Xue, Z. (2019). Gold nanorod etching-based multicolorimetric sensors: Strategies and applications, *J. Mater. Chem. C*, **7** (16), 4610–4621.

25. Cao, J., Sun, T., and Grattan, K.T.V. (2014). Gold nanorod-based localized surface plasmon resonance biosensors: A review, *Sensors Actuators, B Chem.*, **195**, 332–351.

26. He, Y., Wen, C.Y., Guo, Z.J., and Huang, Y.F. (2020). Noble metal nanomaterial-based aptasensors for microbial toxin detection, *J. Food Drug Anal.*, **28** (4), 508–520.

27. Lerdsri, J., Soongsong, J., Laolue, P., and Jakmunee, J. (2021). Reliable colorimetric aptasensor exploiting 72-Mers ssDNA and gold nanoprobes for highly sensitive detection of aflatoxin M1 in milk, *J. Food Compos. Anal.*, **102** (April), 103992.

28. Halas, N.J., Lal, S., Chang, W.S., Link, S., and Nordlander, P. (2011). Plasmons in strongly coupled metallic nanostructures, *Chem. Rev.*, **111** (6), 3913–3961.

29. Chen, W., Hu, H., Jiang, W., Xu, Y., Zhang, S., and Xu, H. (2018). Ultrasensitive nanosensors based on localized surface plasmon resonances: From theory to applications, *Chinese Phys. B*, **27** (10).

30. Medley, C.D., Smith, J.E., Tang, Z., Wu, Y., Bamrungsap, S., and Tan, W. (2008). Gold nanoparticle-based colorimetric assay for the direct detection of cancerous cells, *Anal. Chem.*, **80** (4), 1067–1072.

31. Borghei, Y.S., Hosseini, M., Dadmehr, M., Hosseinkhani, S., Ganjali, M.R., and Sheikhnejad, R. (2016). Visual detection of cancer cells by colorimetric aptasensor based on aggregation of gold nanoparticles induced by DNA hybridization, *Anal. Chim. Acta*, **904**, 92–97.

32. Minopoli, A., Saka, N., Lenyk, B., Mayer, D., Andreas, O., and Della, B. (2020). Sensors and actuators B : Chemical LSPR-based colorimetric immunosensor for rapid and sensitive 17 β -estradiol detection in tap water, **308** (January).

33. Alula, M.T., Karamchand, L., Hendricks, N.R., and Blackburn, J.M. (2018). Citrate-capped silver nanoparticles as a probe for sensitive and selective colorimetric and spectrophotometric sensing of creatinine in human urine, *Anal. Chim. Acta*, **1007**, 40–49.

34. Zhang, Z., Chen, Z., Wang, S., Cheng, F., and Chen, L. (2015). Iodine-mediated etching of gold nanorods for plasmonic ELISA based on colorimetric detection of alkaline phosphatase, *ACS Appl. Mater. Interfaces*, **7** (50), 27639–27645.

35. Lin, T., Zhang, M., Xu, F., Wang, X., Xu, Z., and Guo, L. (2018). Colorimetric detection of benzoyl peroxide based on the etching of silver nanoshells of Au@Ag nanorods, *Sensors Actuators, B Chem.*, **261**, 379–384.

36. Liu, Y., Wang, J., Zhao, C., Guo, X., Song, X., Zhao, W., Liu, S., Xu, K., and Li, J. (2019). A multicolorimetric assay for rapid detection of Listeria monocytogenes based on the etching of gold nanorods, *Anal. Chim. Acta*, **1048**, 154–160.

37. Fang, B., Xu, S., Huang, Y., Su, F., Huang, Z., Fang, H., Peng, J., Xiong, Y., and Lai, W. (2020). Gold nanorods etching-based plasmonic immunoassay for qualitative and quantitative detection of aflatoxin M1 in milk, *Food Chem.*, **329** (May), 127160.

38. Scroccarello, A., Della Pelle, F., Ferraro, G., Fratini, E., Tempera, F., Dainese, E., and Compagnone, D. (2021). Plasmonic active film integrating gold/silver nanostructures for H_2O_2 readout, *Talanta*, **222** (July 2020), 121682.

39. Chen, J., Jackson, A.A., Rotello, V.M., and Nugen, S.R. (2016). Colorimetric detection of *Escherichia coli* based on the enzyme-induced metallization of gold nanorods, *Small*, **12** (18), 2469–2475.

40. Xu, S., Ouyang, W., Xie, P., Lin, Y., Qiu, B., Lin, Z., Chen, G., and Guo, L. (2017). Highly uniform gold nanobipyramids for ultrasensitive colorimetric detection of influenza virus, *Anal. Chem.*, **89** (3), 1617–1623.

41. Li, Y., Ma, X., Xu, Z., Liu, M., Lin, Z., Qiu, B., Guo, L., and Chen, G. (2016). Multicolor ELISA based on alkaline phosphatase-triggered growth of Au nanorods, *Analyst*, **141** (10), 2970–2976.

42. Wang, Y., Zhang, P., Mao, X., Fu, W., and Liu, C. (2016). Seed-mediated growth of bimetallic nanoparticles as an effective strategy for sensitive detection of vitamin C, *Sensors Actuators, B Chem.*, **231**, 95–101.

43. Shrivas, K., Sahu, S., Patra, G.K., Jaiswal, N.K., and Shankar, R. (2016). Localized surface plasmon resonance of silver nanoparticles for sensitive colorimetric detection of chromium in surface water, industrial waste water and vegetable samples, *Anal. Methods*, **8** (9), 2088–2096.

44. Xiong, Y., Pei, K., Wu, Y., Duan, H., Lai, W., and Xiong, Y. (2018). Plasmonic ELISA based on enzyme-assisted etching of Au nanorods for the highly sensitive detection of aflatoxin B1 in corn samples, *Sensors Actuators, B Chem.*, **267**, 320–327.

45. Guo, Y., Wu, J., Li, J., and Ju, H. (2016). A plasmonic colorimetric strategy for biosensing through enzyme guided growth of silver nanoparticles on gold nanostars, *Biosens. Bioelectron.*, **78**, 267–273.

46. Nath, N. and Chilkoti, A. (2002). A colorimetric gold nanoparticle sensor to interrogate biomolecular interactions in real time on a surface, *Anal. Chem.*, **74** (3), 504–509.

47. Lopez, G.A., Estevez, M., Soler, M., and Lechuga, L.M. (2017). Review article Recent advances in nanoplasmonic biosensors: Applications and lab-on-a-chip integration, *Nanophotonics*, **6** (1), 123–136.

48. Kim, D.M., Park, J.S., Jung, S.-W., Yeom, J., and Yoo, S.M. (2021). Biosensing applications using nanostructure-based localized surface plasmon resonance sensors, *Sensors*, **21** (9), 3191.

49. Marinakos, S.M., Chen, S., and Chilkoti, A. (2007). Plasmonic detection of a model analyte in serum by a gold nanorod sensor, *Anal. Chem.*, **79** (14), 5278–5283.

50. Srdjan, S.A., Maria, A.O., Vanesa, S., Johann, B., Jose, L.G.C., Jan, R., Sebastian, J.M., Mark, P.K., and Romain, Q. (2014). LSPR chip for parallel, rapid, and sensitive detection of cancer markers in serum, *Nano Lett.*, **14**, 2636–2641.

51. Oh, S.Y., Heo, N.S., Shukla, S., Cho, H., Vilian, A.T.E., and Kim, J. (2017). Development of gold nanoparticle-aptamer-based LSPR sensing chips for the rapid detection of Salmonella typhimurium in pork meat, *Sci. Rep.*, 1–10.

52. Oh, S.Y., Heo, N.S., Bajpai, V.K., Jang, S., and Ok, G. (2019). Development of a cuvette-based LSPR sensor chip using a plasmonically active transparent strip, *Front. Bioeng. Biotechnol.*, **7**, 1–11.
53. Xie, L., Yan, X., and Du, Y. (2014). Biosensors and bioelectronics an aptamer based wall-less LSPR array chip for label-free and high throughput detection of biomolecules, *Biosens. Bioelectron.*, **53**, 58–64.
54. Kim, J., Yeong, S., Shukla, S., Bok, S., and Su, N. (2018). Heteroassembled gold nanoparticles with sandwich-immunoassay LSPR chip format for rapid and sensitive detection of hepatitis B virus surface antigen (HBsAg), *Biosens. Bioelectron.*, **107**, 118–122.
55. Ki, J., Son, H.Y., Huh, Y., and Haam, S. (2019). Sensitive plasmonic detection of miR-10b in biological samples using enzyme-assisted target recycling and developed LSPR probe, *ACS Appl. Mater. interfacesApplied*, **11**, 18923–18929.
56. Kim, H., Lee, J.U., Song, S., Kim, S., and Sim, S.J. (2018). A shape-code nanoplasmonic biosensor for multiplex detection of Alzheimer's disease biomarkers, *Biosens. Bioelectron.*, **101**, 96–102.
57. Chang, C., Lin, H., Lai, M., Shieh, T., and Peng, C. (2018). Flexible localized surface plasmon resonance sensor with metal – insulator – metal nanodisks on PDMS substrate, *Sci. Rep.*, 1–8.
58. Jin, Y., Wong, K.H., and Granville, A.M. (2016). Developing localized surface plasmon resonance biosensor chips and fiber optics via direct surface modification of PMMA optical waveguides, *Colloids Surfaces A Physicochem. Eng. Asp.*, **492**, 100–109.
59. Jin, Y., Wong, K.H., and Granville, A.M. (2016). Enhancement of localized surface plasmon resonance polymer based biosensor chips using well-defined glycopolymers for lectin detection, *J. Colloid Interface Sci.*, **462**, 19–28.
60. Malinsky, M.D., Kelly, K.L., Schatz, G.C., and Duyne, R.P. Van (2001). Nanosphere lithography: Effect of substrate on the localized surface plasmon resonance spectrum of silver nanoparticles, *J. Phys. Chem. B*, **105** (12), 2343–2350.
61. Bhalla, N., Jain, A., Lee, Y., and Shen, A.Q. (2019). Dewetting metal nanofilms — Effect of substrate on refractive index sensitivity of nanoplasmonic gold, *Nanomaterials*, **9** (1530), 1–14.
62. Austin Suthanthiraraj, P.P. and Sen, A.K. (2019). Localized surface plasmon resonance (LSPR) biosensor based on thermally annealed silver nanostructures with on-chip blood-plasma separation for the detection of dengue non-structural protein NS1 antigen, *Biosens. Bioelectron.*, **132**, 38–46.

63. Caucheteur, C., Guo, T., and Albert, J. (2015). Review of plasmonic fiber optic biochemical sensors: Improving the limit of detection, *Anal. Bioanal. Chem.*, 3883–3897.

64. Jiang, S., Li, Z., Zhang, C., Gao, S., and Li, Z. (2017). A novel U-bent plastic optical fibre local surface plasmon resonance sensor based on a graphene and silver nanoparticle hybrid structure, *J. Phys. D. Appl. Phys.*, **50**, 165105.

65. Chen, J., Shi, S., Su, R., Qi, W., Huang, R., Wang, M., Wang, L., and He, Z. (2015). Optimization and application of reflective LSPR optical fiber biosensors based on silver nanoparticles, *Sensors*, **15** (6), 12205–12217.

66. Lin, H.-Y., Huang, C.-H., Cheng, G.-L., Chen, N.-K., and Chui, H.-C. (2012). Tapered optical fiber sensor based on localized surface plasmon resonance, *Opt. Express*, **20** (19), 21693.

67. García, J.A., Monzón-Hernández, D., Manríquez, J., and Bustos, E. (2016). One step method to attach gold nanoparticles onto the surface of an optical fiber used for refractive index sensing, *Opt. Mater. (Amst).*, **51**, 208–212.

68. Cennamo, N., D'Agostino, G., Donà, A., Dacarro, G., Pallavicini, P., Pesavento, M., and Zeni, L. (2013). Localized surface plasmon resonance with five-branched gold nanostars in a plastic optical fiber for bio-chemical sensor implementation, *Sensors*, **13** (11), 14676–14686.

69. Gowri, A. and Sai, V.V.R. (2016). Development of LSPR based U-bent plastic optical fiber sensors, *Sensors Actuators, B Chem.*, **230**, 536–543.

70. Liu, S., Su, S., Chen, G., and Zeng, X. (2012). Optical fiber sensors based on local surface plasmon resonance modified with silver nanoparticles, *Proc. - 2012 Int. Conf. Intell. Syst. Des. Eng. Appl. ISDEA 2012*, 1444–1447.

71. Tharion, J., Chauhan, S., and Mukherji, S. (2016). Gold Nanoshells coated "U" bend optical fiber for near infra-red LSPR based refractive index sensing, *Procedia Eng.*, **168**, 367–370.

72. Tu, M.H., Sun, T., and Grattan, K.T.V. (2012). Optimization of gold-nanoparticle-based optical fibre surface plasmon resonance (SPR)-based sensors, *Sensors Actuators, B Chem.*, **164** (1), 43–53.

73. Tu, M.H., Sun, T., and Grattan, K.T. V (2014). LSPR optical fibre sensors based on hollow gold nanostructures, *Sensors Actuators B Chem.*, **191**, 37–44.

74. Manoharan, H., Kalita, P., Gupta, S., and Sai, V.V.R. (2019). Plasmonic biosensors for bacterial endotoxin detection on biomimetic C-18 supported fiber optic probes, *Biosens. Bioelectron.*, **129**, 79–86.

75. Divagar, M., Saumey, J., Satija, J., and Sai, V.V.R. (2019). Self-assembled polyamidoamine dendrimer on poly (methyl meth- acrylate) for plasmonic fiber optic sensors, *ChemNanoMat*, **5** (11), 1428–1436.

76. Cheng, S., and Chau, L. (2003). Colloidal gold-modified optical fiber for chemical, *Anal. Chem.*, **75** (1), 16–21.

77. Chau, L.K., Lin, Y.F., Cheng, S.F., and Lin, T.J. (2006). Fiber-optic chemical and biochemical probes based on localized surface plasmon resonance, *Sensors Actuators, B Chem.*, **113** (1), 100–105.

78. Sai, V.V.R., Kundu, T., and Mukherji, S. (2009). Novel U-bent fiber optic probe for localized surface plasmon resonance based biosensor, *Biosens. Bioelectron.*, **24** (9), 2804–2809.

79. Satija, J., Punjabi, N.S., Sai, V.V.R., and Mukherji, S. (2014). Optimal design for U-bent fiber-optic LSPR sensor probes, *Plasmonics*, **9** (2), 251–260.

80. Christopher, C., Subrahmanyam, A., and Sai, V.V.R. (2018). Gold sputtered U-bent plastic optical fiber probes as SPR- and LSPR-based compact plasmonic sensors, *Plasmonics*, **13** (2), 493–502.

81. Kumar, S., Kaushik, B.K., Singh, R., Chen, N.-K., Yang, Q.S., Zhang, X., Wang, W., and Zhang, B. (2019) LSPR-based cholesterol biosensor using a tapered optical fiber structure, *Biomed. Opt. Express*, **10** (5), 2150.

82. Lee, B., Park, J., Byun, J., Heon, J., and Kim, M. (2018). An optical fi ber-based LSPR aptasensor for simple and rapid in-situ detection of ochratoxin A, *Biosens. Bioelectron.*, **102**, 504–509.

83. Nag, P., Sadani, K., Mukherji, S., and Mukherji, S. (2020). Beta-lactam antibiotics induced bacteriolysis on LSPR sensors for assessment of antimicrobial resistance and quantification of antibiotics, *Sensors Actuators, B Chem.*, **311**, 127945.

84. Rithesh Raj, D., Prasanth, S., Vineeshkumar, T. V., and Sudarsanakumar, C. (2015). Surface plasmon resonance based fiber optic dopamine sensor using green synthesized silver nanoparticles, *Sensors Actuators, B Chem.*, **224**, 600–606.

85. Rithesh Raj, D., Prasanth, S., and Sudarsanakumar, C. (2017). Development of LSPR-based optical fiber dopamine sensor using L-tyrosine-capped silver nanoparticles and its nonlinear optical properties, *Plasmonics*, **12** (4), 1227–1234.

86. Rajil, N., Sokolov, A., Yi, Z., Adams, G., Agarwal, G., Belousov, V., Brick, R., Chapin, K., Cirillo, J., Deckert, V., Delfan, S., Esmaeili, S., Fernández-González, A., Fry, E., Han, Z., Hemmer, P., Kattawar, G., Kim, M., Lee, M.C., Lu, C.Y., Mogford, J., Neuman, B., Pan, J.W., Peng, T., Poor, V., Scully, S., Shih, Y., Suckewer, S., Svidzinsky, A., Verhoef, A., Wang, D., Wang, K., Yang, L., Zheltikov, A., Zhu, S., Zubairy, S., and Scully, M. (2020). A fiber optic-nanophotonic approach to the detection of antibodies and viral particles of COVID-19, *Nanophotonics*, **10** (1), 235–246.

87. Kumar, S., Singh, R., Kaushik, B.K., Chen, N.K., Yang, Q.S., and Zhang, X. (2019). Lspr-based cholesterol biosensor using hollow core fiber structure, *IEEE Sens. J.*, **19** (17), 7399–7406.

88. Halkare, P., Punjabi, N., Wangchuk, J., Madugula, S., Kondabagil, K., and Mukherji, S. (2021). Label-free detection of Escherichia coli from mixed bacterial cultures using bacteriophage T4 on plasmonic fiber-optic sensor, *ACS Sensors*, **6** (7), 2720–2727.

89. Jia, S., Bian, C., Sun, J., Tong, J., and Xia, S. (2018). A wavelength-modulated localized surface plasmon resonance (LSPR) optical fi ber sensor for sensitive detection of mercury (II) ion by gold nanoparticles-DNA conjugates, *Biosens. Bioelectron.*, **114**, 15–21.

90. Bharadwaj, R. and Mukherji, S. (2014). Gold nanoparticle coated U-bend fibre optic probe for localized surface plasmon resonance based detection of explosive vapours, *Sensors Actuators B. Chem.*, **192**, 804–811.

91. Wong, R.C. and Y, T.H. (2009). *Lateral Flow Immunoassay*, Springer.

92. Bishop, J.D., Hsieh, H. V., Gasperino, D.J., and Weigl, B.H. (2019). Sensitivity enhancement in lateral flow assays: A systems perspective, *Lab Chip*, **19** (15), 2486–2499.

93. Lou, S., Ye, J., Li, K., and Wu, A. (2012). A gold nanoparticle-based immunochromatographic assay: The influence of nanoparticulate size, *Analyst*, **137** (5), 1174–1181.

94. Kim, D., Kim, Y., Hong, S., Kim, J., Heo, N., Lee, M.-K., Lee, S., Kim, B., Kim, I., Huh, Y., and Choi, B. (2016). Development of lateral flow assay based on size-controlled gold nanoparticles for detection of hepatitis B surface antigen, *Sensors*, **16** (12), 2154.

95. Rong-Hwa, S., Shiao-Shek, T., Der-Jiang, C., and Yao-Wen, H. (2010). Gold nanoparticle-based lateral flow assay for detection of staphylococcal enterotoxin B, *Food Chem.*, **118** (2), 462–466.

96. Panferov, V.G., Safenkova, I.V, Byzova, N.A., Varitsev, Y.A., Zherdev, A.V, Dzantiev, B.B., Panferov, V.G., Safenkova, I.V, Byzova, N.A., and Varitsev,

Y.A. (2017). Silver-enhanced lateral flow immunoassay for highly-sensitive detection of potato leafroll virus, *Food Agric. Immunol.*, **29** (1), 445–457.

97. Qin, Z., Chan, W.C.W., Boulware, D.R., Akkin, T., Butler, E.K., and Bischof, J.C. (2012). Significantly improved analytical sensitivity of lateral flow immunoassays by using thermal contrast, *Angew. Commun.*, **51**, 4358–4361.

98. Yu, Q., Zhang, J., Qiu, W., Li, K., Qian, L., Zhang, X., and Liu, G. (2021). Gold nanorods-based lateral flow biosensors for sensitive detection of nucleic acids, *Microchim. Acta*, **188** (4), 1–8.

99. Hwan, D., Ki, S., Kyoung, Y., Woo, B., Dong, S., Kim, S., Shin, Y., and Kim, M. (2010). A dual gold nanoparticle conjugate-based lateral flow assay (LFA) method for the analysis of troponin I, *Biosens. Bioelectron.*, **25** (8), 1999–2002.

100. Petrakova, A.V, Urusov, A.E., Zherdev, A.V, and Dzantiev, B.B. (2019). Gold nanoparticles of di ff erent shape for bicolor lateral fl ow test, *Anal. Biochem.*, **568**, 7–13.

101. Usha, S.P., Manoharan, H., Deshmukh, R., Alvarez-Diduk, R., Calucho, E., Sai, V.V.R., and Merkoçi, A. (2021). Attomolar analyte sensing techniques (AttoSens): A review on a decade of progress on chemical and biosensing nanoplatforms, *Chem. Soc. Rev.*, **50**, 13012–13089.

102. Thaxton, C.S., Mirkin, C.A., and Nam, J. (2003). Nanoparticle based bio bar codes for the ultrasensitive detection of proteins, *Science (80-.).*, **301**, 1884–1886.

103. Goluch, E.D., Stoeva, S.I., Lee, J.S., Shaikh, K.A., Mirkin, C.A., and Liu, C. (2009). A microfluidic detection system based upon a surface immobilized biobarcode assay, *Biosens. Bioelectron.*, **24** (8), 2397–2403.

104. Ramakrishna, B., Divagar, M., Khanna, S., Danny, C.G., Gupta, S., Janakiraman, V., and Sai, V.V.R. (2020). U-bent fiber optic plasmonic biosensor platform for ultrasensitive analyte detection, *Sensors Actuators, B Chem.*, **321**, 1–6.

105. Divagar, M., Bandaru, R., Janakiraman, V., and Sai, V.V.R. (2020). A plasmonic fiberoptic absorbance biosensor for mannose-capped lipoarabinomannan based tuberculosis diagnosis, *Biosens. Bioelectron.*, **167**, 112488.

106. Divagar, M., Gayathri, R., Rasool, R., Shamlee, J.K., Bhatia, H., Satija, J., and Sai, V.V.R. (2021). Plasmonic fiberoptic absorbance biosensor

(P-FAB) for rapid detection of SARS-CoV-2 nucleocapsid protein, *IEEE Sens. J.*, **21** (20), 22758–22766.

107. George, A., Amrutha, M.S., Srivastava, P., Sunil, S., Sai, V.V.R., and Srinivasan, R. (2021). Development of a U-bent plastic optical fiber biosensor with plasmonic labels for the detection of chikungunya non-structural protein 3, *Analyst*, **146** (1), 244–252.
108. Mcfarland, A.D. and Duyne, R.P. Van (2003). Single silver nanoparticles as real-time optical sensors with zeptomole sensitivity, *Nano Lett.*, **3** (8), 1057–1062.
109. Pei, X., Tao, G., Wu, X., Ma, Y., Li, R., and Li, N. (2020). Nanomaterial-based multiplex optical sensors, *Analyst*, **145** (12), 4111–4123.
110. Sciacca, B. and Monro, T.M. (2014). Dip biosensor based on localized surface plasmon resonance at the tip of an optical fiber, *Langmuir*, **30** (3), 946–954.
111. Ma, J., Zhan, L., Li, R.S., Gao, P.F., and Huang, C.Z. (2017). Color-encoded assays for the simultaneous quantification of dual cancer biomarkers, *Anal. Chem.*, **89** (16), 8484–8489.

Chapter 5

Fiber Gratings–Based Plasmonic Sensors

Maria Simone Soares and Carlos Marques
Physics Department & i3N, University of Aveiro,
3810-183 Aveiro, Portugal
msimone.fsoares@ua.pt, carlos.marques@ua.pt

Plasmonic sensors are normally coated with a metallic layer or nanoparticles to allow the excitation of surface plasmons (SPs) and, consequently, the triggering of surface plasmon resonance (SPR) or localized SPR (LSPR), respectively. Gold is the most commonly used metal since it has advantages in terms of high sensitivity and biocompatibility. Furthermore, it is chemically inert and more stable compared to other metals. Plasmonic sensors were initially developed with planar configurations such as bulky Kretschman prism, nonetheless, they were rapidly replaced by optical fibers that are miniaturized and allow easy light injection and remote operation. Among the different structures that enable the interaction of light with the surrounding, the metal-coated fiber gratings imprinted

Plasmonics-Based Optical Sensors and Detectors
Edited by Banshi D. Gupta, Anuj K. Sharma, and Jin Li

ISBN 978-981-4968-85-0 (Hardcover), 978-1-003-43830-4 (eBook)
www.jennystanford.com

in the fiber core provide modal features that allow the generation of SP at any wavelength. So far, these configurations have been functionalized with many biorecognition elements, being the most applied antibodies and aptamers to turn them into immunosensors and aptasensors, respectively, for analyte sensing. This chapter begins by presenting some points about SPR and LSPR in optical fibers. Afterward, several methods of fiber grating fabrication and the different configurations used for those sensors are explained. A section on gold functionalization with antibodies and aptamers is also described. Finally, some recent applications of these sensors are shown including potential trends for the future.

5.1 Introduction

Optical sensors, more specifically optical fiber sensors present favorable advantages, which justify their widespread use in many applications, such as biomedical diagnostics, life sciences, electrochemistry, environmental safety, and food analysis, among others. The advantages include fast sensing, the possibility of label-free sensing that allows real-time detection, and minimally invasive sensing. Besides, they are flexible, nanoscale, and inexpensive [1, 2].

When light travels through a standard telecommunication fiber, its propagation is confined to the fiber core. Nonetheless, for sensing purposes, light must interact with the surrounding medium. Hence, it is essential to have an optical fiber configuration that allows light coupling toward the surrounding medium. Fiber gratings can be inscribed inside the core ensuring the occurrence of this phenomenon [2, 3]. Fiber gratings can include, for instance, uniform fiber Bragg gratings (FBGs), etched FBGs, tilted FBGs (TFBGs), chirped FBGs, and long period gratings (LPGs). The first referred configuration corresponds to a periodic modification of the refractive index (RI) of the core. In this structure, small reflections are created at each period of the Bragg grating, working as a pass-band filter. Around the Bragg wavelength and for some wavelengths, exists a constructive interference between the forward-propagating and the backward-propagating light waves, leading to an attenuation in the transmitted (reflected) light spectrum. However, the light remains inside the fiber core and, consequently, the interaction of light with the surrounding medium does not exist [2]. If the cladding is totally or partially removed along the grating area, creating etched FBGs, the transmitted light will be influenced by the surrounding RI [4, 5]. Another method is applying gratings with a tilt angle with respect to the fiber axis, which allows the coupling of the core mode with the several cladding modes, and consequently the interaction of the guided light with the surrounding medium. Therefore, the TFBG spectrum exhibits several narrow-bands corresponding to the cladding mode resonances. Each one corresponds to a loss in light core intensity and is characterized by a specific wavelength. According to the tilt angle, the coupling with the cladding modes can occur in backward or forward directions. If the tilt angle is inferior to 16°, the coupling direction is backward and a self-backward core mode coupling is also present; whereas, for tilt angles superior to 45° is forward, and the self-backward core mode coupling is not present [2, 3]. A chirped FBG is similar to a uniform FBG, nonetheless, its grating period changes along its length. As a result, the gating period depends on the propagation axis z since grating periodicity changes along the propagation axis z. Consequently, different Bragg wavelengths are reflected at each different section of the grating, being the spectrum constituted with the spectrum of each grating section. In addition, this configuration enables dispersion compensation due to the dependence of the group delay on the wavelength [6]. In contrast to FBGs that have a period around 500 nm in LPGs, the period can present values between 100 μm and 1 mm. This structure also allows the interaction of the propagating light with the surrounding medium, owing to the forward coupling of the core mode with the various cladding modes. As in TFBG, the cladding modes are sensitive to changes in the surrounding RI [7].

Through the excitation of the SPR, the sensitivity of the sensor can be enhanced [3]. An SP corresponds to an oscillation of the electrons in the metal-dielectric interface. Therefore, to trigger SPR, the fiber surface needs to be covered with a metal layer, usually gold due to good biocompatibility and stainless [2, 3]. The SPR phenomenon is triggered and the phase-matching condition is achieved when the propagation constants of the incident light of the core and the SP waves (SPWs) are equal. Moreover, the incident light wave must be P-polarized in order to excite the SP since SPWs are P-polarized. Depending on the RI of the surrounding medium, light energy is absorbed for a specific wavelength range once SPR is activated, which is visualized in the transmission spectrum [2]. Besides SPR, LSPR is also an adequate optical sensing mechanism, identical to SPR. Instead of a metal layer, this phenomenon is achieved using nanoparticles (NPs), where the resonance is localized to the NPs.

This chapter tries to give a general summary of plasmonic sensors in fiber gratings for the scientific community. This chapter starts with an actual introduction and then is talked about the basics of SPR and LSPR in optical fibers. Then, we try to show the key configurations for biosensors based on optical fiber gratings in a deep way. In addition, we focus on the main applications that research groups are betting heavily and we show potential trends for the future.

5.2 Basics of SPR and LSPR on Optical Fibers

SPR was introduced in 1993 by Liedberg et al. [8]. Since then, optical fiber sensing has been used the most this optical method. This technique presents several advantages, including a high real-time response rate, high sensitivity, and low LOD, and allows label-free monitoring of biomolecular interactions. Moreover, it is sensitive to small changes in RI on the fiber surface. Owing to all these favorable characteristics, SPR is widely applied in sensors and in the field of in situ biomedicine [9, 10].

When fight propagates in the fiber core by total internal reflection (TIR), some light is extended to the surrounding cladding, known as an evanescent wave (EW), which decays exponentially to zero. Penetration depth corresponds to the distance in which the evanescent field decreases to $1/e$ of its initial value at the interface [7, 11]. Therefore, once light propagates in an optical fiber exits the guided field in the core (core modes) and the evanescent field in the cladding (cladding modes) [7].

When a metal film is placed at the fiber surface, the EW is enhanced since penetrates de metal film [12]. Therefore, electrons in the metal surface layer are excited by the EW [13], and quantized collective oscillations of free electrons are generated, which are categorized as SPs [7, 9]. SPWs at the metal-dielectric interface are created when the SPs are excited by EW, with their propagation confined to the referred interface [12, 13]. SPR phenomenon is triggered when the propagation constant of the EW matches that of the SP. Under the resonance condition, the EW transfers energy to the SP [9, 14], and strong absorption of light occurs. Consequently, at the resonance wavelength, the sensor spectrum presents a sharp dip [15]. A displacement of the resonance position or an intensity modification, which occurs due to a change in the SPR propagation constant, and in turn, because of a slight variation of the surrounding RI of the metal layer, can be used as the interrogation method to detect small changes in the RI, which makes this phenomenon a good option for biosensing [16]. As SPWs are P-polarized, having an oscillating electric field normal to the surface, the incident light must be also radially polarized (P-polarized) and present a similar tangential component to that of the plasmon wave to trigger the SPR phenomenon [17].

SPR was firstly developed for planar approaches like Kretschmann configuration using bulky prisms, nonetheless, coupling SPR to optical fiber technology has brought enormous advances such as flexibility, simplicity, cost-effectiveness, miniaturization, in situ monitoring, immunity to electromagnetic interference, and remote operation [1, 4, 6]. Currently, optical sensors based on SPR can be found in different applications in life sciences, electrochemistry, environmental safety, and biomedical diagnostics [9].

In LSPR, NPs are placed on a fiber surface instead of a continuous film. Therefore, the incident light interacts with the conductive NPs, which present dimensions inferior to the wavelength of the incident light. This phenomenon is similar to SPR, nonetheless, the strong oscillations of the surface electrons induced by the interaction of the incident light and the electrons in the conduction band of the metals are localized, since the excitation of the SPs is confined within the conductive NPs [9, 14].

SPR phenomenon involves light losses to the metallic layer leading to oscillation and propagation of SPs along the intersection between optical fiber and metal. In contrast, in the LSPR mechanism, the localized SPs (LSPs) are non-propagating and oscillate locally in the NPs [7, 18, 19]. When an RI variation around the NPs occurs, specific wavelengths are absorbed from the propagating light in the NPs. In this sense, the detection principle using this mechanism is based on monitoring the shift of the resonance wavelength [14].

To improve the performance of plasmonic optical fiber sensors must be taken into account the thickness, uniformity, rugosity, and morphology of the metal. Due to the high sensitivity and biocompatibility, gold and silver are the most common metals used in plasmonic devices [13, 14]. Nonetheless, once silver is exposed to water or air, oxidizes easily. Whereas, gold is chemically inert and more stable compared with silver. As a result, gold is more used in plasmonic sensors for both SPR and LSPR.

5.3 Optical Fiber Gratings Configurations for Biosensing

Fiber gratings are distinguished into short-period fiber gratings, also known as FBGs, or long period gratings, called as LPGs. In the first category, TFBGs are included. All these grating configurations correspond to periodic perturbations of the RI of the core. Other fiber grating configuration consists of chirped FBGs, where the period varies along the grating length, usually, the RI of the core. This configuration consists of aperiodic perturbations of the RI of the core [7, 9]. Usually, an FBG contains a period in the order of 500 nm; whereas, for an LPG, the grating period presents values on the order of 100 micrometers to a millimeter [7]. CFBG contains a grating period similar to that of FBG, nonetheless, the period increases along its length.

5.3.1 Fiber Gratings Fabrication Methods

5.3.1.1 Fiber Bragg gratings fabrication techniques

FBGs were demonstrated by Ken Hill in 1978 [20]. Initially, a visible laser was used to manufacture gratings, in which light from the laser propagated inside the fiber core. Later in 1989, the transverse holographic inscription was demonstrated by Gerald Meltz et al., which consists of a more flexible technique by exposing the core to UV laser, and lighting through the cladding [21]. Using the interference pattern of the UV laser, was possible to perform the periodic structure of the Bragg grating. Over the years, the techniques of gratings fabrication have evolved and, nowadays, it is possible to use femtosecond laser systems to inscribe gratings with high quality [22, 23].

Currently, interferometric and phase mask techniques are the approaches most used for Bragg grating fabrication. In interferometric techniques, a single UV beam is used, which is divided into two and then suffers interference in the fiber. The phase mask approach uses an optical diffractive element for grating fabrication. Therefore, the Bragg gratings can be fabricated by the amplitude-splitting interferometer, wavefront-splitting interferometer, phase mask technique, long FBG inscription, or point-by-point inscription.

Amplitude-splitting interferometer technique was introduced by Gerald Meltz et al. in 1989 as aforementioned and consists of the division of a UV laser into two beams of equal intensity, which are later reflected and intersected by each other on the fiber plane. This process leads to the recombination of beams in order to inscribe an interference pattern. The RI modulation in the fiber core is induced by the interference pattern, presenting the same special periodicity as the interference pattern [24]. Therefore, the Bragg grating presents a period equal to the interference pattern period, given by:

$$\Lambda = \frac{\lambda_{\text{UV}}}{2n_{eff}\sin(\theta)}, \tag{5.1}$$

where λ_{UV} is the wavelength of the UV laser beam and θ is the intersecting angle between the two beams formed due to the splitting of the UV laser beam. With this approach, it is possible to inscribe Bragg gratings at almost any desired wavelength by only changing the intersection angle θ. Nonetheless, this method is easily susceptible to external/environmental interference such as vibrations or air flows, as the optical components are not isolated. These interferences can lead to poor-quality interference patterns [24].

To perform the wavefront-splitting interferometer can be used the prism interferometer or Lloyd's mirror interferometer [24, 25]. In general, the working principle of this technique consists of the incidence of a collimated UV laser beam on the reflecting surface of a mirror or prism edge, depending on the technique applied. Then half of the beam is reflected and its path intersects the path of the incident beam causing interference in the area where the reflected and unreflected beam overlap. Using a cylindrical lens, it is possible to focus the interference pattern on the fiber core [24, 25]. This method compared with the amplitude-splitting interferometer is less subject to environmental influences since it requires fewer optical components. However, the length of the Bragg grating manufactured by this method is restricted to half of the incident UV beam width [24, 25].

Through the phase mask method (Fig. 5.1), it is possible to inscribe an interference pattern using a phase mask, which is a diffractive optical element. In this approach, a UV laser beam is spatially modulated by the phase mask. Then, using the produced pattern, a RI modulation in the photosensitive fiber is photo-imprinted near, parallel, and immediately behind the phase mask [19]. The grooves depth of the phase mask is adjusted to suppress the zero-diffraction order to values of transmitted power below 5%. Whereas, each of the principal beam orders (plus and minus first orders) is maximized to more than 35% of the incident radiation. Owing to the interference of the plus and minus first-order diffraction beams, a fringe interference pattern is created in the fiber core [25]. Therefore, the period of the interference pattern resulting from the beams with orders ±1 is one-half of the phase mask period (Λ_{pm}), given by:

$$\Lambda = \frac{\lambda_{\mathrm{UV}}}{2\sin\frac{\theta_m}{2}} = \frac{\lambda_{\mathrm{UV}}}{2\frac{\lambda_{\mathrm{UV}}}{\Lambda_{pm}}} = \frac{\Lambda_{pm}}{2}. \tag{5.2}$$

Holography and electron beam lithography are the approaches applied in high-quality fused silica to produce phase masks that are transparent to the UV beam. Compared to the electron-beam-formed phase mask, the holographic-formed phase mask does not present stitching errors. Whereas, in the electron-beam-formed phase mask, the stitching error is usually present [25]. Nonetheless, the phase mask method is the most widely used technique for reproducible Bragg gratings fabrication.

As implied in the methods described above, the length of the grating has a value practically equal to the diameter of the UV beam. By the methods presented above, Bragg gratings can have a maximum length of 10 mm (phase mask length) and are not possible to inscribe long FBGs. To overcome this problem and be able to perform long FBG inscription, lasers with a larger diameter could be used, nonetheless, it is difficult to develop one of this kind due to instability problems. Thus, a way to overcome this limitation, which may be important for some applications based on Bragg gratings, is to use a scanning technique, more specifically, laser scanning in the writing plane. Besides, all the optical systems must allow the inscription of a set of sequentially adjacent FBGs in the fiber core until the desired length of the Bragg grating is obtained. Therefore, after the first completed exposition, occurs a translation stage to allow another exposition and Bragg gratings fabrication. This process finishes when the desired Bragg grating length is attained. This technique can be implemented in phase masking and interferometric methods [25].

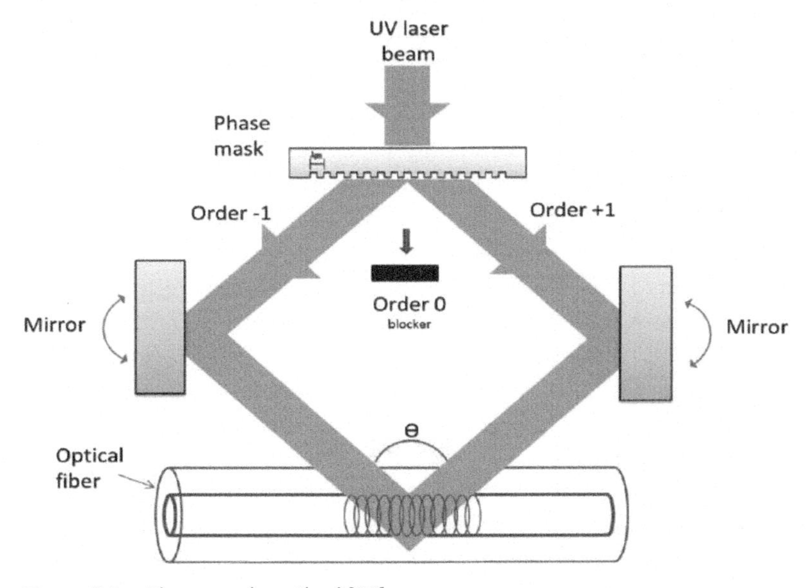

Figure 5.1 Phase mask method [25].

Another method that allows the FBGs fabrication with the desired length is the point-by-point inscription. In this technique, a femtosecond laser or a UV pulse laser is used to inscribe the fringes pattern in the fiber core. In this case, individual grating planes (changes of the RI) are inscribed one step at a time along the core of the fiber. A single pulse of the laser passes through the slit of the mask and is focused onto the optical fiber core. As a result, the core section that was modified presents a locally higher RI. After this process, the optical fiber is translated, in a parallel direction to the fiber axis, to inscribe another grating plane. This process is repeated until the desired grating structure is attained in the core of the fiber. The distance the fiber is translated to perform another grating plane corresponds to the grating period [25].

With this method, it is possible to locally modify the grating period to the desired value and perform long gratings. However, the translation stage movement is very crucial and must be very accurate [25]. This technique is used in Bragg gratings fabrication and was demonstrated for the first time in 1993 by Malo et al. [26], nonetheless, it is more used to produce LPGs.

5.3.1.2 LPG fabrication techniques

LPGs can be fabricated, using an amplitude mask, a phase mask, or an interferometric beam pattern, through the exposition of the photosensitive optical fibers to UV laser radiation [27–31]. This technique is the most commonly used. Besides, LPGs can also be fabricated by changing the RI of the core using, for instance, carbon-dioxide laser irradiation [32], femtosecond laser pulses [33], or by applying electric-arc discharges to inscribe LPGs on a period-by-period basis [34]. Other techniques, such as dopant diffusion into the fiber core, ion implantation through a metal amplitude mask, mechanically induced, and temporary LPGs using acousto-optic modulation are also used [35–38]. Nonetheless, the most used techniques are laser UV light, mechanically induced, and electric-arc discharges [25].

LPGs can be inscribed in optical fibers using UV light in two ways: using the amplitude mask technique (Fig. 5.2a) and the point-to-point technique (Fig. 5.2b) [27, 39]. To use this approach for LPG fabrication, fibers must be photosensitive, by doping the fiber silica core with germanium or boron, or by hydrogen loading [39], for example. The amplitude mask consists of a metal foil milled or etched with rectangular gaps between grooves [40]. More commonly the amplitude mask is made from chrome-plated silica [27]. This mask presents an array of transparent windows to allow the formation of the periodical pattern over the fiber core. Therefore, the mask is placed in contact and perpendicularly to the optical fiber. The LPG inscription occurs by exposing the fiber to the UV beam through the amplitude mask. Peaks in the RI modulation of the photosensitive fiber core are formed, corresponding to alternated bands of maximum and minimum intensity. As for the point-by-point UV technique, this approach was already described. However, more information is given since this method is very applied for LPG fabrication. Besides the information shared, in practical implementations, the fiber can be continuously translated parallel to the fiber axis and a regularly pulsed UV source is used [39], or at each point, the fiber is stopped allowing the illumination of the fiber core in a specific section using a single high energy UV pulse [41].

Mechanically induced long period grating is also a technique used to fabricate LPGs. In general, with this method, periodic stress and/or microbending are/is induced in an optical fiber by squeezing the fiber between a set of grooved plates. Therefore, the fiber is placed on a flat surface and can be pressed with different plates, such as coiled wire [42], grooved plate [43], and graphite rods [44]. Over the years, several techniques of this nature were published, all of which presented the same operating principle already explained.

The electric-arc discharge technique is another approach that can be used to fabricate LPGs. This method is a point-to-point inscription process, in which it is used arc discharges from a commercial fiber fusion splice machine or similar system [45, 46], to inscribe LPGs in the optical fiber core. In contrast to UV-induced LPGs, this technique does not require photosensitive fibers and the spectral properties only depend on the period and the length of the grating. In UV-induced LPGs, the resonant wavelengths depend on the grating strength [47].

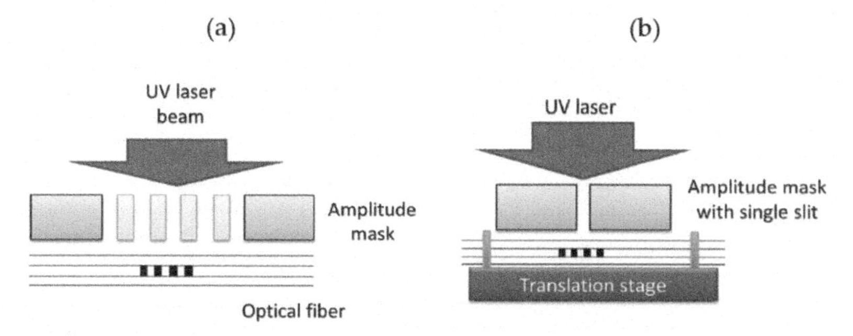

Figure 5.2 Schematic representation of (a) amplitude UV radiation writing and (b) point-to-point UV radiation writing.

5.3.2 Uniform Fiber Bragg Gratings

Uniform FBGs are a type of short-period gratings with gratings fabricated perpendicularly to the light propagation axis, as shown in Fig. 5.3a. In this structure, it is accomplished a coupling between the forward fundamental core mode and its respective counter-propagating mode (backward propagating core mode) [2, 7]. Therefore, the reflected spectrum on an FBG presents a peak, while the transmission spectrum shows an attenuation, as graphically represented in Fig. 5.3b.

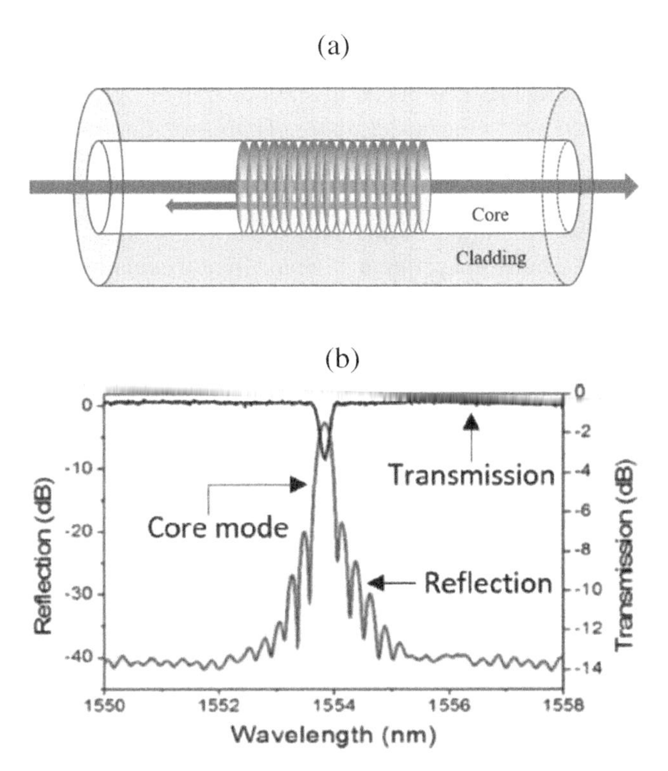

Figure 5.3 (a) Uniform FBG and light mode coupling (propagation direction indicated by the green arrow), and (b) transmitted (black) and reflected (red) spectra of a uniform FBG [1].

The mechanism of an FBG is based on the action of narrow-band reflectors, and the wavelength that is reflected is determined by the grating pitch. In this sense, of many wavelengths reflected at each perturbation, only one experiences constructive interference since the remaining experience destructive interference and disappear along with other reflected signals. Thus, coupling takes place only for a specific wavelength, known as the Bragg wavelength. According to the phase-matching condition, the Bragg wavelength is described as [48]:

$$\lambda_B = 2n_{eff}\Lambda, \tag{5.3}$$

where λ_B is the Bragg wavelength or reflected peak, n_{eff} is the effective RI of FBG, and Λ is the grating period.

Briefly, the operation principle of an FBG is based on the dependence of Bragg wavelength on the effective RI and grating period.

The uniform FBGs present a drawback in the sense that they are insensitive to variations of the surrounding RI, although they are sensitive to temperature, strain, and pressure. This means that the effective RI is not influenced by the external RI [5, 7, 49]. The reason for this phenomenon happen is related to the fact that the light propagation is firmly confined within the core and cannot reach the cladding [7], leading to coupling only between core modes [4]. Therefore, it is not possible to use sensors with this configuration for RI sensing purposes.

5.3.3 Etched Fiber Bragg Gratings

To counteract the barrier of uniform FBGs to sensing purposes, the cladding of FBG can be totally or partially removed along the grating area by the etching process, resulting in an etched FBG and allowing the transmitted light confined in the core to be influenced by the surrounding RI [4, 5]. Cladding reduction can be accomplished by other techniques besides the etching process, such as grinding and fine-drawing-cone [4]. Owing to the structure modification, changes in the external RI influence the effective RI and, consequently, the Bragg wavelength, allowing changes in Bragg wavelength combined with reflection amplitude modulation [5]. As a result, etched FBGs are capable of detecting changes in the measurand and can be used for sensing purposes. For the first time in 2001, an etched FBG was used to manufacture a tunable wavelength filter by Kumazaki et al. [50].

5.3.4 Tilted Fiber Bragg Gratings

Another configuration that can be used instead of uniform FBGs for sensing purposes is TFBG. This structure also presents a periodic modulation of RI along the fiber axis, as in uniform FBGs. However, the gratings are uniformly tilted by an angle θ with respect to the fiber axis [49, 51]. This configuration compared to etched FBG allows the maintenance of the mechanical resistance.

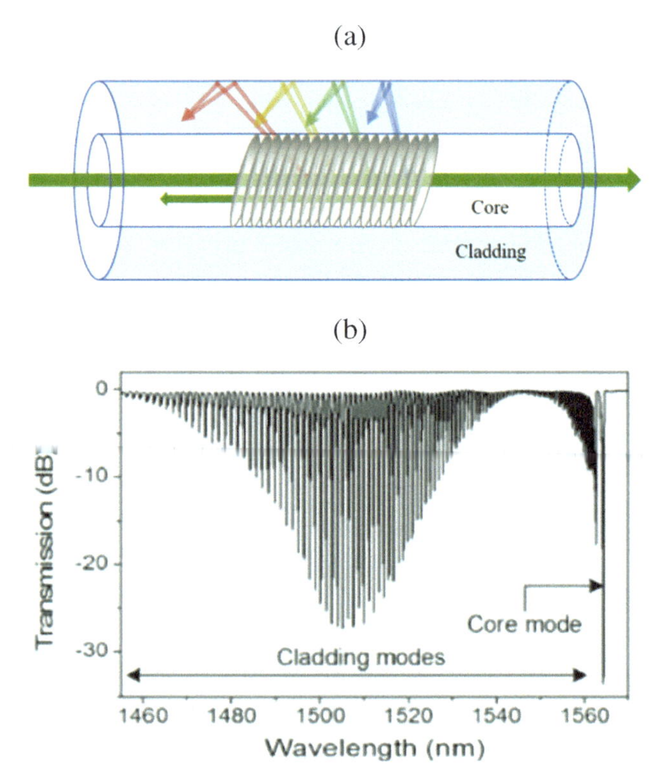

Figure 5.4 (a) Light mode coupling in TFBG with small tilt angle (θ = 10°), and (b) respectively transmitted spectrum [1].

In TFBGs, the core mode couples to several cladding modes, enabling the interaction of light with the surrounding medium. Depending on the tilt angle, the coupled light can have different directions, backward, or forward direction [14, 49]. Therefore, besides the self-backward coupling of the core mode, there is backward coupling between the core mode and several cladding modes, for small θ between 4° and 16° [49] (Fig. 5.4a) [14]. Each cladding mode possesses a unique effective RI, since each mode travels with a different phase velocity [1]. The typical transmission spectrum presents many resonance peaks (narrow attenuation bands), corresponding to different coupled cladding modes, in which each excited cladding mode corresponds to a specific wavelength and loss in the light core intensity [2]. Figure 5.4b represents an example of a transmission spectrum of a TFBG with small θ. This transmission spectrum exhibits narrow attenuation bands (FWHM ≈ 0.1 nm) [14], each one representing the coupling between the core mode and each cladding mode on the smaller wavelength side of the core mode resonance [1]. This core mode resonance is designated the Bragg mode or Bragg peak and appears due to the backward coupling of the core mode, which is sensitive to strain effects and temperature variations [2, 3]. For $\theta > 45°$, coupling only occurs between the core mode and the several cladding modes in the forward direction (Fig. 5.5a), in contrast to small θ. In this case, the transmission spectrum (Fig. 5.5b) also exhibits several attenuation bands corresponding to cladding modes, with FWHM of a few nanometers [1].

The Bragg condition of a TFBG takes into account the resonance wavelengths of the several cladding modes. Thus, for TFBGs, the resonant wavelengths are given by [28, 52]:

$$\lambda_{\mathrm{res}(m)}^{\mathrm{TFBG}} = \left(n_{eff}^{\mathrm{core}} \pm n_{eff}^{\mathrm{clad}(m)}\right)\frac{\Lambda}{\cos(\theta)}, \tag{5.4}$$

where n_{eff}^{core} is the effective RI of the core and $n_{eff}^{\mathrm{clad}(m)}$ is the effective RI of the mth cladding mode. The signs of "+" and "−" in the equation indicate the direction of the mode propagation. "+" corresponds to backward mode propagation and "−" corresponds to forward mode propagation.

Cladding modes are sensitive to the surrounding RI and, the maximum sensitivity is accomplished when the cladding effective RIs are similar to the surrounding RI [4, 53]. When the surrounding RI increases, the resonance wavelengths of the corresponding cladding modes shift to longer values [49]. In 2001, Laffont and Ferdinand reported the first TFBG sensor for surrounding RI detection [54].

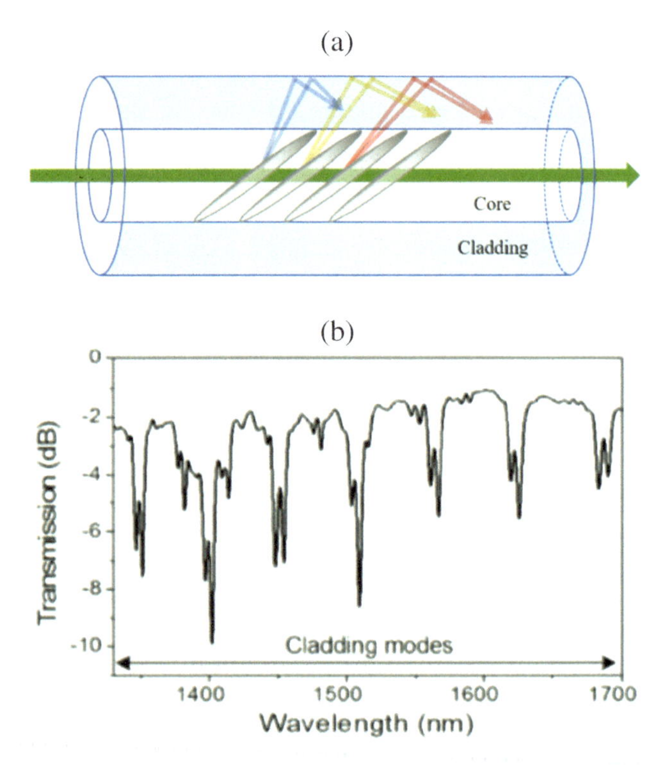

Figure 5.5 (a) Light mode coupling in TFBG with large tilt angle ($\theta > 45°$), and (b) respectively transmitted spectrum [1].

TFBGs allow decorrelation of unwanted temperature, strain variations, and power-level changes from the light source, which make them advantageous for RI sensing since the Bragg resonance is immune to the external medium RI. Hence, this Bragg peak acts as a reference during RI sensing [1, 14]. Although bare silica TFBGs present good features for sensing purposes, the sensitivity can be enhanced even further by applying a metallic coating on the fiber surface over the TFBG region to achieve the SPR effect [55]. The first TFBG-based SPR sensor was reported by Yanina Y. Shevchenko and Jacques Albert in 2007 [56]. As already described, this is possible for P-polarized modes (TM and EH modes) with the electric field polarized mostly radially to couple energy to SPWs. S-polarized modes (TE and HE modes) are unable to trigger SPR as they have electric fields polarized mostly azimuthally, in other words, tangentially to the metal [1, 57]. When the SPR is excited, some light energy with a certain wavelength range is absorbed. This wavelength range of the absorbed light changes according to the surrounding RI. Thus, these spectral variations of the resonance attenuation (SPR signature) are highly sensitive and can be used for sensing purposes. The SPR signature of an Au-coated TFBG is shown in Fig. 5.6.

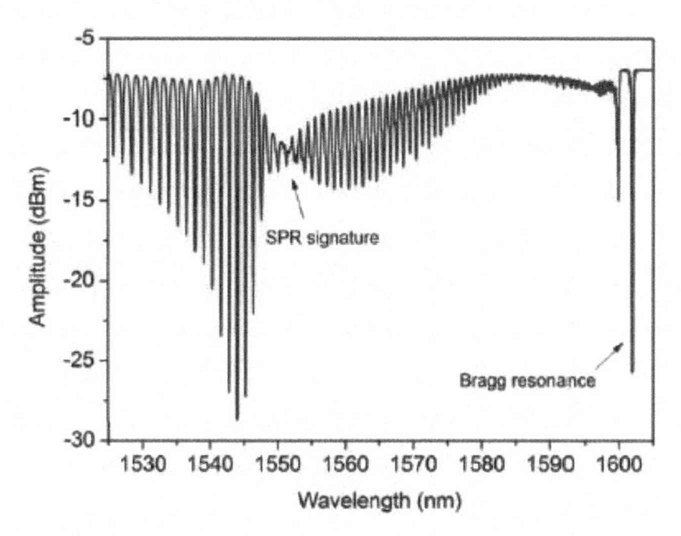

Figure 5.6 Transmission spectrum in salted water of an SPR-TFBG coated with 30 nm gold film [59].

5.3.5 Chirped Fiber Bragg Gratings

All the aforementioned configurations exhibit gratings with spatially uniform periodicity. Once the period of the RI varies along the grating length, Chirped FBGs are created [58]. These structures consist of aperiodic Bragg gratings in which the resonance condition varies along its length, with a different spectrum being reflected by each portion of the grating [6]. Nevertheless, this structure can be considered locally uniform if the variation of the periodicity is small since the different wavelengths are reflected by each part of the grating without affecting each other.

The bandwidth and group delay are two properties affected by the aperiodicity. As already implied, in chirped FBGs, the resonance condition occurs for several wavelengths, resulting in a bandwidth much higher than uniform FBGs. As a result, compared with conventional FBGs, chirped FBGs present a broader reflection spectrum that can range from a few nanometers to tens of nanometers [58]. Another property that is affected is the group delay, in which its characteristics are different from uniform FBGs. In chirped FBGs, the group delay depends on the wavelength since different positions of the fiber are reflected in different wavelengths. The group delay is linear when the aperiodicity is also linear, being the linearly chirped FBGs the most crucial configuration presenting a linear Bragg wavelength variation along the grating length [58].

A chirped FBG is schematized in Fig. 5.7. Its structure is an extension of a uniform FBG. As supra-mentioned, the Bragg wavelength is represented by Eq. (5.3). In chirped FBG, the grating period is given by the function $\Lambda(z)$ since the grating periodicity is not constant, changing along the propagation axis z. As mentioned, a chirped FBG spectrum is constituted with the spectrum of each grating section since different Bragg wavelengths are reflected at each different section of the grating [6].

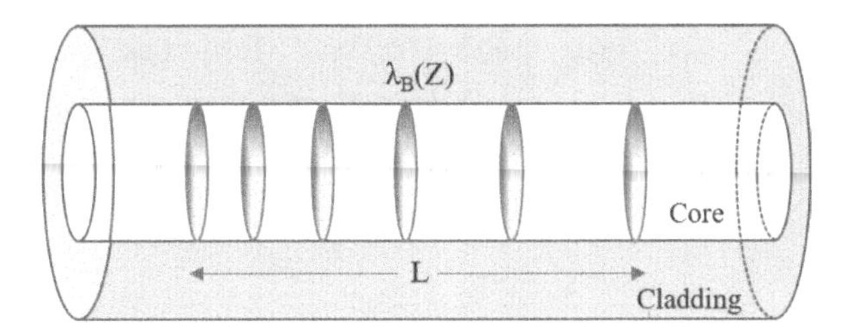

Figure 5.7 Schematic of a chirped FBG.

5.3.6 Long Period Gratings

In contrast to FBGs and as shown in Fig. 5.8a, the coupling in LPGs occurs between the guided mode of light (fundamental core mode) and the several cladding modes in the forwarding direction, owing to the grating period that extends to an order of a fraction of a millimeter. As in TFBG, the transmission spectrum of an LPG also presents several attenuation bands corresponding to resonance peaks related to the coupling of several cladding modes, as represented in Fig. 5.8b. This means that the resonance wavelength for the mth cladding mode is given by [7]:

$$\lambda_{\mathrm{res}(m)}^{\mathrm{LPG}} = (n_{eff}^{\mathrm{core}} - n_{eff}^{\mathrm{clad}(m)})\Lambda \tag{5.5}$$

The cladding modes are also sensitive to changes in the surrounding RI as in TFBGs. Nonetheless, if the surrounding RI increases, the resonance wavelengths shift to lower values [49].

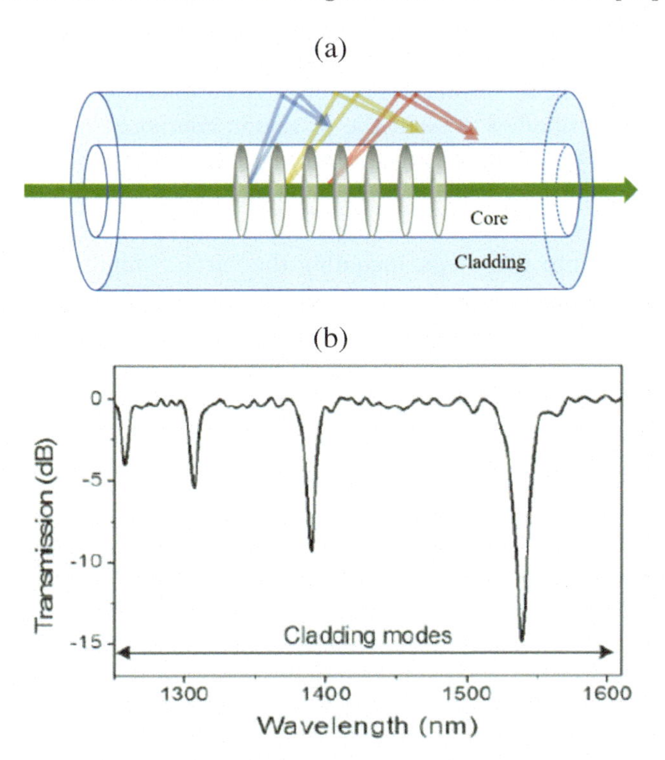

Figure 5.8 (a) LPG and light mode coupling, and (b) respective transmission spectrum [1].

5.4 Surface Functionalization

As already implied, gold and gold NPs are the most chosen metals and structures placed onto silica optical fiber grating surfaces to activate SPR and LSPR phenomena. Immunosensors are one of the main classes of sensors that have been used the most. This type of sensor employs antibodies that recognize and bind to a specific antigen, leading to a highly specific and selective immunoreaction [60]. Thus, it is essential for the immobilization of antibodies on the fiber surface. An antibody belongs to the immunoglobulin (Ig) superfamily and corresponds to a Y-shaped glycoprotein. The antibody is composed of a fragment crystallization (Fc) region and two fragment antigen binding (Fabs) regions where the binding of a specific antigen occurs [61]. This section focuses on the different functionalization approaches that have been applied. These approaches must consider the Fab region composed of the amine terminal, in the sense that this region must be voided to allow antigen binding. Antibodies present the Fc region composed of carboxylic groups that bind covalently to the metallic surface [62]. Besides, the applied functionalization methods must take into consideration the nature of the surface coating. Thus, in general, each approach is divided into two steps, including the surface modification with an intermediary linker that must be chosen according to the used metal, resulting in a surface with, for instance, amines or carboxylic functional groups, being possible to use others, and secondly the linkage of the antibody.

To immobilize the antibody on the Au surface is very used an intermediary linker such as cysteamine. To perform the antibody immobilization employing this molecule, the sensors are firstly immersed in a cysteamine solution, whose thiol (–SH) functional groups bind to the Au surface through a strong affinity interaction. Finally, the carboxylic acid (–COOH) functional groups of antibodies bind to the cysteamine amine (–NH_2) functional groups in the fiber surface, applying 1-ethyl-3-(3-dimethylaminopropyl)carbodiimide hydrochloride (EDC)/ N-hydroxysuccinimide (NHS) bioconjugate reagents. This last step allows the covalent immobilization of the antibodies [63].

Instead of cysteamine, another approach uses 11-mercaptoundecanoic acid (MUA) and protein A covalently bonded to this linker to immobilize antibodies on gold NPs. This strategy using protein A compared to the same without protein A leads to better sensitivity and lower limits of detection. Therefore, the thiol group of MUA strongly binds to gold, and the amine group of protein A links covalently to the carboxylic group of MUA, which is activated by the EDC/NHS chemistry [63].

Besides cysteamine and protein A, dopamine is also used as the cross-linking agent in other studies. As a result, antibodies are immobilized on the fiber surface coated with poly(dopamine) [64].

The reason for EDC is the most popular used bioconjugate reagent is that it allows the activation of the carboxylic groups promoting the formation of amide bonds when they are present. The efficiency of this reaction can be enhanced by using NHS [65].

In this case, the binding between amine groups and the carboxylic groups present in the Fc region of the antibody guarantees an adequate orientation of the antibody [66, 67]. This strategy leads to an available formation of the antibody-antigen complex. Nonetheless, it requires more functionalization steps. As a result, the antibodies are far from the fiber surface, which can lead to sensors with lower sensitivity [55].

After all these steps performed in the functionalization approaches, the last step is the passivation with the aim to reduce non-specific interactions between the analyte and surface, and between interferents present in more complex matrixes. The most common passivating agent is bovine serum albumin (BSA) [68–75].

In addition to antibodies, aptamers have also been chosen as the biorecognition element in many studies since they have the same potential as antibodies in terms of achievements [76]. These aptamers are designed or selected accordingly to the target molecules with high affinity and specificity and can be single-stranded DNA, RNA, or synthetic XNA molecules, meaning that are constituted with nucleotides (or peptides in some specific cases) [77, 78]. Currently, aptamers are engineered using the systematic evolution of ligands by the exponential enrichment (SELEX) method. The trial-and-error experiment will be avoided in the future since the aptamers will be obtained from data learning and through the use of atomic properties to generate computer-assisted sequences in silico. Furthermore, in the near future, the efficiency of aptamers will be improved through synthetic methods that have been carried out in studies, leading to the disuse of SELEX methods to isolate candidate sequences [79, 80]. Over the years, aptamers were used in different sensing platforms for different applications. Nonetheless, the application of aptamers in optical fiber technology generates new challenges in terms of sensing applications.

Therefore, as performed with antibodies, the aptamers immobilization on the gold surface is commonly realized through the chemisorption of thiol groups [76]. One method used for this purpose is self-assembled monolayers (SAMs). This is possible due to the aptamers' small sizes and versatility, which allow efficient immobilization in SAMs. Over the years, several immobilization techniques were applied; however, the chemisorption of thiols to gold is the most used lately [81]. This method based on the formation of SAMs is often used for DNA hybridization on gold surfaces. To reduce steric hindrance and guarantee good conformation of the receptors, thiol-alkanes linked to the aptamer (tail-region) are sometimes coupled with spacers [76]. The formation of the SAM comprises three steps: diffusion-controlled physisorption, chemisorption of the molecules, and the crystallization process. In the first step, only van der Waals interaction is present for the adsorption and is characterized as a gas-like, highly disordered system. In the second step, it accomplishes a strong covalent bond between the sulfur head and three gold atoms. The last step consists of the formation of highly ordered and orientated monomolecular layers through the alignment of the molecules on the gold surface in a parallel configuration. This formation occurs through tail-tail interactions, including van der Waals, repellent, steric, and electrostatic forces [76, 81]. Briefly, the self-assembling process occurs, owing to the alignment of sulfur donor atoms with gold. The SAM formation is not based on covalent bonds [76]. Multiple factors cause several defects in these SAMs, such as molecule disorder in some regions, missing molecules in a small number (pinholes) or in entire rows forming straight or zig-zag lines, or even defects on the gold surface that cause defects on SAM [81].

Therefore, aptamers immobilization can be accomplished via direct thiolation, which is a simple and effective approach. Thiolated aptamers can be found commercially in stable forms of disulfides, it is only necessary to perform a reduction before the immobilization to attain free thiols in solution (-SH). Thus, thiolated aptamers are diluted with, for instance, dithiothreitol (DTT) or tris(2-carboxyethyl) phosphine (TCEP), which are the often used components for this purpose. Afterward, the aptamer solution is placed in contact with the gold surface for aptamer immobilization. In the case of TCEP, it can remain in the solution; whereas, if DTT is chosen, it must be removed from the solution before contact with gold as it can be adsorbed on the gold surface. This extraction can be performed by size-exclusion column or filtration [76]. With thiolated aptamers, the binding not only occurs via Au–S bonds but also presents the non-specifically absorbation via multiple nitrogen atoms. This situation restricts the accessibility of the analyte to the aptamer [81]. Therefore, the surface must be blocked with mercaptohexanol, which corresponds to a small molecule with affinity for the surface. Gold NPs can also be functionalized using this last method by applying the same surface chemistry [76].

Although antibodies and aptamers are the most used biorecognition elements, there are also plasmonic sensors that employ, for instance, enzymes, nucleic acids, and molecularly imprinted polymers as biorecognition elements.

5.5 Applications and New Trends

Over the years, optical fiber grating-based plasmonic sensors have been used in many applications, for instance, in the detection of several analytes, such as viruses and medical biomarkers, among others, as an alternative to traditional methods. In several developed sensors, strategies to enhance the signal response were used, such as the use of highly sensitive coatings to provide SPR or LSPR, for example, gold coating or gold NPs [82].

Epidemics and pandemics are problems that affect the global community. Therefore, accurate and early diagnosis of virus infections is essential to control these problems. Among several works, in 2018, Luo et al. [68] produced an immunosensor using excessively-tilted fiber Bragg grating (ex-TFBG) coated with Au nanospheres (NSs) for Newcastle disease virus (NDV) detection (Fig. 5.9a). In this work, the fiber's surface was modified with 3-mercaptopropyltrimethosysilane (MPTMS) and, then, the MPTMS was coated via Au-S bonds with Au NSs. The next step was the binding of the activated protein A to the NSs through covalent bonds using cysteamine as the intermediate linker. Compared with the fiber without Au NSs, the effect of surrounding RI changes on the fiber cladding mode was enhanced 5–10 times, owing to the presence of NSs and, consequently, due to the LSPR phenomenon. Through the monitorization of the resonance wavelength shifts, it was possible to obtain with this sensor in a linear concentration range of 0–200 pg mL^{-1}, a sensitivity of, approximately, 1.627 pm (pg $mL^{-1})^{-1}$. The LOD presented a value around 25 pg mL^{-1} (see Fig. 5.9b).

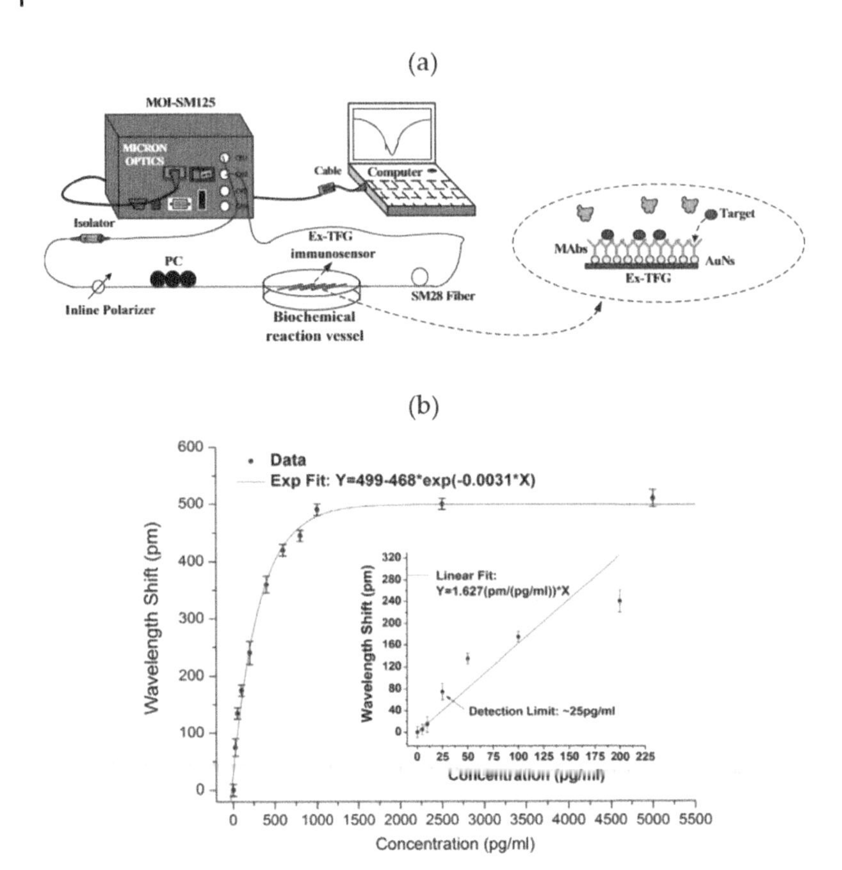

Figure 5.9 (a) Schematic representation of the setup for the development of Ex-TFBG immunosensor coated with Au nanospheres for NDV detection. (b) Experimental fit for resonance wavelength shifts as a response to NDV concentration (Inset presents the linear fit for NDV concentrations of 0~200 pg/mL) [68].

The quantification of cancer biomarkers using portable optical fiber grating sensors allows early screening and diagnosis of the diseases, which significantly improves patient care and leads to higher survival rates. Besides, this type of sensor can be used in POC for real-time detection near the patient. In 2017, Ribaut et al. [83] detected a lung cancer biomarker (CK17) using a TFBG immunosensor coated with Au. TFBGs presented a tilt angle of 7° with 50 nm of Au thickness. In this present work, a self-assembly monolayer by immersing the TFBGs in S_2-PEG_6-COOH was created. Afterward, the TFBGs surface was subjected to AbCK17 antibody immobilization. Gel matrices with encapsulated CK17 were used to simulate tissue samples, in which CK17 detection measurements were taken. This sensing platform was developed with the aim of assessing non-liquid environments. In this situation, the LOD of the immunosensor was 0.1 ng mL^{-1}. This sensor was also used to successfully distinguish samples of healthy tissue from tissue with tumors. With that purpose and based on the same sensing platform, the sensors were applied for ex vivo testing of a human lung biopsy, presenting a LOD that could reach 0.4 nM. In 2019, Luo et al. [84] reported a label-free sensor for soluble programmed death ligand-1 (sPD-L1) detection using ex-TFBG coated with Au nanoshells as represented in Fig. 5.10. The tumor aggressiveness and outcome can be indicated by sPD-L1 level in serum since sPD-L1 represents an essential role in tumor evasion from the immune system. In this work, protein A was implemented to allow the attachment of the anti-sPD-L1 antibodies on the LSPR-based ex-TFBG surface. In buffer solutions, the LOD of this immunosensor was ~1 pg mL^{-1}, and in complex serum media (fetal bovine serum) was 5 pg mL^{-1}. Another cancer biomarker, more specifically, breast cancer HER2 was detected using a plasmonic optical fiber grating sensor in 2020 by Lobry et al. [85]. This study used a TFBG coated with ~35 nm of gold thickness to activate the SPR phenomenon. To detect HER2 breast cancer biomarkers, the Au-coated TFBG was functionalized with thiolated aptamers, which were directly immobilized onto a gold surface. To improve the sensitivity, anti-HER2 antibodies were then immobilized. In this work, for HER2 detection, was employed a demodulation technique based on the SPR signature envelope fit tracking. For that, a peculiar feature (local maximum) of the lower envelope of the cladding mode resonances spectrum was tracked. Comparing the studies reported so far based on the tracking of individual modes of the spectrum, this method showed to be highly sensitive and yielded wavelength shifts several tens of times higher. The aim of this work was not to quantify the limit of detection. The purpose was to strengthen the practicability of a highly sensitive biosensing platform. Therefore, only the HER2 aptamer-protein interaction led to a ~300 pm redshift. Whereas, the detection using anti-HER2 antibodies amplification allowed a high extra shift of 800 pm for 10^{-6} g mL^{-1} HER2 solution in phosphate-buffered saline (PBS).

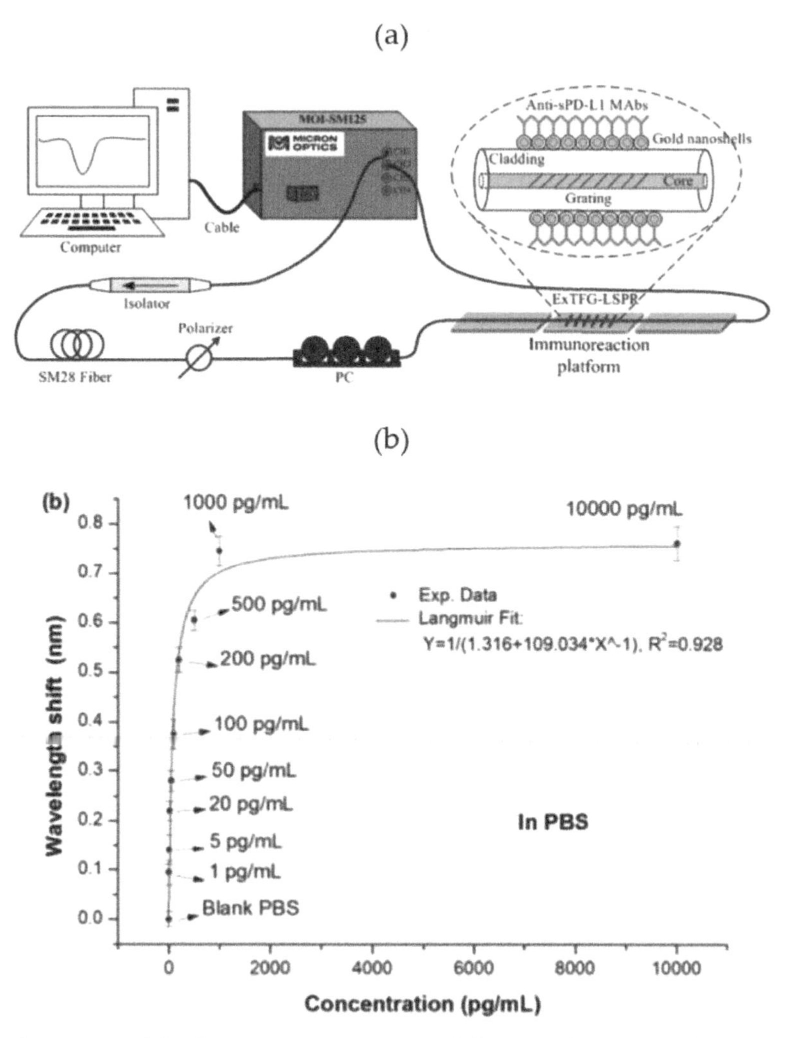

Figure 5.10 (a) Schematic representation of the setup for the development of Ex-TFBG immunosensor coated with Au nanoshells for sPD-L1 detection. (b) Average values of wavelength shifts as a response to sPD-L1 and their fitting curves [84].

Various diseases like depression or cardiovascular diseases can be triggered when stress is persistent and uncontrollable [86]. Besides this area, stress has also a lot of influence on aquaculture production. In recirculating aquaculture systems, when small variations in water chemistry or quality occur, stress is induced, leading to reduced food intake and, consequently, reduced fish growth [87, 88]. It can also lead to mortality when acute or chronic stress is present. Therefore, a technology capable of monitoring stress biomarkers in POC is crucial to prevent these problems. Among several biomarkers, cortisol is considered the main stress biomarker. In 2020, an SPR Au coated-TFBG immunosensor for cortisol detection was reported by Leitão et al. [67], as represented in Fig. 5.11a. For that purpose, TFBG with Au was functionalized with anti-cortisol antibodies employing cysteamine as the intermediate linker. Here, an alternative interrogation method was employed, in which the SPR mode was monitored through the signalization of the local maximum of the plasmonic signature of the lower envelope of the spectra (see Fig. 5.11b). With this, immunosensor was possible to attain a total wavelength shift of 3 nm and a sensitivity of 0.275 ±0.028 nm (ng mL^{-1})$^{-1}$ for a linear cortisol concentration range of 0.1–10 ng mL^{-1}.

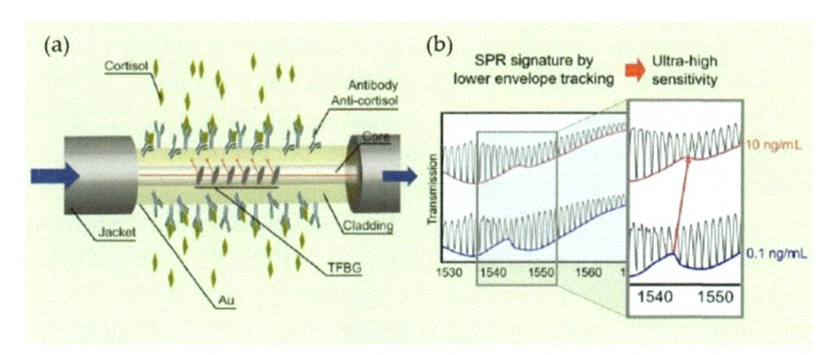

Figure 5.11 SPR-based TFBG immunosensor for cortisol detection as a stress biomarker: (a) schematic representation of the sensing mechanism, and (b) SPR signature spectra variation and respective lower envelope as a response to two cortisol concentrations [67].

Environmental contaminants and pollutants resistant to degradation have been increasingly investigated as they persist in the environment causing harmful effects on the environment and human and animal health. Thus, over the last few years, several methods capable of accurately detecting these substances have been developed, including sensors, which are imperative to protect the environment and health. For instance, in 2019, Tripathi et al. [89] developed an aptasensor for the detection of cyanobacterial toxin, more specifically the environmental toxin microcystin-LR (MC-LR), which is toxic for both humans and animals. For that, a dual-resonance LPG was partially etched, coated with gold, and functionalized with a thiolated aptamer through covalent immobilization (see Fig. 5.12a). Observing the change in the resonance wavelengths, it was possible to measure the MC-LR binding to the aptamer. Thus, for an MC-LR concentration ranging from 5 nM to 250 nM, the resonance wavelength shifted from 0.02 nm to 1.6 nm, as represented in Fig. 5.12b. Besides, the attained LOD of this aptasensor was 5 ng mL^{-1}. In this work, it was also performed atomic force microscopy to confirm the real-time, dynamic, and molecular binding of MC-LR to the sensor surface.

As potential future of biosensors and the way how the researchers will keep progress in this area can be through the use of nanoparticle-doped optical fibers. Although this new generation of optical fiber sensors is still in its infancy, it is starting to show promising results and quite good literature [90–93]. Up to now, these fibers are based on silica glass and it is showing some progress in RI measurements [92]. In addition, progress using such fibers has been done in terms of fiber gratings with the future goal to go for the next step, using them in biosensors [94]. Further, this concept of nanoparticle-doped optical fibers can be extended to other glasses such as chalcogenide [95] or polymer fiber technology [63]. However, there is a clear need to improve the knowledge of researchers on the light-nanoparticles interaction close to the fiber core to engineering periodic structures, for instance. The control of the characteristics of the nanoparticles is far from being simple but possible in a near future. Advanced numerical simulations must be developed to understand the nucleation/growth of phase-separated particles as well as to control the size, position, and shape of such nanoparticles close to the core.

In parallel, recent developments in specialty fibers such as microstructured optical fibers (MOFs) have attracted immense interest, as they demonstrate unique advantages for sensing applications, owing to the easiness to tailor optical and mechanical properties of the fibers. This can be achieved by manipulating the structural design, as well as the material and the added dopants [96]. MOFs that possess a periodic microstructure of air holes along the whole length of the fiber, generally in the cladding region, are known as photonic crystal fibers (PCFs) and can be classified as hollow-

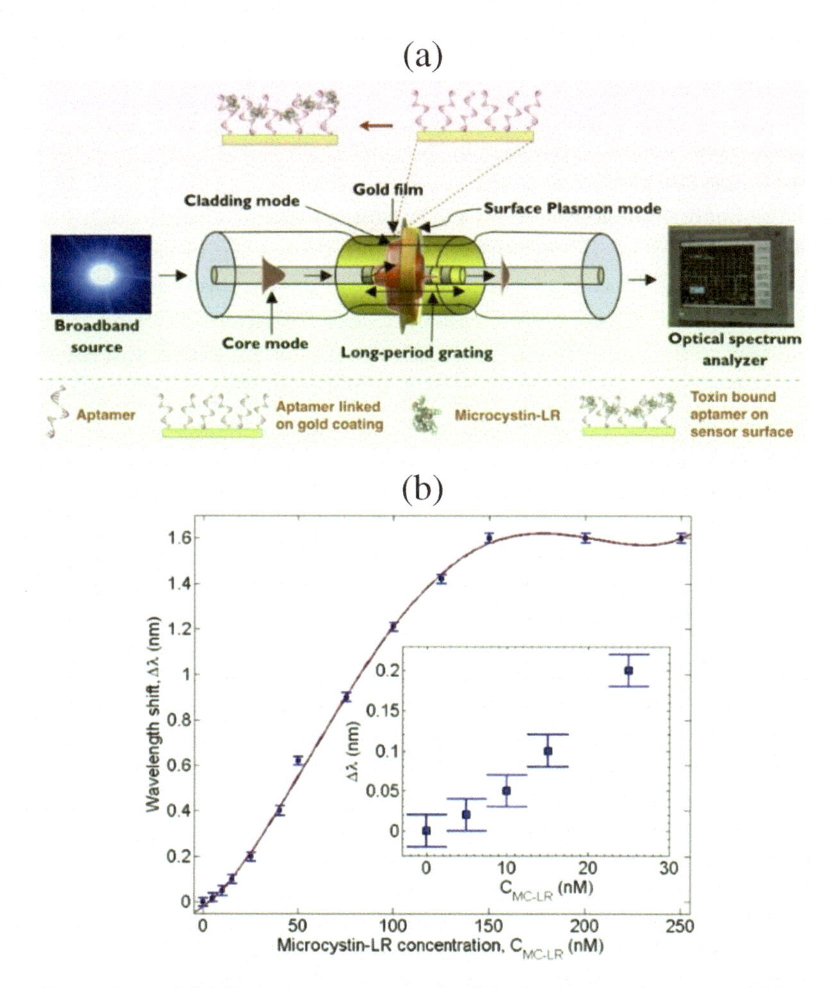

Figure 5.12 (a) Schematic representation of dual-resonance long-period fiber gratings functionalized with a thiolated aptamer used as a bioreceptor for MC-LR toxin detection. (b) Resonance wavelength shifts as a response to MC-LR concentration in the buffer [89].

core or solid-core [96, 97]. For biosensing purposes, the cladding air holes, or the core in the case of hollow-core PCFs, is metal-coated and filled with the analyte (selectively or fully) in order to induce the direct interaction between a part of the light guided in the core and the metal surface. The changes in the sample properties, for example in the RI, will induce a corresponding change in the effective RI of the guided mode. Besides conventional configurations, others have been reported, including multi-core, grapefruit, and with fiber gratings (FBGs, LPGs, and TFBGs) inscribed on MOFs [96], as a means to broaden potential applications and improve sensor efficiency [97].

Lastly, plastic optical fibers (POFs) based on poly(methyl methacrylate) (PMMA) consist of another type of specialty fiber and have been deployed for SPR excitation. Despite the performances reported thus far being inferior compared to those of silica fibers, POFs exhibit increased biocompatibility, an essential feature for biosensors applied in vivo. However, less research has been done on this topic, and it needs more attention and researchers to focus on POF technology with plasmonic. In light of these developments, biosensors incorporating MOFs or POFs must be explored more and more, in future work, as fiber-grating plasmonic detectors of different applications.

References

1. T. Guo, Á. González-Vila, M. Loyez, and C. Caucheteur. (2017). Plasmonic optical fiber-grating immunosensing: A review, *Sensors (Switzerland)*, **17**, pp. 1–20.
2. M. Lobry, D. Lahem, M. Loyez, M. Debliquy, K. Chah, M. David, and C. Caucheteur. (2019). Non-enzymatic D-glucose plasmonic optical fiber grating biosensor, *Biosens. Bioelectron.*, **142**, p. 111506.
3. M. Lobry, M. Loyez, E. M. Hassan, K. Chah, M. C. DeRosa, E. Goormaghtigh, R. Wattiez, and C. Caucheteur. (2020). Multimodal plasmonic optical fiber grating aptasensor, *Opt. Express*, **28**(5), pp. 7539–7551.
4. Y. Zhao, R. jie Tong, F. Xia, and Y. Peng. (2019). Current status of optical fiber biosensor based on surface plasmon resonance, *Biosens. Bioelectron.*, **142**, pp. 1–12.
5. A. Iadicicco, S. Campopiano, A. Cutolo, M. Giordano, and A. Cusano. (2005). Thinned fiber Bragg gratings for sensing applications, *Proc. WFOPC2005 - 4th IEEE/LEOS Work. Fibres Opt. Passiv. Components*, **2005**, pp. 204–209.
6. S. Korganbayev, Y. Orazayev, S. Sovetov, A. Bazyl, E. Schena, C. Massaroni, R. Gassino, A. Vallan, G. Perrone, P. Saccomandi, M. A. Caponero, G. Palumbo, S. Campopiano, A. Iadicicco, and D. Tosi. (2018). Detection of thermal gradients through fiber-optic chirped fiber Bragg grating (CFBG): Medical thermal ablation scenario, *Opt. Fiber Technol.*, **41**, pp. 48–55.

7. M. jie Yin, B. Gu, Q. F. An, C. Yang, Y. L. Guan, and K. T. Yong. (2018). Recent development of fiber-optic chemical sensors and biosensors: Mechanisms, materials, micro/nano-fabrications and applications, *Coord. Chem. Rev.*, **376**, pp. 348–392.
8. B. Liedberg, C. Nylander, and I. Lunström. (1983). Surface plasmon resonance for gas detection and biosensing, *Sensors and Actuators*, **4**, pp. 299–304.
9. M. S. Aruna Gandhi, S. Chu, K. Senthilnathan, P. R. Babu, K. Nakkeeran, and Q. Li. (2019). Recent advances in plasmonic sensor-based fiber optic probes for biological applications, *Appl. Sci.*, **9**, pp. 1–22.
10. N. F. Chiu, C. Du Yang, C. C. Chen, T. L. Lin, and C. T. Kuo. (2018). *Functionalization of Graphene and Graphene Oxide for Plasmonic and Biosensing Applications*, Elsevier Inc.
11. G. P. Anderson and C. R. Taitt. (2008). *Evanescent Wave Fiber Optic Biosensors*, 2nd Ed., Elsevier B.V.
12. Y. Tang, X. Zeng, and J. Liang. (2010). Surface plasmon resonance: An introduction to a surface spectroscopy technique, *J. Chem. Educ.*, **87**(7), pp. 742–746.
13. X. Zhu and T. Gao. (2018). *Spectrometry*, Elsevier Inc.
14. C. Caucheteur, T. Guo, and J. Albert. (2015). Review of plasmonic fiber optic biochemical sensors: Improving the limit of detection, *Anal. Bioanal. Chem.*, **407**, pp. 3883–3897.
15. N. Cennamo, D. Massarotti, L. Conte, and L. Zeni. (2011). Low cost sensors based on SPR in a plastic optical fiber for biosensor implementation, *Sensors*, **11(**12), pp. 11752–11760.
16. W. Gong, S. Jiang, Z. Li, C. Li, J. Xu, J. Pan, Y. Huo, B. Man, A. Liu, and C. Zhang. (2019). Experimental and theoretical investigation for surface plasmon resonance biosensor based on graphene/Au film/D-POF, *Opt. Express*, **27**(3), p. 3483.
17. S. H. Kim and K. Koh. (2006). Functional dyes for surface plasmon resonance-based sensing system, in *Functional Dyes*, S.-H. Kim (ed.), Elsevier B.V., pp. 185–213.
18. Y. E. Monfared. (2020). Overview of recent advances in the design of plasmonic fiber-optic biosensors, *Biosensors*, **10**(7), pp. 1–22.
19. M. Qi, N. M. Y. Zhang, K. Li, S. C. Tjin, and L. Wei. (2020). Hybrid plasmonic fiber-optic sensors, *Sensors*, pp. 1–27.

20. K. O. Hill, Y. Fujii, D. C. Johnson, and B. S. Kawasaki. (1978). Photosensitivity in optical fiber waveguides: Application to reflection filter fabrication, *Appl. Phys. Lett.*, **32**(10), pp. 647–649.

21. G. Meltz, W. W. Morey, and W. H. Glenn. (1989). Formation of Bragg gratings in optical fibers by a transverse holographic method, *Opt. Lett.*, **14**(15), p. 823.

22. T. Geernaert, K. Kalli, C. Koutsides, M. Komodromos, T. Nasilowski, W. Urbanczyk, J. Wojcik, F. Berghmans, and H. Thienpont. (2010). Point-by-point fiber Bragg grating inscription in free-standing step-index and photonic crystal fibers using near-IR femtosecond laser, *Opt. Lett.*, **35**(10), pp. 1647–1649.

23. A. Theodosiou, A. Lacraz, A. Stassis, C. Koutsides, M. Komodromos, and K. Kalli. (2017). Plane-by-plane femtosecond laser inscription method for single-peak bragg gratings in multimode CYTOP polymer optical fiber, *J. Light. Technol.*, **35**(24), pp. 5404–5410.

24. A. Othonos and K. Kalli. (1999). *Fiber Bragg Gratings: Fundamentals and Applications In Telecommunications and Sensing*, Artech House.

25. R. Kashyap. (1999). *Fiber Bragg Gratings*, Academic Press.

26. B. Malo, K. O. Hill, F. Bilodeau, D. C. Johnson, and J. Albert. (1993). Point-by-point fabrication of micro-bragg gratings in photosensitive fiber using single excimer pulse refractive-index modification techniques, *Electron. Lett.*, **29**(18), pp. 1668–1669.

27. M. Vengsarkar, P. J. Lemaire, J. B. Judkins, V. Bhatia, T. Erdogan, and J. E. Sipe. (1996). Long-period fiber gratings as band-rejection filter, *Light. Technol.*, **14**(58).

28. V. Bhatia and A. M. Vengsarkar. (1996). Optical fiber long-period grating sensors, *Opt. Lett.*, **21**(9), pp. 692–694.

29. T. Erdogan. (1997). Fiber grating spectra, *J. Light. Technol.*, **15**(8), pp. 1277–1294.

30. V. Bhatia. (1999). Applications of long-period gratings to single and multi-parameter sensing, *Opt. Express*, **4**(11), pp. 457–466.

31. B. J. Eggleton, C. Kerbage, C. A. White, G. L. Burdge, P. S. Westbrook, and R. S. Windeler. (2000). Cladding-mode-resonances in air-silica microstructure optical fibers, *J. Light. Technol.*, **18**(8), p. 1084.

32. Y. Wang. (2010). Review of long period fiber gratings written by CO_2 laser, *J. Appl. Phys.*, **108**, p. 081101.

33. S. J. Mihailov, D. Grobnic, H. Ding, C. W. Smelser, and J. Broeng. (2006). Femtosecond IR laser fabrication of Bragg gratings in photonic crystal fibers and tapers, *IEEE Photonics Technol. Lett.*, **18**(17), pp. 1837–1839.

34. D. J. Webb, H. Dobb, I. Bennion, J. S. Petrovic, K. Kalli, and V. K. Mezentsev. (2007). Sensitivity of LPGs in PCFs fabricated by an electric arc to temperature, strain, and external refractive index, *J. Light. Technol.*, **25**(5), pp. 1306–1312.

35. M. Fujimaki, Y. Ohki, J. L. Brebner, and S. Roorda. (2000). Fabrication of long-period optical fiber gratings by use of ion implantation, *Opt. Lett.*, **25**(2), pp. 88–89.

36. J. W. Ham, J. H. Lee, J. Y. Cho, H. S. Jang, and K. S. Lee. (2002). A birefringence compensation method for mechanically induced long-period fiber gratings, *Opt. Commun.*, **213**, pp. 281–284.

37. W. J. Bock, J. Chen, P. Mikulic, and T. Eftimov. (2007). A novel fiber-optic tapered long-period grating sensor for pressure monitoring, *IEEE Trans. Instrum. Meas.*, **56**(4), pp. 1176–1180.

38. Q. Li, C. H. Lin, A. A. Au, and H. P. Lee. (2002). Compact all-fibre on-line power monitor via core-to- cladding mode coupling, *Electron. Lett.*, **38**, p. 1013.

39. R. Kashyap. (2010). *Fiber Bragg Gratings*, 2nd Ed., Boston: Academic Press.

40. S. W. James and R. P. Tatam. (2003). Optical fibre long-period grating sensors: Characteristics and application, *Meas. Sci. Technol.*, **14**(5), pp. 49–61.

41. K. O. Hill, M. B., K. A. Vineberg, F. Bilodeau, J. D.C., and I. Skinner. (1990). Efficient mode conversion in telecommunication fiber using externally written gratings, *Electron. Lett.*, **26**, p. 1270.

42. C. B. Probst, A. Bjarklev, and S. B. Andreasen. (1989). Experimental-verification of microbending theory using mode-coupling to discrete cladding modes, *J. Light. Technol.*, **7**, p. 55.

43. S. Savin, M. J. F. Digonnet, G. S. Kino, and H. J. Shaw. (2000). Tunable mechanically induced long-period fiber gratings, *Opt. Lett.*, **25**(10), pp. 710–712.

44. I. B. Sohn and J. W. Song. (2004). Gain flattened and improved double-pass two-stage EDFA using microbending long-period fiber gratings, *Opt. Commun.*, **236**, pp. 141–144.

45. P. Palai, M. N. Satyanarayan, M. Das, K. Thyagarajan, and B. P. Pal. (2001). Characterization and simulation of long period gratings fabricated using electric discharge, *Opt. Commun.*, **193**, p. 181.

46. G. Rego, P. V. S. Marques, J. L. Santos, and H. M. Salgado. (2005). Arc-induced long-period gratings, *Fiber Integr. Opt.*, **24**, pp. 245–259.

47. G. Humbert and A. Malki. (2002). Electric-Arc-induced gratings in nonhydrogenated fibres: Fabrication and high temperature characterizations, *J. Opt. A-Pure Appl. Opt.*, **4**, p. 194.

48. A. Leung, P. M. Shankar, and R. Mutharasan. (2007). A review of fiber-optic biosensors, *Sensors Actuators B Chem.*, **125**, pp. 688–703.

49. F. Chiavaioli, F. Baldini, S. Tombelli, C. Trono, and A. Giannetti. (2017). Biosensing with optical fiber gratings, *Nanophotonics*, **6**, pp. 663–679.

50. H. Kumazaki, Y. Yamada, H. Nakamura, S. Inaba, and K. Hane. (2001). Tunable wavelength filter using a Bragg grating fiber thinned by plasma etching, *IEEE Photonics Technol. Lett.*, **13**(11), pp. 1206–1208.

51. X. Dong, H. Zhang, B. Liu, and Y. Miao. (2011). Tilted fiber bragg gratings: Principle and sensing applications, *Photonic Sensors*, **1**(1), pp. 6–30.

52. T. Erdogan and J. E. Sipe. (1996). Tilted fiber phase gratings, *J. Opt. Soc. Am. A*, **13**(2), pp. 296–313.

53. C. Caucheteur, M. Loyez, Á. González-Vila, and R. Wattiez. (2010). Evaluation of gold layer configuration for plasmonic fiber grating biosensors, *Opt. Express*, **26**(18), pp. 24154–24163.

54. G. Laffont and P. Ferdinand. (2001). Tilted short-period fibre-Bragg-grating- induced coupling to cladding modes for accurate refractometry, *Meas. Sci. Technol.*, **12**, pp. 765–770.

55. M. Loyez, J. Albert, C. Caucheteur, and R. Wattiez. (2018). Cytokeratins biosensing using tilted fiber gratings, *Biosensors*, **8**(3), p. 8.

56. Y. Y. Shevchenko and J. Albert. (2007). Plasmon resonances in gold-coated tilted fiber Bragg gratings, *Opt. Lett.*, **32**(3), pp. 211–213.

57. C. Caucheteur, M. Loyez, Á. González-Vila, and R. Wattiez. (2020). Biofunctionalization strategies for optical fiber grating immunosensors, *Proc. SPIE 11361, Biophotonics Point-of-Care*, **11361**, p. 1136109.

58. D. Tosi. (2018). Review of chirped fiber bragg grating (CFBG) fiber-optic sensors and their applications, *Sensors (Switzerland)*, **18**(7). pp. 1–32.

59. C. Caucheteur, V. Voisin, and J. Albert. (2013). Polarized spectral combs probe optical fiber surface plasmons, *Opt. Express*, **21**(3), pp. 3055–3066.

60. C. Cristea, A. Florea, M. Tertiș, and R. Săndulescu. (2015). Immunosensors, in *Biosensors - Micro and Nanoscale Applications*, T. Rinken (ed.), IntechOpen, pp. 165–202.

61. S. Lara and A. Perez-Potti. (2018). Applications of nanomaterials for immunosensing, *Biosensors*, **8**(4).

62. A. K. Trilling, J. Beekwilder, and H. Zuilhof. (2013). Antibody orientation on biosensor surfaces: A minireview, *Analyst*, **138**(6), p. 1619.

63. M. S. Soares, M. Vidal, N. F. Santos, F. M. Costa, C. Marques, S. O. Pereira, and C. Leitão. (2021). Immunosensing based on optical fiber technology: recent advances, *Biosensors*, **11**(9: 305), pp. 1–35.

64. W. Wang, Z. Mai, Y. Chen, J. Wang, L. Li, Q. Su, X. Li, and X. Hong. (2017). A label-free fiber optic SPR biosensor for specific detection of C-reactive protein, *Sci. Rep.*, **7**(1), pp. 1–8.

65. G. T. Hermanson. (2008). *Bioconjugate Techniques*, 2nd Ed., Amesterdam: Elsevier.

66. C. Leitão, A. Leal-Junior, A. R. Almeida, S. O. Pereira, F. M. Costa, J. L. Pinto, and C. Marques. (2021). Cortisol AuPd plasmonic unclad POF biosensor, *Biotechnol. Reports*, **29**, pp. 1–6.

67. C. Leitao, S. O. Pereira, N. Alberto, M. Lobry, M. Loyez, F. M. Costa, J. L. Pinto, C. Caucheteur, and C. Marques. (2021). Cortisol in-fiber ultrasensitive plasmonic immunosensing, *IEEE Sens. J.*, **21**, pp. 3028–3034.

68. B. Luo, Y. Xu, S. Wu, M. Zhao, P. Jiang, S. Shi, Z. Zhang, Y. Wang, L. Wang, and Y. Liu. (2018). A novel immunosensor based on excessively tilted fiber grating coated with gold nanospheres improves the detection limit of Newcastle disease virus, *Biosens. Bioelectron.*, **100**, pp. 169–175.

69. Q. Wang and B. T. Wang. (2018). Surface plasmon resonance biosensor based on graphene oxide/silver coated polymer cladding silica fiber, *Sensors Actuators, B Chem.*, **275**, pp. 332–338.

70. A. da S. Arcas, F. da S. Dutra, R. C. S. B. Allil, and M. M. Werneck. (2018). Surface plasmon resonance and bending loss-based U-shaped plastic optical fiber biosensors, *Sensors (Switzerland)*, **18**(2), pp. 1–16.

71. R. N. Lopes, D. M. C. Rodrigues, R. C. S. B. Allil, and M. M. Werneck. (2018). Plastic optical fiber immunosensor for fast detection of sulfate-reducing bacteria, *Measurement*, **125**, pp. 377–385.

72. F. Fixe, M. Dufva, P. Telleman, and C. B. Christensen. (2004). Functionalization of poly(methyl methacrylate) (PMMA) as a substrate for DNA microarrays, *Nucleic Acids Res.*, **32**(1), pp. 1–8.

73. B. T. Wang and Q. Wang. (2018). Sensitivity-enhanced optical fiber biosensor based on coupling effect between SPR and LSPR, *IEEE Sens. J.*, **18**(20), pp. 8303–8310.

74. Q. Wang and B. Wang. (2018). Sensitivity enhanced SPR immunosensor based on graphene oxide and SPA co-modified photonic crystal fiber, *Opt. Laser Technol.*, **107**, pp. 210–215.

75. Y. C. Maya, I. Del Villar, A. B. Socorro, J. M. Corres, and J. F. Botero-Cadavid. (2018). Optical fiber immunosensors optimized with cladding etching and ITO nanodeposition, *31st Annu. Conf. IEEE Photonics Soc. IPC 2018*, pp. 4–5.

76. M. Loyez, M. C. DeRosa, C. Caucheteur, and R. Wattiez. (2022). Overview and emerging trends in optical fiber aptasensing, *Biosens. Bioelectron.*, **196**, pp. 1–16.

77. S. Tombelli, M. Minunni, and M. Mascini. (2005). Analytical applications of aptamers, *Biosens. Bioelectron.*, **20**, pp. 2424–2434.

78. M. Janik, E. Brzozowska, P. Czyszczoń, A. Celebańska, M. Koba, A. Gamian, W. J. Bock, and M. Śmietana. (2021). Optical fiber aptasensor for label-free bacteria detection in small volumes, *Sensors Actuators B. Chem.*, **330**, pp. 1–10.

79. J. Im, B. Park, and K. Han. (2019). A generative model for constructing nucleic acid sequences binding to a protein, *BMC Genomics*, **20**, pp. 1–13.

80. C. G. Knight, M. Platt, W. Rowe, D. C. Wedge, F. Khan, P. J. R. Day, A. McShea, J. Knowles, and D. B. Kell. (2009). Array-based evolution of DNA aptamers allows modelling of an explicit sequence-fitness landscape, *Nucleic Acids Res.*, **37**(1), pp. 1–10.

81. F. V. Oberhaus, D. Frense, and D. Beckmann. (2020). Immobilization techniques for aptamers on gold electrodes for the electrochemical detection of proteins: A review, *Biosensors*, **10**(45), pp. 1–50.

82. R. Peltomaa, B. Glahn-Martínez, E. Benito-Peña, and M. C. Moreno-Bondi. (2018). Optical biosensors for label-free detection of small molecules, *Sensors (Switzerland)*, **18**(12), pp. 1–46.

83. C. Ribaut, M. Loyez, J. C. Larrieu, S. Chevineau, P. Lambert, M. Remmelink, R. Wattiez, and C. Caucheteur. (2017). Cancer biomarker sensing using packaged plasmonic optical fiber gratings: Towards in vivo diagnosis, *Biosens. Bioelectron.*, **92**, pp. 449–456.

84. B. Luo, Y. Wang, H. Lu, S. Wu, Y. Lu, S. Shi, L. Li, S. Jiang, and M. Zhao. (2019). Label-free and specific detection of soluble programmed death ligand-1 using a localized surface plasmon resonance biosensor based on excessively tilted fiber gratings, *Biomed. Opt. Express*, **10**(10), pp. 5136–5148.

85. M. Lobry, M. Loyez, K. Chah, E. M. Hassan, E. Goormaghtigh, M. C. DeRosa, R. Wattiez, and C. Caucheteur. (2020). HER2 biosensing through SPR-envelope tracking in plasmonic optical fiber gratings, *Biomed. Opt. Express*, **11**(9), pp. 4862–4871.

86. V. Pinto, P. Sousa, S. O. Catarino, M. Correia-Neves, and G. Minas. (2017). Microfluidic immunosensor for rapid and highly-sensitive salivary cortisol quantification, *Biosens. Bioelectron.*, **90**, pp. 308–313.

87. K. Ogawa, F. Ito, M. Nagae, T. Nishimura, M. Yamaguchi, and A. Ishimatsu. (2011). Effects of acid stress on reproductive functions in immature Carp, Cyprimus Carpio, *Water, Air, Soil Pollut.*, **130**.

88. European Commission, Sustainable Development. [Online]. Available: https://ec.europa.eu/environment/eussd/. [Accessed: 13 Nov 2021].

89. S. M. Tripathi, K. Dandapat, W. J. Bock, P. Mikulic, J. Perreault, and B. Sellamuthu. (2019). Gold coated dual-resonance long-period fiber gratings (DR-LPFG) based aptasensor for cyanobacterial toxin detection, *Sens. Bio-Sensing Res.*, **25**, pp. 1–8.

90. W. Blanc, V. Mauroy, L. Nguyen, B. N. S. Bhaktha, P. Sebbah, B. P. Pal, and B. Dussardier. (2011). Fabrication of rare earth-doped transparent glass ceramic optical fibers by modified chemical vapor deposition, *J. Am. Ceram. Soc.*, **94**(8), pp. 2315–2318.

91. W. Blanc, I. Martin, H. Francois-Saint-Cyr, X. Bidault, S. Chaussedent, C. Hombourger, S. Lacomme, P. L. Coustumer, D. R. Neuville, D. J. Larson, T. J. Prosa, and C. Guillermier. (2019). Compositional changes at the early stages of nanoparticles growth in glasses, *J. Phys. Chem. C*, **123**, pp. 29008–29014.

92. M. Sypabekova, S. Korganbayev, W. Blanc, T. Ayupova, A. Bekmurzayeva, M. Shaimerdenova, K. Dukenbayev, C. Molardi, and D. Tosi. (2018). Fiber optic refractive index sensors through spectral detection of Rayleigh backscattering in a chemically etched MgO-based nanoparticle-doped fiber, *Opt. Lett.*, **43**(24), pp. 5945–5948.

93. X. Wang, R. Benedictus, and R. M. Groves. (2021). Optimization of light scattering enhancement by gold nanoparticles in fused silica optical fiber, *Opt. Express*, **29**(13), pp. 19450–19464.

94. T. Paixão, L. Pereira, R. Min, C. Molardi, W. Blanc, D. Tosic, C. Marques, and P. Antunes. (2020). Bragg gratings and Fabry-Perot interferometers on an Er-MgO-doped optical fiber, *Opt. Laser Technol.*, **123**, pp. 1–6.
95. J. Ballato, H. Ebendorff-heidepriem, J. Zhao, and L. Petit. (2017). Glass and process development for the next generation of optical fibers : A review, *Fibers*, **5**(11), pp. 1–25.
96. Z. Liu, H. Y. Tam, L. Htein, M. L. V. Tse, and C. Lu. (2017). Microstructured optical fiber sensors, *J. Light. Technol.*, **35**(16), pp. 3425–3439.
97. F. Berghmans, T. Geernaert, T. Baghdasaryan, and H. Thienpont. (2014). Challenges in the fabrication of fibre Bragg gratings in silica and polymer microstructured optical fibres, *Laser Photonics Rev.*, **8**(1), pp. 27–52.

Chapter 6

Microstructured and Non-Microstructured Fiber-Based Plasmonic Sensors for High-Performance and Wide-Range Detection of Different Parameters

Vasile A. Popescu[a] **and Anuj K. Sharma**[b]
[a]*Department of Physics, University "Politehnica" of Bucharest, Splaiul Independentei 313, Bucharest 060042, Romania*
[b]*Physics Division, Department of Applied Sciences, National Institute of Technology Delhi, Narela, Delhi 110040, India*
vapopescu@yahoo.com

A short review of recent works in the area of microstructured and non-microstructured fiber plasmonic sensors is presented. The sensing properties of microstructured fiber plasmonic sensors are dependent on the configuration of the devices (hexagonal

Plasmonics-Based Optical Sensors and Detectors
Edited by Banshi D. Gupta, Anuj K. Sharma, and Jin Li

ISBN 978-981-4968-85-0 (Hardcover), 978-1-003-43830-4 (eBook)
www.jennystanford.com

lattice, solid core, partial-solid-core, birefringent single-layer, microchannel incorporated photonic crystal fiber (PCF), two open-ring channels, D-shaped, dual-core D-shaped, honeycomb PCF, truncated honeycomb PCF, etc.). We review only the papers where the plasmonic layer is deposited on the outer surface of the PCF. The sensor parameters of a hexagonal lattice PCF sensor are dependent on the number of layers, thickness, refractive index (RI) of the layers, distribution and number of air holes, number of rings of air holes, thickness of the metal layer, and radius of the core layer. The air holes determine a decrease in the effective RI of the cladding layer, enabling mode guidance in the core mode. Also, the thickness of the metal (gold, silver) layer is very small ($\approx$ 40 nm) to increase the interaction between the core and plasmonic modes. The propagation characteristics of these sensors are analyzed by using a finite element method.

The sensitivity of non-microstructured sensors (Bragg fiber with four layers, and optical fiber with three, four, five, or six layers) is dependent on the number of layers, thickness, and RI of the layers and also on the thickness of the metal layer. The parameters of these sensors are analyzed by using angular and spectral interrogation methods, a finite element method, or an analytical method based on the Bessel functions.

Microstructured fiber plasmonic sensors and non-microstructured fiber plasmonic sensors were applied for the detection of distilled or heavy water, humidity, magnetic field, human blood group, and human-liver tissues, and for the detection of hemoglobin concentration in normal human blood. The best results for the sensitivity (6.25×10^{10} nm/RIU) and resolution (1.6×10^{-10} RIU) for the detection of the RI of a gas medium and magnetic field applied to the graphene layer were obtained with a plasmonic sensor with four layers (germanium, dielectric, graphene, and gas) based on a transverse spin-dependent shift (SDS) of the horizontal photonic spin Hall effect (PSHE) at a given frequency (5 THz).

6.1 Introduction

Microstructured and non-microstructured fiber plasmonic devices are very important for the development of high-sensitivity sensors with applications in the detection of changes in the analytical layer of the sensor. Thus, the optical fiber-based plasmonic sensors have been applied for the detection of distilled or heavy water [1, 2], humidity [3], magnetic field [4], human-liver tissues [5, 6], human blood groups (A, B, and O) [7], hemoglobin concentration in human blood [8, 9], naringin [10], pesticides [11], etc.

This chapter is divided into five parts. The second section of the chapter defines the principal sensor parameters. The third part of the chapter treats microstructured fiber plasmonic sensors. The sensitivity of the sensor is dependent on the number of air holes, number of the rings of air holes, the thickness of the metal layer, and the radius of the fiber core. The role of the air holes is to decrease the effective RI of the cladding layer, enabling mode guidance in the core mode. The propagation characteristics of these sensors are analyzed by using a finite element method.

The fourth part of the chapter treats non-microstructured (without air holes) plasmonic sensors. The sensitivity of the sensor without microstructures is dependent on the number of layers, thickness, and RI of the layers and also on the thickness of the metal layer. The parameters of these sensors are analyzed by using angular and spectral interrogation methods. The results are also analyzed by using a finite element method or an analytical method based on the Bessel functions.

6.2 Principle Operation of Plasmonic Sensors and Sensor Parameters

The properties of the sensor proposed in Refs. [1, 2] are based on a shift of the phase (Fig. 6.1a) or loss (Fig. 6.1b) matching point at a strong interaction between the guided core and plasmon modes when the analyte properties are changed. In the case of the phase matching point, the sensor operates at a resonant wavelength where the difference between the real parts of the effective refractive indices of the guided core modes and plasmon modes is close to zero.

For a loss matching point, the imaginary parts of the effective refractive indices of the core and plasmon modes are very close to the resonant wavelength (Fig. 6.2).

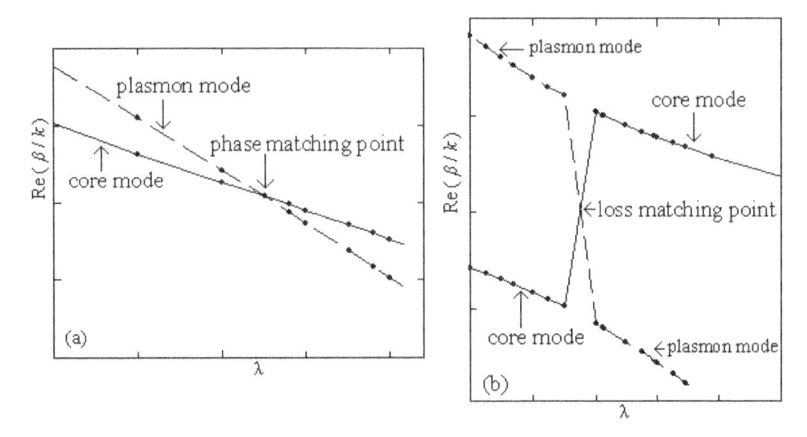

Figure 6.1 Wavelength dependence of the real parts of the effective indices β/k near the phase (a) or loss (b) matching point.

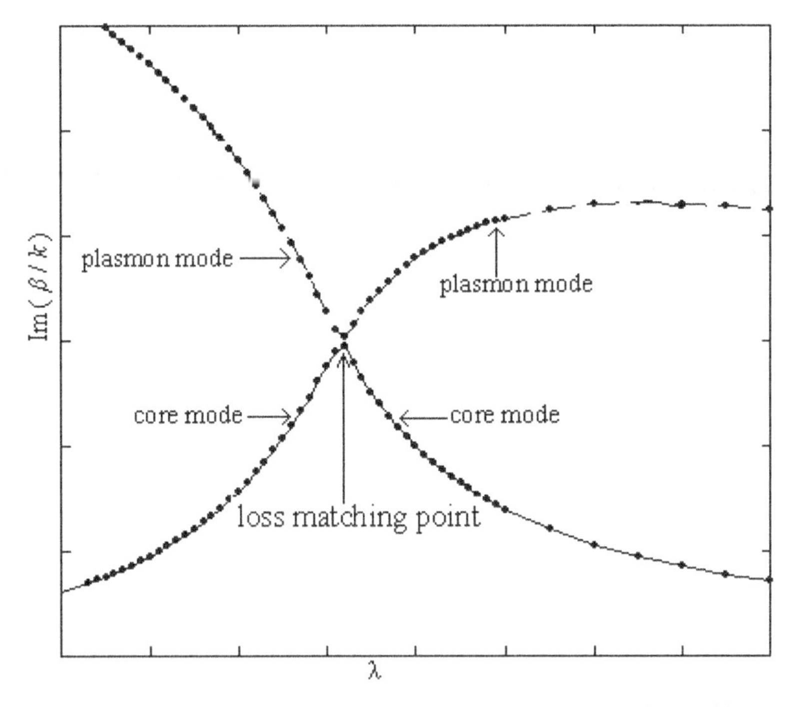

Figure 6.2 Wavelength dependence of the imaginary parts of the effective indices β/k near the loss matching point.

The real and imaginary parts of the effective index at the resonant wavelength are very sensitive to the analyte RI. For a loss matching point, there is an anti-crossing point between the plasmonic mode and core-guided mode and then the modes of each other are exchanged after the anti-crossing point. Figure 6.3 shows a wavelength dependence of the imaginary part of the effective indices β/k for a core mode in a microstructured optical fiber with a RI of the analyte n_a and for a fixed change (0.001 RIU) in the value of n_a, (i.e., n_a + 0.001) for a small part of the spectrum.

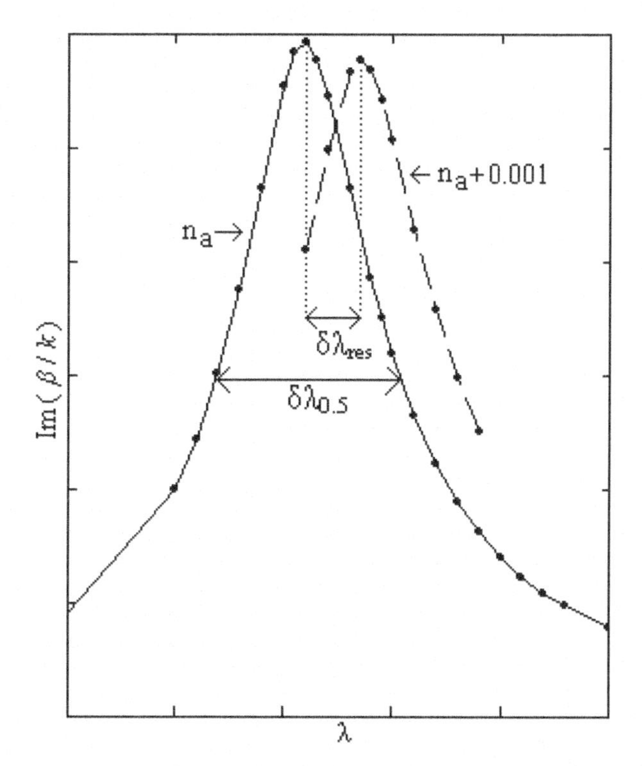

Figure 6.3 Wavelength dependence of the imaginary part of the effective indices β/k for a core mode in a microstructured optical fiber with a RI of the analyte n_a and for a fixed change (0.001 RIU) in the value of n_a, (i.e., n_a + 0.001) for a small part of the spectrum.

The phase or the loss matching point is shifted with $\delta\lambda_{\text{res}}$ toward higher wavelengths when the analyte RI is increased with δn_a = 0.001 RIU. Also, the resonance spectral width computed at the full width at half maximum (FWHM) of the loss spectra is $\delta\lambda_{0.5}$.

The wavelength sensitivity is:

$$S_{\lambda} = \frac{\delta\lambda_{\text{res}}}{\delta n_a}, \ \delta n_a = 0.001\ \text{RIU}. \quad (6.1)$$

The corresponding spectral resolution is:

$$SR_{\lambda} = \frac{\delta_{\min}}{S_{\lambda}}, \ \delta_{\min} = 0.1\ \text{nm}, \quad (6.2)$$

if it is assumed that a $\delta_{\min}$ = 0.1 nm change in the position of a resonance peak can be detected reliably. Also, the resonance spectral width computed at the FWHM of the loss spectra is $\delta\lambda_{0.5}$ and the corresponding signal-to-noise ratio (SNR) is:

$$\text{SNR} = \frac{\delta\lambda_{\text{res}}}{\delta\lambda_{0.5}}. \quad (6.3)$$

The corresponding figure of merit (FOM) is:

$$\text{FOM} = \frac{\delta\lambda_{\text{res}}}{\delta n_a \times \delta\lambda_{0.5}}. \quad (6.4)$$

The amplitude sensitivity is:

$$S_A = \frac{1}{\alpha}\frac{\delta\alpha}{\delta n_a}, \quad (6.5)$$

where

$$\alpha = \frac{400000\pi}{\lambda \ln(10)} \times \text{Im}\left(\frac{\beta}{k}\right) \quad (6.6)$$

is the loss corresponding to optical power in dB/cm, λ is the wavelength in μm, and β/k is the effective index of the core mode. The corresponding amplitude resolution is:

$$SR_A = \frac{\delta_A}{S_A}, \ \delta_A = 0.01, \quad (6.7)$$

where it is assumed that a δ_A = 1% = 0.01 change in the transmitted intensity can be detected.

Some plasmonic sensors [3] are analyzed by using the surface plasmon resonance (SPR) angular interrogation method where the effective indices of the evanescent wave of the incident p-polarized light and the surface plasmon wave (SPW) become identical and there takes place a transfer of energy from the incident light to SPW at a given wavelength, and a maximum of the power loss appears at the resonance incidence angle (θ_{res}). The value of θ_{res} depends on the RI of sensing the medium (n_a). The real part of the SPW effective index ($n_{eff} = \beta/k$) is given by the relation (valid for the case where the layers are semi-infinite):

$$\mathrm{Re}(\beta/k) = n_1 \sin\theta_{\mathrm{res}} = \sqrt{\frac{\varepsilon_{mr} n_a^2}{\varepsilon_{mr} + n_a^2}} \tag{6.8}$$

In Eq. (6.8), n_1 is the RI of the optical fiber core, θ_{res} is the resonance angle, ε_{mr} is the real part of the gold dielectric function, and n_a is the analyte RI. The angular sensitivity is:

$$S_\theta = \frac{\delta\theta_{\mathrm{res}}}{\delta n_a},\ \delta n_a = 0.001\ \mathrm{RIU}. \tag{6.9}$$

The corresponding SNR is:

$$\mathrm{SNR} = \frac{\delta\theta_{\mathrm{res}}}{\delta\theta_{0.5}}. \tag{6.10}$$

The corresponding angular resolution is:

$$SR_\theta = \frac{\delta_{\theta_a}}{S_\theta},\ \delta\theta_a = 0.0001^\circ, \tag{6.11}$$

if it is assumed that angular resolution is $\delta\theta_a = 0.0001^\circ$ [12]. The FOM of the sensor is defined as:

$$\mathrm{FOM} = \frac{\delta\theta_{\mathrm{res}}}{\delta n_a \times \delta\theta_{0.5}} \tag{6.12}$$

where $\delta\theta_{\mathrm{res}}$ is the shift in resonance angle corresponding to small variation δn_a in the analyte RI and $\delta\theta_{0.5}$ is the angular width of power loss (PL) vs θ spectrum. The maximum of local power sensitivity (S_P) calculated for an increase of n_a with 0.001 RIU is given by:

$$S_P = \frac{\delta PL}{\delta n_a},\ \delta n_a = 0.001\ \mathrm{RIU}, \tag{6.13}$$

where

$$\mathrm{PL} = 10\log_{10}\left(\frac{1}{P_p}\right), \tag{6.14}$$

where P_p is the normalized output power transmitted from the optical fiber.

$$P_p = R_p\left(\frac{L}{D\tan\theta}\right), \tag{6.15}$$

where L is the length of the sensing part of the fiber, D is the core thickness, θ is the angle inside the fiber (L/D = 25), and R_p is the intensity reflection coefficient given as:

$$R_p = |r_p|^2. \tag{6.16}$$

Here, r_p is the amplitude reflection coefficient and can be calculated using the transfer matrix method (TMM) as (for a structure with three layers):

$$r_p = \frac{(M_{11} + M_{12}q_3)q_1 - (M_{21} + M_{22}q_3)}{(M_{11} + M_{12}q_3)q_1 + (M_{21} + M_{22}q_3)}, \tag{6.17}$$

where M_{11}, M_{12}, M_{21}, and M_{22} are the elements of corresponding 2 × 2 transfer matrix given as:

$$M_{11} = M_{22} = \cos\beta_2,\ M_{12} = -\frac{i\sin\beta_2}{q_2},\ M_{21} = -iq_2\sin\beta_2 \tag{6.18}$$

with

$$\beta_2 = \frac{2\pi d_2}{\lambda}\sqrt{n_2^2 - n_1^2\sin^2(\theta)} \tag{6.19}$$

and

$$q_1 = \frac{\sqrt{n_1^2 - n_1^2\sin^2(\theta)}}{n_1^2},\ q_2 = \frac{\sqrt{n_2^2 - n_1^2\sin^2(\theta)}}{n_2^2},\ q_3 = \frac{\sqrt{n_3^2 - n_1^2\sin^2(\theta)}}{n_3^2}. \tag{6.20}$$

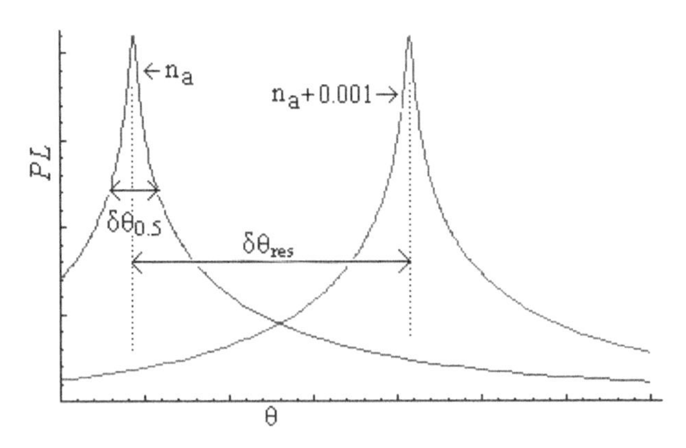

Figure 6.4 Power loss PL versus the incident angle θ on a fiber for a TM mode for n_a and n_a + 0.001.

Figure 6.4 shows an angular dependence of the power loss *PL* with the incident angle and for a fixed change (0.001 RIU) in the value of n_a, (i.e., $n_a + 0.001$).

The maximum of the power loss is shifted with $\delta\theta_{\text{res}}$ toward higher values of θ when the analyte RI is increased with $\delta n_a = 0.001$ RIU. Also, the resonance angular width computed at the FWHM of the power loss is $\delta\theta_{0.5}$.

6.3 Microstructured Fiber Plasmonic Sensors

A microstructured optical fiber is also called a PCF when the air holes are arranged in the regular pattern in 2D arrays. The propagation characteristics of a microstructured optical fiber-based plasmonic sensor are determined by using the finite element method. In the numerical simulations, Re(β/k) and Im(β/k) are almost constant for a large value of the analyte layer thickness.

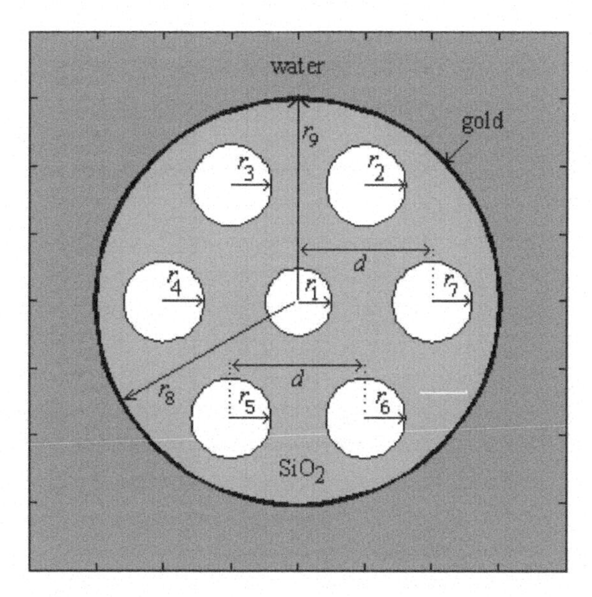

Figure 6.5 Schematic cross-section of a microstructured optical fiber (adapted from Ref. [1]).

In the case of the work [1] proposed by Popescu et al., a microstructured optical fiber (Fig. 6.5) was made by a small air hole (radius r_1) in the center of the structure, six air holes (radii $r_2 = r_3 =$

$r_4 = r_5 = r_6 = r_7$), which are placed at the vertices of a hexagon with vertice-to-vertice distance d and are inserted in a SiO_2 core (radius r_8), which is surrounded by a gold layer and a very thickness distilled water layer.

This sensor contains a single ring of air holes and the analyte layer is in the exterior part of the sensor. The fabrication process is simplified by using a single circular gold and analyte layer outside the fiber structure. The structure is designed to have the phase matching point corresponding to the maximum in the water and the gold layers and a minimum in the glass layer of the power fraction for a core-guided mode. Also, this point corresponds to a maximum in the glass layer and a minimum in the water layer of the power fraction for a plasmon mode. At the phase matching point, there is a strong interaction between the core and plasmon modes, causing a splitting in the real part of the propagation constant. The phase matching point is shifted with $\delta\lambda_{res}$ = 4 nm toward higher wavelengths with a sensitivity of S_λ = 4000 nm/RIU when the analyte RI is increased with δn_a = 0.001 RIU. The corresponding spectral resolution is $SR_\lambda = 2.5 \times 10^{-5}$ RIU if it is assumed that a δ_{min} = 0.1 nm change in the position of a resonance peak can be detected reliably. Also, the resonance spectral width computed at the FWHM of the loss spectra is $\delta\lambda_{0.5}$ = 31 nm and the corresponding signal-to-noise ratio is SNR = 0.13.

Another proposed structure [2] was made by a silica core with a small air hole in the center, and surrounded by six air holes placed at the vertices of a hexagon, four smaller air holes between some large air holes, and further enclosed by gold and water layers. The difference (0.000019) between the real parts of the effective indices of core (Fig. 6.6a) and plasmon (Fig. 6.6b) modes is the smallest at the phase matching point λ = 0.618 μm where β/k = 1.434439 + 0.000561i for the core mode and β/k = 1.434420 + 0.002459i for the plasmon mode, and there is a maximum of the power fraction of the core mode in the gold layer.

The advantages of another configuration [2] with five supplementary small air holes between five pairs of large air holes and with an increase of the radius of the silica core of 40 nm are a smaller value of the FWHM parameter ($\delta\lambda_{0.5}$ = 22 nm) and a higher amplitude sensitivity S_A = 3941.5 RIU^{-1} where α = 950.3 dB/cm is the loss corresponding to optical power in dB/cm, λ = 0.6507 μm is the wavelength and β/k is the effective index of the core mode. The corresponding amplitude resolution is $SR_A = 2.5 \times 10^{-6}$ RIU where it is assumed that $\delta_A = 1\%$ change in the transmitted intensity can be detected.

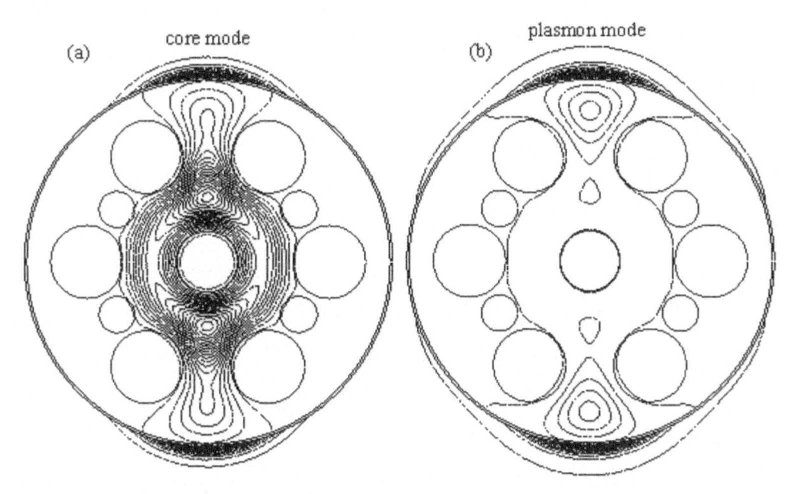

Figure 6.6 Schematic contour plot of the z-component $S_z\,(x, y)$ of the Poynting vector for the core guided (a) and plasmon (b) modes in a microstructured optical fiber at the resonant wavelength (λ = 0.618 μm) of a microstructured optical fiber (adapted from Ref. [2]).

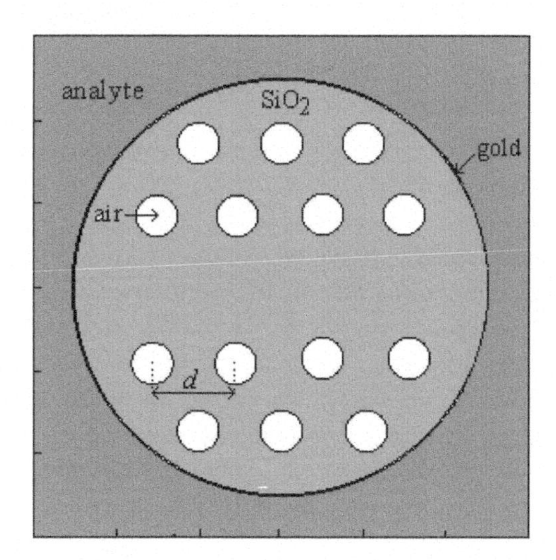

Figure 6.7 Schematic cross-section of a birefringent solid-core microstructured optical fiber (adapted from Ref. [13]).

A finite element method was applied [13] to microstructured optical fiber-based SPR sensors where five central horizontal air holes from a two-ring hexagonal lattice of air holes in silica fiber are omitted (Fig. 6.7). In this fiber structure for n_a = 1.39, the FWHM of the loss spectra is small ($\delta\lambda_{0.5}$ = 12.2 nm) for the first core mode I where the main electric fields are oriented in the same x-direction in substrate and gold layers.

The same research group proposed another plasmonic biosensor [14], which is based on a birefringent partial-solid-core microstructured optical fiber (Fig. 6.8) where a single core mode can interact resonantly with the corresponding plasmon mode.

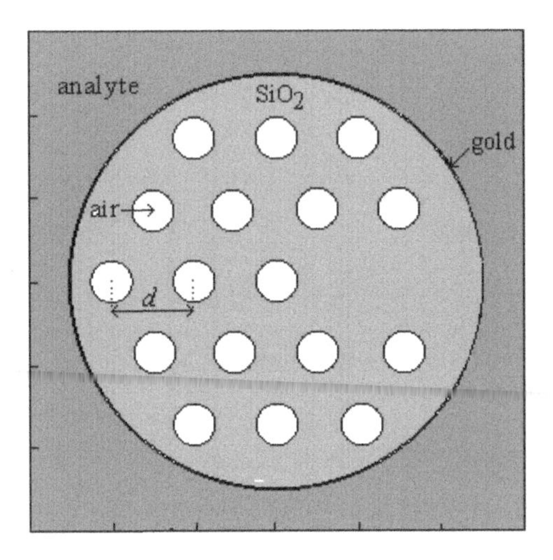

Figure 6.8 Schematic cross-section of a birefringent partial-solid-core microstructured optical fiber (adapted from Ref. [14]).

This structure can be scaled to a larger radius of the gold layer with a simultaneous increase in the number of rings and thus of the holes. For a RI of the analyte n_a = 1.34, a very small resonance spectral width (11.3 nm) is obtained for the silica layer radius (7 μm), three rings of the air holes, and 35 air holes.

Liu et al. proposed a birefringent single-layer coating PCF-based SPR sensor [15] with only eight air holes (Fig. 6.9), where the maximum wavelength sensitivity is 15180 nm/RIU and the amplitude sensitivity can reach up to 498 RIU^{-1}. Some examples of microstructured optical fiber biosensors have been reviewed in

Ref. [16]. In the same paper, a plasmonic biosensor with only six air holes (Fig. 6.10) was obtained from a fiber with 14 small air holes, in which 12 small air holes were replaced with four large air holes while maintaining the same surface.

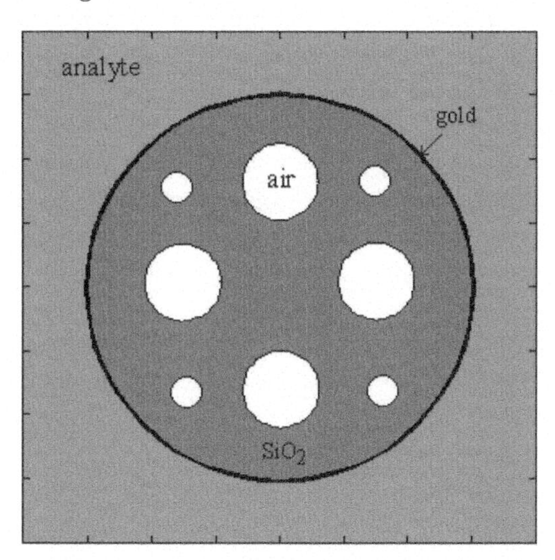

Figure 6.9 Schematic cross-section of a birefringent single-layer coating PCF-based SPR (adapted from Ref. [15]).

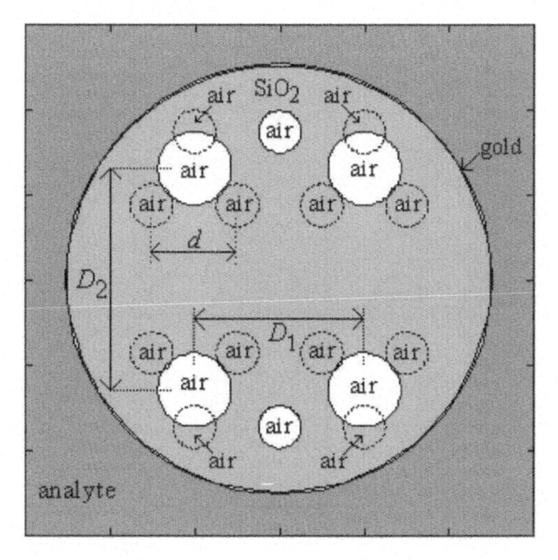

Figure 6.10 Schematic cross-section of the fiber sensor structure with six air holes in which 12 small air holes (dashed line, radius $r = 0.5$ μm) are replaced with four larger air holes (radius $R = 0.866025$ μm) while maintaining the same surface. Adapted from Ref. [16].

The main advantages of this biosensor in comparison with a structure with 14 air holes [13] are the higher value of maximum amplitude sensitivity (960.4 RIU^{-1}) for core mode I ($\lambda = 0.7921$ μm) at $n_a = 1.395$, higher loss (1123.6 dB/cm) for core mode II ($\lambda = 0.6504$ μm) at $n_a = 1.36$ and a smaller number of air holes that facilitates the fabrication. The corresponding FOM = $\delta\lambda_{res}/(\delta n_a \times \delta\lambda_{0.5}) = 7.2/(0.001 \times 13.4) = 537.3$ RIU^{-1} for the core mode I and FOM = $2.8/(0.001 \times 13.6) = 205.9$ RIU^{-1} for the core mode II.

Islam et al. proposed a highly sensitive PCF-based SPR sensor [17] where a high-birefringence fiber is obtained by means of an array of air holes at the center of the fiber. For this device, $S_\lambda = 25000$ nm/RIU, $S_A = 1411$ RIU^{-1}, and FOM = 502 RIU^{-1} for a RI range of 1.33–1.38 were obtained.

Haque et al. proposed a microchannel incorporated PCF-based SPR sensor [18] for the detection of low RI at the near-infrared wavelength where the gold and analyte layers are placed outside the fiber and a thin layer of TiO_2 is employed as an adhesive layer between the gold and silica layers (Fig. 6.11). For this device, the maximum wavelength and amplitude sensitivities $S_\lambda = 51000$ nm/RIU for $n_a = 1.36$ and $S_A = 1872$ RIU^{-1} for $n_a = 1.37$ were obtained. Also, this sensor shows a minimum FOM of 6 RIU^{-1} for $n_a = 1.22$ and a maximum FOM of 566 RIU^{-1} for $n_a = 1.36$.

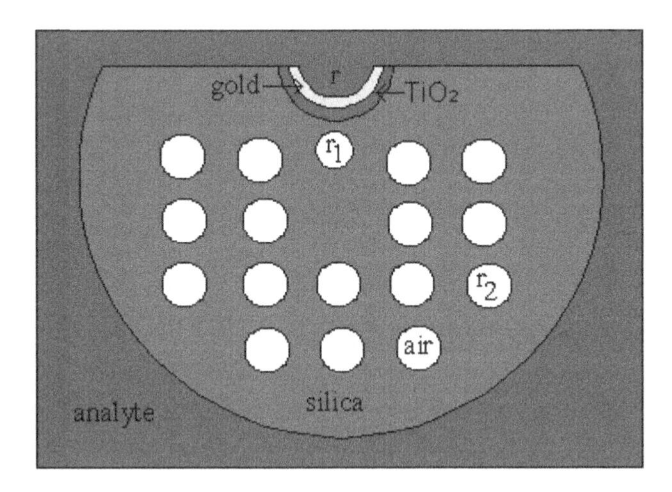

Figure 6.11 Cross-section of a microchannel incorporated PCF-based SPR (adapted from Ref. [18]).

Liu et al. proposed an SPR sensor [19] with two open-ring channels based on a PCF (Fig. 6.12), which detects low RI between 1.23 and 1.29 with the wavelength in the mid-infrared region between 2550 nm and 2900 nm. The spectral sensitivity and amplitude sensitivity of the sensor can reach 5500 nm/RIU and 333.8 RIU^{-1}, respectively. Also, the maximum wavelength resolution of 7.69×10^{-6} RIU can be obtained.

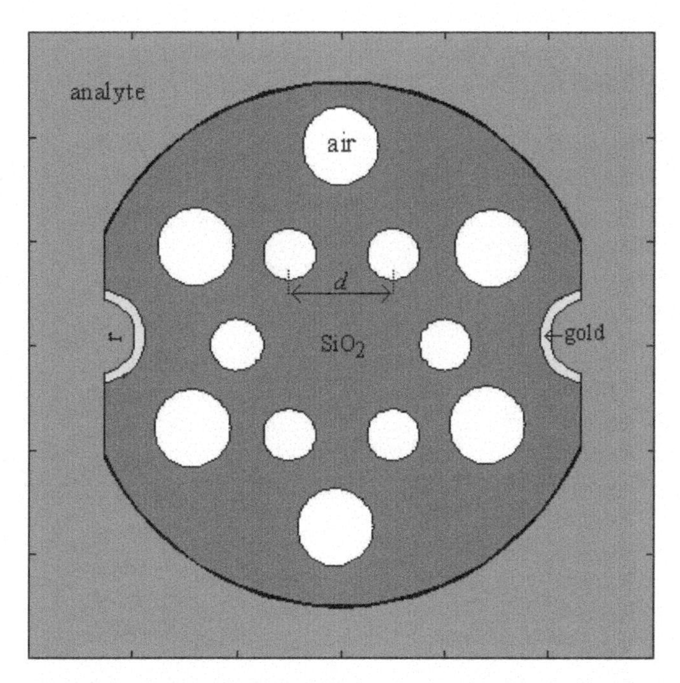

Figure 6.12 Cross-section of an SPR sensor with two open-ring channels based on a PCF (adapted from Ref. [19]).

Another research group [20] proposed a new type of resonant interaction between two core modes in a plasmonic biosensor based on a birefringent solid-core microstructured optical fiber (Fig. 6.13). In this sensor, the transmission losses of the lateral core modes are larger than those of the corresponding central core modes for the *x*- and *y*-polarization components. The maximum value S_A (II, *x*) = 227.7 RIU^{-1} of the amplitude sensitivity for the second (lateral) core mode (*x*-polarization) is at a wavelength $\lambda = 0.744$ μm, which is close to the resonant wavelength λ (I, *x*) = 0.747 μm of the first (central) core mode when the RI of the analyte is $n_a = 1.37$. Also, the maximum value S_A (II, y) = 271.0 RIU^{-1} of the amplitude sensitivity for the second core mode (y-polarization) is at a wavelength λ = 0.734 μm, which is close to the resonant wavelength λ (I, y) = 0.735 μm of the first core mode. The local spectral sensitivities ($S_\lambda = \Delta\lambda_{res}/\Delta n_a$), calculated for an increase of n_a with 0.001, are increased with the RI of the analyte n_a. Thus, S_λ (I, x) = S_λ (I, y) = 2000 nm/RIU, S_λ (II, x) = 1000 nm/RIU, S_λ (II, y) = 2000 nm/RIU for n_a = 1.33, and S_λ (I, x) = 11000 nm/RIU, S_λ (I, y) = 10000 nm/RIU, S_λ (II, x) = 12000 nm/RIU, and S_λ (II, y) = 11000 nm/RIU for n_a = 1.39.

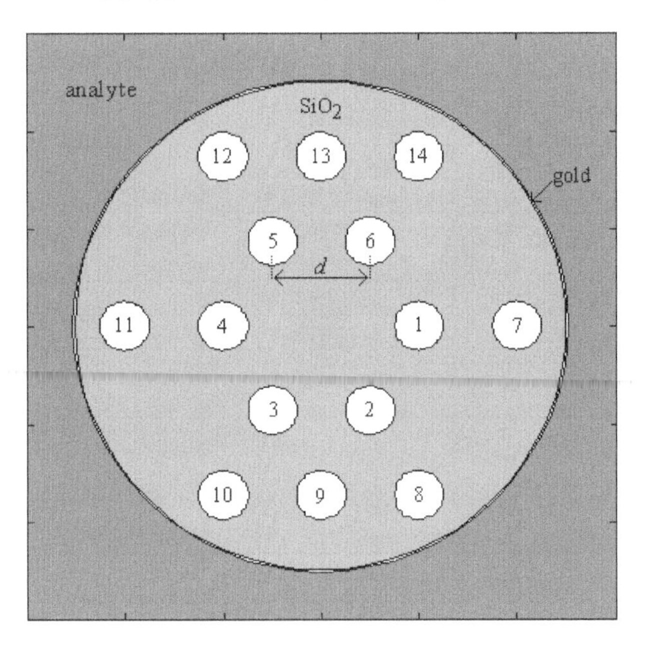

Figure 6.13 Cross-section of a fiber consisting of 14 air holes (d = 2 μm) with radius r = 0.5 μm, which are inserted in a SiO_2 core (radius r_g = 5 μm) surrounded by a gold layer (thickness t_g = 40 nm) and an analyte layer (adapted from Ref. [20]).

The sensor proposed in Ref. [21] is based on a resonant interaction between a core mode and two complementary supermodes in a honeycomb PCF reflector-based SPR sensor. Figure 6.14 shows a schematic cross-section of a honeycomb PCF made up of a SiO_2 glass with a small air hole (diameter d_1 = 0.2695 μm) in its center, surrounded by the cladding holes (diameter d_2 = 0.4235 μm) where the center-to-center distance between adjacent cladding holes is d = 0.77 μm, a gold layer (thickness t_g = 40 nm and radius r_g = 3.48425 μm), and a distilled water layer.

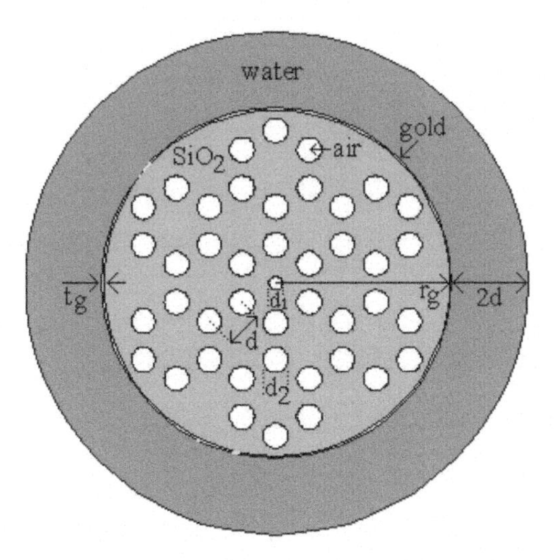

Figure 6.14 Schematic cross-section of a honeycomb PCF (adapted from Ref. [21]).

The light in the core mode is confined by the bandgap of the honeycomb reflector. For a wavelength λ = 1.3593 μm, the losses of the supermodes are close to the maximum of the loss of the core mode indicating that a strong interaction is between the core mode and these supermodes. A very important observation is that the local spectral sensitivity of the core mode is very sensitive to the radius of the gold layer and for a wavelength detection limit $\Delta\lambda_L$ = 0.1 nm [3], the sensor resolution is $SR_\lambda = 2.0 \times 10^{-5}$ RIU, a better value than in Ref. [22] where $SR_\lambda = 3.6 \times 10^{-5}$ RIU, or in Ref. [23] where $SR_\lambda = 2.5 \times 10^{-5}$ RIU. In Ref. [4], a PCF-based magnetic field sensor was simulated by using a honeycomb structure without the first ring of air holes where the core air hole is filled with a magnetic fluid, which is surrounded by a thin reflector gold layer. Figure 6.15 shows a schematic cross-section of a truncated honeycomb PCF made by a SiO_2 glass (radius r = 3.48425 μm) with MF filled into the air hole (radius r_1 = 1 μm) located at the center of the photonic bandgap-guiding core, surrounded by a thin gold layer (thickness t_g = 40 nm) and the cladding holes (diameter $d_2 = 0.4235$ µm) where the center-to-center distance between adjacent cladding holes is $d = 0.77$ µm and the first ring of the air holes is absent.

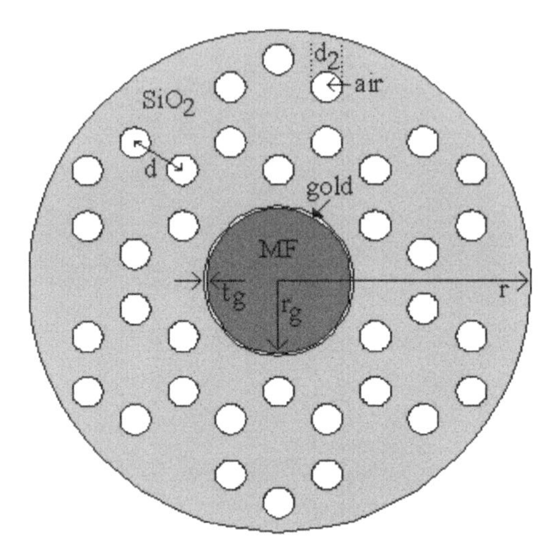

Figure 6.15 Schematic cross-section of a truncated honeycomb PCF (adapted from Ref. [4]).

This sensor was based on the variation of the RI with the applied magnetic field at a fixed wavelength (1557 nm) for three different temperatures ($t = 24.3$ °C, 40 °C, and 60 °C), and for different values of the volume fractions ($c = 1.21\%$, 1.48%, 1.52%, and 1.93%) of Fe_3O_4 particles in a magnetic fluid solution. The sensor was also based on the variation of the transmission loss of optical power with the magnetic field for the fundamental mode HE_{11} when the length of PCF is $L = 1$ cm. For a given range of the magnetic field, the slope of the linear fitting line of the transmission loss versus the magnetic field gives the average sensitivity of the magnetic field. For the average sensitivity of the magnetic field $S_H = (\Delta\alpha_L/\Delta \mathrm{H})_S$, the magnetic field resolution of the sensor is defined as $SR_H = (\Delta \mathrm{H}/\Delta\alpha_L)_S \times (\Delta\alpha)_a$ where the transmission loss α_L in dB is defined as $\alpha_L = \alpha L$, L is the length (in centimeters) of the PCF and the amplitude resolution is $(\Delta\alpha_L)_a = 0.01$ dB [24]. The best sensitivity (4.3763 dB/Oe) and resolution (2.2850×10^{-3} Oe) are obtained for a volume fraction of 1.93% and for a range of the magnetic field between 28.58 Oe and 192.45 Oe. In Ref. [24], the maximum sensitivity is only 0.042 dB/Oe (resolution

0.24 Oe) is achieved at the magnetic field of 589 Oe where L (10 cm) is 10 times larger than in our case.

A finite element method was used to calculate the effective indices β/k for HE_{11}, TM_{01}, and the first order of plasmon modes (Fig. 6.16) for a truncated honeycomb PCF-based magnetic field plasmonic sensor at the wavelength $\lambda = 1.557$ μm [25].

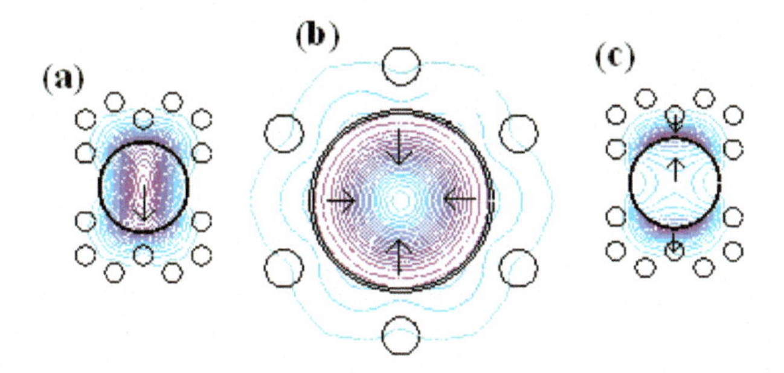

Figure 6.16 Typical sections of the contour plots of the z-component S_z (x, y) of the Poynting vector of a truncated honeycomb PCF-based magnetic field plasmonic sensor at the wavelength $\lambda = 1.557$ μm for core mode (a), for a TM_{01} mode (b) and for a plasmonic mode (c) (adapted from Ref. [26] and Ref. [4]).

The resolution $SR_H = 0.019$ Oe for the magnetic field detection in a truncated honeycomb structure calculated with the finite element method for a TM_{01} mode is finer than the value (0.080 Oe) for the first-order plasmon mode and the one (0.241 Oe) for a fundamental HE_{11} mode, when the volume fraction of Fe_3O_4 particles in a magnetic fluid solution is $c = 1.48\%$. In addition, the resolution $SR_H = 0.0028$ Oe for the magnetic field detection in a truncated honeycomb structure calculated with the finite element method for the first-order plasmon mode is very close to a reported [4] value where $SR_H = 0.0023$ Oe for an HE_{11} mode when $c = 1.93\%$.

A very sensitive D-shaped PCF sensor based on SPR was proposed in Ref. [26]. This sensor is composed (Fig. 6.17) of three layers of air holes with a triangular lattice where the small radius is $r_1 = 0.5$ μm, the large radius is $r = 1$ μm, the radius of other air holes is $r_2 = 0.6$ μm, and the lattice spacing is 2 μm. The thickness of the gold layer is very small (20 nm). The maximum sensitivity can reach $S_\lambda = 31000$ nm/RIU for a wide analyte RI detection range (from 1.36 to 1.40). The corresponding spectral resolution is $SR_\lambda = \delta\lambda_{min}/S_\lambda = 0.1/31000 = 3.2 \times 10^{-6}$ RIU where $\delta\lambda_{min} = 0.1$ nm.

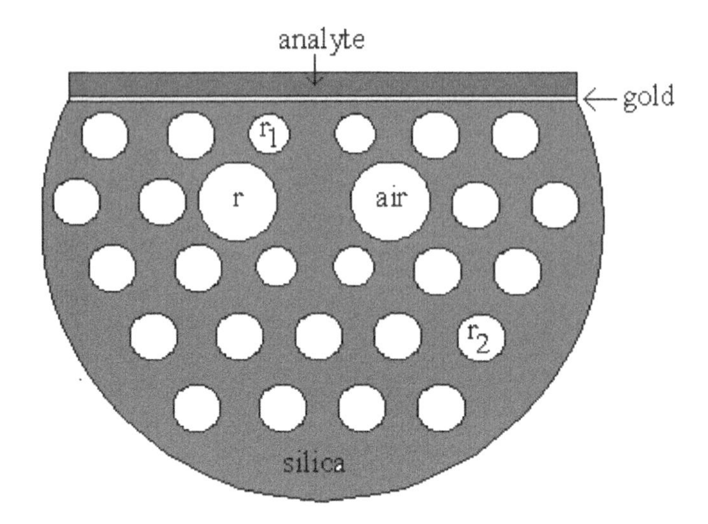

Figure 6.17 Schematic cross-section of a D-shaped microstructured optical fiber sensor (adapted from Ref. [26]).

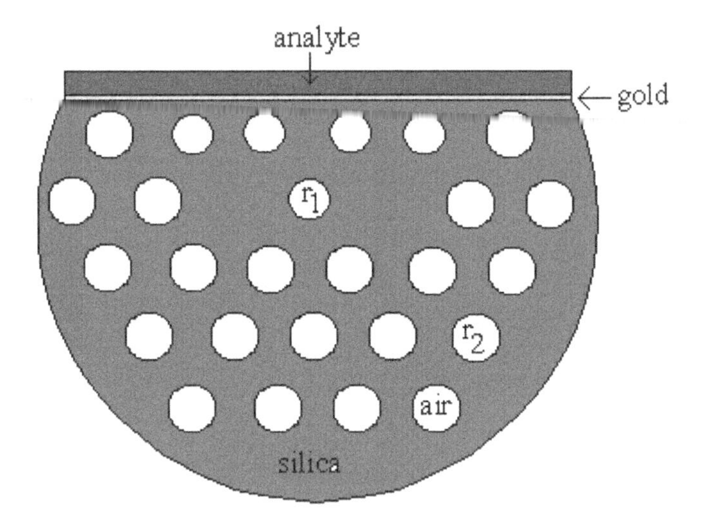

Figure 6.18 Schematic cross-section of a dual-core D-shape microstructured optical fiber sensor (adapted from Ref. [27]).

Another sensitive dual-core D-shaped PCF sensor based on SPR was proposed in Ref. [27]. This sensor is composed (Fig. 6.18) of three layers of air holes where the small radius is r_1 = 0.475 μm, the large radius is r_2 = 0.95 μm, and the lattice spacing is 1.9 μm. The thickness of the gold layer is 30 nm. The maximum spectral sensitivity is S_λ = 8000 nm/RIU for n_a = 1.48 RIU and λ = 1.3 μm. The corresponding spectral resolution is $SR_\lambda = 1.25 \times 10^{-5}$ RIU. The amplitude sensitivity is 700 RIU^{-1}, the corresponding amplitude resolution is 1.78×10^{-5} RIU and FOM = 138 RIU^{-1}.

A highly sensitive D-shaped PCF sensor based on SPR was proposed in Ref. [28]. This sensor is composed (Fig. 6.19) of 10 air holes with the radius r = 0.825 μm and the lattice spacing is 3.3 μm. The thickness of the gold layer is 70 nm. The maximum spectral sensitivity is S_λ = 216000 nm/RIU for n_a = 1.41 RIU in infrared. The corresponding spectral resolution is $SR_\lambda = 4.63 \times 10^{-7}$ RIU. The amplitude sensitivity is 1680 RIU^{-1} and FOM = 1200 RIU^{-1}.

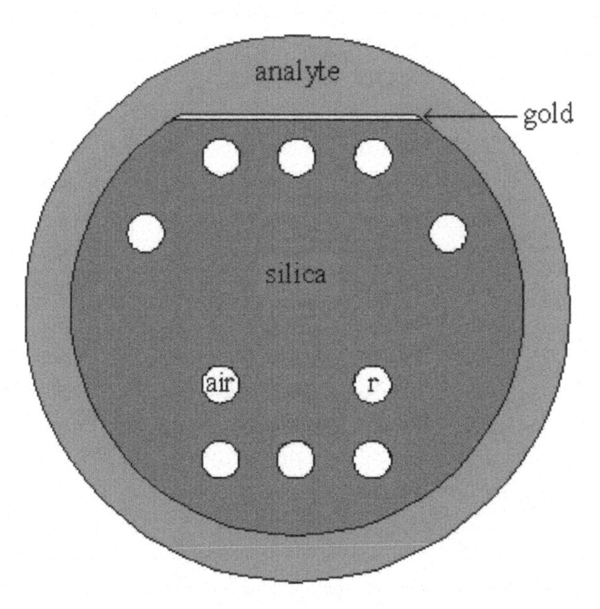

Figure 6.19 Schematic cross-section of a D-shaped microstructured optical fiber sensor (adapted from Ref. [28]).

The sensitivity of a D-shaped sensor is higher than other PCF sensors, due to a stronger interaction between the core and plasmon modes since the plasmonic gold is deposited near the core and the evanescent field can easily reach the gold surface. The sensitivity is large for high values of the analyte RI since with the increase of analyte RI, the RI contrast between the core mode and SPP mode reduces and there is a strong coupling between the core and plasmon modes.

6.4 Non-Microstructured Fiber Plasmonic Sensors

In Ref. [29], a theoretical analysis of the sensitivity and SNR of a step-index fiber optic SPR sensor with bimetallic layers has been carried out. For a ratio Ag/Au of 0:1, 1:1, and 1:0, the wavelength sensitivity is 5973 nm/RIU, 5120 nm/RIU, and 5067 nm/RIU, respectively. The corresponding SNR is 1.0090, 1.7143, and 1.9592, respectively.

A theoretical analysis of a fiber optic remote sensor based on SPR for temperature detection has been carried out [30]. The SPR curve width increases with the rise in temperature because the imaginary part of the metal-dielectric function increases with temperature.

The influence of dopants on the performance of a fiber optic SPR sensor has been carried out [31]. The spectral sensitivity for pure (without dopant) SiO_2 at n_a = 1.343 is 2900 nm/RIU when the resonance wavelength is 0.5296 μm. The spectral sensitivity for silica doped with B_2O_3 with a molar concentration (mole percentage) of 5.2 at n_a = 1.343 is 3100 nm/RIU when the resonance wavelength is 0.538 μm. Thus, the doping of B_2O_3 increases the sensitivity of the sensor.

The effect of taper ratio and taper profiles on the sensor sensitivity was studied [32]. The maximum sensitivity of the sensor was observed for an exponential-linear taper profile with a high taper ratio. The minimum sensitivity was obtained for a parabolic profile.

Some examples of SPR-based fiber optics sensors have been reviewed in Ref. [33]. In this review, reference has presented the principle of the SPR technique for sensing, various designs (bimetallic coating, tapered probe, U-shaped, and side polished fiber) of the fiber optic for the enhancement of the sensitivity of the sensor, and some applications (detection of the temperature [30], naringin [10], and pesticides [11]).

An analytical method based on Bessel functions was applied to the calculation of transmission loss, average sensitivity, and magnetic field resolution for the TM_{01} and the first order of two-fold degenerate plasmon modes in an optical fiber structure with three layers (silicon, gold, and magnetic fluid) [25]. The radius of the core is 1 μm and the thickness of the gold layer is 51 nm. In this method, the electromagnetic field is represented by a Bessel function of the first kind in the core layer (silicon), a linear combination of the Hankel functions in the gold layer, and a modified Bessel function of the second kind in the magnetic fluid layer. The values of the effective indices calculated by using the finite element method are in agreement with our analytical method.

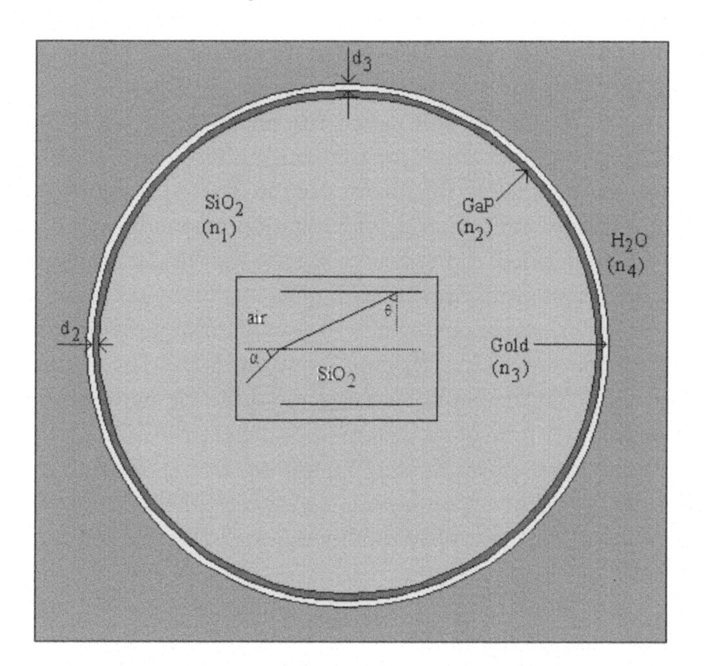

Figure 6.20 Schematic of a Bragg fiber with four layers (SiO_2, GaP, gold, and H_2O). The light is incident under the angle α from an air medium in a SiO_2 core of the fiber and the angle inside the fiber is θ (inset). Adapted from Ref. [34].

The angular and spectral interrogation methods were applied [34] for the calculation of the reflectivity, power loss, and spectral and amplitude sensitivities for the *TE*, *TM*, and hybrid modes in a Bragg fiber-based plasmonic sensor with four layers (SiO_2, GaP, gold, and H_2O). Figure 6.20 shows a Bragg fiber with four layers where n_1, n_2, n_3, and n_4 are the refractive indices of the core (SiO_2), GaP, gold, and H_2O, respectively. The thicknesses of the GaP and gold layers are d_2 and d_3, respectively. The light is incident under the angle α from an air medium in a SiO_2 core of the fiber and the angle inside the fiber is θ.

The relation between these angles (Fig. 6.20) is given by Snell's law: $\sin \alpha / \sin (90 - \theta) = n_1/n_{air}$, where $n_{air} = 1$ is the RI of the air. The results calculated with the finite element method are in agreement with another analytical method [34], where the electromagnetic field is represented by a Bessel function of the first kind in the core region (SiO_2), a linear combination of Bessel functions of the first and second kinds in the dielectric interior layer (GaP), a linear combination of the Hankel functions in the gold region, and a modified Bessel function of the second kind in the outermost region (H_2O). The amplitude sensitivity for the core modes near the maximum power loss point for the Bragg fiber with a gold layer is increased for optimized thicknesses of the GaP and gold layers.

The angular and spectral interrogation methods were applied [35, 36] for the calculation of the power loss, FOM, and spectral and amplitude sensitivities for the *TM* modes in a fiber-based plasmonic sensor with four layers using a very thin aluminum oxide insulator as the interior or exterior cladding layer (Fig. 6.21). The values of the effective indices calculated by using the finite element method are in agreement with an analytical method where the electromagnetic field is represented by a Bessel function of the first kind in the core region (SiO_2), a linear combination of the Hankel functions in the gold and Al_2O_3 regions, and a modified Bessel function of the second kind in the outermost region (H_2O).

For some values of the core (SiO_2) radius, there is a loss matching point between HE_{11} and HE_{12} modes and between TM_{01} and TM_{02} modes when the imaginary parts of the effective indices of the HE_{11} (TM_{01}) and HE_{12} (TM_{02}) modes calculated with an analytical method based on the Bessel functions are very close. The angular and spectral interrogation methods show that for an optical fiber-based plasmonic sensor with a thin Al_2O_3 insulator layer, the best FOM and the maximum amplitude sensitivity in the infrared region near the laser wavelength $\lambda = 1.064$ µm are obtained when the gold layer is in the exterior part of the cladding region of the fiber for the distilled water as the analyte layer. For a fiber with the gold layer in the exterior part of the cladding region, the limit of detection (LOD) in intensity interrogation mode (7.78×10^{-8} RIU) is better than in angular interrogation mode (8.41×10^{-6} RIU).

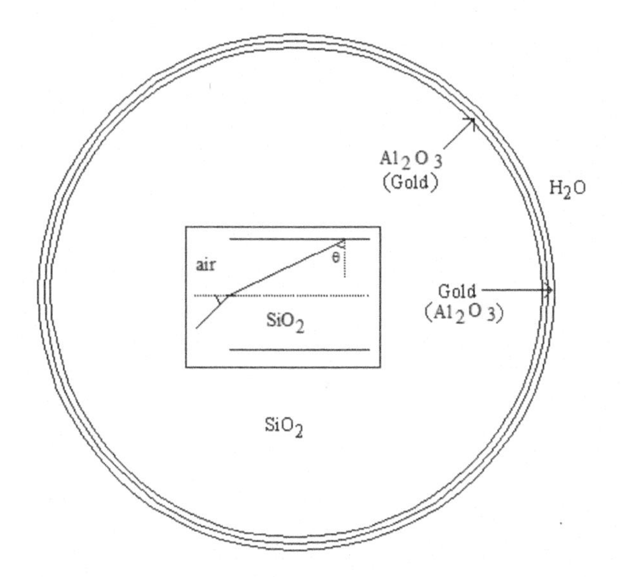

Figure 6.21 Schematic of an optical fiber with four layers (SiO_2, Al_2O_3, gold, and H_2O) where the Al_2O_3 (gold) layer is in the interior (exterior). The light is incident from an air medium in a SiO_2 core of the fiber and the angle inside the fiber is θ (inset). Adapted from Ref. [35].

The ability of plasmonic structures based on metal-2D material junction has been explored [3] for simulation and analysis of fiber optic relative humidity sensing. The influence of various 2D materials/heterostructures has been analyzed. The sensor structure is simplified in Fig. 6.22.

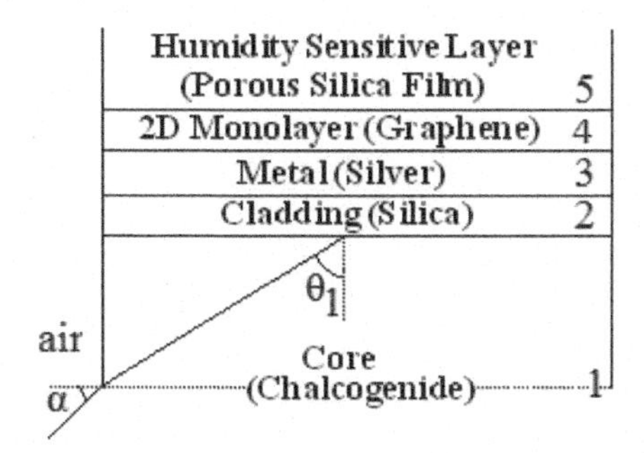

Figure 6.22 Schematic of half of a fiber with five layers (adapted from Ref. [3]).

In Ref. [3], angular interrogation and intensity interrogation methods have been used. The graphene-based sensor probe shows the highest complete performance factor (CPF) (171.11 dB/% RH). This sensing performance factor is defined as:

$$\text{CPF} = \frac{\delta\theta_{\text{res}}}{\delta RH} \times \frac{\delta PL_{\text{max}}}{\delta\theta_{0.5}}, \ \text{CPF} = \frac{\delta\alpha_{\text{res}}}{\delta RH} \times \frac{\delta PL_{\text{max}}}{\delta\alpha_{0.5}} \tag{6.21}$$

where $\delta\theta_{\text{res}}$ ($\delta\alpha_{\text{res}}$) is the shift in peak power loss angle for a given RH variation (δRH, in percentage) with reference to 0% RH, $\delta\theta_{0.5}$ ($\delta\alpha_{0.5}$) represents the angular width of the power loss spectrum, and δPL_{max} is the differential peak power loss corresponding to any arbitrary RH value (with RH = 0% as reference). Also, with an intensity interrogation method, the LOD achievable with graphene-based and WS_2-based sensor probes is as fine as 6.8×10^{-4}% RH, whereas the LOD for the angular interrogation method is 4.1×10^{-3}% RH.

Three types of plasmonic sensors [6] were applied for the detection of five types of human-liver tissues (normal: N, metastatic: MET, non-cancerous metastatic: NMET, hepatocellular carcinoma: HCC, and non-cancerous hepatocellular carcinoma: NHCC).

The first type of fiber with seven layers is made by a ZBLAN core surrounded by NaF, silicon, silver, graphene, Al_2O_3, and liver layers. The second type of fiber with six layers (Fig. 6.23) is similar to the first one but without the graphene layer. The third type of fiber with five layers is made by a SiO_2 core surrounded by GaP, gold, Al_2O_3, and liver layers. We can distinguish among the N, NMET, NHCC, HCC, and MET liver tissues by analyzing the values of the resonant parameters. Thus, for a decrease of the real part of the RI from 1.3625 (N human-liver) to 1.3439 (MET human-liver), the resonance angular width is increased from 0.152° (0.050°) to 1.026° (0.975°) for the second (third) type of the fiber. In addition, the FOM is decreased from 1197.6 RIU^{-1} (2716.4 RIU^{-1}) to 165.7 RIU^{-1} (124.6 RIU^{-1}), and the amplitude sensitivity S_A is decreased from 2067.8 RIU^{-1} (1453.1 RIU^{-1} to 293.2 RIU^{-1} (212.5 RIU^{-1}) for the second (third) type of sensor. For optimized thicknesses of the interior layers at a laser wavelength (1.53 μm), one obtains large values of the FOM and amplitude sensitivity at an angle very close to the resonance angle. The advantage of the third (second) type of fiber for normal liver is a large FOM (maximum amplitude sensitivity) in comparison with that for the second (third) type of sensor. For all variants of the sensors, the FOM and maximum amplitude sensitivity are decreasing in the same order (N, MET, NMET, NHCC, and HCC) for the liver tissues.

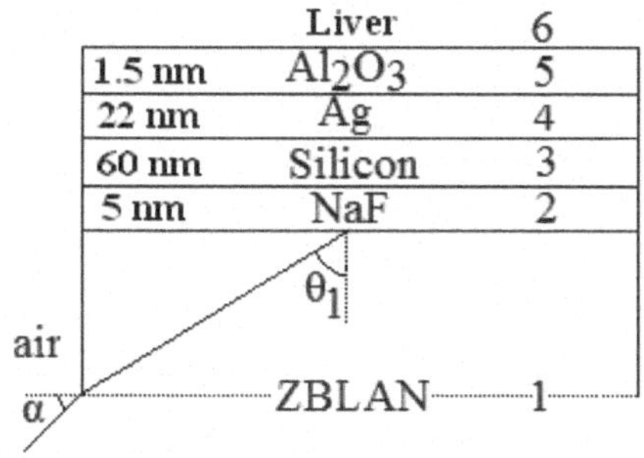

Figure 6.23 Schematic of half of a fiber with six layers with the thickness of the interior layers (adapted from Ref. [6]).

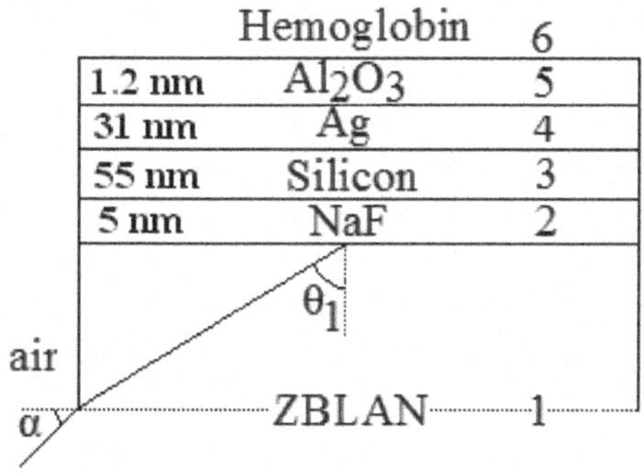

Figure 6.24 Schematic of half of a fiber with six layers with the thickness of the interior layers (adapted from Ref. [9]).

Two new types of plasmonic biosensors were applied for the detection of hemoglobin concentration in normal human blood by using the angular interrogation method [9]. The first type of fiber with five layers is made by a SiO_2 core surrounded by GaP, gold, Al_2O_3, and hemoglobin layers. The second type of fiber with six layers is made by a ZBLAN core surrounded by NaF, silicon, silver, Al_2O_3, and hemoglobin layers (Fig. 6.24). For some thicknesses of the interior layers and a hemoglobin concentration close to the mean value of a male human at a laser wavelength (1.53 μm), a large FOM and maximum value of the amplitude sensitivity are obtained at an angle very close to the resonance angle. For the second type of sensor, 0.0036 g/dl resolution is obtained, which is 10% finer than the estimated detection limit (0.004 g/dl) reported very recently by using an electrochemical method. Also, our resolution is better than the previously reported value (0.006 g/dl if the angular resolution is 0.001°) by using a gold-aluminum-based SPR sensor.

A fiber-based plasmonic sensor with three layers (Fig. 6.25) was studied and the analytical method based on Bessel functions was applied to the calculation of the loss, FOM, and amplitude and power sensitivities for the HE_{12} mode in a resonant interaction with a TM_{01} mode [37]. In this method, the electromagnetic field is represented by a Bessel function J of the first kind in the core region (SiO_2), a linear combination of the Hankel functions H_1 and H_2 in the TiO_2/PSS (poly(styrene sulfonate)) region, and a modified Bessel function K of the second kind in the analyte region.

The values of the effective indices calculated by using the finite element method are in agreement with the analytical method. The results and subsequent analysis indicate that the FWHM of the loss and maximum value of the loss P for HE_{12} mode in resonant interaction with a TM_{01} mode tends to decrease when the radius of the fiber core is increased. In addition, the shift $\delta\lambda_{res}$ toward longer wavelengths of the maximum loss P for an increase Δn_a of the analyte RI by 0.001 RIU, the FOM and the maximum amplitude S_A and power S_P sensitivities increase with fiber core radius.

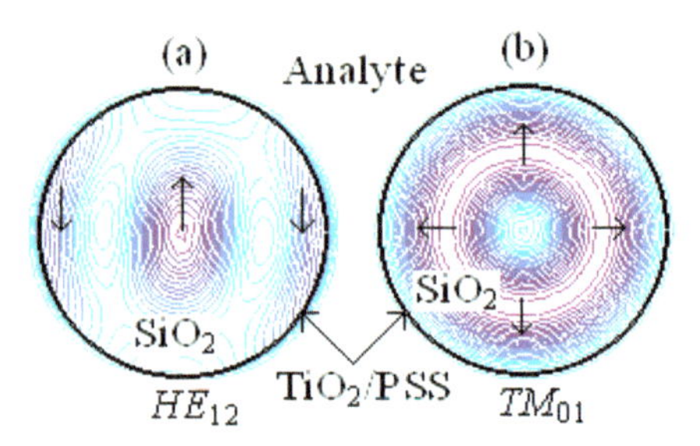

Figure 6.25 Schematic of an optical fiber with three layers (SiO_2, TiO_2/PSS, and analyte) and the contour plot of the z-component $S_z(x, y)$ of the Poynting vector at the resonance (λ = 0.8494 μm) between the HE_{12} (a) and TM_{01} (b) modes when the fiber core radius is r_1 = 5 μm, the thickness of TiO_2/PSS lossy layer is d_2 = 70 nm, and the analyte RI is n_a = 1.333. The arrows show the electrical field orientation. Adapted from Ref. [37].

The analytical method based on the Bessel functions shows that for an optical fiber-based lossy mode resonance sensor with only three layers, the best values of the local spectral sensitivity S_{λ}, the angular width of the resonance spectrum $\delta\lambda_{0.5}$, the figure of merit FOM, maximum of the amplitude sensitivity S_A, maximum of the local power sensitivity S_P, RI resolution $S_{R\lambda}$, and power resolution S_{RP} for the HE_{12} mode in resonant interaction with the TM_{01} mode, are obtained when the radius of the fiber core is increased from r_1 = 2.5 μm to r_1 = 10 μm. Thus, when r_1 is increased (2.5 μm, 5 μm, 7.5 μm, and 10 μm), the local spectral sensitivity S_{λ} increases (1600 nm/RIU, 2100 nm/RIU, 2300 nm/RIU, and 2500 nm/RIU), the full width $\delta\lambda_{0.5}$ at half maximum of the loss decreases (371 nm, 208 nm, 135 nm, and 91 nm), the FOM increases (4.31 RIU^{-1}, 10.09 RIU^{-1}, 17.04 RIU^{-1}, and 27.47 RIU^{-1}), the maximum of the amplitude sensitivity S_A increases (13.3 RIU^{-1}, 34.7 RIU^{-1}, 50.5 RIU^{-1}, and 112.6 RIU^{-1}), the maximum of the local power sensitivity S_P increases (7454.6 dB cm^{-1} RIU^{-1}, 8348.6 dB cm^{-1} RIU^{-1}, 10241.0 dB cm^{-1} RIU^{-1}, and 17334.0 dB cm^{-1} RIU^{-1}), the power resolution (LOD) S_{RP} gets finer (1.34×10^{-6} RIU, 1.20×10^{-6} RIU, 9.76×10^{-7} RIU, and 5.77×10^{-7} RIU) and the resonance wavelength undergoes a redshift (0.8098 μm, 0.8494 μm, 0.8583 μm, and 0.8600 μm), respectively. Also, the spectral resolution $S_{R\lambda}$ is 4.35×10^{-7} RIU for r_1 = 7.5 μm and 4.0×10^{-7} RIU for r_1 = 10 μm.

In recent times, SPR-based optical sensors have attained very high levels of performance. For example [38], in an SPR-based fiber optic sensor with six layers (ZBLAN fluoride core, 5 nm thick NaF clad, 60 nm thick amorphous Si layer, 45.3 nm thick Ag layer, 11 nm thick Al_2O_3 interlayer, and heavy water as an analyte) and 938.7 nm wavelength, an extremely large FOM of 31806.65 RIU^{-1} was obtained.

A transverse SDS of the horizontal PSHE in the waveguide at a given wavelength (1557 nm) for a *TM* mode was analyzed [39]. The sensor structure is analyzed in two ways. In the angular method, the results indicate a maximum FOM of 4007.0 RIU^{-1} for an optimum thickness (51 nm) of gold layer and this value is significantly greater than the corresponding FOM (413.7 RIU^{-1}) reported earlier for 50 nm thick gold layer.

The experimental device (Fig. 6.26a) includes a linearly polarized light (1557 nm) from a laser source (LS), a half-wave plate (HWP) for adjusting the intensity of the light, a short-focus lens (L_1), a polarizer (P_1) to select the incident polarization angle, a waveguide (W), or a silicon prism (SP) where the reflected light splits into left-handed and right-handed components and generate displacement in both directions perpendicular and parallel to the incident plane, a post-selection polarizer (P_2) to adjust the post-selection angle Δ, a long-focus lens (L_2), and a recorded by a charge-coupled device (CCD) camera (Coherent Laser-Cam HR).

In the spin Hall effect, the maximum sensitivity was 6.25×10^7 μm/RIU for an amplified angle $\Delta = 0.1^\circ$ and 1.09×10^6 μm/RIU for $\Delta = 0.1$ rad in the conventional weak measurements. The corresponding finest possible resolution of magnetic field detection is 1.5×10^{-6} Oe in the conventional weak measurements for $\Delta = 0.1^\circ$ and 7.8×10^{-8} Oe in the modified weak measurements for $\Delta = 0.5$ rad when the volume fraction of Fe_3O_4 particles in the magnetic fluid is 1.93%. These values are significantly finer than those existing in the related state of the art.

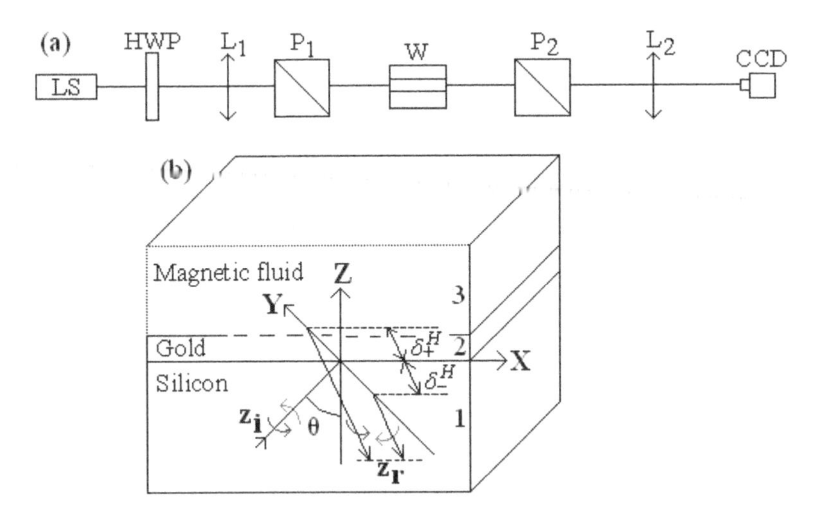

Figure 6.26 Schematic of the (a) experimental device and (b) a plasmonic waveguide with three layers (silicon, gold, and magnetic fluid) where δ_-^H and δ_+^H are the transverse displacements of left circularly polarized (LCP) and right circularly polarized (RCP) components of light, respectively (adapted from Ref. [39]).

A plasmonic sensor with four (germanium, dielectric, graphene, and gas) layer waveguide structure (Fig. 6.27) for the detection of the RI of a gas medium and magnetic field applied to the graphene layer was proposed and analyzed [40]. We consider the transverse SDS of the horizontal PSHE at a given frequency (5 THz). The RI and thicknesses of the constituted layer of a plasmonic sensor (Fig. 6.26): $n_1 = 4$ (germanium), $n_2 = 1.5$, $d_2 = 51$ nm (organic layer), $n_3 = n_g$, and $d_3 = 0.34$ nm (graphene) are used to find a RI resolution of the gas medium when the RI of the gas medium is changed from $n_4 = 1$ RIU to $n_{4a} = 1.1$ RIU for a fixed magnetic field applied to the graphene layer.

The same method is also applied when $n_1 = 4$, $n_2 = 1.5$, $d_2 = 51$ nm, $n_3 = n_g$ (B = 0), $n_{3a} = n'_g$, (B ≠ 0), and $d_3 = 0.34$ nm are used to find a magnetic field resolution when the magnetic field applied to the graphene layer is changed from 0 T to 1 T for a fixed RI of the gas medium ($n_4 = 1$).

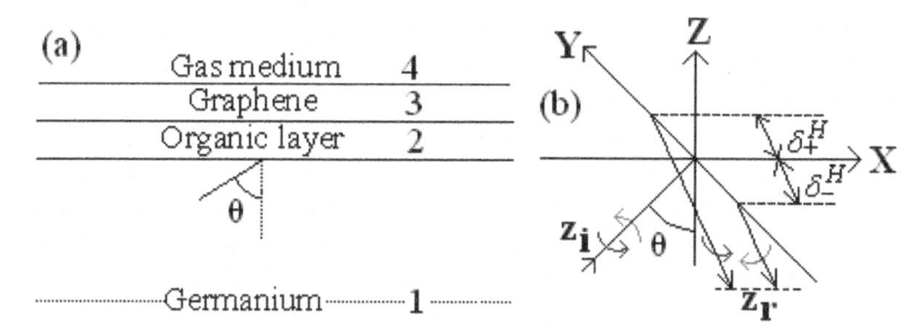

Figure 6.27 Schematic of (a) an optical structure with four layers (germanium, organic layer, graphene, and gas medium), and (b) a plasmonic sensor with PSHE, where δ_-^H and δ_+^H are the transverse displacements of LCP and RCP components of light, respectively (adapted from Ref. [40]).

The sensor structure is analyzed in two ways. Firstly, in the conventional weak measurements, for a magnetic field $B = 0$ T and for an amplified angle $\Delta = 0.1°$ ($\Delta = 0.1$ rad), a RI resolution of 1.22×10^{-11} RIU (6.99×10^{-10} RIU) was obtained for the gas medium when the RI of the gas is changed from 1 RIU to 1.1 RIU. Also, in the modified weak measurements, for an amplified angle $\Delta = 0.1°$ ($\Delta = 0.1$ rad), a RI resolution of 3.59×10^{-15} RIU (6.24×10^{-17} RIU) was obtained for the gas medium when the RI of the gas is changed with 0.1 RIU. Secondly, in the conventional weak measurements and for an amplified angle $\Delta = 0.1°$ ($\Delta = 0.1$ rad), a magnetic field resolution of 0.0146 μT (0.84 μT) when the RI of the gas is 1 RIU and the magnetic field applied to the graphene layer is changed from 0 T to 0.1 T. Also, in modified weak measurements, for an amplified angle $\Delta = 0.1^{\circ}$ ($\Delta = 0.1$ rad), a magnetic field resolution of 2.06×10^{-3} μT (3.59×10^{-6} μT) was obtained. Our results are significantly finer than those existing (5×10^{-9} RIU [44] and 0.7 μT [45]) in the related state of the art. Table 6.1 makes a comparison of the parameters n_a, λ, $\delta\lambda_{0.5}$, S_{λ}, SR_{λ}, S_A, and SR_A for some optical fiber sensors.

Table 6.1 Values of n_a, λ [μm], $\delta\lambda_{0.5}$ [nm], S_{λ} [nm/RIU], SR_{λ} [RIU], S_A [RIU^{-1}], and SR_A [RIU] for optical fiber sensors

n_a	λ	$\delta\lambda_{0.5}$	S_{λ}	SR_{λ}	S_A	SR_A
Distilled water	0.623 [1]	31	4000	2.5×10^{-5}	–	–
Distilled water	0.6507 [2]	22	5229	1.9×10^{-5}	3941.5	2.5×10^{-6}
Distilled water	0.62909 [41]	19.4	3236	3.1×10^{-5}	3708.8	2.7×10^{-6}
Distilled water	0.5986 [42]	18.0	3419	2.9×10^{-5}	4257.2	2.3×10^{-6}
1.33	0.629 [12]	34.2	2000	5.0×10^{-5}	118.1	8.5×10^{-5}
1.33	0.863 [20]	32.0	10000	1.0×10^{-5}	560.6	1.8×10^{-5}
1.343	0.538 [31]	–	3100	3.2×10^{-5}	–	–
1.40–1.43	0.904 [15]	–	15180	5.7×10^{-6}	498	2.0×10^{-5}
1.37, x-pol.	0.850 [17]	–	16100	6.2×10^{-6}	837	1.2×10^{-5}
1.37, y-pol.	0.880 [17]	–	25000	4.0×10^{-6}	404	2.5×10^{-5}
1.22	1.65 [18]	–	1000	1.0×10^{-4}	–	–
1.23–1.29	2.55–2.90 [19]	–	5500	7.69×10^{-6}	333.8	–
1.36–1.40	2.1–3.3 [26]	–	30000	3.33×10^{-6}	–	–
1.48	1.3 [27]	–	8000	1.25×10^{-5}	700	1.8×10^{-5}
1.36	2.22 [18]	–	51000	1.96×10^{-6}	–	–
1.41	1.32–3.85 [28]	–	216000	4.63×10^{-7}	1680	5.95×10^{-6}
1.33 (δn_a = 0.005)	0.633 [43]	–	1.09×10^{8}	9.0×10^{-8}	–	–
1.4671	1.557 [40]	–	6.25×10^{10}	1.6×10^{-10}	–	–

6.5 Summary

In this chapter, we describe some examples of microstructured and non-microstructured fiber plasmonic sensors. Two types of plasmonic sensors have been reviewed in the above sections: microstructured fiber plasmonic sensors and non-microstructured fiber plasmonic sensors. For the first type of the sensor, the sensitivity of the sensor is dependent on the configuration of the devices (hexagonal lattice, solid core, partial-solid-core, birefringent single-layer, microchannel incorporated PCF, two open-ring channels, D-shaped, dual-core D-shaped, honeycomb PCF, truncated honeycomb PCF, etc.). The sensor parameters of a hexagonal lattice PCF biosensor are dependent on the distribution and number of air holes, the number of rings of air holes, the thickness of the metal, and the radius of the core layer. The role of the air holes is to decrease the effective RI of the cladding layer, enabling mode guidance in the core mode. The metal layer is very small ($\approx$ 40 nm) to increase the interaction between the core and plasmonic modes. The sensitivity of a D-shaped sensor is higher than other PCF sensors, due to a stronger interaction between the core and plasmon modes since the plasmonic gold is deposited near the core and the evanescent field can easily reach the gold surface. The sensitivity of non-microstructured fiber plasmonic sensors (Bragg fiber with four layers, and optical fiber with three, four, five, or six layers) is dependent on the number of layers, thickness, and RI of the layers, and also on the thickness of the metal layer.

Microstructured fiber plasmonic sensors and non-microstructured fiber plasmonic sensors (fiber optic SPR sensors) were applied for the detection of distilled or heavy water, humidity, magnetic field, human blood group, and human-liver tissues, and for the detection of hemoglobin concentration in normal human blood. The best results for the sensitivity (6.25×10^{10} nm/RIU) and resolution (1.6×10^{-10} RIU) for the detection of the RI of a gas medium and magnetic field applied to the graphene layer were obtained with a plasmonic sensor with four layers (germanium, dielectric, graphene, and gas) based on a transverse SDS of the horizontal PSHE at a given frequency (5 THz).

Acknowledgment

Anuj K. Sharma gratefully acknowledges the sponsored research project grants "03(1441)/18/EMR-II" and "CRG/2019/002636" funded by the Council of Scientific & Industrial Research (India) and Science & Engineering Research Board (India), respectively.

References

1. Popescu, V. A., Puscas, N. N., and Perrone, G. (2012). Power absorption efficiency of a new microstructured plasmon optical fiber, *JOSA B*, **29,** pp. 3039–3046.
2. Popescu, V. A., Puscas, N. N., and Perrone, G. (2014). Strong power absorption in a new microstructured holey fiber-based plasmonic sensor, *JOSA B*, **31,** pp. 1062–1070.
3. Sharma, A. K., Kaur, B., and Popescu, V. A. (2020). On the role of different 2D materials/heterostructures in fiber-optic SPR humidity sensor in visible spectral region, *Opt. Mater.*, **102,** 109824.
4. Sharma, A. K. and Popescu, V. A. (2021). Magnetic field sensor with truncated honeycomb photonic crystal fiber: Analysis under the variations in magnetic fluid composition and temperature for high performance in near infrared, *OQEL.*, **53,** 145.
5. Popescu, V. A. (2018). Application of a plasmonic biosensor for detection of human-liver tissues, *Plasmonics*, **13,** pp. 575–582.
6. Popescu, V. A. and Sharma, A. K. (2019). Simulation and analysis of different approaches towards fiber optic plasmonic sensing for detection of human-liver tissues, *OQEL.*, **51,** 290.
7. Popescu, V. A. (2017). Application of a plasmonic biosensor for detection of human blood groups, *Plasmonics*, **12,** pp. 1733–1739.
8. Popescu, V. A. (2018). Simulation of some plasmonic biosensors for detection of hemoglobin concentration in human blood, *Plasmonics*, **13,** pp. 1507–1511.
9. Popescu, V. A. and Sharma, A. K. (2020). New plasmonic biosensors for determination of human hemoglobin concentration in blood, *Sens Imaging*, **21,** 5.
10. Rajan, Chand S. and Gupta, B. D. (2006). Fabrication and characterization of a surface plasmon resonance based fiber-optic sensor for bittering component-Naringin, *Sens. Actuator B*, **115,** pp. 344–348.

11. Rajan, Chand S. and Gupta, B. D. (2007). Surface plasmon resonance based fiber-optic sensor for the detection of pesticide, *Sens. Actuator B*, **123,** pp. 661–666.

12. Sharma, A. K. and Nagao, T. (2014). Design of a silicon-based plasmonic optical *sensor* for *magnetic field* monitoring in the infrared, *Appl. Phys. B*, **117**, pp. 363–368.

13. Popescu, V. A., Puscas, N. N., and Perrone, G. (2017). Simulation of the sensing performance of a plasmonic biosensor based on birefringent solid-core microstructured optical fiber, *Plasmonics*, **12,** pp. 905–911.

14. Popescu, V. A., Puscas, N. N., and Perrone, G. (2017). Plasmonic biosensor based on a birefringent partial-solid-core microstructured optical fiber, *J. Opt.*, **19,** 075004

15. Liu, M., Yang, X., Shum, P., and Yuan, H. (2018). High-sensitivity birefringent and single-layer coating photonic crystal fiber biosensor based on surface plasmon resonance, *Appl. Opt.*, **57,** pp. 1883–1886.

16. Perone, G., Popescu, V. A., and Puscas, N. N. (2019). Enhanced plasmonic biosensors based on microstructured optical fibers, *Opt. Eng.(SPIE)*, **58,** 072013.

17. Islam, MD. S., Cordeiro, C. M. B. Sultana, J., Aoni, F. A., Feng, S., Ahmed, R., Dorraki, M., Dinovitser, A., Wai-Him Ng, B., and Abbott, D. (2018). A hi-bi ultra-sensitive surface plasmon resonance fiber sensor, *IEEE Access*, **7,** pp. 79085–79094.

18. Haque, E., Hossain, Md. A., Namihira, Y., and Ahmed, F. (2019). Microchannel-based plasmonic refractive index sensor for low refractive index detection, *Appl. Opt.*, **58,** pp. 1547–1554.

19. Liu, C., Yang, L., Lu, X., Liu, Q., Wang, F., Lv, J., Sun, T., Mu, H., and Chua, P. K. (2017). Mid-infrared surface plasmon resonance sensor based on photonic crystal fibers, *Opt. Express*, **25,** pp. 14227–14237.

20. Popescu, V. A. and Sharma, A. K. (2018). Resonant interaction between two core modes in a plasmonic biosensor based on a birefringent solid-core microstructured optical fiber, *OSA Continuum*, **1,** pp. 496–505.

21. Popescu, V. A., Sharma, A. K., and Marques, C. (2021). Resonant interaction between a core mode and two complementary supermodes in a honeycomb PCF reflector-based SPR sensor, *Optik*, **227,** 166121.

22. Hassani, A., Gauvreau, B., Fehri, M. F., Kabashin, A., and Skorobogatiy, M. (2008). Photonic crystal fiber and waveguide-based surface plasmon resonance sensors for application in the visible and near-IR, *Electromagnetics*, **28,** pp. 198–213.

23. Liu, C., Wang, L., Liu, Q., Wang, F., Sun, Z., Sun, T., Mu, H., and Chu, P. K. (2018). Analysis of a surface plasmon resonance probe based on photonic crystal fibers for low refractive index detection, *Plasmonics*, **13**, pp. 779–784.

24. Gao, R. and Jiang, Y. (2013). Magnetic fluid-filled microhole in the collapsed region of a photonic crystal fiber for the measurement of a magnetic field, *Opt. Lett.*, **38**, pp. 3181–3184.

25. Popescu, V. A. and Sharma, A. K. (2021). On the performance of plasmonics-based optical fiber magnetic field sensor: Comparison between analytical and finite element methods, *Rom. Rep. Phys.*, **73**, 406.

26. Wu, J., Li, S., Wang, X., Shi, M., Feng, X., and Liu, Y. (2018). Ultrahigh sensitivity refractive index sensor of a D-shaped PCF based on surface plasmon resonance, *Appl. Opt.*, **57**, pp. 4002–4007.

27. Sakib, Md. N., Hossain, Mb. B., Al-tabatabaie, K. F., Mehedi, I. M., Hasan, Md. T., Hossain, Md. A., and Amiri, I. S. (2019). High performance dual core D-shape PCF-SPR sensor modeling employing gold coat, *Results Phys.*, **15**, 102788.

28. Haque, E., Noman, A. A, Hossain, Md. A., Hai, N. H., Namihira, Y., and Ahmed, F. (2021). Highly sensitive D-shaped plasmonic refractive index sensor for a broad range of refractive index detection, *IEEE Photon. J.*, **13**, 4800211.

29. Sharma, A. K. and Gupta, B. D. (2005). On the sensitivity and signal to noise ratio of a step-index fiber optic surface plasmon resonance sensor with bimetallic layers, *Opt. Commun.*, **245**, pp. 159–169.

30. Sharma, A. K. and Gupta, B. D. (2006). Theoretical model in a fiber optic remote sensor based on surface plasmon resonance for temperature detection, *Opt. Fiber Technol.*, **12**, pp. 87–100.

31. Sharma, A. K. and Gupta, B. D. (2007). Influence of dopants on the performance of a fiber optic surface plasmon resonance sensor, *Opt. Commun.*, **274**, pp. 320–326.

32. Verma, R. K., Sharma, A. K., and Gupta, B. D. (2008). Surface plasmon resonance based tapered fiber optic sensor with different taper profiles, *Opt. Commun.*, **281**, pp. 1486–1491.

33. Gupta, B. D. and Verma, R. K. (2009). Surface plasmon resonance-based fiber optic sensors: Principle, probe designs, and some applications, *J. Sens.*, **979761.**

34. Popescu, V. A. (2019). Interrogation methods for Bragg fiber-based plasmonic sensors, *Rom. Rep. Phys.*, **71**, 408.

35. Popescu, V. A., Puscas, N. N., and Perrone, G. (2015). Efficient light absorption in a new Bragg fiber-based plasmonic sensor, *JOSA B*, **32**, pp. 473–478.

36. Popescu, V. A. (2020). Optical fiber-based plasmonic sensors using aluminium oxide insulator, *Rom. Rep. Phys.*, **72,** 407.

37. Popescu, V. A. and Sharma, A. K. (2020). Theoretical understanding of the resonant interaction between TM_{01} and HE_{12} modes in a lossy mode fiber-based plasmonic sensors, *Rom. Rep. Phys.*, **72,** 414.

38. Sharma, A. K., Dominic, A., Kaur, B., and Popescu, V. A. (2019). Fluoride fiber sensor with huge performance enhancement via optimum radiative damping at Ag–Al_2O_3–Graphene heterojunction on silicon, *J. Light. Technol.*, **37**, pp. 5641–5646.

39. Popescu, V. A., Prajapati, Y. K., and Sharma, A. K. (2021). Highly sensitive magnetic field detection in infrared region with photonic spin Hall effect in silicon waveguide plasmonic sensor, *IEEE Trans. Magn.*, **57**, pp. 1–10.

40. Popescu, V. A., Sharma, A. K., and Prajapati, Y. K. (2021). Graphene-based plasmonic detection with photonic spin Hall effect in terahertz region: Application to magnetic field and gas sensing, *Journal of Electronic Materials (JEMS),* accepted for publication.

41. Popescu, V. A., Puscas, N. N., and Perrone, G. (2016). Sensing performance of the Bragg fiber-based plasmonic sensors with four layers, *Plasmonics*, **11,** pp. 1183–1189.

42. Popescu, V. A., Puscas, N. N., and G. Perrone, G. (2014). Supermode resonances in a multi-core holey fiber-based sensor, *Eur. Phys. J. D.*, **68,** 229.

43. Zhou, X., Sheng, L., and Ling, X. (2018). Photonic spin Hall effect enabled refractive index sensor using weak measurements, *Sci. Rep.*, **8,** 1221.

44. Pevec, S. and Donlagic, D. (2018). Miniature fiber-optic Fabry-Perot refractive index sensor for gas sensing with a resolution of 5×10^{-9} RIU, *Opt. Express*, **26,** pp. 23868–23882.

45. Zhong, S., Guan, T., Xu, Y., Zhou, C., Shi, L., Guo, C., Zhou, X., Li, Z., He, Y., and Xing, X. (2021). Simultaneous sensing axial and radial magnetic fields based on weak measurement, *Opt. Commun.*, **486**, 12677.

Chapter 7

Microstructured and Nanostructured Fiber Plasmonic Sensors

Qi Wang and Zi-Han Ren
College of Information Science and Engineering,
Northeastern University, 110819, Shenyang, Liaoning, China
wangqi@ise.neu.edu.cn

Optical sensors have multiple detection capabilities, are immune to electromagnetic interference, require no labeling, and are easy to integrate with microfluidic devices. Optical fibers have received widespread attention for their tiny structure, rapid response, ease of modification, and integration. The surface plasmon resonance (SPR) phenomenon has attracted widespread interest, due to its ability to allow label-free, highly sensitive, and rapid detection of biomarkers. Focusing on research on the application of the SPR phenomenon in bioassays, this chapter revisits SPR biosensors modified with different optical fiber structures and surface materials over the past five years, outlining microstructured and nanostructured fiber optic sensors directly related to SPR sensing. Finally, this chapter gives

Plasmonics-Based Optical Sensors and Detectors
Edited by Banshi D. Gupta, Anuj K. Sharma, and Jin Li

ISBN 978-981-4968-85-0 (Hardcover), 978-1-003-43830-4 (eBook)
www.jennystanford.com

an outlook on the sensitivity enhancement of fiber optic biosensors with microstructures and nanostructures, new demodulation mechanisms, etc.

7.1 Introduction

As technology advances, there is a growing expectation for sensors that are portable and quick to detect. Integratable sensors for microstructures and nanostructures have received a lot of attention in applications such as physical (pressure [1], temperature [2]), chemical (photocatalysis [3]), and biomedical (antigen [4], protein [5], heavy metal ions [6]) parameter sensing. Thanks to the development of micro-/nanofabrication technologies, miniature sensors can be realized on increasingly smaller scales. As a popular miniature sensor platform, optical fiber-based sensors have been widely studied due to their small size, lightweight, flexibility, interference immunity, and remote sensing capability. Fiber optic sensors can modulate the intensity, phase, polarization, wavelength, or propagation time of light in an optical fiber by the physical quantity to be measured for detection purposes. Recently, various fiber sensors based on fiber Bragg grating (FBG) [7], SPR [8], Mach-Zehnder interference [9], and Fabry-Pérot interference [10] have been reported and demonstrated.

Among these, sensors based on the SPR mechanism have received wide attention for their high sensitivity, label-free characteristics, simple operation, and fast and convenient detection. SPR is a phenomenon in which the incident light is strongly absorbed by the surface plasma (SP) at a specific incident wavelength or angle, resulting in a resonance spectrum. When the refractive index (RI) and other parameters of the biological sample attached to the sensor surface change, the resonance wavelength or angle will shift, thus enabling the detection of a specific biological sample. Fiber optic SPR biosensors are easier to integrate and process, less expensive, and have stronger resistance to temperature interference than traditional prism and grating sensors [11]. In addition, with advances in nanofabrication technology, the integration of plasmonic nanostructures into optical fibers has been realized, leading to the evolution of a new technology known as "fiber optic lab". Fiber optic lab technology is the integration of nano-sized functionalized materials and structures on optical fibers, aiming to produce a novel all-fiber miniaturized probe for a wide range of applications in environmental monitoring, biomedical, instant diagnostics, and photonic device integration. Although fiber optic SPR sensors have the above existing advantages, they also have some shortcomings for improvement; for example, the detection limits are not sufficiently low compared to traditional biological assays, and specific adsorption of biological samples needs to be pre-treated. In order to realize the portable integration of the SPR bio-detection method, improve the sensitivity of the sensor, and lower the minimum detection limit, researchers proposed methods such as changing the structure of the fiber optic substrate and modifying the fiber optic surface.

Exploring new fiber optic structures for high-accuracy detection of sensors has been a popular research area in recent years. Designing single kind of fiber substrate structures based on high-order cladding modes, fusion spliced fiber substrate structures based on the core offset to stimulate the SPR phenomenon; and designing photonic crystal fibers (PCF) structures based on different internal structures to achieve sensor sensitivity enhancement, very narrow linewidth, and low detection limits, are popular development directions.

The detection of biomarkers currently relies on changes in the resonance spectrum when the concentration of a biological sample changes. Therefore, the treatment of the sensing region is critical. Modifying high RI dielectric materials on the surface of SPR biosensors increases the electric field strength of the sensor surface and the interaction area between the biomarker and the sensor to improve the sensitivity of the sensor and reduce the detection limit of the biomarker. The modification of metal micro-nanostructures on the sensor surface increases the local electric field strength through the coupling between local localized surface plasmon polariton (LSPP) and surface plasmon polariton (SPP) to suppressing the resonance spectrum from excessively broadening while enhancing the sensitivity of the sensor. Surface modification by applying polymers and biomaterials enhances the adsorption of specific biomass and improves resistance to contamination, allowing SPR bioassays to be applied for in situ detection, and skipping biological sample processing steps while reducing detection limits.

In this chapter, Section 7.1 points out the advantages and disadvantages of the microstructured and nanostructured fiber optic SPR bio-detection methods and the main research directions of researchers in the last five years. Section 7.2 reviews the sensing characteristics of fiber-coupled SPR biosensors with different structures in recent years. Section 7.3 summarizes that researchers have enhanced the sensor performance such as sensitivity by surface modification of the sensing region. Section 7.4 summarizes promising research directions in terms of new demodulation mechanisms, etc.

7.2 Microstructured and Nanostructured SPR Fiber Sensors Based on Fiber Shaping

Compared with other substrate-structured SPR sensing systems with multiple devices, large size, complex structure, long detection time, and large modification area, the optical fiber SPR sensing system has gained wide attention for its simple operation, easy implementation, small size, and low price. In recent years, due to the fabrication process has advanced, optical fiber biosensors have been developed quickly. In this section, we first introduce fiber grating SPR biosensors, and then briefly classify fiber SPR biosensors into several categories: single-mode fiber (SMF), multimode fiber (MMF), and PCF. The development of optical fiber SPR sensors in the past five years is outlined from the structural level.

7.2.1 Fiber Grating SPR Biosensor

Fiber grating is an optical fiber device with the permanent periodic change of RI on the fiber core, which has the advantages of polarization insensitivity, low additional losses, good coupling, wide bandwidth, and easy to use and maintain. Fiber optic grating sensors with SPR have achieved good results in measuring biomass. However, the structure of fiber grating is mostly fixed, in the last five years mostly for FBG and other structures. The new structure of the grating is still mostly in the simulation stage. The trade-off between the complexity of the structure and the ease of processing and manufacturing as well as the sensor performance improvement effect still needs to be studied. Its performance improvement in this chapter is mainly reflected in the section on surface modification, this section will not be repeated.

7.2.2 Single-Mode Fiber

A single species of fiber is tapered, bent, and polished to excite higher-order cladding modes to produce the SPR phenomenon. Therefore, SPR biosensors based on tapered fiber, D-shaped fiber, and U-shaped fiber have been developed. Aitor Urrutia et al. [12] proposed tapered SMF (Fig. 7.1A) for the detection of biological proteins. By coating the fiber surface with gold film in different ways, such as layer-by-layer (LBL) self-assembly and sputtering, the limit of detection (LOD) of the sensors was 271 pM and 806 pM, respectively, for the same taper diameter. The strong evanescent field of the tapered fiber improved the detection sensitivity. But the loss of the tapered region introduced some noise so the signal-to-noise ratio (SNR) of the sensor is reduced. H. Ahmad et al. [13] used a numerical control machine to fabricate tapered fibers with uniform tapered waist diameters with a loss of less than 0.1 dB (Fig. 7.1C). The increased SNR improves sensor recognition of small signal changes. The sensor has multiple resonance valleys in the range of 1.52–1.55 μm, which enables the sensor performance to be enhanced by selecting the detection point with the highest sensitivity when performing the detection of biological samples.

As the outer cladding on the fiber is removed, the core is micrometer level, the mechanical strength of the fiber becomes low, and surface modification becomes difficult. SPR sensors based on U-shaped optical fibers achieve increased sensitivity while improving mechanical strength. Hang Song et al. [8] proposed a U-shaped multichannel biosensor that cascaded MMF with SMF (Fig. 7.1B) to detect human immunoglobulin G (IgG). A gold film of the same thickness was applied to both sensing zones, and the U-shaped MMF part was used as the detection channel, while the straight SMF part was used as a reference channel to improve the interference resistance. In complex samples, the sensor LOD can reach 0.104 μg/mL.

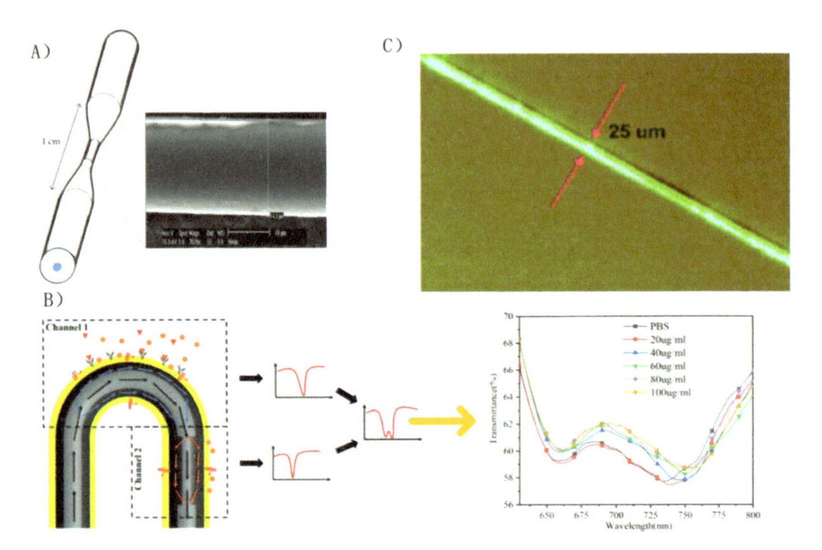

Figure 7.1 (A) Tapered fiber and its SEM image (Reprinted with permission from Ref. [12]). (B) U-shaped dual-channel SPR biosensor is resistant to temperature interference and reflectance spectra at different IgG concentrations (Reprinted with permission from Ref. [8]). (C) Low-loss tapered fiber with uniform waist (Reprinted with permission from Ref. [13]).

SMFs are widely studied as sensor probes in SPR biosensors because of their high plasticity and low light transmission loss rate compared to fibers of other structures. For example, SMFs are tapered to produce more intense evanescent fields, and U-shaped fibers are used to achieve multiple channels. However, as SMF has very low mechanical strength when the cladding is removed and is susceptible to interference from factors such as pressure, adding mechanical structures (composite RI sensor for abrupt wave fiber loop attenuation [14]) to the fiber is the key to solving the problem.

7.2.3 Multimode Fiber

Similar to SMF, a single MMF structure is also unable to excite the SPR phenomenon, so by changing the side coating or by incorporating a hetero-core fiber, MMFs are beginning to be used in fiber optic sensing. Yuzhi Chen et al. [15] proposed the detection of urine specific gravity (USG) by coating only one side of the MMF with Au films (Fig. 7.2A). This is the first time that USG detection has been achieved through an optical fiber. This structural treatment makes the sensor easy to be produced production. The sensitivity of the sensor for the USG detection was 856.09 nm/SG (wavelength modulation) and 183.31%/SG (intensity modulation), respectively. This structure excels in the detection of proteinuria with a wavelength sensitivity of 12645 nm/SG and an intensity sensitivity of 1410%/SG.

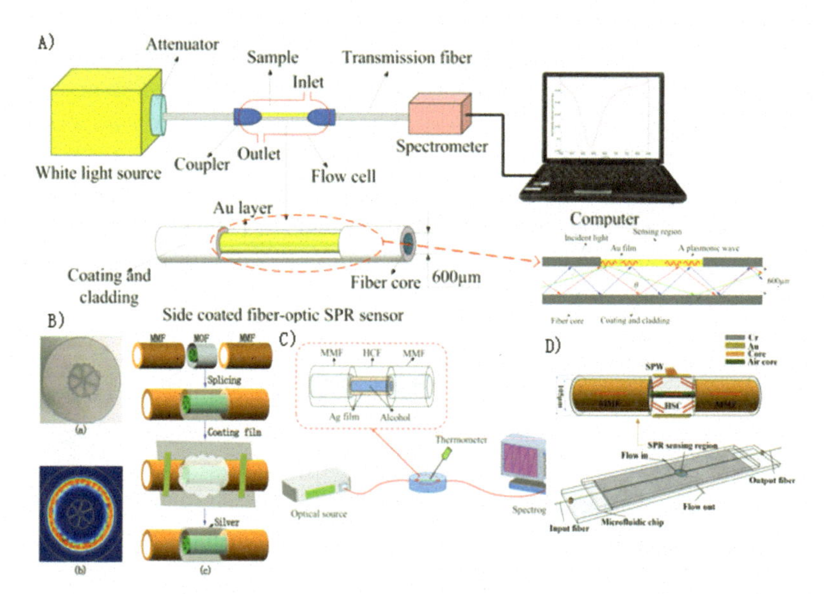

Figure 7.2 (A) Side-coated fiber optic USG sensor and side-coated fiber optic detail picture (Reprinted with permission from Ref. [15]). (B) MMF-MOF-MMF structure diagram (Reprinted with permission from Ref. [17]). (C) MMF-HCF-MMF SPR sensor and experimental setup of the temperature sensing system (Reprinted with permission from Ref. [18]). (D) MMF-HSC-MMF (Reprinted with permission from Ref. [16]).

The fusion of MMFs with other structured fibers (coreless fibers, microstructured fibers, and capillaries), constituting hetero-core fibers to excite the SPR phenomenon is also a popular direction for development in recent years. The hetero-core fibers embedded in MMF are usually slightly smaller in diameter than MMFs. Core mismatch excites the high-order cladding mode. In bioassay experiments, temperature affects the rate quantity of binding of biological samples to the modification zone. Zhao Yang et al. [16] (Fig. 7.2D) proposed the structure of MMF-hollow silica capillary

(HSC)-MMF, and the experiment demonstrated that the sensor sensitivity was up to 7225.63 nm/RIU. Moreover, the temperature can shift the resonance spectrum, so to improve the resistance to temperature interference. Shuguang Li et al. [17] proposed incorporating microstructured optical fibers (MOF) into MMFs (Fig. 7.2B) to achieve high sensitivity (3223 nm/RIU) for biomass detection. Due to the low-temperature dependence of MOF, the sensor exhibits excellent stability within a temperature variation of 21 °C (25–46 °C). Tonglei Cheng et al. [18] proposed the structure of MMF-ethanol-filled hollow-core fiber (HCF)-MMF (Fig. 7.2C). The structure is simple and easy to integrate. Due to the high thermo-optical coefficient of alcohol, the sensor is highly resistant to ambient temperature interference. Haofeng Hu et al. [19] proposed incorporating a no-core fiber (NCF) in the middle of an MMF, and optimized parameters of hetero-core fiber improve the sensitivity (11792 nm/RIU).

In recent years, structural changes and the formation of hetero-core fibers have been widely used for MMF. The SPR phenomenon is stimulated by coating the material on only one side of the light, simplifying the sensor fabrication steps and facilitating the fixation of the biological aptamer. By incorporating multiple fiber structures into an MMF, the SPR phenomenon is stimulating. This method enables the detection of biological samples while improving the sensor's resistance to temperature interference. Although the hetero-core fiber structure has the advantage of simple fabrication, the choice of fiber is very important because the small core and thick cladding of hetero-core fiber limit the leakage of evaporative waves.

7.2.4 Photonic Crystal Fibers

In the late 1920s, various air hole structures were proposed to be constructed within optical fibers to improve the detection accuracy of sensors [20–24]. PCFs are widely used for sensor probes due to their structural flexibility, adjustable dispersion, and large mode field area [20]. This section summarizes the changes in the arrangement of the air holes and the external structure of the PCF in recent years. Since the diameter of the optical fibers is on the order of microns, structural changes inside the fibers are extremely demanding for the fabrication process, so the arrangement of air holes in PCFs is dominated by simple circles, rectangles, and hexagons, and is mostly at the simulation stage. To make the detection of biological samples easy, asymmetric fiber structures have been proposed.

Due to the circular shape of the fiber itself and the highly symmetrical geometric nature of the circle, the air holes arranged in circular arrays are expected to be realized by the existing fabrication process [25]. Hasan et al. [26] have implemented a PCF sensor with a double-layer circular air hole arrangement through simulation. After optimizing the thickness of the gold film and other parameters, the simulation results show that the structure has a maximum wavelength sensitivity of 2200 nm/RIU and a maximum amplitude sensitivity of 266 RIU^{-1} at refractive indices in the range 1.33–1.36. The resolution of the sensor is 3.75×10^{-5}. Khalek et al. [27] proposed a kind of perfectly circular lattice of PCF. Compared to the lattice structure proposed by Md. Rabiul Hasan, the sensor has a wavelength sensitivity of 9000 nm/RIU, and its maximum amplitude sensitivity is 318 RIU^{-1}. The detection range is 1.34–1.37 and the maximum wavelength resolution in this interval is 1.11×10^{-5}. Jahan et al. [28] proposed a simple circular array structure with air holes of different diameters. Improved sensitivity and a higher level of resolution compared to the two structures mentioned above. The values are 1380 RIU^{-1} (amplitude), 20000 nm/RIU (wavelength), and 5.26×10^{-6} RIU. Although the circular shape is easier to produce than the other air hole arrangements, the reduced accuracy of the sensor due to attrition needs more consideration.

A PCF based on an array of periodic circular air holes transmits two orthogonal modes with the same mode effective RI. However, in practice, non-circular symmetry factors inevitably arise in optical fibers. So that the two orthogonal modes are coupled during propagation, thus limiting the optical transmission rate. A high level of birefringence is required to reduce polarization coupling, and high birefringence (HB) PCF has been achieved in recent years mainly by constructing non-circularly symmetric defect structures, such as multi-diameter air holes, anisotropic lattice, non-circularly-symmetric-fiber-core design, polygonal air holes [29], etc. Islam et al. [27] constructed a rectangular air hole arrangement inside the fiber, and by optimizing parameters such as the diameter and arrangement of the air holes, the coupling effect between the fiber core and the plasma resonance mode was enhanced to improve parameters such as sensor sensitivity. Simulation results showed that the sensitivity was 58000 nm/RIU (X-polarization direction, wavelength modulation) and 62000 nm/RIU (Y-polarization direction, wavelength modulation), which is the maximum value of sensitivity that can be achieved under wavelength modulation in recent years. Due to the complexity of the coating process inside the air holes, the outside of the base rectangular structure is coated with a metal oxide to improve the resolution of the sensor. Liu et al. [24] designed a rectangular array of air hole structures with different diameters and coated the outside of the fiber core with a metal oxide material (ITO) to enhance the resolution of the sensor. The numerical results are 2.86×10^{-6} RIU (wavelength modulation) and 8.92×10^{-6} RIU (amplitude modulation), respectively.

Compared to circular arrays, hexagons have a reduced degree of symmetry. The arrangement of air holes can be denser in hexagons than in rectangular arrays. This reduces the losses caused by polarization coupling while enhancing the coupling between the core and the plasma mode. The full-width half-maximum (FWHM) is reduced while the sensitivity of the sensor is improved. Firoz Haider et al. [30] propose a hexagonal array composed of double-diameter air holes to further reduce the loss due to polarization coupling and improve sensitivity. The numerical results are 2843 RIU^{-1} (amplitude modulation) and 3.5×10^{-6} RIU (amplitude modulation). This is the highest value for a PCF-SPR sensor with a hexagonal air hole arrangement in amplitude modulation. Mahfuz et al. [22] proposed a hexagonal array of multi-diameter air holes. Numerical results show that the sensitivity and resolution of the sensor can reach 6829 RIU^{-1} and 5×10^{-6} RIU (amplitude modulation), and 28000 nm/RIU and 3.57×10^{-6} RIU (wavelength modulation). It is worth mentioning that the sensor has a FOM of up to 2800 RIU^{-1}. This is the highest value achieved by a PCF sensor in simulation this year.

The presence of the micro air holes makes it more difficult to fill the PCF with the object to be measured. In addition to the complexity of the fabrication process itself, the complexity of the bio-adaptor modification needs to be considered in order to use PCF-SPR sensors for real bioassays. To solve this problem, the researchers proposed changing the external structure of the fiber, such as the D-type PCF, to reduce the difficulty of detecting the sample to be tested while taking advantage of the strong evanescent field created by the structure. Paul et al. [21] proposed a side opening in the PCF with a hexagonal air hole arrangement to facilitate the filling of the substance to be tested. The simulation results show that the sensitivity and resolution of the sensor in the high RI solution range (1.42–1.43) are 11700 nm/RIU and 8.55×10^{-6} RIU. Zhang et al. [31] constructed a D-shaped fiber based on a hexagonal air hole arrangement, which was forming symmetrical micro-openings on the end face. The presence of the micro-openings allows the sensor to be released from the substance to be measured over a larger area, improving the sensor resolution. Experimental results show that the maximum sensitivity and resolution of the sensor is 11750 nm/RIU and 8.51×10^{-6} RIU (RI range of 1.31–1.37). Haque et al. [32] optimized the polishing position, polishing depth, and other values of the D-shaped fiber. The simulation results show that the sensitivity of the sensor is 20000 nm/RIU (wavelength modulation) and 1054 RIU^{-1} (amplitude modulation). The resolution of the sensors is 16.7×10^{-6} RIU and 5×10^{-6} RIU, respectively.

Changing the internal air hole arrangement and optimizing the parameters of the air holes is an effective way to improve the sensitivity and resolution of the PCF-SPR sensor. The circular air-hole arrangement makes processing possible because of its high symmetry, but also because of its symmetry leads to the inevitable poor performance of the sensor. The rectangular arrangement and hexagonal arrangement, on the other hand, have been studied extensively because of the small mode field loss that can be achieved, and the structure, although more complex than circular, has the possibility of fabrication with the upgrading of the machining process. The sensitivity and FOM of the rectangular arrangement have been greatly improved compared to the circular arrangement, while the hexagonal arrangement enhances the energy coupling between the plasma mode and the fundamental mode, due to the denser air holes, which further improves the sensitivity and other parameters of the sensor. Compared with optical fiber sensors with complete cores, D-type PCFs have attracted attention because they enable simple replacement of biological samples, but problems, such as higher fabrication process requirements, have yet to be solved. The PCF-SPR sensors are summarized in Table 7.1.

Table 7.1 Summary of PCF

Air hole arrangement	Schematic	Maximum wavelength sensitivity (nm/RIU)	Maximum amplitude sensitivity (RIU^{-1})	Amplitude resolution (RIU)	Ref.
Circle	Fused silica, Gold, Liquid, PML, Air	2200	266	3.75×10^{-5}	[26]
	Gold, Analyte, Air hole, Fused Silica, PML	9000	318	1.11×10^{-5}	[27]
		20000	1380	5.26×10^{-6}	[28]
Rectangle		62000	1293	1.6×10^{-6}	[33]
	PML, Analyte, Silica, ITO, Air	35000	1120.73	2.86×10^{-6}	[24]
Hexagonal	PML Layer, Liquid, Gold Layer, Air, Fused Silica	18000	2843	3.5×10^{-6}	[30]
	Air Hole, Fused Silica, Gold Layer, Analyte, PML Layer	22000	N/A	4.55×10^{-6}	[23]

Air hole arrangement	Schematic	Maximum wavelength sensitivity (nm/RIU)	Maximum amplitude sensitivity (RIU^{-1})	Amplitude resolution (RIU)	Ref.
Hollow core PCF		11700	N/A	8.55×10^{-6}	[24]
Double micro-opening D-type PCF		11750	N/A	8.51×10^{-6}	[31]

N/A: there are no relevant data in reference.

In summary, changes in the substrate structure of fiber optic SPR sensors can be broadly classified into three categories. The SPR phenomenon is achieved by pulling taper, bending, and side coating of a single fiber structure to increase the mode field area to achieve biological detection. By designing an MMF–hetero-core fiber-MMF structure to achieve temperature compensation, high accuracy detection is finally achieved. High sensitivity detection is achieved by designing different air hole arrangements of PCFs.

7.3 Sensor Performance Enhancement Using Micromaterials and Nanomaterials

Since the sensitivity of SPR biosensors is proportional to the overlapping integral of the electromagnetic field of the mode with the environmental medium. Therefore, sensing area processing is very important to improve the performance of sensors (especially biosensors). Due to the development of material science and the advancement of fabrication processes, a variety of novel materials have been used to modify the surface layer to achieve specific adsorption of biological samples, enhance the sensitivity of the sensor, and reduce the detection limit to achieve high accuracy measurements of the sensor. Among those that have been applied to sensors in recent years are high-RI dielectric materials, nanomaterials, polymeric materials, and specific bioreceptors. In this section, the research progress of surface-modified materials applied to SPR sensors in the past five years is summarized.

7.3.1 High RI Dielectric Film Modification

Since materials that excite the SPR phenomenon such as gold and silver have low bio-affinity, it is the development trend to assemble high-RI dielectric membrane materials on their surfaces to enhance bio-affinity while improving sensitivity for high-precision detection of biological samples. Based on the research in the past five years, the effects of materials such as graphene and composite membranes composed of multilayer materials on sensors are mainly summarized. The applications of graphene fiber-coupled SPR sensors (improved sensitivity, interference resistance, etc.) are presented; the effects of metal oxide materials (indium tin oxide (ITO), indium molybdenum oxide (IMO) and ZnO) on the resolution, sensitivity, FOM, and LOD of the sensors are reviewed. Finally, a multilayer membrane composite structure is presented to improve the sensor performance under the coupling effect of multiple materials.

(a) Graphene and its derivatives

So far, among the 2D materials under investigation, graphene and its derivatives are widely used in SPR sensors on optical fiber substrates due to their superior properties such as bio-affinity, large surface area, high electron mobility, and rich conjugated structure. Qi Wang et al. [34] proposed a graphene oxide/silver-plated polymer-clad fiberoptic SPR (Fig. 7.3C) sensor with high sensitivity (3311 nm/RIU) and low detection limit (0.04 μg/mL) for the detection of IgG proteins. This sensor has a detection sensitivity of 0.4985 nm/(μg/mL) after surface functionalization modification. Graphene oxide both improves the sensor sensitivity and increases the stability of the silver film. They [35] fused PCF to two MMF segments to form a hetero-core fiber and modified graphene oxide and biofunctionalized the sensing area (Fig. 7.3A) to achieve IgG detection. The sensor takes advantage of the strong bio-affinity and large surface area of graphene oxide in order to bind more biomolecules and enhance the interaction between the gold film and the external medium, ultimately achieving highly sensitive detection of biological samples. The sensitivity of the sensor reached 4649.8 nm/RIU and the LOD for IgG was 10 ng/mL, which further reduced the detection limit. Among the grating structures, Qi Wang et al. [36] coated the same things on the outside of a TFBG (Fig. 7.3B) for the detection of IgG. The experimental results showed that the sensitivity and LOD of the biosensor were 0.096 dB/(g/mL) and 0.5 dB/(g/mL), respectively.

(b) Metal oxide

Since the concentration and mobility of free electrons can be adjusted by changing the oxygen content, the metal oxide can be coated on the surface of the metal, which can excite the SPR phenomenon to expand the detection range of the sensor; since the metal oxide is more stable than some precious metals, it can be deposited on the surface of the active metal, making it possible for the active metal to form the SPR sensor.

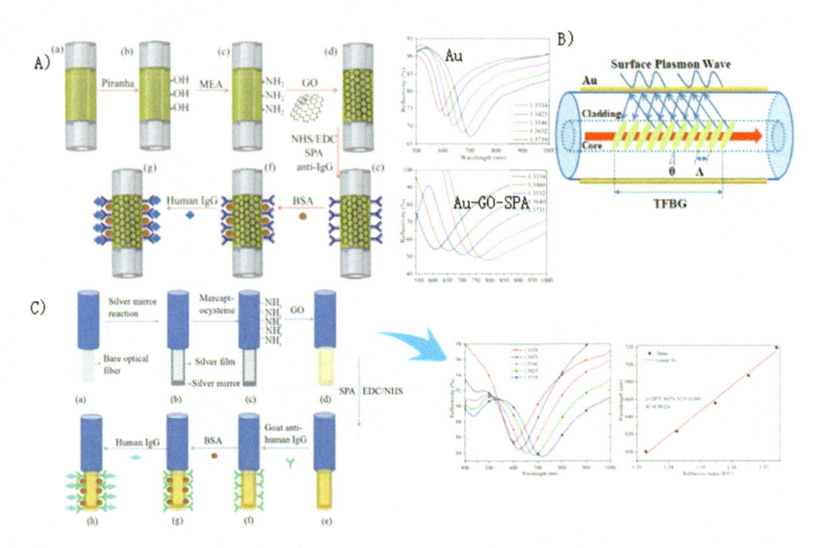

Figure 7.3 (A) Reflectance spectra of SPR biosensors with surface modification using graphene oxide and SPA and conventional SPR and modified sensors (Reprinted with permission from Ref. [35]). (B) TFBG-SPR biosensor (Reprinted with permission from Ref. [36]). (C) Graphene oxide/silver-plated polymer-clad fiber optic biosensor and reflection spectra at different RI solutions (Reprinted with permission from Ref. [34]).

Due to the high electron mobility and stable chemistry of IMO, IMO enables both multi-channel measurements by exciting the SPR phenomenon on the dielectric surface and covering the precious metal surface as a protective layer. First, the possibility of IMO as a plasma resonance film was theoretically verified by Aray et al. [37]. Simulation results show that IMO can be used as a coating layer for the sensing area and that sensors can achieve detection over a wide range of wavelengths. Wang Qi et al. [38] covered the ITO layer on the surface of the copper film (Fig. 7.4B), with a 61.6% improvement in FOM compared to the sensors without ITO. When testing bovine serum protein solutions, the sensors' sensitivity could be reached 1.907 nm/(mg/mL) and LOD of 5.70×10^{-7} mg/mL. The sensor sensitivity can reach sub-micro levels. Since dual-channel can enhance the anti-interference characteristics of sensors, the use of metal oxides as reference channels is the realization of dual-channel measurement is a hot development in recent years. Wei Peng et al. [39] proposed to sputter ITO and gold films on the fiber surface to form a dual resonant optical fiber SPR sensor (Fig. 7.4A). The presence of a dual resonance spectrum makes the sensor more resistant to interference. The sensor has shown promise in areas such as the real-time monitoring of biological samples.

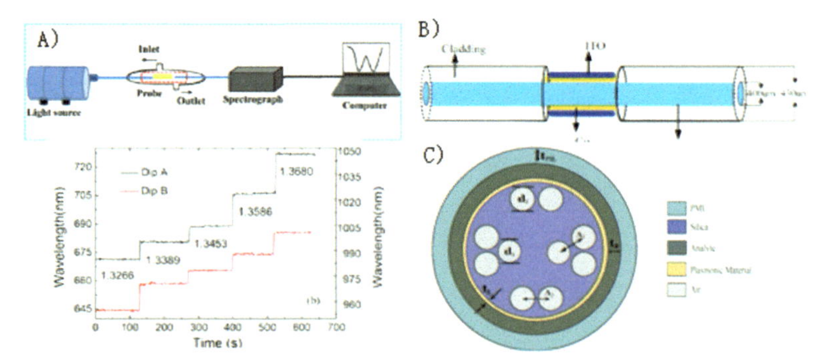

Figure 7.4 (A) SPR sensor with dual resonant (Reprinted with permission from Ref. [38]). (B) Cu/ITO coated uncladded fiberoptic SPR biosensor (Reprinted with permission from Ref. [39]). (C) AZO-modified PCF-SPR biosensor (Reprinted with permission from Ref. [40]).

The presence of non-centro-symmetry due to charge transfer at the ZnO-metal interface, the zinc-metal interface also exerts phonon interactions and flattening of the energy band structure leading to its enhanced light aggregation capability. Therefore, the coating of ZnO in the optical fiber structure to enhance the detection accuracy of the sensor is a promising research direction. Mohammad Rakibul Islam et al. [40] coated a PCF with gold or AZO (aluminum-doped zinc oxide) to excite SPR (Fig. 7.4C), the FOM is improved by 5.36 times by data simulation comparison.

Among the precious metals, although gold films are more stable than silver films, silver has a higher sensitivity, so to achieve a silver-covered SPR biosensor, covering its surface with a chemically stable film is a proven method. Du Bobo et al. [41] numerically simulated the three-layer structure of coated alumina-silver-alumina in a D-type fiber optic sensor, and the results showed that the RI sensitivity could reach 6558 nm/RIU and a resolution of 1.5×10^{-6} RIU. After performing the sensor fabrication, it was found that the sensitivity of the sensor in bovine serum protein detection was 0.4357 nm/μM and the LOD was 23 nM.

(c) Other materials

In recent years, new materials have emerged, such as hyperbolic metamaterials, transition metal disulfides, carbon nanotubes, barium titanate [42], etc., whose large surface area and layered structure, identical to that of graphene, have attracted the attention of researchers. Wang et al. [43] took advantage of the large surface area and high biomolecular compatibility of carbon nanotubes, so that more biomolecules could be immobilized. It covered the PCF-SPR surface to reduce the LOD of bovine serum proteins and accelerate the response rate. Usually, the material is covered on the surface of the sensing area, but Wang Qi et al. [44] added the material between the gold film and the optical fiber (Fig. 7.5A), using MoS_2, a band-gap semiconductor, to absorb light energy to facilitate electron transfer, enhancing the electric field strength and improving the sensitivity of the sensor.

(d) Multi-material composite film

Since high RI dielectric materials are richer in free electrons, which increase the electric field strength, and thus lead to more obvious changes in the resonance spectrum, it has become a hot topic of research whether multi-layer material composite films can make the charge density rise further and make the sensitivity of the sensor improve further. Anuj K. Sharma et al. [45] tested the attachment of a variety of 2D materials to a silver-based SPR optical fiber sensor (Fig. 7.5C) for the detection of salivary cortisol. The experimental results show that the bio-affinity of $Ti_3C_2O_2$ (MXenes material) and its high stability are more favorable for the detection of biological samples, with LODs up to 15.7 fg/mL.

Table 7.2 Summary of high RI dielectric film materials

Type of specific recognition	Trim Materials	Type of coupling	Sensitivity (nm/RIU)	LOD	Ref.
Graphene and its derivatives	Graphene oxide	Fiber optic	3311	0.04 μg/mL	[34]
	Graphene oxide	Grating	0.096 dB/(μg/mL)	0.5 mg/mL	[36]
	Graphene oxide	Grating	1782	N/A	[46]
	Graphene oxide (Dual Channel)	Fiber optic(first)	1942	N/A	[47]
		Fiber optic(second)	2315.4	N/A	
	Graphene oxide	Fiber optic	4649.8	10 ng/mL	[35]
Metal Oxide	AZO	Fiber optic	1700	N/A	[40]
	ITO	Fiber optic	4583.4	5.70×10^{-7} mg/mL	[38]
	Al_2O_3	Fiber optic	6558	23 nM	[41]
Other materials	CNTs	Fiber optic	8.18 nm/(mg/mL)	2.5 μg/mL	[43]
	MoS_2	Fiber optic	6184.4	19.7 ng/mL	[44]
Multi-material composite film	Polymer clad-silver (Ag) - $Ti_3C_2O_2$	Fiber optic	N/A	15.7 fg/mL	[45]

Increased sensitivity requires greater electric field enhancement at the interface, and high RI dielectric films are mostly of wide interest due to their high number of free electrons. Coating the fiber with different materials can increase the strength of the electric field coupling while achieving a larger surface area, increased sensitivity, lower LOD, and more stable sensor surfaces. Coating multi-material composite film further increases the electric field enhancement effect of the material and then allows the sensitivity of the sensor to be further improved. Table 7.2 summarizes the current state of development of 2D materials and metal oxide films for sensors as well as multilayer structures.

7.3.2 Metallic Micromaterials and Nanomaterials Modifications

Applying the LSPR phenomenon generated by nanomaterials to a basic SPR fiber sensor to enhance the sensitivity of the sensor through electric field enhancement is a widely used surface modification method today. In this section, precious metal nanoparticles and metal oxide nanoparticles are introduced as examples of increasing dimensionality.

(a) Precious metal nanoparticles

Precious metals (gold, platinum) are currently more widely studied because they are easy to process and purchase. Wang et al. [48] biofunctionalized the sensing zone of a PCF-SPR biosensor stacked with gold nanospheres for the detection of human IgG. Experimental results showed that the sensitivity of the sensor (3915 nm/RIU) was improved by 1.6 times compared to conventional PCF-SPR sensors, and the LOD was 37 ng/mL for the detection of IgG. Vikas et al. [49] sputtered gold films on MMFs and reassembled gold nanoparticles to study the coupling phenomenon between LSPR and SPR. The results show an increase in sensitivity of 291% for biological samples. Various forms of nanomaterials (nanoclusters, nanostars, nano sea urchins) are used in SPR biosensors because of the more significant electric field enhancement effect at the tips of the nanoparticles. Wang et al. [50] proposed an optical fiber SPR sensor composed of multiple layers of gold nanorods. Its RI sensitivity can reach 25642 nm/RIU, which is 10 times higher than that of ordinary SPR sensors and 1.6 times higher than that of SPR sensors covered by gold nanospheres. In addition, further enhancement of the local electric field by a combination of both nanomaterials and substrate structure changes is one way to improve the detection accuracy of the sensors. Wang et al. [51] proposed to link gold nanoparticles and gold film through chemical bonding and cover them on D-type optical fiber (Fig. 7.5B), the chemical bonding makes a gap between the gold nanoparticles and the sensor, the simulation results show that the electric field strength (at this time) is 4–5 times of the traditional gold film SPR, and the RI sensitivity is 3074 nm/RIU. Lidiya et al. [52] proposed treating gold nanoparticles firstly, filling the surface of the particles with titanium dioxide, and then assembling them on a D-type PCF-SPR sensor for glucose detection, which was achieved by optimizing the parameters of the nanoparticles and the thickness of the titanium dioxide, etc. to detect blood glucose in biological samples.

(b) Oxide nanoparticles

In addition to precious metal nanoparticles, oxide nanoparticles can also stimulate the LSPR effect and have a higher bio-detection accuracy compared to, for example, gold and silver. Ta_2O_5 nanoparticles have attracted the attention of researchers due to their high RI. Kant et al. [53] proposed the detection of xanthines by XO-coated Ta_2O_5 nanoparticles. After experimenting with various forms of oxide nanoparticles, it was found that Ta_2O_5 nanofibers exhibited excellent sensing properties with a minimum detection limit of 0.0127 μM. Kant et al. [54] proposed a multilayer structure of the silver film and tantalum oxide for SPR sensors. The presence of the oxide both makes the silver layer more stable and enhances the local electric field using its LSPR effect. After optimizing the structure of the sensor for acetylcholine detection, a sensitivity of 8.709 nm/μM nm and a LOD of 38 nM were achieved.

(c) Other forms of nanomaterials

The fabrication of nanopore arrays on-chip SPR biosensors is a way to instrument and smarten up the sensors. In 2015, Monteiro et al. [55] proposed a nanopore array SPR biosensor combined with a microfluidic system and showed that the sensor has high sensitivity with a minimum detection limit of 30 ng/mL for human epidermal receptor protein-2 (HER2), which is the first attempt to use an optical device to detect calmodulin.

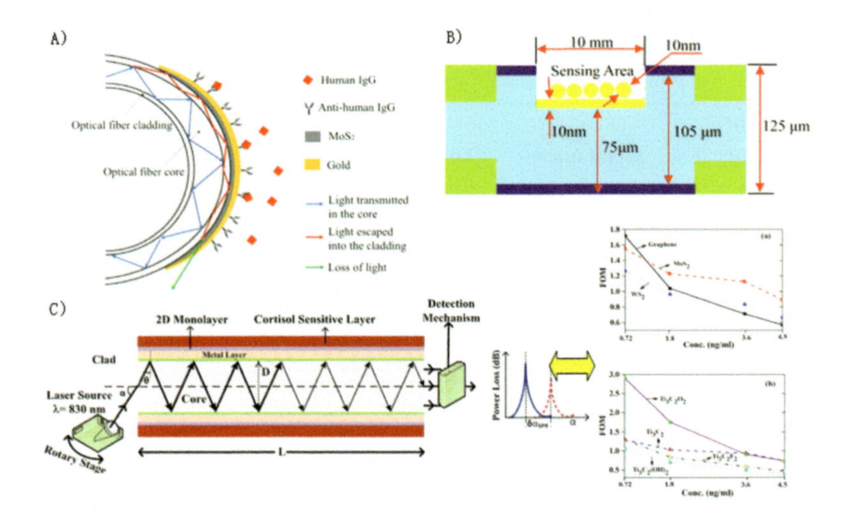

Figure 7.5 (A) Addition of MoS_2 between fiber and gold film for IgG detection (Reprinted with permission from Ref. [44]). (B) Longitudinal section of AuNPs SPR sensor assembled on D-type optical fiber (Reprinted with permission from Ref. [51]). (C) SPR biosensor consists of a sulfur glass core, a polymer cladding, a silver layer, and a 2D material. FOM values for different 2D and MXene materials are shown on the right (Reprinted with permission from Ref. [45]).

In the last five years, the introduction of the SPP-LSPP electric field coupling effect using the LSPR phenomenon has been widely investigated to enhance the sensitivity of sensors by surface electric field enhancement. The introduction of multiple shapes of noble metal nanoparticles and multi-layer stacking of nanoparticles has enhanced the sensitivity of the sensors; the introduction of oxide nanoparticles has amplified the signal intensity by exploiting their magnetization effect and bio-affinity, thus lowering the LOD. The introduction of nanowire and nanopore arrays reduces the FWHM spreading caused by material loss, thus allowing the sensor to be multi-channel and integrated while having high accuracy. The current status of research on SPR fiber optic biosensors based on micromaterial and nanomaterial modifications is shown in Table 7.3.

Table 7.3 Summary of metal microstructured and nanostructured sensor performance

Type of specific recognition	Metal micro-nanoparticles	Adhesive film	Sensitivity (nm/RIU)	LOD	Ref.
Noble metal nanoparticles	AuNP	Gold film	3915	37 ng/mL	[48]
	Gold nanorod	Gold film	25642	4.6 ng/mL	[50]
D-type optical fiber-AuNPs	AuNP	Gold film	3074	N/A	[51]
	AuNP	TiO_2 layer	0.83 nM/g/L	N/A	[52]
Metal oxide nanoparticles	Ta_2O_5 nanofibers	Silver film	26.2 nm/μM	0.0127 μM	[53]
	Ta_2O_5 nanoflakes	Silver film	8.709 nm/μM	38 nM	[54]
Nanopore arrays	Nanohole arrays	Gold film	304.5	30 ng/mL	[55]

7.3.3 Surface Anti-fouling Material Finishing

For specific biomass detection, the collected bio-samples need to be pre-processed, but the pre-processing requires complex operations and long response times by specialized personnel. Therefore, in order to achieve bedside detection and immediate diagnosis, the sensor needs to achieve direct detection of the original bio-sample, so specific adsorption as well as improving the interference resistance of the sensor are issues that need to be addressed by the SPR biosensor [56].

7.3.3.1 Polymer films modification

Since biomolecules such as proteins have randomly distributed positively or negatively charged residues on their surfaces, it is important that the sensor has a corresponding electrical charge in order to immobilize proteins on the surface of the optical fiber. Polymer materials with polyhydrophobic and polyamphiphilic groups are therefore modified on the sensor surface to immobilize the biomolecules. Polymers with a surface hydration layer enable a physical and energetic barrier to be formed in the sensing zone, preventing the adsorption of non-specific proteins and improving the resistance of the sensor to interference [57].

Firstly, hydrophilic materials, such as polyethylene glycol, are introduced, which are widely used in biosensors due to their high affinity for specifically adsorbed biosamples, low toxicity, strong repulsion for non-specifically adsorbed biosamples, and low price. Jeroen Pollet et al. [58] combined DNA aptamer with polyethylene glycol on a gold membrane to detect DNA hybridization and DNA-protein interactions, and the proposed sensor not only achieved quantitative DNA testing but also revealed the binding kinetics of DNA and protein that occurred on the sensor surface.

The super-hydrophilic nature of the amphiphilic ions enables an increased resistance to interference and, thanks to its charge balance, among other things, makes it possible to build antifouling surfaces by means of layer-by-layer self-assembly at the nanoscale. Yu-Sin Wang et al. [59] proposed to modify two kinds of thiols on the incoming surface to form nanoscale self-assembled monolayers. The experimental results showed that the sulfobetaine-thiols (SB-thiols) film enhanced the stacking density of biological samples and the hydrophilicity of the sensor surface; while the zwitterionic carboxybetaine-thiols (CB-thiols) enhanced the bio-affinity and facilitated the modification of biological samples in the sensing area. The nanofilms were prepared by doping and other means to reduce the detection limit, making it possible for the direct detection of biological samples.

Although the polymer film enables enhanced specific adsorption of biological samples, the attenuation characteristics of the evanescent wave and the losses associated with the polymer film cause the FWHM of the sensor to widen as the thickness of the polymer film increases, reducing the monitoring accuracy of the sensor. There is therefore a trade-off between improving the immunity of the sensor and parameters such as the thickness of the polymer film. Roberta D'Agata et al. [60] prepared a new surface coating with a thickness of less than 2 nm using a triglyceride (PEG (3))-pentyl trimer carboxybetaine system. The thickness of the polymer film is much less than the penetration depth of SP, enabling high-precision detection of biological samples.

7.3.3.2 Biomaterial modification

For the detection of biological samples, specific biological materials, such as antigens, antibodies, and DNA aptamers, need to be modified in the sensing region. Therefore, researchers have focused on finding functionalized aptamers to enhance the adsorption capacity of sensors for specific biological samples in recent years. However, as biological aptamers are specific, the main focus of current research is on finding aptamers that form low-fouling surfaces, allowing the detection limit of the sensor to be reduced. Husun Qian et al. [61] proposed a sandwich structure, which was consisting of aptamer-functionalized AuNPs, platelet-derived growth factor (PDGF-BB), and growth factor aptamer to detect PDGF-BB. When applied to diluted human serum samples, the sensor exhibits good adsorption properties for specific biomarkers.

To achieve high resolution of complex biological samples, it is necessary that the sensing zone is able to achieve specific adsorption even at low concentrations, so the sensor should be highly resistant to fouling, which is made possible by polymer films and biomaterials. Coating modified polymer films on top of the basic SPR biosensor to achieve increased adsorption of specific proteins while blocking fouling and facilitating LOD reduction; coating with different bio-adapters to achieve high affinity for specific biological samples while rejecting non-specific biological samples.

7.4 Summary, Perspectives, and Outlook

The detection of specific substances in the natural environment or biological samples places a high demand on the resolving power of sensors. Microstructured and nanostructured fiber plasmonic sensors have a promising future in environmental monitoring and biomedical applications due to their high sensitivity, no pre-processing, and real-time monitoring of markers in line with the trend toward integration and intelligence. However, SPR sensors have a high detection limit for markers and cannot be used for direct detection of the originally collected samples. To address these two issues, this chapter reviews the research progress in the past five years, mainly from the aspects of optical fiber structure change and sensing zone modification.

Changes to the substrate structure are inevitable in order to increase the accuracy of the sensor and achieve integration. The treatment of single-type fibers, the incorporation of hetero-core fibers into fibers, and the rearrangement of air holes within the PCF are the main ways to achieve high-performance sensors. The modification of the sensing zone is inevitable in order to build sensors with high sensitivity and strong specificity for adsorption. The use of high RI dielectric materials, nanoparticles, polymeric materials, and biomaterials for electric field enhancement and forming antifouling surfaces are the main ways to achieve low detection sensors.

Reviewing the development of SPR biosensors in recent years, researchers have proposed a number of approaches to achieve high performance in fiber optic biosensors, which fall into two main areas: changes in the substrate structure and modification of the sensing region. These methods have certain side effects while achieving certain performance improvements in the sensor. For example, the treatment of optical fibers reduces the mechanical strength of the sensor and makes the process more complex. The use of optical fiber and lab-on-chip to achieve portable high-precision sensor production is in line with the direction of the development trend. In terms of sensing zone modification, the coating material makes the radiation loss greater and although the sensitivity is increased, the FWHM of the sensor is significantly wider, which is contrary to the extremely narrow FWHM that is sought today. The use of structures to realize optical effects (Fano resonance) and combining them with materials is a viable research direction to solve the problem of FWHM.

In addition, considering today's demand for biosamples detection, exploring new signal processing methods, and optical sensing mechanism to enhance the detection performance of optical sensors and improve the detection accuracy of biosamples is an exciting attempt in the field of biosensing. The new signal processing allows for more subtle observation of changes in the resonance spectrum, enabling a wider range and more dimensions of detection. For example, the use of surface plasmon resonance imaging combines SPR with image processing to enable real-time detection of the binding of biological samples. A new optical mechanism can enable sensors to be sensitive to higher orders of magnitude, such as lossy mode resonance, and if the width of their resonance spectrum is reduced, the detection accuracy of optical devices as biosensors rises to a new level.

In conclusion, optical fiber SPR biosensors excel in sensor integration and in achieving high-precision detection of biological samples. With the maturity of the fabrication process and the progress of artificial intelligence, optical plasma sensors technology will be widely used in telemedicine, environmental monitoring, and intelligent diagnosis and treatment [62].

References

1. Yang, M., Zhu, Y., An, R. (2021) Underwater fiber-optic salinity and pressure sensor based on surface plasmon resonance and multimode interference. *Appl. Opt.*, 60 (30), 9352–9357.
2. Wei, Y., Li, L., Liu, C., Hu, J., Su, Y., Wu, P., Zhao, X. (2021) Cascaded dual-channel fiber SPR temperature sensor based on liquid and solid encapsulations. *Chin. Phys. B*, 30 (10).
3. Li, X., Jiang, H., Ma, C., Zhu, Z., Song, X., Wang, H., Huo, P., Li, X. (2021) Local surface plasma resonance effect enhanced Z-scheme ZnO/Au/g-C_3N_4 film photocatalyst for reduction of CO_2 to CO. *Appl. Catal. B*, 283.
4. Mai, Z., Zhang, J., Chen, Y., Wang, J., Hong, X., Su, Q., Li, X. (2019) A disposable fiber optic SPR probe for immunoassay. *Biosens. Bioelectron.*, 144, 111621.
5. Liu, Y., Li, P., Zhang, N., Chen, S., Liu, Z., Guang, J. (2019) A compact biosensor for binding kinetics analysis of protein–protein interaction. *IEEE Sens.*, 19 (24), 11955–11960.
6. Maruthupandi, M., Vasimalai, N. (2021) Nanomolar detection of L-cysteine and Cu2+ ions based on Trehalose capped silver nanoparticles. *Microchem. J.*, 161.
7. Wang, Q., Guo, M., Zhao, Y. (2017) A sensitivity enhanced microdisplacement sensing method improved using slow light in fiber Bragg grating. *IEEE Trans. Instrum. Meas.*, 66 (1), 122–130.
8. Wang, Q., Song, H., Zhu, A., Qiu, F. (2021) A label-free and anti-interference dual-channel SPR fiber optic sensor with self-compensation for biomarker detection. *IEEE Trans. Instrum. Meas.*, 70, 1–7.
9. Wang, Q., Wang, B.-T., Kong, L.-X., Zhao, Y. (2017) Comparative analyses of bi-tapered fiber mach–zehnder interferometer for refractive index sensing. *IEEE Trans. Instrum. Meas.*, 66 (9), 2483–2489.

10. Li, X., Warren-Smith, S.C., Xie, L., Ebendorff-Heidepriem, H., Nguyen, L.V. (2020) Temperature-compensated refractive index measurement using a dual fabry–perot interferometer based on C-fiber cavity. *IEEE Sens.,* 20 (12), 6408–6413.

11. Kim, H.J., Sohn, Y.S., Kim, C.D., Jang, D.H. (2016) Surface plasmon resonance sensing of a biomarker of Alzheimer disease in an intensity measurement mode with a bimetallic chip. *J. Korean Phys. Soc.*, 69 (5), 793–797.

12. Urrutia, A., Bojan, K., Marques, L., Mullaney, K., Goicoechea, J., James, S., Clark, M., Tatam, R., Korposh, S. (2016) Novel highly sensitive protein sensors based on tapered optical fibres modified with Au-based nanocoatings. *J. Sens.,* 2016, 1–11.

13. Ahmad, H., Amiri, I.S., Soltanian, M.R.K., Narimani, L., Zakaria, R., Ismail, M.F., Thambiratnam, K. (2017) High sensitivity surface plasmon resonance (SPR) refractive index sensor in 1.5 μm. *Mater. Express.,* 7 (2), 145–150.

14. Wu, D., Zhao, Y., Wang, Q. (2015) SMF Taper evanescent field-based RI sensor combined with fiber loop ring down technology. *IEEE Photonics Technol. Lett.,* 27 (17), 1802–1805.

15. Chen, Y., Yu, Y., Li, X., Zhou, H., Hong, X., Geng, Y. (2016) Fiber-optic urine specific gravity sensor based on surface plasmon resonance. *Sens. Actuators, B.,* 226, 412–418.

16. Yang, Z., Xia, L., Li, S., Qi, R., Chen, X., Li, W. (2019) Highly sensitive refractive index detection based on compact HSC-SPR structure in a microfluidic chip. *Sens. Actuator A Phys.,* 297.

17. Wang, Y., Li, S., Guo, Y., Zhang, S., Li, H. (2021) Surface plasmon polariton high-sensitivity refractive index sensor based on MMF-MOF-MMF structure. *Infrared Phys. Technol.*, 114.

18. Zhou, X., Li, S., Li, X., Yan, X., Zhang, X., Wang, F., Cheng, T. (2020) High-sensitivity SPR temperature sensor based on hollow-core fiber. *IEEE Trans. Instrum. Meas.,* 69 (10), 8494–8499.

19. Hu, H., Song, X., Han, Q., Chang, P., Zhang, J., Liu, K., Du, Y., Wang, H., Liu, T. (2020) High sensitivity fiber optic SPR refractive index sensor based on multimode-no-core-multimode structure. *IEEE Sens. J.,* 20 (6), 2967–2975.

20. Rifat, A.A., Ahmed, R., Yetisen, A.K., Butt, H., Sabouri, A., Mahdiraji, G.A., Yun, S.H., Adikan, F.R.M. (2017) Photonic crystal fiber based plasmonic sensors. *Sens. Actuator A Phys.,* 243, 311–325.

21. P Paul, A.K., Samiul Habib, M., Hai, N.H., Abdur Razzak, S.M. (2020) An air-core photonic crystal fiber based plasmonic sensor for high refractive index sensing. *Opt. Commun.*, 464.

22. Mahfuz, M.A., Hossain, M.A., Haque, E., Hai, N.H., Namihira, Y., Ahmed, F. (2020) Dual-core photonic crystal fiber-based plasmonic RI sensor in the visible to near-RI operating band. *IEEE Sens. J.*, 20 (14), 7692–7700.

23. Mahfuz, M.A., Hasan, M.R., Momota, M.R., Masud, A., Akter, S. (2019) Asymmetrical photonic crystal fiber based plasmonic sensor using the lower birefringence peak method. *OSA Continuum,* 2 (5).

24. Liu, Q., Sun, J., Sun, Y., Ren, Z., Liu, C., Lv, J., Wang, F., Wang, L., Liu, W., Sun, T., Chu, P.K. (2020) Surface plasmon resonance sensor based on photonic crystal fiber with indium tin oxide film. *Opt. Mater.,* 102.

25. Wieduwilt, T., Tuniz, A., Linzen, S., Goerke, S., Dellith, J., Hubner, U., Schmidt, M.A. (2015) Ultrathin niobium nanofilms on fiber optical tapers: A new route towards low-loss hybrid plasmonic modes. *Sci. Rep.*, 5, 17060.

26. Hasan, M.R., Akter, S., Rifat, A.A., Rana, S., Ali, S. (2017) A highly sensitive gold-coated photonic crystal fiber biosensor based on surface plasmon resonance. *Photonics,* 4 (1).

27. Khalek, M.A., Chakma, S., Paul, B.K., Ahmed, K. (2018) Dataset of surface plasmon resonance based on photonic crystal fiber for chemical sensing applications. *Data Brief,* 19, 76–81.

28. Jahan, N., Rahman, M.M., Ahsan, M., Based, M.A., Rana, M.M., Gurusamy, S., Haider, J. (2021) Photonic crystal fiber based biosensor for pseudomonas bacteria detection: A simulation study. *IEEE Access,* 9, 42206–42215.

29. Kim, S., Kee, C.S., Lee, C.G. (2009) Modified rectangular lattice photonic crystal fibers with high birefringence and negative dispersion. *Opt. Express,* 17 (10), 7952–7957.

30. Haider, F., Aoni, R.A., Ahmed, R., Miroshnichenko, A.E. (2018) Highly amplitude-sensitive photonic-crystal-fiber-based plasmonic sensor. *J. Opt. Soc. Am. B,* 35 (11).

31. Zhang, S., Li, J., Li, S., Liu, Q., Wu, J., Guo, Y. (2018) Surface plasmon resonance sensor based on D-shaped photonic crystal fiber with two micro-openings. *J. Phys. D: Appl. Phys.*, 51 (30).

32. Bhardwaj, S.K., Basu, T. (2018) Study on binding phenomenon of lipase enzyme with tributyrin on the surface of graphene oxide array using surface plasmon resonance. *Thin Solid Films*, 645, 10–18.

33. Islam, M.S., Sultana, J., Rifat, A.A., Ahmed, R., Dinovitser, A., Ng, B.W., Ebendorff-Heidepriem, H., Abbott, D. (2018) Dual-polarized highly sensitive plasmonic sensor in the visible to near-IR spectrum. *Opt. Express,* 26 (23), 30347–30361.

34. Wang, Q., Wang, B.-T. (2018) Surface plasmon resonance biosensor based on graphene oxide/silver coated polymer cladding silica fiber. *Sens. Actuators, B.,* 275, 332–338.

35. Wang, Q., Wang, B. (2018) Sensitivity enhanced SPR immunosensor based on graphene oxide and SPA co-modified photonic crystal fiber. *Opt. Laser Technol.,* 107, 210–215.

36. Wang, Q.J., J. Y., Wang, B.T. (2019) Highly sensitive SPR biosensor based on graphene oxide and staphylococcal protein A co-modified TFBG for human IgG detection. *IEEE Trans. Instrum. Meas.,* 68 (9), 3350–3357.

37. Aray, A., Ranjbar, M. (2019) SPR-based fiber optic sensor using IMO thin film: Towards wide wavelength tunability from visible to NIR. *IEEE Sens. J.,* 19 (7), 2540–2546.

38. Wang, Q., Sun, B., Hu, E., Wei, W. (2019) Cu/ITO-coated uncladded fiber-optic biosensor based on surface plasmon resonance. *IEEE Photonics Technol. Lett.,* 31 (14), 1159–1162.

39. Li, L., Liang, Y., Guang, J., Cui, W., Zhang, X., Masson, J.F., Peng, W. (2017) Dual Kretschmann and Otto configuration fiber surface plasmon resonance biosensor. *Opt. Express,* 25 (22), 26950–26957.

40. Islam, M.R., Iftekher, A.N.M., Hasan, K.R., Nayen, J., Islam, S.B., Khan, M.M.I., Chowdhury, J.A., Mehjabin, F., Islam, M., Islam, M.S. (2021) Design and analysis of a biochemical sensor based on surface plasmon resonance with ultra-high sensitivity. *Plasmonics,* 16 (3), 849–861.

41. Du, B., Yang, Y., Zhang, Y., Yang, D. (2018) SPR label-free biosensor with oxide-metal-oxide-coated d-typed optical fiber: A theoretical study. *Plasmonics,* 14 (2), 457–463.

42. Wang, Q., Niu, L.-Y., Jing, J.-Y., Zhao, W.-M. (2020) Barium titanate film based fiber optic surface plasmon sensor with high sensitivity. *Opt. Laser Technol.,* 124.

43. Jing, J.-Y., Wang, Q., Wang, B.-T. (2018) Refractive index sensing characteristics of carbon nanotube-deposited photonic crystal fiber SPR sensor. *Opt. Fiber Technol.,* 43, 137–144.

44. Song, H., Wang, Q., Zhao, W.-M. (2020) A novel SPR sensor sensitivity-enhancing method for immunoassay by inserting MoS_2 nanosheets between metal film and fiber. *Opt. Laser Technol.,* 132.

45. Sharma, A.K., Kaur, B., Marques, C. (2020) Simulation and analysis of 2D material/metal carbide based fiber optic SPR probe for ultrasensitive cortisol detection. *Optik,* 218.

46. Sadeghi, Z., Shirkani, H. (2020) Highly sensitive mid-infrared SPR biosensor for a wide range of biomolecules and biological cells based on graphene-gold grating. *Physica E,* 119.

47. Tong, K., Guo, J., Wang, M., Dang, P., Wang, F., Zhang, Y., Wang, M. (2017) An optical fiber surface plasmon resonance biosensor for wide range detection. *Eur. Phys. J.: Appl. Phys.,* 80 (1).

48. Wang, B.-T., Wang, Q. (2018) Sensitivity-enhanced optical fiber biosensor based on coupling effect between SPR and LSPR. *IEEE Sens. J.,*18 (20), 8303–8310.

49. Vikas, Gupta, S., Tejavath, K., Verma, R.K. (2020) Urea detection using bio-synthesized gold nanoparticles: An SPR/LSPR based sensing approach realized on optical fiber. *Opt Quant Electron,* 52 (6).

50. Xia, F., Song, H., Zhao, Y., Zhao, W.-M., Wang, Q., Wang, X.-Z., Wang, B.-T., Dai, Z.-X. (2020) Ultra-high sensitivity SPR fiber sensor based on multilayer nanoparticle and Au film coupling enhancement. *Measurement,* 164.

51. Niu, L.-Y., Wang, Q., Jing, J.-Y., Zhao, W.-M. (2019) Sensitivity enhanced D-type large-core fiber SPR sensor based on gold nanoparticle/Au film co-modification. *Opt. Commun.,* 150, 207–295.

52. Lidiya, A.E., Raja, R.V.J., Pham, V.D., Ngo, Q.M., Vigneswaran, D. (2019) Detecting hemoglobin content blood glucose using surface plasmon resonance in D-shaped photonic crystal fiber. *Opt. Fiber Technol.,* 50, 132–138.

53. Kant, R., Tabassum, R., Gupta, B.D. (2018) Xanthine oxidase functionalized Ta2O5 nanostructures as a novel scaffold for highly sensitive SPR based fiber optic xanthine sensor. *Biosens. Bioelectron.,* 99, 637–645.

54. Kant, R., Gupta, B.D. (2018) Fiber-optic SPR based acetylcholine biosensor using enzyme functionalized Ta_2O_5 nanoflakes for Alzheimer's disease diagnosis. *J. Lightwave Technol.* 36 (18), 4018–4024.

55. Monteiro, J.P., de Oliveira, J.H., Radovanovic, E., Brolo, A.G., Girotto, E.M. (2015) Microfluidic plasmonic biosensor for breast cancer antigen detection. *Plasmonics,* 11 (1), 45–51.

56. Moro, G., Chiavaioli, F., Liberi, S., Zubiate, P., Del Villar, I., Angelini, A., De Wael, K., Baldini, F., Moretto, L.M., Giannetti, A. (2021) Nanocoated

fiber label-free biosensing for perfluorooctanoic acid detection by lossy mode resonance. *Results Opt.*, 5.

57. Chen, S.F., Li, L.Y., Zhao, C., Zheng, J. (2010) Surface hydration: Principles and applications toward low-fouling/nonfouling biomaterials. *Polymer,* 51 (23), 5283–5293.
58. Pollet, J., Delport, F., Janssen, K.P., Jans, K., Maes, G., Pfeiffer, H., Wevers, M., Lammertyn, J. (2009) Fiber optic SPR biosensing of DNA hybridization and DNA-protein interactions. *Biosens. Bioelectron.,* 25 (4), 864–869.
59. Wang, Y.S., Yau, S., Chau, L.K., Mohamed, A., Huang, C.J. (2019) Functional biointerfaces based on mixed zwitterionic self-assembled monolayers for biosensing applications. *Langmuir,* 35 (5), 1652–1661.
60. D'Agata, R., Bellassai, N., Giuffrida, M.C., Aura, A.M., Petri, C., Kogler, P., Vecchio, G., Jonas, U., Spoto, G. (2021) A new ultralow fouling surface for the analysis of human plasma samples with surface plasmon resonance. *Talanta,* 221, 121483.
61. Qian, H., Huang, Y., Duan, X., Wei, X., Fan, Y., Gan, D., Yue, S., Cheng, W., Chen, T. (2019) Fiber optic surface plasmon resonance biosensor for detection of PDGF-BB in serum based on self-assembled aptamer and antifouling peptide monolayer. *Biosens. Bioelectron.*, 140, 111350.
62. Wang, Q., Ren, Z., Zhao, W.-M., Wang, L., Yan, X., Zhu, A., Qiu, F., Zhang, K.-K. (2021) Research advances on surface plasmon resonance biosensors. *Nanoscale,* 2022, 14 (3), 564.

Chapter 8

Surface-Enhanced Raman Scattering–Based Plasmonic Sensors

Sachin Kumar Srivastava[a] **and Ibrahim Abdulhalim**[b]
[a]*Department of Physics, Indian Institute of Technology Roorkee, Roorkee 247667, Uttarakhand, India*
[b]*Electrooptics and Photonics Engineering Unit, School of Electrical and Computer Engineering and Ilse Katz Institute of Nanoscale Science and Technology, Ben Gurion University of the Negev, Beer Sheva 84105, Israel*
sachin.srivastava@ph.iitr.ac.in

Starting from the basics of plasmonics and Raman scattering, the present chapter will introduce surface-enhanced Raman spectroscopy (SERS), where the role of electromagnetic enhancement due to plasmonic coupling will be discussed. Further, the role of plasmonic nanostructure optimization in maximizing the SERS signal will be presented through an example of plasmonic nanosculptured thin films (nSTFs). Further, various modalities of SERS sensing with examples will be provided. At last, an insight into the introduction and utility of machine learning (ML) techniques for

Plasmonics-Based Optical Sensors and Detectors
Edited by Banshi D. Gupta, Anuj K. Sharma, and Jin Li

ISBN 978-981-4968-85-0 (Hardcover), 978-1-003-43830-4 (eBook)
www.jennystanford.com

SERS analysis and sensing of species in complex analyte matrix will be presented. The discussions will be substantiated with examples from relevant literature.

8.1 Surface-Enhanced Raman Spectroscopy

8.1.1 Plasmons

Plasmons are the quanta of collective longitudinal oscillations of free electrons in (generally) bulk metals. Let us consider a noble metal as a sea of free electrons. When an external stimulation leads to local displacement in the position of a collection of electrons, the mutual electrostatic repulsion among the displaced electrons sets up a restoring force, which leads to collective longitudinal oscillations. This type of wave can be thought analogous to the waves emanating by dropping a pebble in calm water. The waves in calm water traveling away from the center of excitation are longitudinal in nature while the ones on the surface are complex in character. They possess both longitudinal and transverse characters. Such kinds of waves on surfaces can be termed surface plasmon waves and are partially transverse in nature, and can be excited using electromagnetic waves, such as light.

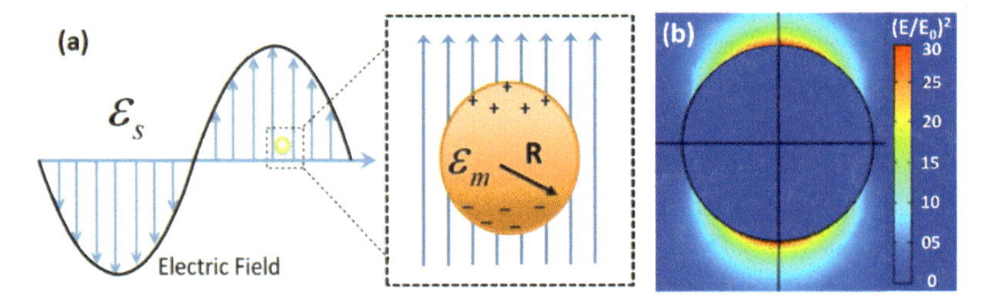

Figure 8.1 (a) Schematic of LSPR excitation. (b) Electromagnetic field enhancement in a 40 nm Ag sphere, simulated using COMSOL Multiphysics v5.6.

When the dimensions of noble metals are reduced to a few or tens of nanometers, even if there are plasmonic excitations, the structure cannot withstand a wave. Such a kind of plasmonic excitation is termed localized surface plasmon resonance (LSPR) (Fig. 8.1a) [1]. In such a case, the electromagnetic field gets confined in structures much smaller than the wavelength of incident light (Fig. 8.1b). When a molecule is brought closer to such a plasmonic nanostructure, it experiences a high electromagnetic field [2], which, in turn, leads to enhanced spectroscopic signals from the molecule.

8.1.2 Enhancement of Raman Signal by Plasmons

Molecules can be subdivided, in general, into two categories; the first possesses permanent dielectric moment, while the second does not possess any dipole moment. When the molecules of the first category are subject to any incident electromagnetic field, there is an interaction of dipoles of the molecule, which leads to a class of optical phenomena such as fluorescence, etc. In the case of molecules with no dielectric moment, when subjected to the electromagnetic field, such molecules get polarized, thereby creating induced dipoles. The induced dipole moment of such molecules depends on the polarizability of the molecule. Such a class of molecules possesses scattering processes, which are predominantly elastic in nature. Elastic scattering means that these molecules predominantly scatter the radiation of a frequency similar to that of the incident one. This is termed as Rayleigh scattering. A very small fraction of the incident photons on such molecules, say one in 10 million gets scattered inelastically. To say, the scattered radiation has a frequency either larger or smaller than that of the incident radiation. This phenomenon is called Raman scattering. The scattered radiation with a smaller frequency is called Stokes, while the one with a larger frequency is called anti-Stokes radiation. The small shift in the frequency of the incident radiation is a characteristic of the molecular bonds and the Raman bands of every molecule being unique, this is called molecular fingerprinting. As stated earlier, the probability of Raman scattering is very low, 1:10,000,000; hence the spectroscopic throughput of the phenomenon is very low. Various techniques for the enhancement of Raman signal from the molecules of interest have been invented. One of the most popular techniques is surface-enhanced Raman scattering/spectroscopy (SERS). SERS is the enhancement of Raman signals in the vicinity of plasmonic nanostructured surfaces [3–5]. The induced dipole moment, $\vec{p}$ of the molecule, is given as:

$$\vec{p} = \alpha \vec{E} \tag{8.1}$$

where α is the polarizability of the molecule and $\vec{E}$ is the applied electric field. Out of the two factors on which $\vec{p}$ depends, the contribution of α, known as chemical contribution to SERS [6], is of the order of 10^2. The contribution of SERS enhancement due to enhancement in the electric field $\vec{E}$ is termed electromagnetic contribution and is of the order of magnitude of 10^6–10^{10} [7, 8]. Such enhancement in Raman intensities enables to achieve single molecule limit of detection [9].

8.1.3 Estimation of Electromagnetic Enhancement of Stokes Lines

Since most of the molecules remain in the ground state prior to excitation, the probability of Stokes transitions is about 10^4 times higher than that for anti-Stokes. Let us try to estimate the extent of SERS enhancement due to plasmonic nanoparticles. For this purpose, let us assume a small molecule is kept at a distance d from the surface of a plasmonic nanosphere of radius R, as shown in Fig. 8.2.

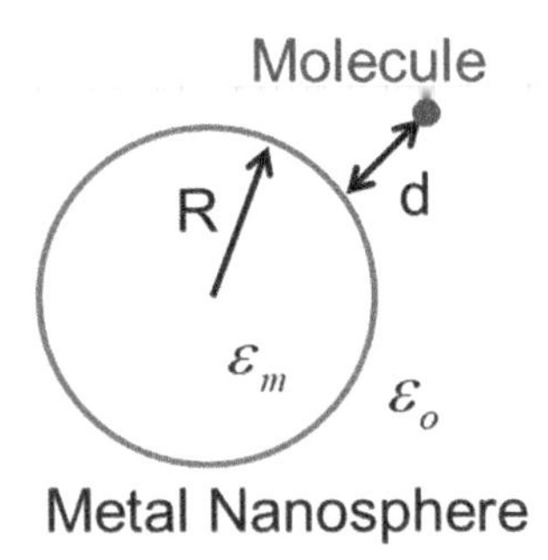

Figure 8.2 Schematic of a molecule placed at a small distance near a plasmonic nanosphere.

Let us denote the dielectric function of the nanoparticle as ε_m and that of the medium surrounding it as ε_0. Let E_0 be the magnitude of an incident in the electric field and E_{plasmon} that of the plasmons. If the molecule plus the nanosphere is illuminated with a laser of electric field E_0, the total electric field experienced by the molecules will be

$$E_{\text{Molecule}} = E_0 + E_{\text{Plasmon}}. \tag{8.2}$$

E_{Plasmon} due to the metallic nanosphere is given by

$$E_{\text{Plasmon}} = \frac{\varepsilon_m - \varepsilon_0}{\varepsilon_m + 2\varepsilon_0}\frac{R^3}{(R+d)^3}E_0. \tag{8.3}$$

The field enhancement factor, A can be defined as

$$A = \frac{\text{Field at the position of molecule}}{\text{Incident field}}. \tag{8.4}$$

Hence, if ν is the frequency of the incident laser,

$$A(\nu) = \frac{E_{mol}(\nu)}{E_0(\nu)} \sim \frac{\varepsilon_m(\nu) - \varepsilon_0}{\varepsilon_m(\nu) + 2\varepsilon_0}\frac{R^3}{(R+d)^3}. \tag{8.5}$$

Therefore, the total electromagnetic enhancement factor for Stokes SERS power,

$$G_{em}(\nu_S) = \left|A(\nu_L)\right|^2 \left|A(\nu_S)\right|^2$$

$$\sim \left[\frac{\varepsilon_m(\nu_L) - \varepsilon_0}{\varepsilon_m(\nu_L) + 2\varepsilon_0}\frac{\varepsilon_m(\nu_S) - \varepsilon_0}{\varepsilon_m(\nu_S) + 2\varepsilon_0}\right]^2 \left(\frac{R}{R+d}\right)^{12}, \tag{8.6}$$

where ν_S is the Stokes frequency. From Eq. (8.6), it can be concluded that the SERS enhancement in power is roughly of the order of E^4. However, this enhancement is inversely proportional to $\sim d^{12}$, which means that it is a very small-range phenomenon and is effective only when the molecule is very close to the plasmonic surface. That is why it is termed SERS. Further, if the size of the analyte is big, say ~100 nm, the plasmonic nanostructures may not be able to enhance the Raman signal from the whole molecule.

Bringing a molecule near a nanostructured plasmonic surface leads to about 10^6–10^8 enhancement in the SERS power. Since the Raman signal of each molecule is a unique signature and it gets highly enhanced in SERS, SERS can be used for highly sensitive and selective detection of various molecules of interest. A number of bio and chemical sensor studies based on SERS, ranging from clinical diagnostics to water sensing, etc. have been reported in the literature [10–15]. It can also be observed from Eq. (8.6) that for optimum enhancement, the plasmon resonance must coincide with the laser wavelength and the Stokes line [16]. The nanospheres of about 60–70 nm size, which can be prepared with good control on size distribution possess plasmon resonances around 500–600 nm wavelength, a

532 nm laser can be a good choice for SERS spectroscopy. However, the fluorescence signals possessed by certain moieties dominate over the Raman signal at such wavelengths. Therefore, it is preferred to work on near infra-red (NIR) wavelengths to avoid any fluorescence signals. The cost paid for that is low throughput power. It is also difficult to make nanospheres plasmonically resonant at NIR wavelengths. Moreover, the dipole approximation-based analysis is only approximate and does not accurately predict the actual SERS enhancement. In general, for 785 nm excitation, a nanorod-like structure is employed, which works as a nanoantenna and is one of the high-performance tools for SERS. This nanorod-shaped nanoantenna can be fabricated using lithographic techniques, but it is very costly to fabricate them. A relatively cost-effective technique for the fabrication of high-performance nanorod-like structures called nSTFs is the glancing angle deposition (GLAD) technique under physical vapor deposition (PVD) [17].

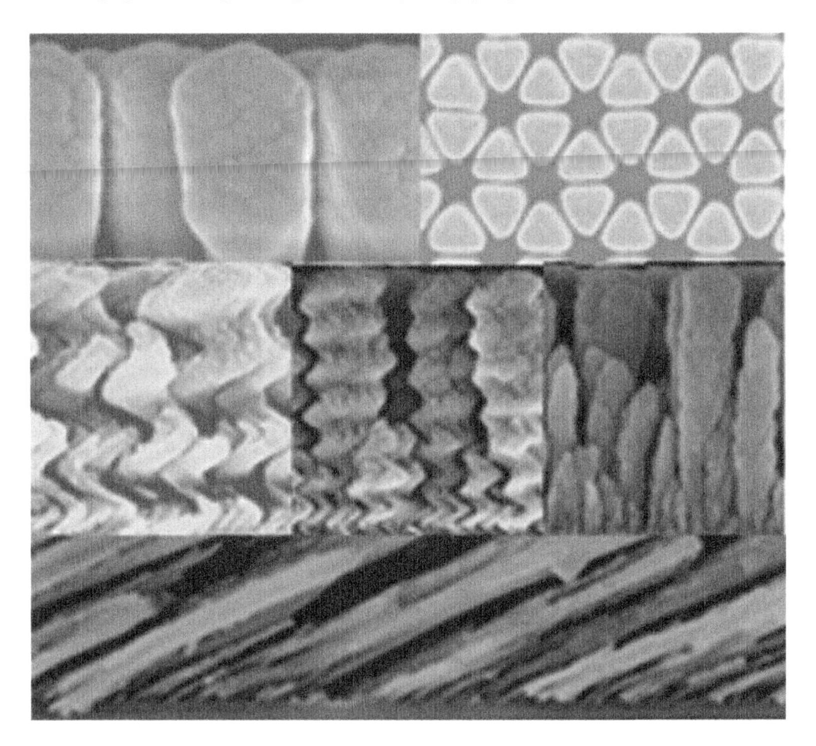

Figure 8.3 SEM image of various nSTFs (Adapted with permission from Ref. [18]).

8.1.4 Nanosculptured Thin Plasmonic Films and GLAD

Plasmonic nSTFs are basically porous films of plasmonic metals, which possess generally nanorod-like structures. A scanning electron microscope (SEM) image of various nSTFs has been shown in Fig. 8.3. These films are usually grown by the GLAD technique, as mentioned in the earlier paragraph. Pre-templating gives a rather defined topography of the grown nanosculptures, as evident from the SEM images. A schematic of the GLAD setup inside a vacuum chamber has been shown in Fig. 8.4a. The substrate to be coated with the nSTF is kept at glancing angles to the collimated metal plume directed toward it. It has provisions for the control of tilt, online rotation and heating of the substrate, an influx of desired gases during deposition, etc., which enable one to fabricate desired, rod-like, zigzag, and helical porous nanosculptured films. Figure 8.4b illustrates the basic mechanism of the formation of

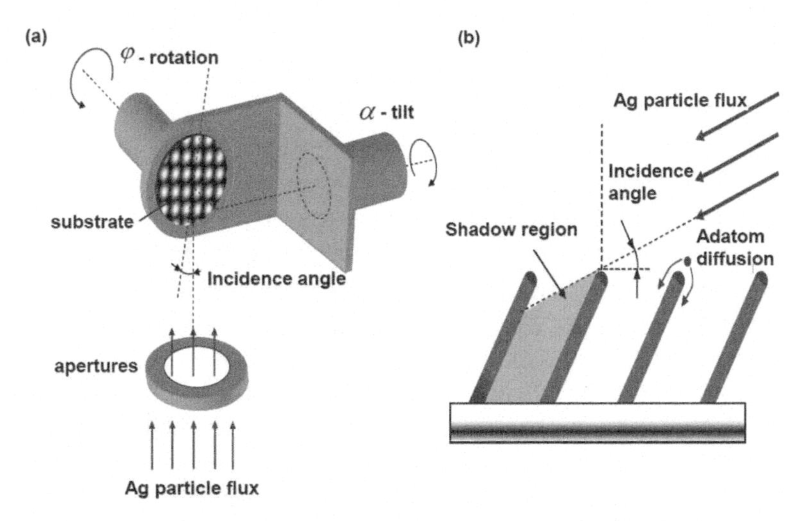

Figure 8.4 (a) Schematic of GLAD setup and (b) mechanism of nSTF formation (Adapted with permission from Ref. [19]).

columnar films. During the initial stage of deposition, certain nucleation sites of the incoming vapor flux are grown on the substrate. These nucleation sites form a shadowing region for the later incoming vapor flux, thereby making it a columnar, porous film [20]. Pre-templated substrates grow columnar structures in a regular fashion, as the templates play the same role as that of the nucleation sites, which were randomly formed. At later stages, the high-speed adatoms get deposited near the tip and diffuse toward the bottom to provide relatively smoothers and homogeneous films. The tilt of the nanorods is a function of the tilt angle. However, vertical nanorods can be achieved by adequate rotation of the substrate under suitable conditions [21]. By controlling the temperature of the substrate, deposition rate, and gas influx rate, one may control the porosity of deposited films.

8.1.4.1 Performance optimization

Optimization with respect to nSTF height

Though Raman scattering is a wavelength-independent phenomenon, SERS enhancement depends on the resonant excitation of the plasmonic nanostructures along with the Stokes frequencies. Hence, the choice of both the plasmonic films and the laser wavelengths becomes crucial. For a wavelength of 785 nm of the excitation laser, hence, it is important to optimize the nSTFs of silver such that they are resonant with the laser wavelength. For this purpose, the nSTFs with rod lengths ranging from 100 to 100 nm having fixed (optimum) porosity (to be discussed later in this section) were fabricated on Si using GLAD. The incubation of these nSTFs in 1% (W/W) 4-aminothiophenol (4-ATP) in ethanol solution for about an hour resulted in the formation of a self-assembled monolayer (SAM) of 4-ATP over the Ag nSTF. The SAM forms due to the bonding of the thiol (-SH) group in 4-ATP with Ag, while the NH_2 group of 4-ATP offers avenues for further bonding over it. After the SAM formation, the nSTFs are taken out and cleaned in running ethanol to remove any unbound 4-ATP molecules from the nSTF surface. The SERS spectra from the 4-ATP functionalized Ag nSTFs were recorded and studied using a custom-made fiber optic Raman spectrometer system. The schematic of the fiber optic Raman system has been shown in Fig. 8.5. Light of wavelength of 785 nm launched into the excitation fiber of a fiber optic Raman probe, equipped with suitable laser line and long-pass filters, lenses, and dichroic and folding mirrors, was focused on the SERS chip and the Raman signal scattered off, and it was collected from the same spot by using the same lens. The SERS signal accessed from the sample was fed to the fiber optic Raman spectrometer via collection fiber. The Raman spectrometer was further interfaced with a computer to assess and analyze the SERS spectra. SERS spectra from 4-ATP functionalized Ag nSTFs of varying heights using the aforementioned setup were studied by Shalabney and Abdulhalim [22] and have been plotted in Fig. 8.6. It can be observed that for a closed film of about 50 nm thickness (with 0% porosity), no SERS enhancement was observed. However, for porous nSTFs, SERS peaks characteristic of 4-ATP were observed. This is due to about one order of magnitude larger electromagnetic field enhancements in localized surface plasmons, as compared to extended or propagating surface plasmons.

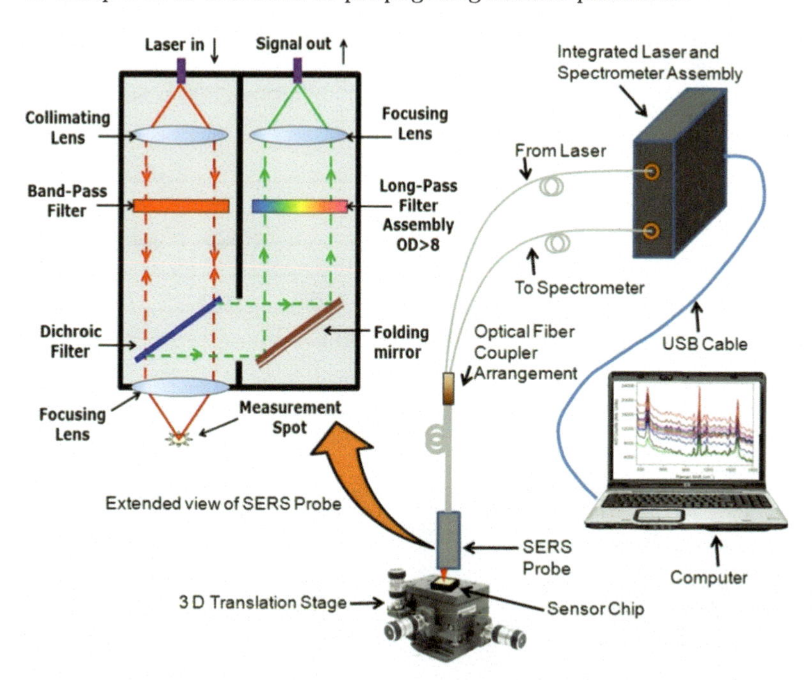

Figure 8.5 Schematic of a custom-made fiber optic Raman spectroscopy setup (Adapted with permission from Ref. [23]).

It can be observed that with an increase in the height of the nSTF from 100 to 400 nm, the SERS enhancement increased, became maximum around 300 nm height, and then decreased again at 400 nm height. It was concluded that the optimum height of nSTF for maximum SERS enhancement is about 300 nm. It can be noted that this height is not exactly at $\lambda/2$ for 785 nm wavelength, as sought in nanoantenna resonances. This is because the plasmonic wavelength corresponding to 785 nm is slightly smaller than $\lambda/2$ and has been explained in Ref. [24].

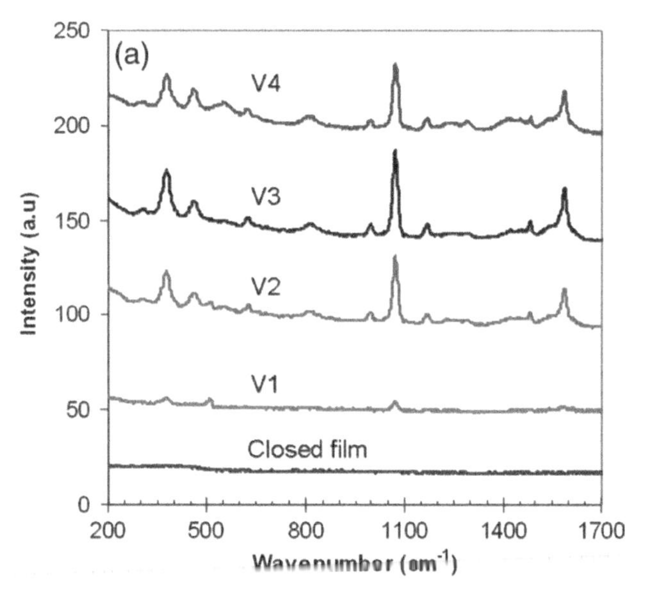

Figure 8.6 SERS spectra from 4-ATP functionalized Ag nSTFs, where $V_1 = 100$, $V_2 = 200$, $V_3 = 300$, and $V_4 = 400$ nm (Adapted with permission from Ref. [22]).

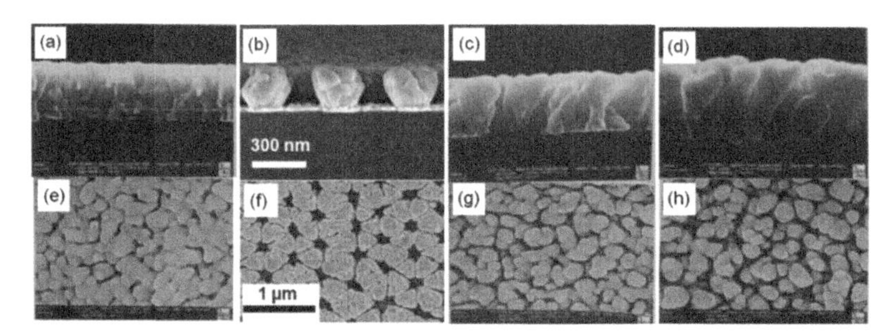

Figure 8.7 SEM images of Ag STFs of 300 nm height on Si (100) substrate having varying porosity. Figures (a), (b), (c), and (d) represent the cross-sectional views of nSTFs with 20%, 30%, 40%, and 60% porosities, respectively, and (e), (f), (g), and (h) represent the respective top views (Adapted, with permission from Ref. [19]).

Optimization of nSTF porosity

The nSTFs with an optimal height of 300 nm and varying porosities ranging from 0% to 60% were fabricated using the same GLAD setup discussed in Fig. 8.4. SEM images of nSTFs having different porosities have been shown in Fig. 8.7. The porosity is defined as the percentage of open area in the total film comprising the open, coated portion. The SEM images could be processed using ImageJ software from NIH [25] to estimate the respective porosities.

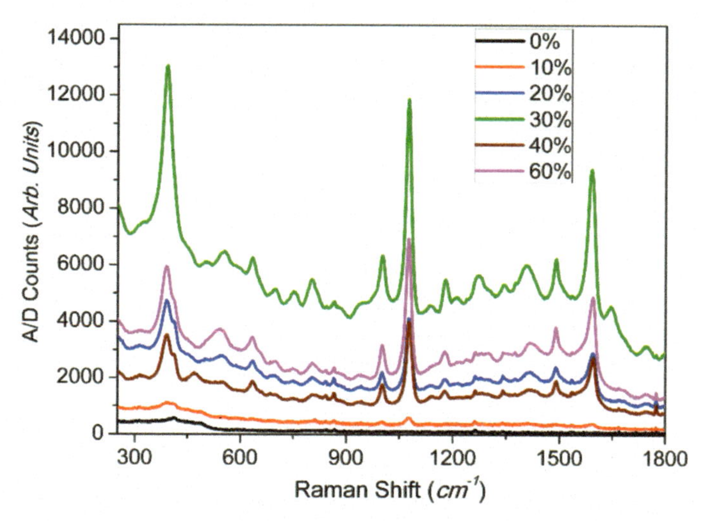

Figure 8.8 SERS spectra of 4-ATP adsorbed on Ag nSTFs of different porosities (Adapted with permission from Ref. [19]).

The SERS spectra from 4-ATP functionalized Ag nSTFs of different porosities have been plotted in Fig. 8.8. Prima facie, it can be observed that with an increase in porosity from 0% to 60%, the SERS power increases with an increase in porosity from 0% to 30% porosity and then it starts decreasing with further increase in porosity from 30% to 60%. An explanation for the aforementioned observation is presented as follows: The EM field in the gap between two closely spaced plasmonic nanostructures is much stronger than that on individual plasmonic nanostructures. Further, the EM field intensity decreases rapidly with an increase in the gap between the plasmonic nanostructures [26]. Accordingly, when the porosity of the nSTF is small, the EM field in the gap among the nanocolumns is very strong as the spacing is relatively small. However, the number of 4-ATP molecules in the gaps is also very small due to obvious reasons. When the porosity of the nSTF is increased, the EM field intensity in the gap decreases rapidly due to the decoupling of plasmonic fields of individual nanocolumns; thereby leading to a relatively weaker SERS signal from individual molecules. However, the number of 4-ATP molecules increases in the gaps, leading to a rise in the SERS power. Hence, there is a trade-off when the total SERS signal from the molecules on the nanocolumns and that from the pores sums up to become maximum. In the experimental window of the article discussed here, the maximum enhancement was achieved at an optimum porosity of about 30%. Hence, the Ag nSTFs of about 300 nm height and 30% porosity were found to possess maximum SERS enhancement.

Estimation of SERS enhancement factor

The measured enhancement factor (EF) is defined as the ratio of SERS intensity per adsorbed molecule to the Raman intensity per molecule in a bulk sample [22]. The common expression is denoted as

$$EF = (I_{\text{SERS}} / I_{\text{bulk}}) \cdot (M_{\text{bulk}} / M_{\text{ads}}). \tag{8.7}$$

Here, I_{SERS} is the SERS intensity of a specific Raman band of the sample under study adsorbed over a nanoplasmonic surface and I_{bulk} is the intensity of the same Raman band from a bulk chunk of the same sample. For the solid phase, M_{bulk} is the number of molecules in the entire volume exposed to the laser beam, which is calculated using the size of the focused laser spot on the sample, the penetration depth of the beam inside the sample, and the density of the molecules in the sample. For the SERS spectrum, the number of the adsorbed molecules M_{ads} at the nanoplasmonic surface can be calculated by

$$M_{\text{ads}} = C \cdot \left\{A_{\text{laser}} \cdot P_{\text{den}}\right\} \cdot \left\{S_p / S_m\right\}. \tag{8.8}$$

P_{den} is the density of the plasmonic nanostructures and is given by the number of such nanostructures per unit area, S_p is the upper surface area of a single nanostructure, and S_m is the area occupied by a single molecule when it is adsorbed to the plasmonic nanostructures. The coefficient C expresses the concentration of the molecular solution in which the nanoplasmonic chip is incubated. By applying the formulae mentioned in Eqs (8.7) and (8.8), the EF over the nSTFs from the prominent bands of SERS spectra of the 4-ATP functionalized optimal nSTF was estimated. The diameter of the laser spot focused on the sample, penetration depth of the laser beam in a bulk chunk of 4-ATP, density, and the molecular weight of 4-ATP used in the calculations were 200 μm, 1 mm, 1.18 g/cm^3, and 125.19 g/mole, respectively. The area occupied by a single 4-ATP molecule was considered to be 0.2 nm^2, and the concentration of the 4-ATP, C was 1%. Using Eq. (8.7), an EF of the order of 10^7 was estimated for the optimized nSTF.

From the discussions mentioned above, it can be concluded that the proper design of the plasmonic nanostructured thin films is an important step for resonant excitation of localized plasmons around the laser wavelength, which leads to ultra-high enhancements of the Raman signal.

8.2 SERS-Based Sensing Mechanisms

As was concluded by Eq. (8.6), SERS is a very short-range phenomenon. If the distance between the analyte and the plasmonic nanostructure becomes large, the SERS enhancement decreases rapidly. Also, the molecule of interest shall be small enough to assess the SERS signal from the whole molecule. If the size of the molecule is small and the SERs signal is assessed directly from the analyte molecule, this mode of sensing is called direct detection/recognition of the analyte (Fig. 8.9a). The assessment of SERS signal from analytes bigger in size is not possible due to the aforementioned reasons and hence the sensing of bigger analytes remains challenging. However, if the binding of the big analyte on the crosslinker/molecular recognition element (MRE) leads to substantial changes in the SERS signal of the cross-linker/MRE, the presence of even bigger analytes can also be sensed. Such a scheme is called indirect or non-direct mode of sensing (Fig. 8.9b). The indirect mode of sensing also depends on turning ON (Fig. 8.9c) or turning OFF (Fig. 8.9d) of the SERS signal, due to certain transformations taking place in the vicinity of the plasmonic nanostructure. For turn ON kind of detection, the enhancement of the SERS signal from the inter-particle plasmonic hotspots of plasmonic complexes is formed due to the binding of the ligands. Similarly, the turn OFF of the SERS signal takes place when competitive binding of an analyte (say, heavy metal ions) leads to the removal of SERS probes off the plasmonic nanostructures.

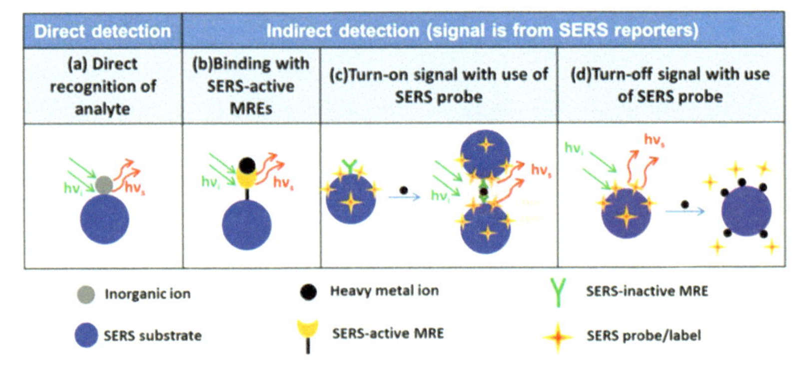

Figure 8.9 Illustration of different SERS sensor design strategies for inorganic ion detection. (a) Direct recognition of analyte; indirect detection of analyte using (b) binding with SERS-active MREs; (c) turn-on signal with the use of SERS probe; and (d) turn-off signal with the use of SERS probe (Adapted with permission from Ref. [27]).

In the following sub-sections, the discussion of sensors based on both the direct and indirect modes of sensing is presented.

8.2.1 Direct Detection: Ultra-Sensitive Detection of Anions in Water Using Ag Nanoparticles

Similar to heavy metal ion contaminations, inorganic ion contaminations too are very dangerous for living organisms and it is important to detect them even in small traces, as these can be severely harmful. For example, ClO_4^- can disrupt iodide intake in the thyroid gland, thereby leading to hyperplasia. This further leads to hypothyroidism.

Colloidal sols of both positively and negatively charged silver nanoparticles (Ag^+ and Ag^-, respectively) were obtained via UV-assisted reduction of $AgNO_3$. These Ag^+ nanoparticles were further immobilized for the development of SERS-based sensors and then used for the detection of various anions in aqueous solutions, while Ag^- nanoparticles supported control experiments.

8.2.1.1 Sensor chip fabrication

Positively charged spherical Ag nanoparticles (Ag^+) were obtained via UV-assisted reduction of $AgNO_3$ using branched poly(ethyleneimine) (BPEI) and 4-(2-hydroxyethyl)-1-pierazineethanesulfonic acid (HEPES). The Ag^+ nanoparticles were then immobilized onto glass slides for the development of the sensor chips. Prior to the immobilization, the glass slides were cleaned by UV exposure for several hours and Nochromix solution in concentrated H_2SO_4 overnight. Then the slides were washed with a copious amount of Mill-Q water to remove any impurities as well as unused chemicals from the glass surface. As illustrated in Fig. 8.10, three types of substrates were developed.

The first type of substrate illustrated in Fig. 8.10a was developed by the adsorption of Ag^+ nanoparticles on the glass substrate by incubating it in Ag^+ colloidal solution at pH 5 for 4 hours. After the incubation was over, the substrate was rinsed thoroughly with a copious amount of Milli-Q water.

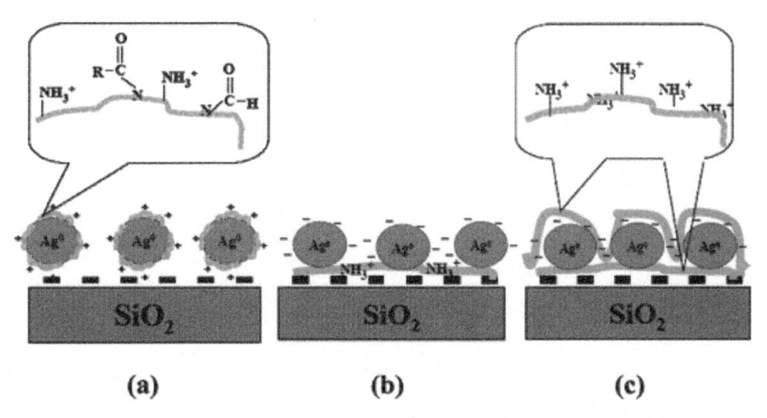

Figure 8.10 Schematic Illustration of (a) Type I, (b) Type II, and (c) Type III substrates for direct sensing of different anions (Adapted with permission from Ref. [28]).

The second type of substrate was prepared by immobilizing Ag^- nanoparticles on BPEI-modified glass surface (discussed thoroughly in Ref. [28]) by incubating the BPEI-modified glass slides in Ag^- colloidal solution in HEPES buffer at pH 7 for 4 hours and then washing it with a copious amount of Milli-Q water.

The third type of substrate was prepared by the introduction of an additional BPEI layer over the second type of substrate to mimic the behavior of the first type of substrate. The concentrations of the reagents were chosen such that all the substrates possessed a comparable nanoparticle coverage density of ~30 particles/μm^2.

8.2.1.2 Anion sensing experiment

Aqueous solutions of ClO_4^-, CN^-, SCN^-, and SO_4^{2-} ranging from micromolar to nanomolar concentrations were prepared and the SERS spectra for each sample solution over the substrate of type I were recorded. The SERS spectra for the abovementioned four anions with varying concentrations have been shown in Fig. 8.11. It can be observed that all the different anions possess qualitative signatures of Raman bands specific to them. Furthermore, with an increase in concentration, in each spectral window, a rise in SERS intensity is observed, which leads to quantitative prediction as well. The limits of detection (LoDs) of 1 ppb, 8 ppb, 7 ppb, and 4 ppb were achieved for ClO_4^-, CN^-, SCN^-, and SO_4^{2-}, respectively.

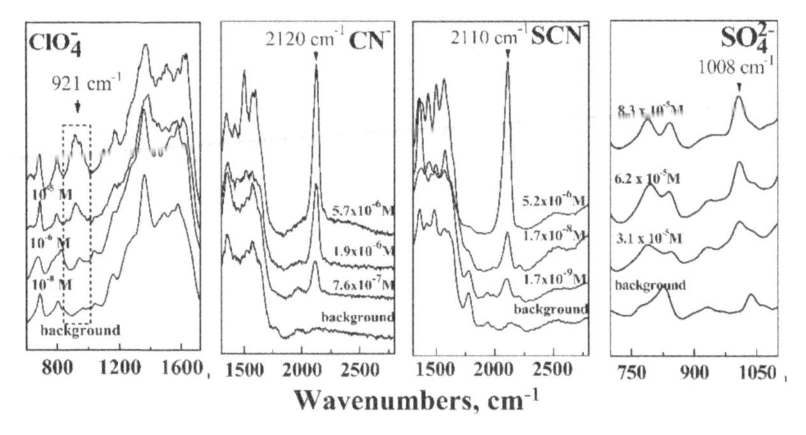

Figure 8.11 SERS spectra of aqueous solutions of ClO_4^-, CN^-, SCN^-, and SO_4^{2-} using the first type substrate (Adapted with permission from Ref. [28]).

A comparative study of the performance of the three substrates for a fixed concentration of SCN^- was made to assess the optimal sensing capabilities of the three sensors. In Fig. 8.12a, the SERS spectra assessed from 5 ppb SCN^- in an aqueous medium over I, II, and III types of sensor substrates have been plotted. It can be observed that among all the types of sensors, type I has the best performance. The substrate with Ag^- nanoparticles has the worst performance toward anions sensing among the three types of sensors. In type II, though positively charged BPEI maintains a similar charged environment as that of type I sensor, the distance between the anions and the plasmonic surface becomes relatively larger, thereby leading to a relatively weaker SERS signal. In Fig. 8.12b, the absorption isotherm shows a similar behavior, which is explained as previously.

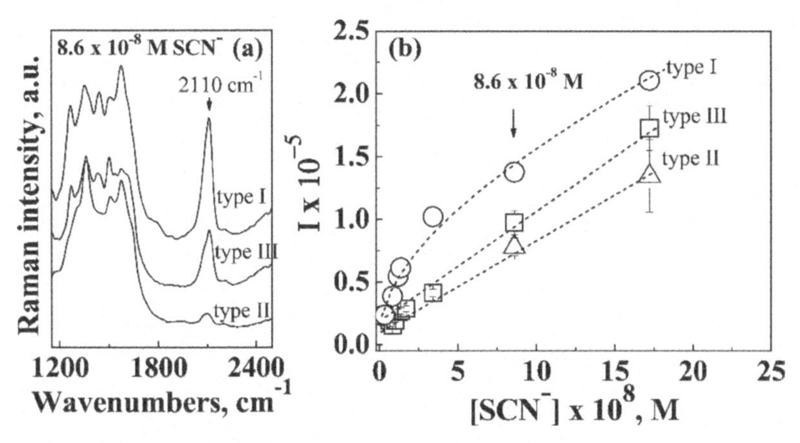

Figure 8.12 (a) SERS spectra for all the three types (I, II, and III) of substrates at 5 ppb concentration of SCN^-. (b) Adsorption isotherm of SCN^- for all the three substrates (Adapted with permission from Ref. [28]).

8.2.2 Indirect Detection: *Escherichia coli* Detection Using nSTFs

Escherichia coli (*E. coli*) is a rod-shaped, gram-negative, facultative anaerobic bacterium, which is considered to be an indicator of fecal contamination (fecal coliforms). Some strains of *E. coli* are highly pathogenic and lead to food or waterborne gastrointestinal diseases. Consumption of *E. coli* contaminated water and/or food may lead to hemolytic-uremic syndrome (HUS), especially in elderly children. Prolonged HUS may cause hemolysis and kidney failure, which may result in seizures, strokes, or even death [29]. According to the World Health Organization (WHO) survey reports, approximately two billion people annually suffer from gastrointestinal diseases caused by *E. coli* [30].

8.2.2.1 Sensor chip development

A stepwise illustration of sensor chip development has been presented in Fig. 8.13. Firstly, an Ag nSTF chip was incubated in 4-ATP solution in ethanol (1% w/w) for 1 hour. This led to the formation of a self-assembled monolayer (SAM) of 4-ATP over the Ag nSTF surface. The nSTF chip was rigorously washed with ethanol and then a copious amount of Milli-Q water to remove any remnants and dried using a blow of nitrogen (N_2) gas after the SAM formation. The chip was then incubated in an aqueous solution of glutaraldehyde (5% v/v), a cross-linker, for 1 hour to substantiate receptor binding. After binding of cross-linker on the sensor chip, it was again washed with Milli-Q water to remove any unbound molecules and blow-dried with N_2 gas. The chip was further incubated in T4 bacteriophage solution for 4 hours to form an *E. coli* specific receptor layer. After this, the chip was incubated in an aqueous solution of BSA, an antifouling agent, of 1 mg/ml concentration in 50 mM PBS buffer for 1 hour to enable the prevention of putative non-specific bindings on the sensor surface. This step leads to increased specificity, thereby improving the overall performance of the sensor while reducing the foul signal from other bacteria present in the sample. After BSA immobilization, the sensor chip was taken out, washed rigorously in a copious amount of Milli-Q water and PBS, and blow-dried with N_2. The chip was stored in a refrigerator at 4 °C and was taken out only to perform SEM, atomic force microscope (AFM), and SERS studies.

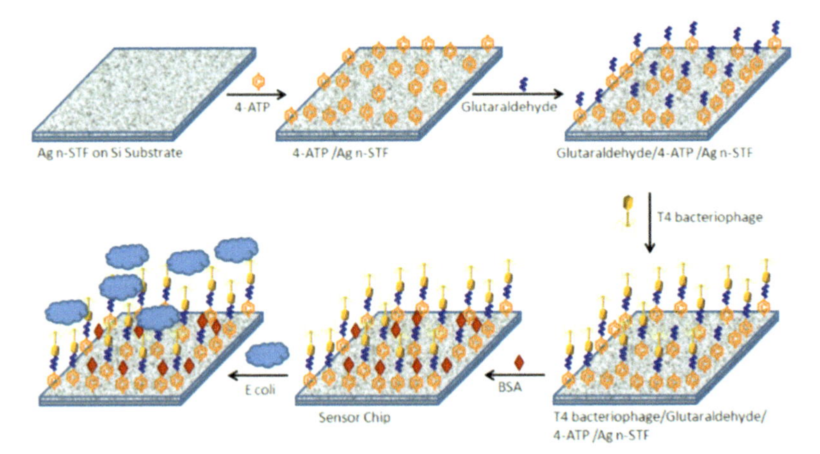

Figure 8.13 Illustration of sensor chip development for detection of *E. coli* (Adapted with permission from Ref. [23]).

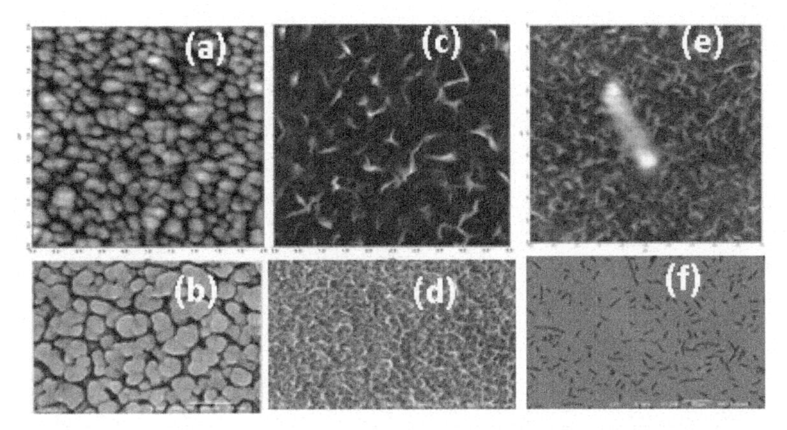

Figure 8.14 AFM and SEM images of the nSTF sensor: (a), (b) before and (c), (d) after functionalization, respectively; (e), (f) SEM images of *E. coli B* attached to the sensor surface (Adapted with permission from Ref. [23]).

The AFM and SEM images of the *E. coli* sensor chip before and after the nSTF functionalization have been shown in Fig. 8.14a–d, respectively. In both the AFM (Fig. 8.14c) and the SEM (Fig. 8.14d) images, the leg-like parts of the attached T4-bacteriophages are visible. Figure 8.14e,f shows the AFM and SEM images of a number of *E. coli* (rod-like microstructures) captured at the sensor surface, respectively.

The sample solutions of different concentrations ranging from 1.5×10^2 to 1.0×10^5 cfu/mL were prepared in PBS buffer from the stock solutions of *E. coli B, E. coli µX, P. aeruginosa, CV026,* and *P. dentrificans*. Bacterial strains other than *E. coli B* were chosen for negative control experiments to be detailed later in Fig. 8.15. The sensor chips interacted with the sample solutions for 10 minutes. SERS spectra from the chip before and after the interaction were assessed for each sample solution to ensure intensity referencing. After the assessment of SERS spectra, the sensor chip was washed with 20 mM aqueous solution of NaOH to detach bacteria from the surface, thereby regenerating the sensor surface. The sensor surface was then washed twice with PBS to remove the remnants of the NaOH solution, blow-dried with N_2 gas, and then used for other solutions.

The SERS spectra for different concentrations of *E. coli B, E. coli µX, P. dentrificans, P. aeruginosa,* and *CV026* bacteria were recorded.

The peak SERS intensity of the most prominent peak at 1077 cm^{-1} Raman shift was considered for quantitative analysis. The variation of peak SERS intensities @1077 cm^{-1} with bacterial concentrations has been plotted in Fig. 8.15. The X-axis representing concentration has been chosen logarithmic as it spans a large dynamic range. The background SERS signal from the bare sensor chip was subtracted from the SERS signal of samples over the chip to correct the baseline for all the measurements. The Y-axis, therefore, represents the A/D counts for the chip with sample minus that from the bare sensor chip ($I_{Sample} - I_{Sensor}$) @1077 cm^{-1}.

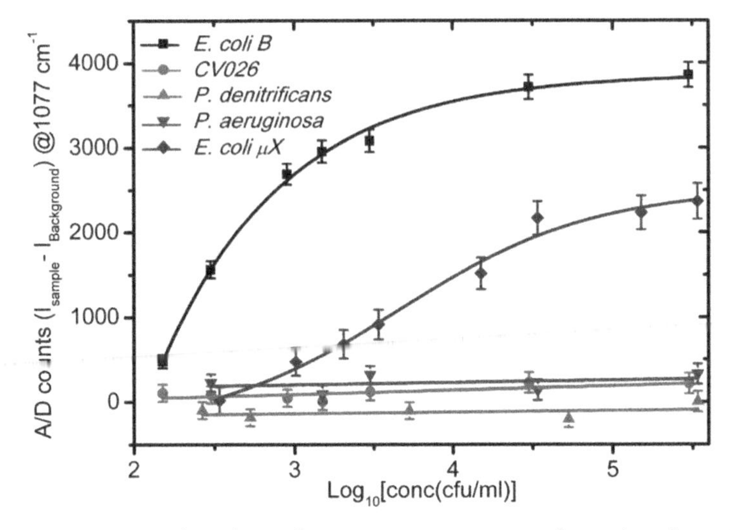

Figure 8.15 SERS-based *E. coli* sensor response curves for various bacteria: Negative control (Adapted with permission from Ref. [23]).

The processed signal was termed differential Raman enhancement and has been used so throughout the chapter. The symbols represent the differential SERS Intensities at varying concentrations of studied bacteria, while the lines through symbols are the best curve fits. With an increase in the concentration of bacterial concentrations except for the two kinds of *E. coli*, the differential SERS intensity does not show any change. For both the strains of *E. coli*, *E. coli B,* and *E. coli μX,* differential SERS intensity first increases at small concentrations and then becomes nearly constant, as all the receptor sites get consumed in binding. This kind of response confirms that the sensor possesses high specificity for *E. coli*. Approximately 10 μL volumes of the bacterial samples were used for SERS sensor studies. It can be estimated that taking 10 μL from a well-mixed sample of 1.5×10^2 cfu/mL will nearly have only one bacterial cell; thereby reaching the ultimate limit of detection of a single bacterium.

8.2.2.2 Indirect detection: DNA hybridization-based sensors

Gold nanostructures connected to single-stranded DNA (ss-DNA) can be useful as micro-SERS probes for reaching ultra-low limits of detection. Here, we briefly discuss a couple of schemes for the detection of heavy metal ion Hg^{2+}, which leads to the hybridization of ss-DNA into double-stranded DNA (ds-DNA), thereby changing the SERS signal. Two such schemes have been illustrated in Fig. 8.16a,b. These schemes have been taken from Refs. [31] and [32] and a detailed discussion of these could be found there, respectively. As shown in Fig. 8.16a, the micro-SERS probe contains two interconnected Thymine (T) rich ss-DNA strands, one of which acts as a probe and the other, complementary. One end of the probe DNA has the thiol (-SH) group to facilitate binding with Au nanoshell, while the other end has tetramethylrhodamine (TAMRA) to provide an enhanced SERS signal. When Hg^{2+} is introduced to this kind of micro-SERS sensor, the Hg^{2+} leads to hybridization of the T-bases, thereby bringing TAMRA, close to the Au-nanoshell surface. This leads to the enhancement of the SERS signal, as SERS is a highly distance-dependent phenomenon.

Figure 8.16b illustrates the hybridization of two gold nanostar immobilized ss-DNA strands when Hg^{2+} is introduced. The hybridization leads to enhanced SERS signal, thereby providing non-direct information about the presence of Hg^{2+}.

These sensing mechanisms provide highly sensitive detection of analytes of interest. However, when an analyte is present in a complex matrix of molecules/species having almost similar or overlapping Raman bands, it becomes extremely difficult to accurately predict the presence of a particular analyte. As an analogy, one may find it as difficult as finding a needle in hay. To overcome this difficulty, in recent times, artificial neural networks (ANNs) equipped with machine learning/deep learning (ML/DL) algorithms have been employed. In the next sub-section, starting with the preface of the methods and various steps involved in these algorithms, the results of a recent article as a case study using one kind of DL algorithm for SERS-based detection will be discussed.

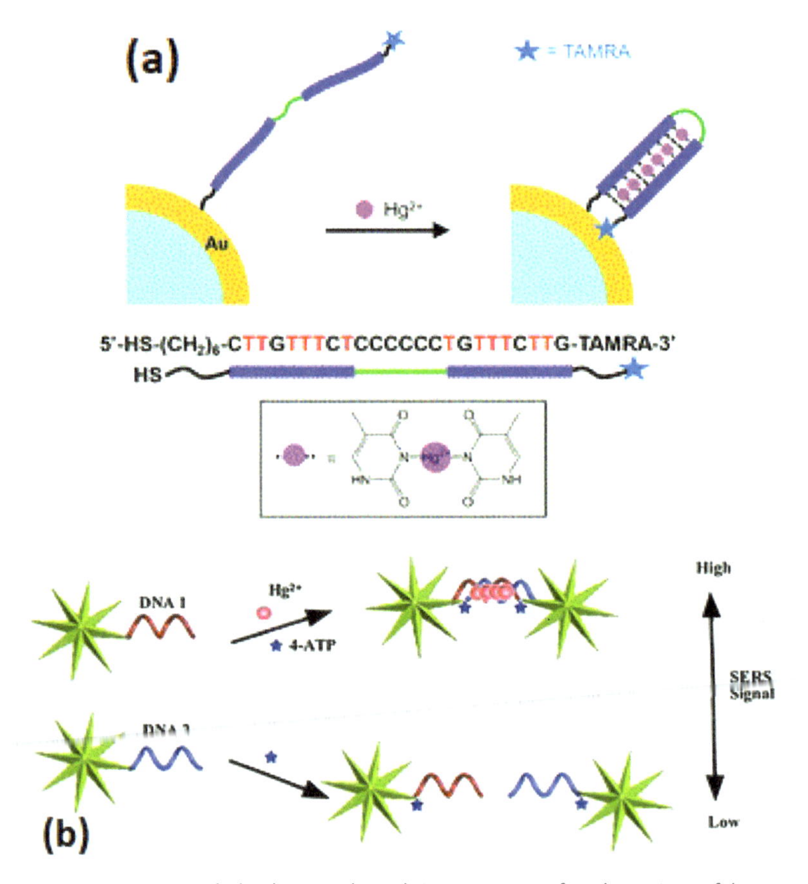

Figure 8.16 DNA hybridization-based SERS sensors for detection of heavy metal ion (Hg^{2+}) using (a) gold nanoshells (Adapted with permission from Ref. [31]), (b) gold nanostars (Adapted with permission from Ref. [32]).

8.2.3 Machine Learning–Enabled SERS Sensors

Raman spectroscopy is a non-invasive method to characterize molecules by retrieving vibrational information from different chemical bonds and using this information to identify components in the given samples. However, it comes with certain challenges associated with it, due to the complex nature of the technique, such as it is still a very hard problem to find different hidden molecular components in Raman spectra of mixtures. To overcome this, researchers are always searching for new advanced data processing techniques to extract certain meaningful features and analyze complex Raman spectra. For problems with relatively simple Raman signals of pure components, a plethora of techniques are readily available such as Euclidean distance, linear regression, and multivariate data analysis algorithms to recognize and categorize simpler molecular vibrational spectra of pure components with relatively high accuracy. But in practical applications, usually, multiple components are present in Raman spectra of complex samples. And for complex Raman spectra of mixtures of entities of similar structure/functional groups, it still is a challenging task to qualitatively analyze and identify components in Raman spectra of mixtures with high accuracy and sensitivity.

For quantitative analysis of Raman spectra of mixtures, many methods have been suggested and used in the past, such as statistics, chemometrics, and various search-based methods. Most of the proposed methods are based on the characterization of peaks, distance, and similarities between spectra and dataset-searching algorithms [33–36]. Moreover, these methods are required to have a spectral data preprocessing step, which can cause some information loss at the cost of increasing variance and thus can introduce errors. Also, one has to choose the searching algorithm and similarity criteria for these methods; hence these methods are not versatile and thus are fit for only certain types of Raman spectra analysis.

Due to the advancement in computer hardware technology, there has been a renewed interest in artificial intelligence (AI), where the algorithms are developed over ANNs and trained via ML/DL to extract accurate information from complex or big datasets. In SERS-based sensors, it could be helpful in the assessment of a large dataset of vibrational spectra of complex chemical mixtures. Moreover, it has a large number of applications in various fields such as computer vision, natural language processing, chemistry, biology, sensors, and even in spectroscopy. It is similar to the training of a brain via repeated exposure of the sensory organs to various features of different systems, as illustrated in Fig. 8.17. Once trained, the brain gets acquainted with the systems and can easily identify them. Similarly, when a DL algorithm gets trained to a dataset via exposure through a number of epochs/iterations, it can then test and validate the same dataset through certain features. This can both be used for regression and classification. DL basically extracts features from the

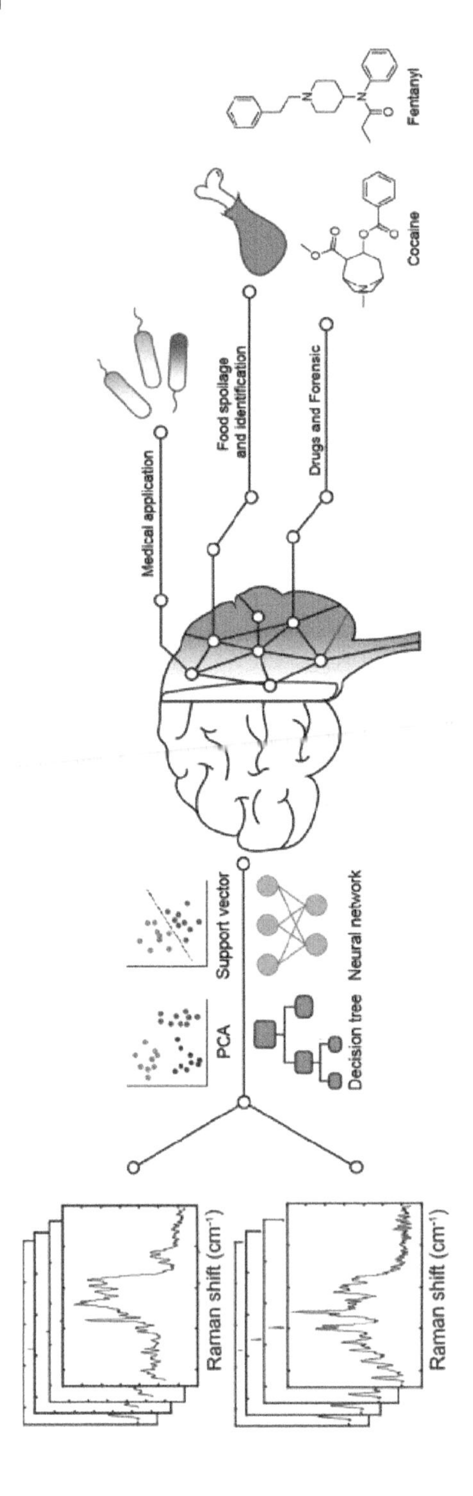

Figure 8.17 Illustration of training and classification using AI through ML (Adapted with permission from Ref. [39]).

data, then learns and models complex relationships. Convolutional neural networks (CNNs) are a class of deep neural networks that are mostly used in computer vision but have much wider applications in different fields and also can be used with different classes of neural network architectures. A few notable models with CNN architectures are ResNet developed by Microsoft, VGGNet [37] developed by Visual Geometry Group, and Inception Net [38] developed by Google, etc.

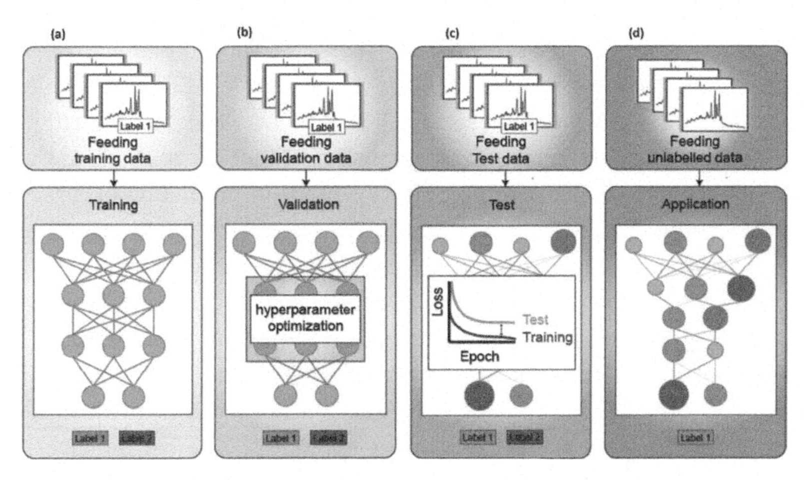

Figure 8.18 General scheme presenting the generation pipeline of an ANN applied to SERS spectroscopy (Adapted with permission from Ref. [39]).

At first, the experimentally acquired Raman/SERS spectral dataset is divided into three parts, 80%, 15%, and 5%, respectively, named training, validation, and test datasets and preprocessed identically. Each spectrum composing the abovementioned datasets is then labeled to a certain corresponding identity, defined as a feature or label. For example, a SERS spectrum can be converted into a 1D 1000:1 vector, having 1000 features, which need to be fed for the training of an untrained ANN algorithm, as shown in Fig. 8.18a. An ANN is composed of a number of artificial neurons/nodes, arranged and layered to create a complex network, which is capable of extracting information from the input. The nodes are the basic computational units of the ANN and have been shown as grey spheres in this figure. After the spectrum is converted into features, each feature is assigned certain weights depending on the importance of the feature. For example, a SERS peak is given high weightage as compared to the baseline because it helps determine the molecular fingerprint of the associated bonds. In ANN, weight is the learning parameter. Herein weights have been represented as green lines connecting the various nodes of ANN. While training the algorithm, the predicted output is compared to the true features of the input spectra, and then the error in prediction is evaluated. During the training phase, several metrics, such as mean squared error (MSE), recall, and differentiable F1 score function, may be employed. The monitored metric needs to be judicially chosen depending on the input data and kind of analysis. For example, MSE is generally chosen for 1D CNN analysis of SERS spectra. The ML algorithm then minimizes or maximizes the selected metric to improve the performance of ANN through improvement in prediction accuracy by running the program in a number of epochs. As a first step, one may minimize MSE in order to improve the accuracy of ANN by using ADAM optimizer, a commonly used learning algorithm. Apart from optimization of accuracy, it is also important to optimize the number of nodes and various layers, types of layers, etc. For this purpose, the validation dataset is presented to the ANN to optimize hyperparameters (such as number of nodes, number of layers, types of layers, e.g., convolutional or densely connected, regulation coefficient, learning rate, optimizer, etc.), as shown in Fig. 8.18b. The validation may be performed by using Bayesian optimization or grid/random search kinds of algorithms to find the optimal set of hyperparameters, which leads to the maximum performance of the ANN. Typically, the hidden layers imply the most important changes during the validation process. Further, the optimally trained and validated ML algorithm is composed of an optimal number of nodes, layers, and weights assigned to the features, and it is employed on the test dataset to predict the output, as represented in Fig. 8.18c. If after running the code for a number of epochs, the loss functions (errors) of both the training and test datasets converge to the same value, and the performance of the algorithm is said to be sufficient for predictions, depending upon the aforementioned convergence. In Fig. 8.18d, the use of a fully trained, validated, tested, and optimized ML algorithm for the assessment of an unlabeled SERS dataset is shown.

Based on the previously mentioned approach, we present a study of the classification of 30 pathogenic bacteria by employing DL algorithms on their SERS spectra by the Jennifer Dionne group at Stanford University [40].

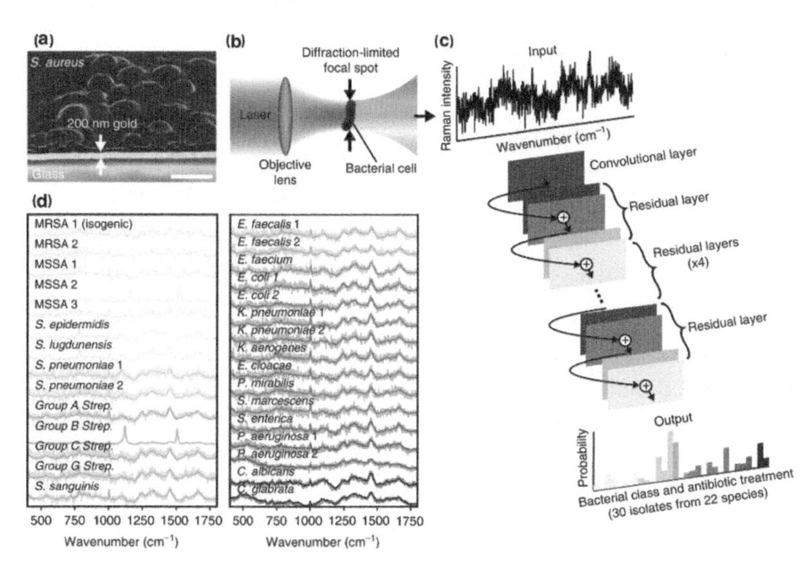

Figure 8.19 (a) SEM image of a cross-section of the sample with a monolayer of bacteria (Scale bar is 1 μm), (b) Conceptual schematic: Acquisition of Raman signal from single cells by focusing the excitation laser to a diffraction-limited spot size, (c) Using a 1D residual network with 25 total convolutional layers. (d) SERS spectra averaged over 2000 spectra from 30 isolates shown in bold and overlaid on representative examples of noisy single spectra for each isolate. Spectra are color-grouped according to antibiotic treatment. These reference isolates represent over 94% of the most common infections seen at Stanford Hospital in the years 2016–1739 (Adopted with permission from Ref. [40]).

Firstly, 2000 SERS spectra each from 30 classes of pathogenic bacteria were recorded to make a dataset for the DL algorithm, as shown in Figs. 8.19 a, b, and d. The SERS spectra (Fig. 8.19d) were recorded by focusing a laser light (Fig. 8.19b) on a gold-coated SERS chip, which had the bacterial samples on it (Fig. 8.19a). The SERS spectra were then divided into training, validation, and test datasets. Further, the SERS spectra were converted into a vector of 1000:1 feature, each having different weights. Figure 8.19c shows a representation of a CNN for processing the SERS spectra. The same algorithm was also tested for antibiotic susceptibility and was found to provide treatment identification accuracies of about 100%. Before we substantiate the findings of the DL algorithm for the above-mentioned algorithm, the algorithm and the performance parameters need to be explained. The algorithm shown in Fig. 8.19c has got 26 layers in total, 25 convolutional layers, and 1 pooling layer. Apart from these, it has skipped connections between layers such that learning from one layer can be passed to another layer skipping intermediate layers along with the learning of the intermediate layer. Due to this approach of adding skip connections, one can enable the model to also learn and fit residual mapping, instead of only learning underlying mapping. This can also be described mathematically; let us say the mapping without the skipped connection is $M(x)$, where x is the output from the previous layer. Now with the skipped connections, the final output becomes $M(x) = F(x) + x$, where $F(x)$ is the residual mapping as shown in Fig. 8.20.

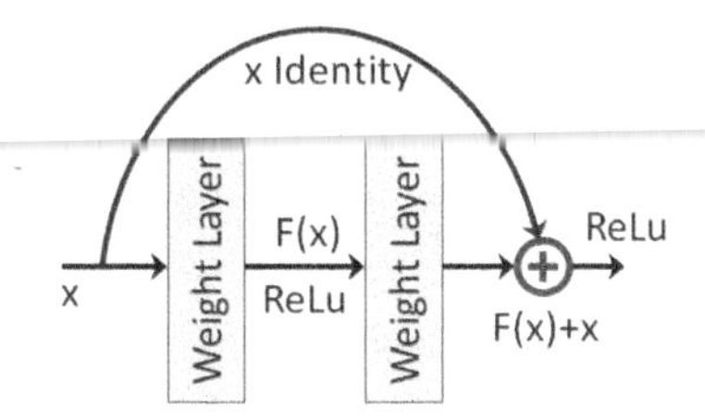

Figure 8.20 Schematic of a residual block.

8.2.3.1 Evaluation matrices (confusion matrix)

Before further discussion of results is made, let us introduce the evaluation matrices, which are required for analysis of the output results of the DL algorithm. The confusion matrix as given in Table 8.1 is used to evaluate, characterize, and validate the performance of the model or classifier on a given testing dataset for which the labels or the true values are known [41]. While testing the model on a test dataset, the number and predicted class of correct and incorrect predictions is counted for each class.

Table 8.1 Predicted label types

Positive (P)	Positive observation
Negative (N)	Negative observation
True Positive (TP)	Predicted value is positive and observation is positive
True Negative (TN)	Predicted value is negative and observation is negative
False Positive (FP)	Predicted value is positive and observation is negative
False Negative (FN)	Predicted value is negative and observation is positive

Based on Table 8.1, there are four performance parameters defined as follows:

Accuracy: It is the most basic measure of performance and is calculated by dividing correctly predicted observation labels by the total observation labels. Accuracy cannot be considered as the single most important measure to evaluate a model. Accuracy can only be considered as a good verdict if the model is trained on a balanced dataset having equal no of positive and negative examples of each class. It is very important to consider other metrics to estimate the performance of a model.

$$\text{Accuracy} = \frac{TP + TN}{TP + TN + FP + FN} \tag{8.9}$$

Recall: It is defined as the ratio of the number of correct positive results to the number of all the examples that must be predicted as positive. Recall defines how many of the actual positive 1 models capture by labeling it as positive (true positive). A recall of 1 signifies that one has no false negatives.

$$\text{Recall} = \frac{TP}{TP + FN} \tag{8.10}$$

Precision: It can be expressed as the number of correct positive results divided by the number of all positive results returned by the classifier. In other words, it is the ratio of correctly predicted positive observations to the total predicted positive observations. Precision is a good measure to identify when the cost of a false positive is high. A precision of 1 means that one has no false positives.

$$\text{Precision} = \frac{TP}{TP + FP} \tag{8.11}$$

Specificity: It is defined and calculated by dividing the number of correct negative predictions by the total number of negative samples. Specificity is also defined as a true negative rate (TNR). The best specificity is 1.0, whereas the worst is 0.0.

$$\text{Specificity} = \frac{TN}{TN + FP} \tag{8.12}$$

A confusion matrix illustrated in Fig. 8.21 helps estimate the aforementioned four evaluation parameters.

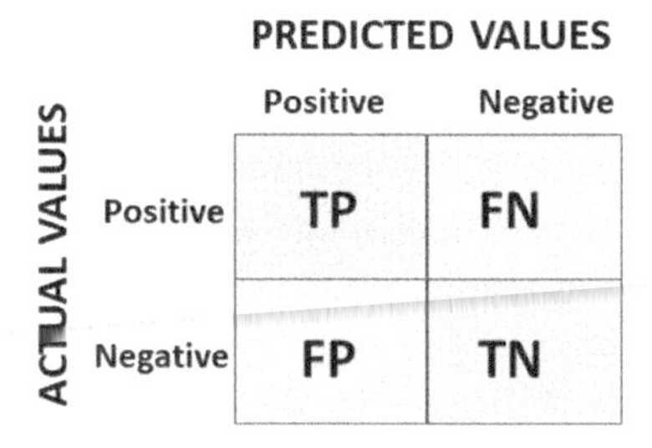

Figure 8.21 Confusion matrix.

The ResNet algorithm (CNN algorithm with residual blocks) was trained on 30 classifications of bacterial strains, where the algorithm output is in terms of probability distribution across all the 30 bacterial classes. The maximum probability is considered as the predicted class.

A confusion matrix, in line with the previously mentioned definitions and discussions, has been plotted in Fig. 8.22a for each individual class of 30 bacterial strains. Units corresponding to i, j box represents the predicted percentage as jth strain out of 100 test spectra fed to CNN for true labels of ith strain; entries along the diagonal represent the accuracies for each class. Misclassifications are mostly within antibiotic groupings, indicated by different colored squares, and barely affect the treatment outcome. Values below 0.5% were not shown, and matrix entries covered by figure insets were all below 0.5% aside from a 2% misclassification of MRSA 2 as *P. aeruginosa* 1 and a 1% misclassification of Group B Strep as *K. aerogenes*. Predictions can be combined into antibiotic groupings also to estimate treatment accuracy, as shown in Fig. 8.22b. TZP = piperacillin-tazobactam. All values below 0.5% were not shown.

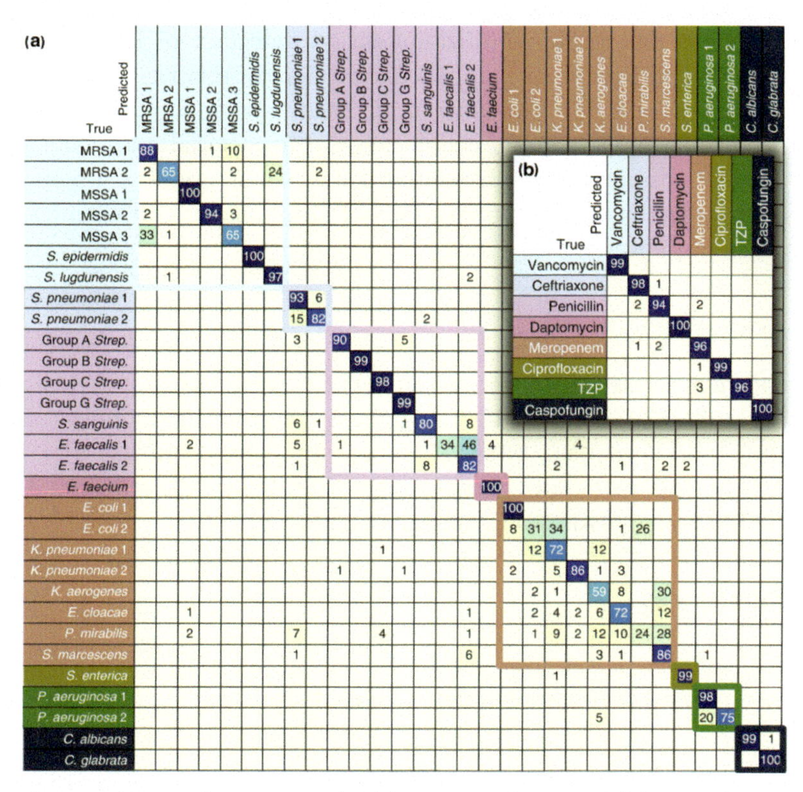

Figure 8.22 Confusion matrix for (a) 30 strain classes and (b) 08 antibiotics (Adapted with permission from Ref. [40]).

8.3 Summary

Plasmons being a versatile tool for EM enhancements can be utilized for the manifold enhancement of Raman signals; the phenomenon termed SERS. Plasmonic nSTFs prepared by the GLAD technique turned out to be cost-effective high throughput SERS substrates.

It was observed that SERS enhancement factors of about 10^7–10^8 per molecule could be achieved using optimal plasmonic nSTFs. Further, the optimized nSTFs were shown to work as highly sensitive and selective sensors. Pros and cons of plasmonic nanostructure optimization were discussed and optimal structure was used to demonstrate non-direct sensing of *E. coli* bacteria. Direct detection of anions and DNA hybridization-based detection of Hg^{2+} were discussed. ML techniques for the assessment of relatively complex matrices of analytes were introduced. Their functionality along with recent development for the classification of 30 bacterial strains along with 08 antibiotics was presented. This chapter presents a blend of both the basics and applications of SERS for bio and chemical sensing applications.

Acknowledgments

Sachin K. Srivastava gratefully acknowledged financial support from the MoE-STARS grant, SERB Start-up research grant, FIG IIT Roorkee, DST: Indo-Korea JNC, DST-INSPIRE Faculty Award, and DST-BDTD grants. This work was partially supported by the EU MSCA-RISE H2020 project IPANEMA grant agreement N° 872662.

References

1. Hutter, E. and Fendler, J. H. (2004). Exploitation of localized surface plasmon resonance. *Adv. Mater.*, **16**, pp. 1685–1706.
2. Hao, E. and Schatz, G. C. (2004). Electromagnetic fields around silver nanoparticles and dimers. *J. Chem. Phys.*, **120**, pp. 357–366.
3. Fleischmann, M., Hendra, P. J., and McQuillan, A. J. (1974). Raman spectra of pyridine adsorbed at a silver electrode. *Chem. Phys. Lett.*, **26**, pp. 163–166.
4. Jeanmaire, D. L. and van Duyne, R. P. (1977). Surface Raman spectroelectrochemistry: Part I. Heterocyclic, aromatic, and aliphatic amines adsorbed on the anodized silver electrode. *J. Electroanal. Chem. and Interfac. Electrochem.*, **84**, pp. 1–20.
5. Moskovits, M. (2005). Surface-enhanced Raman spectroscopy: A brief retrospective. *J. Raman Spec.*, **36**, pp. 485–496.
6. Valley, N., Greeneltch, N., van Duyne, R. P., and Schatz, G. C. (2013). A look at the origin and magnitude of the chemical contribution to the

enhancement mechanism of surface-enhanced Raman spectroscopy (SERS): Theory and experiment. *J. Phys. Chem. Lett.*, **4**, pp. 2599–2604.

7. le Ru, E. C., Blackie, E., Meyer, M., and Etchegoint, P. G. (2007). Surface enhanced Raman scattering enhancement factors: A comprehensive study. *J. Phys. Chem. C*, **111**, pp. 13794–13803.
8. Ding, S. Y., You, E. M., Tian, Z. Q., and Moskovits, M. (2017). Electromagnetic theories of surface-enhanced Raman spectroscopy. *Chem. Soc. Rev.*, **46**, pp. 4042–4076.
9. Nie, S. and Emory, S. R. (1997). Probing single molecules and single nanoparticles by surface-enhanced Raman scattering. *Science*, **275**, pp. 1102–1106.
10. Pang, S., Yang, T., and He, L. (2016). Review of surface enhanced Raman spectroscopic (SERS) detection of synthetic chemical pesticides. *Tr. Anal. Chem.*, **85**, pp. 73–82.
11. Pilot, R., Signorini, R., Durante, C., Orian, L., Bhamidipati, M., and Fabris, L. (2019). A review on surface-enhanced Raman scattering. *Biosensors*, **9**, pp. 9020057.
12. Sharma, B., Frontiera, R. R., Henry, A. I., Ringe, E., and van Duyne, R. P. (2012). SERS: Materials, applications, and the future. *Mater. Today*, **15**, pp. 16–25.
13. Shvalya, V., Filipič, G., Zavašnik, J., Abdulhalim, I., and Cvelbar, U. (2020). Surface-enhanced Raman spectroscopy for chemical and biological sensing using nanoplasmonics: The relevance of interparticle spacing and surface morphology. *Appl. Phys. Rev.*, **7**, pp. 031307
14. Cialla, D., März, A., Böhme, R., Theil, F., Weber, K., Schmitt, M., and Popp, J. (2012). Surface-enhanced Raman spectroscopy (SERS): Progress and trends. *Anal. Bioanal. Chem.*, **403**, pp. 27–54.
15. Fateixa, S., Nogueira, H. I. S., and Trindade, T. (2015). Hybrid nanostructures for SERS: Materials development and chemical detection. *Phys. Chem. Chem. Phys.*, **17**, pp. 21046–21071.
16. McNay, G., Eustace, D., Smith, W. E., Faulds, K., and Graham, D. (2011). Surface-enhanced Raman scattering (SERS) and surface-enhanced resonance Raman scattering (SERRS): A review of applications. *Appl. Spec.* **65**, pp. 825–837.
17. Liedtke, S., Grüner, C., Lotnyk, A., and Rauschenbach, B. (2017). Glancing angle deposition of sculptured thin metal films at room temperature. *Nanotech.*, **28**, pp. 385604.
18. Abdulhalim, I. (2014). Plasmonic sensing using metallic nanosculptured thin films. *Small*, **10**, pp. 3499–3514.

19. Srivastava, S. K., Shalabney, A., Khalaila, I., Gru¨ner, C., Rauschenbach, B., and Abdulhalim, I. (2014). SERS biosensor using metallic nano-sculptured thin films for the detection of endocrine disrupting compound biomarker vitellogenin. *Small,* **10**, pp. 3579–3587.

20. Robbie, K. and Brett, M. J. (1998a). Sculptured thin films and glancing angle deposition: Growth mechanics and applications. *J. Vac. Sci. Tech. A: Vac., Surf. Films,* **15**, pp. 1460.

21. Gall, D. D. and Gall, D. (2011). Sculptured thin films: Nanorods, nanopipes, nanosmiles. *Proc. SPIE,* **8104**, pp. 158–164.

22. Shalabney, A. H., Khare, C., Bauer, J., Rauschenbach, B., and Abdulhalim II, I. (2012). Detailed study of surface-enhanced Raman scattering from metallic nanosculptured thin films and their potential for biosensing. *J. Nanophoton,* **6**, pp. 061605.

23. Srivastava, S., Hamo, H., Kushmaro, A., Marks, R., Gru¨ner, C., Rauschenbach, B., and Abdulhalim, I. (2015). Highly sensitive and specific detection of E. coli by a SERS nanobiosensor chip utilizing metallic nanosculptured thin films. *Analyst,* **140**, pp. 3201–3209.

24. Novotny, L. (2007). Effective wavelength scaling for optical antennas. *Phys. Rev. Lett.,* **98**, 26, pp. 266802.

25. Schneider, C. A., Rasband, W. S., and Eliceiri, K. W. (2012). NIH image to ImageJ: 25 years of image analysis. *Nat. Meth.,* **9**, pp. 671–675.

26. Chung, T., Lee, S.-Y., Song, E. Y., Chun, H., and Lee, B. (2011). Plasmonic nanostructures for nano-scale bio-sensing. *Sensors,* **11**, pp. 10907–10929.

27. Tang, H., Zhu, C., Meng, G., and Wu, N. (2018). Review—surface enhanced Raman scattering sensors for food safety and environmental monitoring. *J. Electrochem. Soc.,* **165**, pp. B3098–B3118.

28. Tan, S., Erol, M., Sukhishvili, S., and Du, H. (2008). Substrates with discretely immobilized silver nanoparticles for ultrasensitive detection of anions in water using surface-enhanced Raman scattering. *Langmuir,* **24**, pp. 4765–4771.

29. MicroSEQ® E.coli O157:H7 Detection Kit (2014), *http://tools.lifetechnologies.com/content/sfs/brochures/CO13764.pdf,* Accessed January 21, 2022.

30. Kalele, S. A., Kundu, A. A., Gosavi, S. W., Deobagkar, D. N., Deobagkar, D. D., and Kulkarni, S. K. (2006). Rapid detection of Escherichia coli by using antibody-conjugated silver nanoshells. *Small,* **2**, pp. 335–338.

31. Han, D., Lim, S. Y., Kim, B. J., Piao, L., and Chung, T. D. (2010). Mercury (ii) detection by SERS based on a single gold microshell. *Chem. Comm.*, **46**, 30, pp. 5587–5589.

32. Ma, W., Sun, M., Xu, L., Wang, L., Kuang, H., and Xu, C. (2013). A SERS active gold nanostar dimer for mercury ion detection. *Chem. Comm.*, **49**, pp. 4989–4991.

33. Lee, S., Lee, H., and Chung, H. (2013). New discrimination method combining hit quality index based spectral matching and voting. *Anal. Chim. Acta*, **758**, pp. 58–65.

34. Khan, S. S. and Madden, M. G. (2012). New similarity metrics for Raman spectroscopy. *Chemomet. Intel. Lab. Sys.*, **114**, pp. 99–108.

35. Rodriguez, J. D., Westenberger, B. J., Buhse, L. F., and Kauffman, J. F. (2011). Quantitative evaluation of the sensitivity of library-based Raman spectral correlation methods. *Anal. Chem.*, **83**, pp. 4061–4067.

36. Vandenabeele, P., Hardy, A., Edwards, H. G., and Moens, L. (2001). Evaluation of a principal components-based searching algorithm for Raman spectroscopic identification of organic pigments in 20th century artwork. *Appl. Spec.*, **55**, pp. 525–533.

37. Simonyan, K. and Zisserman, A. (2014). Very deep convolutional networks for large-scale image recognition. *3rd International Conference on Learning Representations, ICLR 2015 - Conference Track Proceedings.*

38. Szegedy, C., Liu, W., Jia, Y., Sermanet, P., Reed, S., Anguelov, D., Erhan, D., Vanhoucke, V., and Rabinovich, A. (2015). Going deeper with convolutions. In *Proc. IEEE Conf. Comp. Vis. Patt. Recog.*, pp. 1–9.

39. Lussier, F., Thibault, V., Charron, B., Wallace, G. Q., and Masson, J.-F. (2020). Deep learning and artificial intelligence methods for Raman and surface-enhanced Raman scattering. *Tr. Anal. Chem.*, **124**, pp. 115796.

40. Ho, C.-S., Jean, N., Hogan, C. A., Blackmon, L., Jeffrey, S. S., Holodniy, M., Banaei, N., Saleh, A. A., Ermon, S., and Dionne, J. (2019). Rapid identification of pathogenic bacteria using Raman spectroscopy and deep learning. *Nat. Commun.*, **10**, pp. 1–8.

41. Hossin, M. and Sulaiman, M. N. (2015). A review on evaluation metrics for data classification evaluations. *Int. J. Data Min. and Knowled. Managem. Proc.*, **5**, pp. 1–11.

Chapter 9

Effective Enhancement of Raman Scattering, Fluorescence, and Near-Field Imaging

Raisa Siqueira Alves, Italo Odone Mazali, and Diego Pereira dos Santos
Institute of Chemistry, University of Campinas, P.O. Box 6154, ZIP Code 13084-970, Campinas, Brazil
santosdp@unicamp.br

9.1 Introduction

The technological potential of plasmonic materials is enormous, especially when it comes to nanostructured systems. Primarily, plasmonic nanoparticles have been employed to improve the optical signals of molecules for spectroscopic detection and imaging. This signal enhancement comes from the modified optical response of nearby molecules due to excitations of localized surface plasmons in the particle.

Plasmonics-Based Optical Sensors and Detectors
Edited by Banshi D. Gupta, Anuj K. Sharma, and Jin Li

ISBN 978-981-4968-85-0 (Hardcover), 978-1-003-43830-4 (eBook)
www.jennystanford.com

Plasmonics enables to achieve the ultimate limit of detection, which is the impressive power to capture the signal emitted from a single molecule. Together with the use of near-field optics, plasmonic nanosystems are used to image a single molecule with subdiffraction resolution. These breakthroughs have made plasmonics an extremely attractive field in the past few decades. In this context, a deep knowledge of the enhancement mechanisms is crucial for designing highly efficient plasmonic materials and devices according to the desired application. Thus, this chapter is devoted to discussing the main electromagnetic phenomena responsible for plasmonic signal enhancement. Particularly, spectroscopic detection and near-field optical imaging via plasmon-enhanced fluorescence and Raman scattering will be covered. Ultimately, selected examples of single-molecule detection are presented to illustrate the power of each technique.

9.2 Optical Properties of Plasmonic Systems

The localized surface plasmons (LSP) are quantized oscillations of the free electron gas confined to conductor-dielectric boundaries of sub-wavelength particle systems. Plasmonic particles act as nanosized antennas, interacting efficiently with light, within an effective area much larger than the own physical size of the nanoparticle. Therefore, plasmonic nanoparticles are capable of focusing far-field radiation into confined nanosized regions much smaller than the wavelength of light, overcoming Abbe's diffraction limit [1–3].

Plasmonic resonances appear in a wide range of optical frequencies. For instance, nanostructured Na, Mg, Al, and Rh show LSP resonances (LSPR) mostly in the ultraviolet region [4]. Colloidal nanocrystals of heavily-doped semiconductors and metal oxides display LSPR in the near-infrared region (NIR) [5]. The majority of works on the LSPR effect employ nanoparticles of coinage metals such as copper, silver, and gold. Due to their optimum dielectric properties in the visible range and high free electron densities, those metals exhibit excellent plasmonic efficiencies [4, 6].

The resonant excitation of LSPs leads to a coupling between the incident field and the field produced by the charge oscillations. Consequently, the electromagnetic field at the particle vicinities is redistributed, resulting in regions of strong field concentration and depletion at the particle surface (Fig. 9.1). The enhanced local field shows a strong spatial and frequency dependence with the incident field. The field intensity rapidly decreases with increasing distance from the surface. Also, the magnitude of field enhancement raises rapidly as the incident wavelength becomes resonant with the plasmon oscillation.

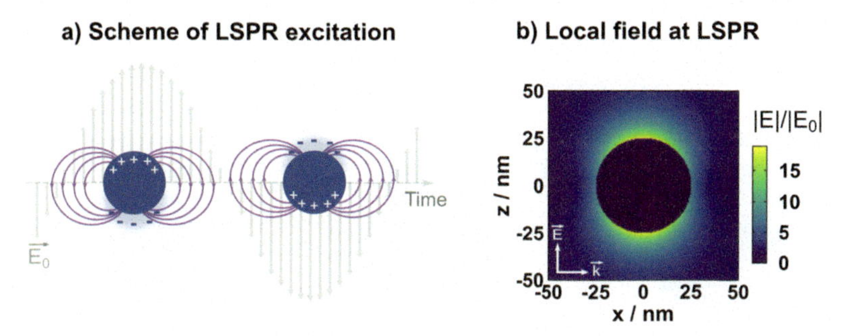

Figure 9.1 (a) Illustration of localized surface plasmon resonances. (b) Electric field mapping at the particle surroundings for resonant plasmon excitation of a 50 nm Ag sphere embedded in a vacuum. The simulation was performed using the Mie theory calculated with the GMM-field package [16]. The optical data for silver from the compilation of Johnson and Christy [18] was used for modeling.

The near-field properties of plasmonic systems reflect in the far field as increased light extinction toward plasmon oscillation frequencies. The major contribution to extinction comes from elastic scattering, due to the induced charge oscillations. But ohmic losses are inevitable, and the plasmon energy is partially dissipated through absorption processes. This is particularly important for metals that show interband transitions at optical frequencies, such as gold and copper. Most of the early research on plasmonic systems was directed toward the design of nanomaterials with high Q-factor for applications in photonics and sensing. Although absorption losses were undesirable at first, they have proved useful for hyperthermia and photocatalysis.

The magnitude of amplification and the spatial distribution of the local electromagnetic field is extremely sensitive to the nature of the particle boundaries. This is because the boundary conditions define what types of plasmon oscillations will arise according to the frequency of the incoming light. For small particles below 20 nm, the incident field is considered spatially uniform so that only charge oscillations of dipolar character arise. But, as the particle grows, retardation effects become relevant, and higher-order resonances become possible. Higher-order plasmon modes can also arise by changing the particle shape or aspect ratio. Therefore, the optical features of plasmonic particles can be tuned across a wide spectral region.

Modification of size, geometry, composition, and dielectric environment are examples of parameters that can be controlled for that purpose (Fig. 9.2) [3, 7].

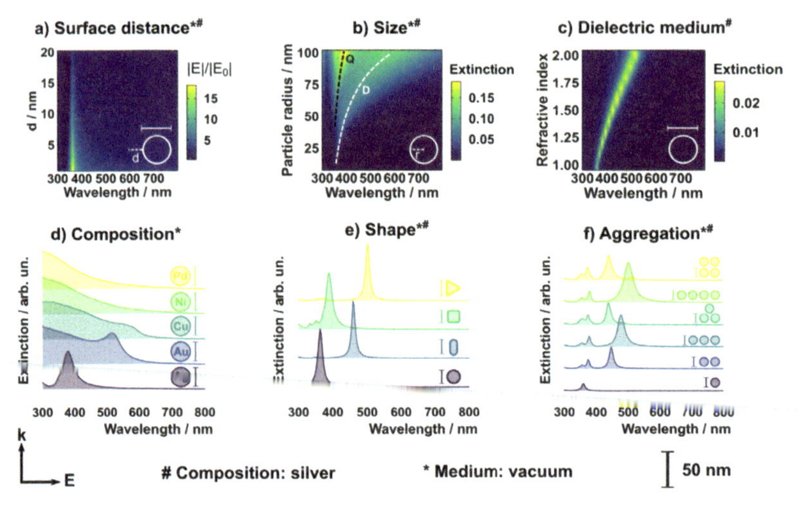

Figure 9.2 Spectral changes of plasmonic particles by varying the parameters of the nanosystem. (a) Dependence of electric field magnitude according to the distance from the surface. (b) Spectral mapping for nanospheres with variable radii, highlighting the dipolar (D) and quadrupolar (Q) plasmonic modes. (c) Spectral shifts for a nanosphere by changing the dielectric medium. (d) Extinction spectra for nanospheres of different compositions, (e) shapes, and (f) aggregation geometries at a fixed gap of 1 nm. The calculations with spheres were obtained with the GMM-field [16] program and other shapes were calculated with the MNPBEM17 package [15] using the tabulated data from Johnson and Christy [17, 18] in both cases.

In fact, with the rapid progress of new synthesis routes, metal nanoparticles with a countless number of shapes have been reported in the literature, e.g., spheres [8], rods [9], wires [10], stars [11], flowers [12], cubes [13], and so forth. Also, advances in numerical calculations have enabled the prediction of the optical features of particle systems. Several methods are available nowadays, e.g., the discrete dipole approximation (DDA), the finite-difference time-domain (FDTD), the generalized Mie theory (GMT), and the boundary element method (BEM) [14–16]. Usually, tabulated dielectric functions such as Johnson and Christy's reference data [17, 18] are used as input in the simulations. With these advances, the development of theoretical calculations has strongly contributed to the rational design of plasmonic materials [19–21].

Furthermore, when two or more closely spaced plasmonic particles are resonantly excited, the near-field intensity is further enhanced at the interparticle gap. Due to the enormous field amplification, these regions are termed hot spots. Hot spots originate wherever highly dense, localized plasmon modes can arise. In this sense, not only particle junctions, but also plasmonic cavities, nanostructures with sharp corners (e.g., triangles), and nanosized metal tips can generate hot spots under light excitation. In the latter case, the formation of hot spots relates to a lightning rod effect [22, 23].

For particle aggregates, the formation of hot spots is qualitatively treated in terms of a plasmon hybridization model [24]. This model is often evoked to explain the charge distributions and the spectral changes observed for particle clusters compared to the spectra of the corresponding isolated particles. The plasmon hybridization picture provides an intuitive view of plasmon resonances in particle aggregates and other complex structures, e.g., core-shell systems. Considering a particle dimer as an example, the wavefunctions of two individual particle plasmons superimpose, causing in-phase and out-of-phase interferences. The in-phase interference leads to a bonding plasmon mode of lower energy. On the other hand, the out-of-phase interference generates an anti-bonding plasmon state of higher energy. Additionally, the orientation of the incident field relative to the interparticle axis plays an important role in the plasmon coupling. The field orientation determines what single-particle plasmons will be excited and, hence, the overlapping profile between them. Therefore, different couplings can be expected for parallel and perpendicular field incidence (Fig. 9.3).

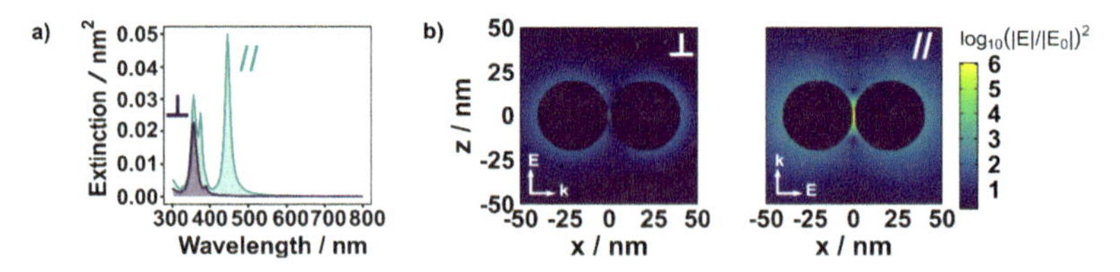

Figure 9.3 Optical response of a 20 nm silver dimer embedded in a vacuum. (a) Extinction spectra and (b) local field distribution for incident polarization parallel and perpendicular to the dimer axis at the main resonance wavelengths (357 nm for perpendicular and 447 nm for parallel incidence). The above examples were calculated with the GMM-field program [16] using the data from Ref. [18].

Hot spots can be understood as local regions of high Q-factor, where the energies of the incoming and scattered fields of a plasmon are accumulated [25]. Thus, from a quantum perspective, hot spots are associated with an increase in the local density of photon states (LDOS). The LDOS at a given position in space directly relates to the probability of spontaneous decay of a quantum emitter (e.g., a molecule or an atom) placed at that position. The higher the LDOS at the frequency emitted by the quantum system, the more favorable the spontaneous decay of this emitter. In fact, the pioneering work of E.M. Purcell showed that placing an emitter inside a cavity in resonance with a quantum transition leads to an enhancement of the spontaneous emission rate [26]. Since hot spots are conceptually similar to open cavities, they induce the Purcell effect, providing enormous enhancement of secondary radiation by molecules.

The formation of hot spots enabled achieving ultrasensitive detection of analytes at the single-molecule level in surface-enhanced spectroscopic techniques like SERS and SEF. On this basis, nanoparticle engineering toward control over hot spot formation has been intensively investigated to produce high-efficiency substrates for sensing [27, 28].

9.3 Emission of Secondary Radiation by Molecules

When light encounters a quantum system, e.g., a molecule, the incoming photon can either be transmitted without interacting or couple with the system in various ways. Typically, molecules absorb radiation at frequencies that match the energy spacing between two quantum levels. But incident photons of mismatching energy can cause a weak perturbation in the energy levels, still opening channels for quantum transitions. This perturbation is often treated as a "virtual" absorption in opposition to a real absorption process, which requires resonant excitation.

Excited quantum systems can spontaneously dissipate their energies through the formation of a second photon. Secondary radiation can occur via elastic (Rayleigh) scattering, inelastic (Raman) scattering, or spontaneous emission, also known as luminescence in the case of relaxation between electronic states. In most cases, secondary photons differ in phase, directionality, and polarization compared to the excitation photons.

Despite this common feature, there are major differences between scattering and luminescence. Scattering occurs instantaneously to the excitation of the system due to the virtual absorption of a photon. In contrast, luminescence is part of a multi-step decay process triggered by real photon absorption. Because of that, luminescence occurs within a time delay after excitation [23].

Additionally, luminescence can be further divided into fluorescence or phosphorescence, depending on the spin nature of the excited states prior to emission. Generally, emitted photons from luminescence have lower energies than incoming light. Historically, those processes are termed Stokes luminescence. Otherwise, if the emitted photons show higher energies, one has an anti-Stokes luminescence. This nomenclature is equivalently applied for treating the energy differences in Raman scattering [25]· An overview of scattering and luminescence processes is comparatively illustrated by Jablonski diagrams in Fig. 9.4.

Every quantum state transition leads to a change in the electron densities within the molecule in response to the redistribution of the electron energies. Thereby, electric and magnetic multipoles arise from these light-induced transitions. Among all possible multipoles, the electric dipole moment is the most relevant to describe linear optical processes. Therefore, spectroscopic transitions of any kind are treated as resultant of a dipolar coupling between light and matter. Based on that, luminescence and light scattering can be treated using the same formalism, namely the time-dependent perturbation theory.

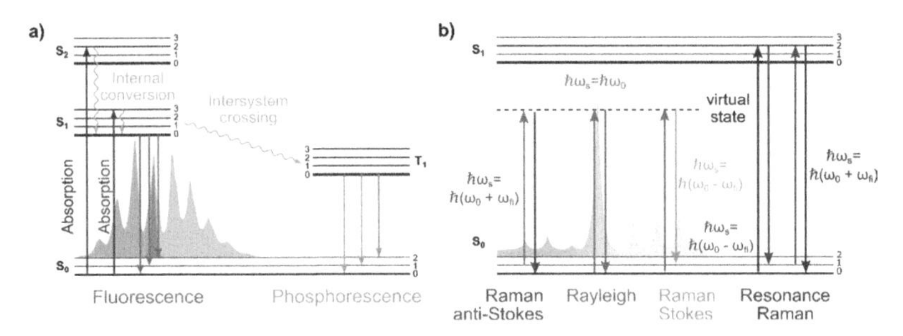

Figure 9.4 Jablonski diagram representing the spectroscopic transitions involved in (a) luminescence and (b) scattering.

9.3.1 Basic Formalism for Direct Spectroscopic Transitions

For spectroscopic transitions involving real photon absorption, such as luminescence, the time-dependent perturbation theory enables describing the initial and final states of the quantum system by stationary, or time-independent, wavefunctions. Following this theory, a direct spectroscopic transition occurs when the respective transition dipole moment $(\boldsymbol{p})_{fi}$ connecting the initial and final states of an electron is non-zero:

$$(\boldsymbol{p})_{fi} = \left\langle \Psi_f | \hat{\boldsymbol{p}} | \Psi_i \right\rangle = \int \Psi_f^* \hat{\boldsymbol{p}} \Psi_i \mathrm{d}\tau = e \int \Psi_f^* \hat{\boldsymbol{r}} \Psi_i \mathrm{d}\tau, \tag{9.1}$$

where $\hat{\boldsymbol{p}}$ is the dipole moment operator, $\hat{\boldsymbol{r}}$ is the position operator, and e is the electron charge. Ψ_i and Ψ_f are the stationary wavefunctions for the initial and final states, respectively, and Ψ_f^* is the complex conjugate of the final state wavefunction.

The square modulus of the dipole transition moment gives the transition probability $|(\boldsymbol{p})_{fi}|^2$. The transition probability is directly proportional to the intensities of secondary radiation, thus representing the starting point for defining spectroscopic selection rules. Allowed transitions show a non-zero transition probability. On the other hand, if $|(\boldsymbol{p})_{fi}|^2$ is zero, the transition is forbidden. A non-zero transition probability physically means that the incident photon can effectively couple with the system and drive the transition.

Selection rules rely on finding the appropriate pairs of electronic wavefunctions for which the transition dipole integral does not vanish. This condition is met whenever the wavefunctions of initial and final states show different parities but the same spin multiplicities.

9.3.2 Fluorescence Emission

Luminescence can be described through a timeline of events, illustrated in Fig. 9.4a. Consider a molecule initially in the ground electronic singlet state S_0. Upon photon absorption, the molecule is promoted to a higher vibrational level of an excited electronic state, commonly S_1 or S_2. The molecule rapidly reaches thermal equilibrium and relaxes to the lowest vibrational level of S_1. This non-radiative relaxation pathway occurs within a few picoseconds and is termed internal conversion. After that, the molecule can decay back to the ground state S_0 by either emitting a photon or by any other non-radiative relaxation process. The radiative decay pathway from a singlet excited state is a rapid spin-allowed process with a typical lifetime of 10 ns, characterizing a fluorescence emission.

Alternatively, the molecule can decay radiatively or non-radiatively from S_1 to a triplet excited state T_1. This process is known as an intersystem crossing. The radiative decay from T_1 to S_0 is forbidden due to the change in spin multiplicities. Therefore, this type of decay, named phosphorescence, is a much slower event compared to fluorescence, with lifetimes in the range between milliseconds to seconds.

The emission spectra of fluorophores are usually a red-shifted mirrored image of the absorption spectra relative to the transition $S_0 \rightarrow S_1$. Due to the low thermal energy at room temperature, the population of excited vibrational states is usually negligible. Therefore, almost all fluorophores in the condensed phase undergo internal conversion. As a result, the brightest emissions lines originate from the ground vibrational state of S_1.

Given the stochastic nature of spontaneous emission, the fluorescence efficiency is measured in terms of probabilistic variables, such as decay rates. Those variables help define two key quality parameters in fluorescence, namely, the lifetime (τ) and the quantum yield (Q). The lifetime corresponds to the average time the molecule stays in an excited electronic state before damping occurs. It is the reciprocal of the sum of the radiative (Γ_{rad}) and non-radiative

(k_{NR}) decay rates:

$$\tau = \frac{1}{\Gamma_{\mathrm{Rad}} + k_{NR}}. \tag{9.2}$$

The quantum yield is a measurement of the brightness of a given fluorescence emission. It is defined as the ratio between the number of emitted to the number of absorbed photons, expressed by the decay rates:

$$Q = \frac{\Gamma_R}{\Gamma_R + k_{NR}}. \tag{9.3}$$

Different processes can quench fluorescence emission, drastically reducing the quantum yield of a fluorophore. Quenching commonly results from the interaction of the fluorophore with another chemical species that acts as a quencher and deactivates the fluorophore. For instance, luminescence can be quenched by Förster or fluorescence resonance energy transfer (FRET) [29]. FRET occurs when the emission frequency matches the absorption of a closely located molecule, resulting in non-radiative energy transfer from the excited fluorophore. Besides the extent of spectral overlap, FRET also varies with the distance r between the donor and acceptor. The FRET efficiency falls with r^6 and depends on the relative dipolar orientation between the emitter and the absorber.

9.3.3 Raman Scattering

The energy transfer model helps illustrate light scattering by molecules based on a fully quantum description (Fig. 9.4b) [30]. In this picture, an incident photon of arbitrary energy $\hbar\omega_0$ interacts with the molecule and causes a perturbation in the electronic states upon virtual absorption. The initial photon $\hbar\omega_0$ gets annihilated in the process and a new photon of energy $\hbar\omega_s$ is created and scattered in all directions via the promotion to an intermediate virtual state. This virtual state does not have a defined energy value since it is not a solution to the time-independent Schrödinger's equation.

Commonly, the energy difference between the incident and scattered photons corresponds to the energy $\hbar\omega_{fi}$ of a vibrational transition $i \rightarrow f$ within the ground electronic level of the molecule, i.e., $\omega_s = \omega_0 \pm \omega_{fi}$. If $\omega_s = \omega_0 - \omega_{fi}$ the process is termed Stokes Raman scattering, and the molecule is promoted from the ground to an excited vibrational level. Conversely, if $\omega_s = \omega_0 + \omega_{fi}$ the anti-Stokes Raman scattering takes place. In this case, the molecule is initially at an excited vibrational state and, after perturbation, decays to a lower vibrational level. Finally, the condition $\omega_s = \omega_0$ describes the resonant Rayleigh scattering.

Given these, the intensities of Stokes (I_S) and anti-Stokes (I_{AS}) Raman scattering will depend, respectively, upon the populations of the ground and excited vibrational states at the ground electronic state. The relative population of states varies according to the temperature (T) of the system, following the Boltzmann distribution. Being k_B the Boltzmann constant, the ratio between anti-Stokes (I_{AS}), and Stokes (I_S) scattering intensities can be calculated from Refs. [30, 31]:

$$\frac{I_{AS}}{I_S} = \left[\frac{\omega_0 + \omega_{fi}}{\omega_0 - \omega_{fi}}\right]^4 \exp\left(\frac{-hc\omega_{fi}}{k_B T}\right). \tag{9.4}$$

The mathematical treatment of scattering requires the second-order correction to the perturbation theory [30, 32]. Then, the total transition dipole moment $(\boldsymbol{p})_{fi}$ for scattering becomes [30]:

$$(\boldsymbol{p})_{fi} = \left\langle \Psi'_f | \hat{\boldsymbol{p}} | \Psi'_i \right\rangle, \tag{9.5}$$

where Ψ'_i and Ψ'_f denote the time-dependent perturbed wavefunctions of the initial and final molecular states, respectively. A series expansion can be used to express Ψ'_i and Ψ'_f as a sum of each unperturbed counterpart $\Psi^{(0)}_{i,f}$ and nth-order perturbations $\Psi^{(n)}_{i,f}$. The first-order term for the $(\boldsymbol{p})_{fi}$ is sufficient to describe linear Raman scattering, which leads to a transition dipole that varies linearly with the incident electric field $\boldsymbol{E}_0$:

$$\left(\boldsymbol{p}_\rho(\omega_s)\right)_{fi} = \left(\alpha_{\rho\sigma}\right)_{fi} \boldsymbol{E}_{\sigma 0}(\omega_0), \tag{9.6}$$

where ρ and σ denote the x, y, or z direction and $(\alpha_{\rho\sigma})_{fi}$ is a component of the general Raman polarizability tensor $(\alpha)_{fi}$. The tensor components $(\alpha_{\rho\sigma})_{fi}$ relate to the Raman scattering efficiency. In a sample containing N scattering molecules, the total intensity of Raman line related to the transition $i \rightarrow f$ is given by:

$$I_{fi} \propto N\omega_s^4 |E_0|^2 \sum_{\rho\sigma} \left|\left(\alpha_{\rho\sigma}\right)_{fi}\right|^2, \tag{9.7}$$

where $|E_0|$ is the magnitude of the incident electric field. Therefore, the Raman intensity varies strongly with the frequency of the mode. Besides, the measured Raman intensities are also dependent on experimental parameters, such as the detection angle relative to the incident light. From the perturbation theory, the solution to the tensor $(\alpha_{\rho\sigma})_{fi}$ is [33]:

$$\left(\alpha_{\rho\sigma}\right)_{fi} = \frac{1}{\hbar}\sum_{r \neq i,f}\left\{\frac{\langle\psi_f|\hat{p}_\rho|\psi_r\rangle\langle\psi_r|\hat{p}_\sigma|\psi_i\rangle}{\omega_{ri} - \omega_0 - \mathrm{i}\Gamma_r} + \frac{\langle\psi_f|\hat{p}_\rho|\psi_r\rangle\langle\psi_r|\hat{p}_\sigma|\psi_i\rangle}{\omega_{ri} + \omega_0 + \mathrm{i}\Gamma_r}\right\}. \tag{9.8}$$

The previous expression is the Kramers-Heisenberg-Dirac (KHD) dispersion equation, which depends only on stationary wavefunctions ψ_i, ψ_f, and ψ_r of the initial, the final, and any other state r of all possible electronic states of the molecule. The term ω_{ri} indicates the energy frequency associated with the electronic transition $i \rightarrow r$. The equation for $(\alpha_{\rho\sigma})_{fi}$ involves the summation of all r possible states of the molecule besides the initial and final states [33]. Assuming that $\omega_r > \omega_i$ and $\omega_r > \omega_f$, the numerators $\langle\psi_f|\hat{p}_\rho|\psi_r\rangle$ and $\langle\psi_r|\hat{p}_\sigma|\psi_i\rangle$ could be understood as a virtual absorption related to $i \rightarrow r$ and an emission involving the transition $r \rightarrow f$. In other words, $(\alpha_{\rho\sigma})_{fi}$ depends upon all possible pathways connecting $i \rightarrow r$ and $r \rightarrow f$ that lead to a non-zero transition electric dipole moment $(\boldsymbol{p})_{fi}$. This condition is what defines the Raman selection rules. Whenever $(\boldsymbol{p})_{fi}$ vanishes, the respective Raman line is forbidden.

The $\omega_{ri} - \omega_0$ term in the denominator means that a resonance condition can be fulfilled as $\omega_0 \rightarrow \omega_{ri}$. In this case, the polarizability, and hence, the Raman scattering intensity is strongly enhanced. For that reason, this phenomenon is termed resonance Raman scattering, as schemed in Fig. 9.4b.

Raman scattering is a much less efficient process compared to fluorescence. For instance, the differential Raman scattering cross-section of rhodamine 6G (RH6G), a common dye used in Raman experiments, lies around 10^{-27} cm^2. In contrast, the fluorescence cross-section of RH6G at a 514 nm excitation is 11-fold larger [23, 34]. Thus, from a practical perspective, Raman spectroscopy may be unfeasible for routine analysis, due to its low sensitivity. This is particularly important when it comes to the detection of non-chromophores substances, for which the resonance Raman condition cannot be achieved. Due to this limitation, several approaches to increase the detection efficiency of Raman spectroscopy have emerged in the past few decades. Among them, surface-enhanced Raman spectroscopy (SERS) is one of the most explored nowadays [35].

9.4 Mechanisms of Modified Secondary Radiation by Plasmonics

The excitation of plasmonic resonances opens up several channels to alter the spectroscopic response of nearby molecules, mostly by enhancing the emission of secondary radiation. This effect is mainly due to interactions of electromagnetic nature between the plasmonic antenna and the molecular dipole.

The emission of secondary radiation can be treated as a problem of point dipole emission. A molecule placed near an excited plasmonic particle is subjected to the action of the enhanced local field at that position. As a result, a dipole moment $\boldsymbol{p}(\omega)$ is created in the molecule from this interaction. This dipole emission can strongly differ from the emission in free space, due to the change in the local dielectric environment caused by the presence of the metal. Additionally, the field created from the radiative dipole $\boldsymbol{p}(\omega)$ can, in turn, re-excite the plasmon, transferring part of its energy to the metal without involving a photon emission. On this basis, the electromagnetic origin of modified secondary radiation of molecules by metal particles can be broken down into three main effects, namely, the local field enhancement, the radiation enhancement, and a non-radiative enhancement [23, 36].

9.4.1 Local Field Enhancement

As the excitation frequency ω_0 approaches the natural frequency of plasmon oscillation, sharp field gradients occur near the particle surface. As a consequence, the magnitude and orientation of the local field $\boldsymbol{E}_{Loc}(\omega_0)$ can differ dramatically from the incident field $\boldsymbol{E}_0(\omega_0)$ depending on the position (r). For linear spectroscopic transitions, the signal intensity is proportional to the square of the electric field. Thus, one defines the local field intensity enhancement factor F_{Loc}

to quantify the contribution to signal enhancement provided by the change in the local field distribution as follows:

$$F_{Loc} = \frac{\left|\boldsymbol{E}_{Loc}\left(\omega_0, r\right)\right|^2}{\left|\boldsymbol{E}_0\left(\omega_0, r\right)\right|^2}. \tag{9.9}$$

Notice that the equation for F_{Loc} does not take into account changes in the electric field polarization. Due to the field confinement, F_{Loc} represents the change in the photon density at a certain point due to the presence of the plasmonic structure compared to the photon density at the same point in free space. It should be clear that this enhancement does not violate the energy conservation principle. Although the field redistribution leads to areas of local field enhancement, this is compensated by areas of field depletion so that the energy balance remains unchanged.

The primary contribution of the local field intensity enhancement factor is the excitation rate of molecular dipoles placed in the particle vicinities. For absorbing species, F_{Loc} also describes the increase in the absorption cross-section σ_{Abs} and, hence, in the absorption rate Γ_{Abs} of the molecule.

Extremely intense local fields can be found in hot spots such as interparticle gaps. For instance, the F_{Loc} can reach up to 10^5 orders of magnitude in these regions. (Fig. 9.5a). Usually, the local field enhancement is inversely proportional to the gap size, due to the weakening of plasmonic coupling between the particles with increasing separation.

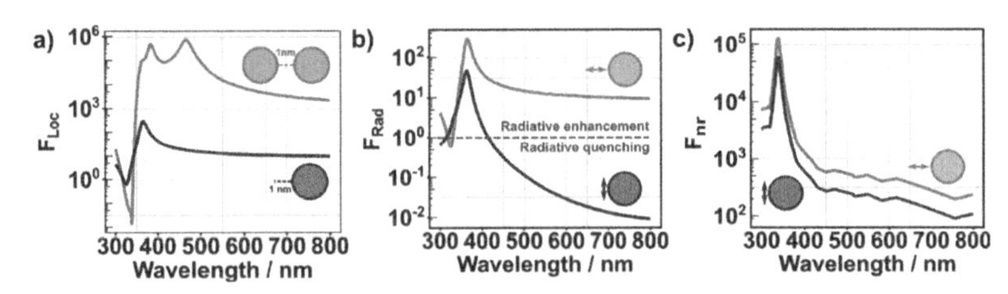

Figure 9.5 (a) Calculated local field enhancement factor at 1 nm from the particle surface and at the center of the dimer hot spot with an interparticle gap of 1 nm. (b) Radiative and (c) non-radiative enhancement factors for a dipole located at 1 nm from the particle surface with parallel and perpendicular relative orientation. The above examples were calculated using the GMM-field [16] package with the compiled data from Johnson and Christy [18].

9.4.2 Radiation Enhancement

Besides local field enhancement, plasmonic antennas will induce the Purcell effect on nearby molecular dipoles. The change in the dielectric environment provided by the particle strongly alters the dipole emission. In this case, both the directional profile and the total emitted power of the dipole will be affected. The latter effect can be quantified by the radiation enhancement factor F_{Rad}. By defining P_{Rad} as the total radiative power extracted from the dipole in free space and P_{Rad}^{M} as the total emitted power modified by the metal, F_{Rad} is given by:

$$F_{Rad} = \frac{P_{Rad}^{M}}{P_{Rad}}. \tag{9.10}$$

The F_{Rad} relates to the re-emission of radiation by the dipole after excitation. This is separated from the contribution to the dipole excitation, F_{Loc}, presented in the previous section. A comparison between the F_{Loc} and F_{Rad} calculation schemes is given in Fig. 9.6b.

The expression for F_{Rad} depends on the dipole emitted power, which can be correlated with the quantum picture by the direct proportionality to the radiative decay rate, i.e., $\left(P_{Rad}^{M} / P_{Rad}\right) = \left(\Gamma_{Rad}^{M} / \Gamma_{Rad}\right)$. The radiative decay rate follows Fermi's golden rule, originally deduced by Dirac [37, 38]. The radiative decay rate in free space Γ_{Rad} increases as:

$$\Gamma_{Rad} \propto \left|(\boldsymbol{p})_{fi}\right|^{2} D\left(\omega_{\mathrm{Dip}}\right). \tag{9.11}$$

On this basis, the spontaneous emission rate is directly proportional to the transition probability $|(\boldsymbol{p})_{fi}|^2$ and to the local density of states $D(\omega_{\mathrm{Dip}})$ at the dipole emission frequency ω_{Dip}. For scattering, the same proportionality with the LDOS holds, but at the scattered frequency [25]. While $|(\boldsymbol{p})_{fi}|^2$ is a property of the emitter, $D(\omega_{\mathrm{Dip}})$ can be tuned by shaping the electromagnetic boundary conditions. The maximum achievable enhancement, f, in the spontaneous emission rate is determined by [26]:

$$f = \frac{\Gamma_{Rad,\,cav}}{\Gamma_{Rad,\,vac}} = \frac{3Q}{4\pi^2}\frac{\lambda^3}{V}, \tag{9.12}$$

where V is the volume of the cavity relative to λ^3. The quality factor Q is the ratio between the amount of energy stored to the energy dissipated by a resonator in each oscillation. Equivalently, Q can be determined by the linewidth $\Delta\omega$ of the resonance. For a resonance frequency ω, $Q = \omega/\Delta\omega$. So, the above equation expresses the emission enhancement that arises due to the discretization and increased density of the resonant mode caused by the confinement of the cavity. For treating open cavities of plasmonic nanosystems, V is replaced by the effective volume of the plasmon mode, V_{eff} [39]. Thus, F_{Rad} relates to the change in the LDOS, due to plasmonic excitation at the frequency of the corresponding secondary radiation emitted from the dipole, ω_{Dip}.

Except for nanospheres, the calculation of F_{Rad} is usually difficult for particles of complex shapes. The value of F_{Rad} is frequently approximated based on the local field properties by applying the optical reciprocity theorem (ORT) and the plane wave approximation (PW). Then F_{Rad} can be reformulated to [23]:

$$F_{Rad} = \frac{P_{Rad}^{M}}{P_{Rad}} \approx \frac{\left|\boldsymbol{E}_{loc}\left(\omega_{\mathrm{Dip}}, r\right)\right|^{2}}{\left|\boldsymbol{E}_{0}\left(\omega_{\mathrm{Dip}}, r\right)\right|^{2}}. \tag{9.13}$$

Thus, the modified dipolar emission in the far field could be estimated in terms of the local field enhancement at the position r and the frequency of the radiating dipole, ω_{Dip}. This will lead to important results on the enhancement of Raman scattering, as discussed later herein. However, the above approximation must be used with caution, since its validity is limited to very specific conditions. The agreement of the approximation is limited to small spherical particles dispersed inhomogeneous media. This is because the approximation fails when dealing with particles supported on substrates and when the excitation of higher-order plasmon modes occurs. Therefore, the particles should be small enough to sustain only dipolar oscillations. Also, the approximation implies optical measurements in backscattering configuration so that other measurement schemes would strongly diverge from predictions using the previous relation [36, 39–41].

9.4.3 Non-radiative Enhancement

In contrast to the effect of local field enhancement, secondary radiation from a dipole is not always enhanced by a plasmonic antenna. The radiative decay of a dipole can be suppressed in some cases as a result of increasing non-radiative damping channels provided by the close interaction with the metal particle. By calculating the total power emitted by the dipole in non-radiative form P_{NR}^{M}, the non-radiative enhancement factor F_{NR} of a plasmonic antenna can be obtained from Ref. [23]:

$$F_{NR} = \frac{P_{NR}^{M}}{P_{Rad}}. \tag{9.14}$$

In scattering, the re-emission of the photon is instantaneous and occurs simultaneously with the virtual photon absorption. For this reason, radiative decay is independent of any other non-radiative decay that may occur in scattering. In other words, non-radiative decay does not significantly compromise the radiative efficiency of a scattering molecule. For this reason, the enhancement of non-radiative decay becomes irrelevant for most applications involving metal-enhanced scattering. Considering non-radiative energy transfer between the metal particle and the molecule, a possible channel is when the Raman frequency corresponds to the frequency of subradiant plasmon modes. The metal then absorbs the Raman photons, which are in turn undetectable in the far-field [23].

On the other hand, non-radiative decay is crucial in luminescence, leading to a substantial decrease in quantum efficiency. Spontaneous emission is one of many steps that can take place after the real excitation of a quantum emitter. In this case, there exists a competitive interplay between radiative and non-radiative decay channels. Metal particles can favor the non-radiative decay of luminescent centers mainly due to resonance energy transfer (RET). RET can occur when the metal absorbs at the emission frequency of the fluorophore. The RET to the metal is a result of non-radiative dipolar coupling between the dipolar donor and the particle acceptor. This means that the observed non-radiative enhancement near a metal particle does not involve a direct photon emission. Similar to FRET, luminescence quenching by a metal shows a strong distance dependence between the molecular donor and the particle acceptor. Thus, the non-radiative enhancement will be prominent at small distances from the particle surface.

9.4.4 Overview

The whole range of effects discussed so far that a nanoantenna can induce to change secondary emission is schematically represented in a three-level Jablonski diagram in terms of the corresponding decay rates (Fig. 9.6). The magnitudes of each enhancement factor are comparatively shown in Fig. 9.5. In Fig. 9.6, the superscript *M* refers to the changes in absorption and decay rates of the emitter by the plasmonic particle. Although real quantum transitions are represented, the scheme is adequate for understanding virtual excitations and, thus, modified scattering. But the relative importance of each of these contributions differs in fluorescence and Raman scattering. These will be addressed in the next sessions by discussing the influence of each mechanism on the enhancement factor calculations for SERS and SEF.

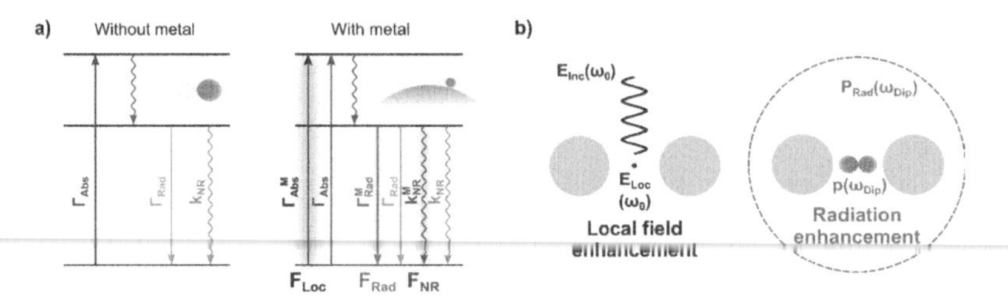

Figure 9.6 (a) Jablonski diagram illustrating the emission of secondary radiation by molecules in free space and in the presence of a plasmonic particle. (b) Calculation parameters for the electromagnetic enhancement factors.

9.5 Plasmon-Enhanced Spectroscopies

The discovery of plasmonic enhancement of secondary radiation can be traced back to almost 50 years ago. Those findings resulted from intense research on molecular monolayers on solid substrates. The starting point was the paper by Fleischmann and co-workers in 1974 [42]. They observed increased Raman scattering of pyridine once adsorbed on roughened Ag electrodes. Initially, the result was associated with improved molecular adsorption, due to the higher surface area provided by the roughened substrate. But soon the contradictions in this interpretation were published [43, 44]. Later, the electromagnetic origin of the phenomenon was recognized as caused by the resonant excitation of surface plasmons [45]. By the time, the same effect was reported for luminescent probes under similar conditions [46–50].

Since then, plasmonic systems have been extensively studied to increase the detection capabilities of spectroscopic techniques. Originally called surface-enhanced spectroscopies [51, 52], this nomenclature has evolved to comprise a large scope of derived techniques that emerged from advances in colloidal nanosynthesis and nanofabrication methods. Now, the so-called plasmon-enhanced spectroscopies refer to several techniques that differ in the measurement schemes and the type of plasmonic substrates used. Among these, three groups of techniques can be highlighted, namely, traditional surface-enhanced, shell-isolated nanoparticle-enhanced, and tip-enhanced spectroscopies. Despite their relevance, this chapter will not be dedicated to discussing the specifics of each technique. Instead, the interested reader is referred to the appropriate references [35, 53–56].

The design of plasmonic substrates for enhanced spectroscopies has been focused mostly on nanocavity engineering envisioning high-energy, reproducible hot spot formation with low energy loss. Combined with advanced optical setups, plasmonic hot spots have pushed the detection limit of spectroscopic methods down to the single-molecule (SM) level. The detection capability of ultralow concentrations makes plasmonic systems very attractive tools, especially for biosensing [57]. Besides that, plasmon-enhanced single-molecule spectroscopy has found several application niches, ranging from super-resolution imaging [58, 59] and optical trapping [60–62] to nanophotonics [63, 64]. Some of these applications are highlighted in this section.

9.5.1 Signal Enhancement in SERS

By definition, the SERS enhancement factor F_{SERS} is the ratio between the SERS (σ_{SERS}) and the Raman (σ_{Raman}) cross-sections at the same experimental conditions. This relative quantity can be expressed in terms of the enhancement factors previously defined. The excitation of surface plasmons enhances the Raman signal in two ways. First, by enhancing the excitation of the Raman dipole due to the strong local field. Second, by changing the emitted power of the Raman dipole. Therefore, the definition of F_{SERS} must include both the local field F_{Loc} and the radiative F_{Rad} enhancement factors presented earlier. Considering a single molecule on a plasmonic substrate, the observed SERS signal of this molecule integrated overall measurement directions will be enhanced by [34, 65]:

$$F_{\text{SERS}} = \frac{\sigma_{\text{SERS}}}{\sigma_{\text{Raman}}} = F_{Loc}\left(\omega_0\right)F_{Rad}\left(\omega_{\text{R}}\right), \tag{9.15}$$

where ω_{R} is the emission frequency of the Raman dipole. Based on the discussion in the previous section, the non-radiative term F_{NR} can be omitted from the calculation since non-radiative decay is independent of radiative emission for scattering. The expression for F_{SERS} can be extended to account for all the possible positions of a molecule by taking the surface average of each enhancement factor.

An important consequence of the F_{SERS} formulation is that the expression justifies the surface selection rules in SERS. The magnitude of radiative enhancement depends strongly on the dipole orientation relative to the particle surface. As presented in Fig. 9.5b, perpendicular dipoles lead to higher radiative enhancements than parallel dipoles. Therefore, the factor $F_{Rad}(\omega_R)$ helps us understand why vibrational modes perpendicular to the surface are preferentially enhanced. Frequently, this results in changes in the relative intensities of the modes when comparing the SERS and normal Raman spectra [65–67].

Additionally, the scattered intensity varies with the direction of observation. Consider a SERS signal measured in a specific direction d within a solid angle Ω with respect to the incident beam. In this situation, the expression for F_{SERS} should be replaced by the corresponding differential enhancement factor at that angle $F_{\text{SERS}}^{d}\left(\Omega\right)$ by taking the directional radiative enhancement $F_{Rad}^{d}\left(\Omega,\omega_R\right)$.

In backscattering measurements, the excitation and detection directions are the same. When one applies the plane-wave approximation and the ORT, $F_{Rad}^{d}\left(\omega_R\right)$ becomes approximately equal to $F_{Loc}(\omega_R)$. Also, the frequency shift between the incident and the Raman scattered photon is usually small compared to the linewidth of the plasmon band. Thus, it is reasonable to assume that $\omega_R \approx \omega_0$ for vibrational modes of low frequencies. Given these, the SERS enhancement factor can be approximated to:

$$F_{\text{SERS}} \approx F_{Loc}(\omega_0)F_{Loc}(\omega_{\text{R}}) \approx F_{Loc}(\omega_0)F_{Loc}(\omega_0) = \frac{|\boldsymbol{E}_{Loc}(\omega_0)|^4}{|\boldsymbol{E}_0(\omega_0)|^4}. \tag{9.16}$$

The above expression is known as the $|E|^4$-approximation to SERS. Provided that the conditions of validity are fulfilled, the $|\text{E}|^4$ approximation provides a simplified but accurate measurement of F_{SERS} (Fig. 9.7).

9.5.2 Signal Enhancement in SEF

Fluorescence absorption is affected by the local field enhancement, while the emission depends on the radiation enhancement, which, in turn, affects the decay rates. Contrary to the SERS case, non-radiative enhancement plays an important role in the calculation of fluorescence enhancement factors.

The absorbed power of a fluorophore scales as the square of the excitation field, $|\boldsymbol{E}_{Loc}(\omega_0)|^2$. Because the absorption cross-section (σ_{Abs}) is the ratio between the absorbed and the incident power, the same proportionality applies. When considering a molecule randomly oriented near a plasmonic particle, the absorption enhancement factor F_{Abs} corresponds to the local field enhancement factor:

$$F_{Abs}(\omega_0) = \frac{\sigma_{Abs}^M}{\sigma_{Abs}} = \frac{|\boldsymbol{E}_{Loc}(\omega_0, r)|^2}{|\boldsymbol{E}_0(\omega_0, r)|^2} = F_{Loc}(\omega_0). \tag{9.17}$$

To account for the changes in the fluorescence emission rates due to the metal, the quantum yield and the lifetime must be modified to:

$$Q^M = \frac{\Gamma_{Rad} + \Gamma_{Rad}^M}{\Gamma_{Rad} + \Gamma_{Rad}^M + k_{NR} + k_{NR}^M} \tag{9.18}$$

$$\tau^M = \frac{1}{\Gamma_{Rad} + \Gamma_{Rad}^M + k_{NR} + k_{NR}^M}. \tag{9.19}$$

The expression for the modified quantum yield Q^M reveals that the increase in radiative decay rate also increases the quantum yield of a molecule. However, since Q^M is a relative quantity that has a limit value of 1, this effect is pronounced only for weak emitters. That is, strong fluorophores benefit only slightly from the radiative enhancement contribution because Q is already close to 1. Nevertheless, the lifetime follows an inverse trend with Γ_r^M.

It is also useful to rewrite the Q^M in terms of the electromagnetic enhancement factors discussed in the previous section. Using Eq. (9.3) and the relations $F_{Rad} = \Gamma_{Rad}^M / \Gamma_{Rad}$ and $F_{NR} = \Gamma_{NR}^M / \Gamma_{NR}$, the expression for Q^M becomes:

$$Q^M = \frac{F_{Rad}}{F_{Rad} + F_{NR} + Q^{-1} - 1}. \tag{9.20}$$

The enhancement factor for surface-enhanced fluorescence (F_{SEF}) is defined as the ratio between the fluorescence cross-sections in SEF and non-SEF conditions. Considering a fluorophore that emits at a frequency ω_F, one gets to the following relation to F_{SEF} [23]:

$$F_{\mathrm{SEF}} = \frac{\sigma_{\mathrm{SEF}}}{\sigma_{\mathrm{Fluor}}} = \frac{\sigma_{Abs}^M Q^M}{\sigma_{Abs} Q} = \frac{F_{Loc}(\omega_0)(F_{Rad}(\omega_F))}{Q\left[F_{Rad}(\omega_F) + F_{NR}(\omega_F)\right] + 1 - Q}. \tag{9.21}$$

The analysis of the F_{SEF} shows that the plasmonic enhancement in SEF will always be lower than that observed in SERS (Fig. 9.7). An important observation is that the SEF signal scales as the square of the local field, $|\boldsymbol{E}_{\boldsymbol{Loc}}(\omega_0, r)|^2/|\boldsymbol{E}_{\boldsymbol{0}}(\omega_0, r)|^2$, since F_{SEF} is directly proportional to F_{Loc}. On the other hand, SERS enhancements follow a 4th power dependence with the local field, assuming the $|E|^4$-approximation. The reason is that SERS enhancements benefit strongly from the radiation field. But this effect is not so pronounced in SEF due to the competition between radiative and non-radiative decay channels. Typical SEF enhancements occur in the range of 10–10^2. Hot spots can deliver higher SEF enhancements, reaching 10^3 orders of magnitude [68, 69]. In contrast to SERS where the Raman signal is always enhanced by the metal, the F_{SEF} expression confirms the possibility of enhanced fluorescence quenching (Fig. 9.7).

As discussed previously, fluorescence quenching dominates at small distances from the particle surface due to favored non-radiative energy transfer between the molecule and the metal. But this distance should not be too large so that the molecule still profits from radiative enhancement due to the enhanced local field. The first experimental proof of this distance dependence was provided by Anger and co-workers [70].

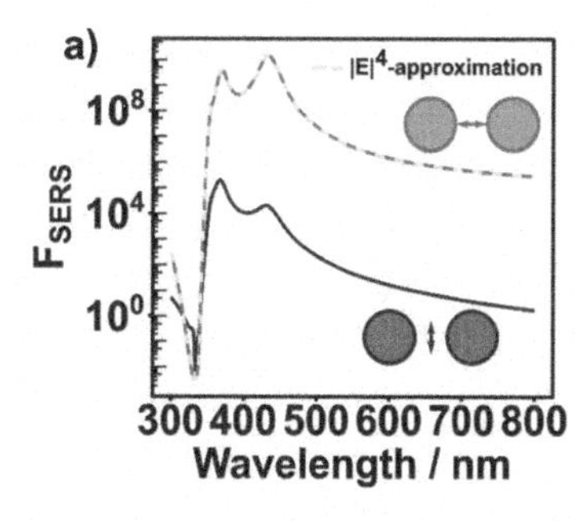

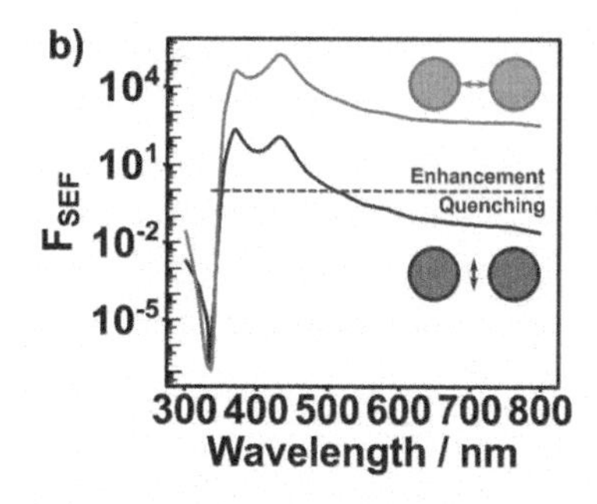

Figure 9.7 Comparative enhancement factors for SERS and SEF for a molecule located at the hot spot center of a 25 nm silver homodimer embedded in a vacuum. The parameters are calculated considering a molecule oriented parallel and perpendicular to the dimer axis. The SERS enhancement factor is also compared to the $|E|^4$ approximation. The above enhancement factors were simulated using the GMM-field [16] package using the tabulated dielectric data of Ag from reference [18].

9.5.3 Applications on Single-Molecule Detection

Experimental measurements of a large number of molecules offer an averaged response of the molecular ensemble contained in the probed volume. But this averaged signal often hides important information about the analyzed sample. On the other hand, knowing the individual response of each molecule enables calculating the probability distribution of the measured value. This allows for obtaining a deep understanding not only of the molecular dynamics but also of the local environment around the molecule.

Single-molecule experiments usually require an extremely small probed volume obtained with very diluted samples ($< 10^{-9}$ mol L^{-1}) to guarantee that the signal comes from a single molecule at the time. Also, a high signal-to-noise ratio (SNR) is crucial since the numerous surrounding solvent molecules can mask the signal. These conditions can be satisfied when the signal originates from a molecule located at a plasmonic hot spot.

9.5.3.1 Sensing

An important aspect of single-molecule measurements is the occurrence of spectral shifts and signal fluctuations, both temporal and spatial [71]. This stochastic behavior is a universal feature found in all single-molecule systems [72]. Several different effects can contribute to those fluctuations, such as molecular adsorption and desorption, changes in molecular orientation, and surface diffusion, among others. Nonetheless, the signal fluctuations represent a bottleneck for analytical applications because there is no direct correlation between signal intensity and concentration. Thus, it becomes challenging to construct calibration curves based solely on spectral intensities.

To tackle this issue, Brolo's group developed a digital protocol of quantification using single-molecule SERS (SM-SERS). The premise was that, in the single-molecule regime, only molecules located at highly efficient hot spots contribute to the signal. Then, the number of SM-SERS events was considered proportional to the analyte concentration rather than the intensities. For that, they analyzed two emerging water pollutants, enrofloxacin (ENRO) and ciprofloxacin (CIPRO), at ultralow concentrations below the nanomolar range. Bi-analyte experiments with isotopologues were performed to prove the single-molecule regime in the studied concentrations. The SM-SERS events were counted by spatially mapping the substrate containing the analytes. The chosen plasmonic substrate was a thin film of gold nanoparticles immobilized on a glass slide. This method provided a robust strategy for performing quantitative SERS, achieving a limit of quantification in the picomolar range.

Compared to SERS, single-molecule detection by SEF is a harder task, due to quenching effects when the molecule locates too closely to the particle surface. In this sense, it is important to develop strategies to control the position of the fluorophore to favor signal enhancement over quenching. A common approach is to use dielectric spacers between the particle and the molecule, usually in a core-shell particle shape. Measurements in this configuration compose a technique known as shell-isolated nanoparticle-enhanced fluorescence (SHINEF) [68, 73].

Another way to control the molecule-particle spacing is by DNA origami technique. In this technique, DNA structures are employed for self-assembling plasmonic particles with fluorescent dyes at plasmonic hot spots with controlled distances. This method was developed by Tinnefeld's group in 2012 and proved to be a valuable tool for single-molecule SEF studies. The developed DNA origami nanoantennas provided reproducible hot spots and precise control over the molecule location [74]. A recent improvement in DNA-origami technology enabled the detection of single fluorophores using a custom-made smartphone microscope [75]. The so-called DNA NanoAntennas with Cleared HOtSpots (NACHOS) were developed and used to detect DNA specific to antibiotic-resistant bacteria by fluorescent labeling. The NACHOS were capable to deliver up to 461-fold fluorescence enhancement and could be easily used for other bioassays.

9.5.3.2 Super-resolution imaging

Super-resolution imaging techniques have revolutionized biology and material sciences by overcoming the diffraction limit of light and offering the possibility to image individual molecules. In this context, plasmonics establishes a symbiotic relationship with the field of super-resolution imaging [59]. Primarily, plasmonic systems serve as localized excitation sources in the near-field using far-field radiation. They provide excitation volumes of subwavelength dimensions below the diffraction limit, due to the ability of strong field confinement. Additionally, the large extinction cross-sections and high photostability of plasmonic particles enable their use as contrast agents in optical microscopy.

The signal fluctuations of single molecules near plasmonic structures open the possibility to track plasmonic hot spots by localization-based super-resolution techniques. This was first reported by Stranahan and Willets in 2010 [76]. The measured SERS signals of an RH6G solution at 10^{-9} nM dispersed onto silver nanoparticle aggregates. They observed the temporal fluctuations in the SERS signals of RH6G. After performing a 2D Gaussian fit, they localized the spatial origins of the SERS signals and constructed intensity maps with a 10 nm nanometer resolution. The well-defined intensity gradients in the maps suggested molecular diffusion over the nanoparticle surface and indicated possible hot spot positions. They corroborated these results in a second study by correlating the intensity maps with scanning microscopy images of the aggregates [77].

However, care should be taken when imaging fluorophores in the presence of plasmonic particles by super-resolution approaches. The emission profile of fluorophores is strongly altered by the plasmonic particle, creating image distortions. This change is also strongly sensitive to the geometry of the local environment. Therefore, the recorded emission pattern is not an accurate representation of the point-spread function of the microscope and a simple Gaussian fit leads to mislocalization effects. As a consequence, the precision of the position of the molecule might actually be lost in plasmon-enhanced fluorescence [78]. Given these, approaches have been developed to study the mislocalization effects based on computational and analytical methods [79, 80].

Recent studies from Brolo and co-workers [81] on SM-SERS imaging have led to exciting findings on SERS. They investigated SM-SERS signal fluctuations on silver nanoshells by using super-resolution imaging with ultrafast acquisition rates. A monolayer of 5,5'-dithiobis-(2-nitrobenzoic acid) (DTNB) was used as a SERS probe to reduce the possible contribution of surface diffusion to signal fluctuations. Although the silver particles were fully covered with DTNB molecules, the ultrafast measurements demonstrated the same signal fluctuation behavior observed at ultralow concentrations. The results suggested that the signal fluctuations originated from transient hot spot generation, due to random surface reconstruction. The same trend was observed for different particle geometries and configurations [82]. They were able to show that the signal fluctuation could be in fact a universal phenomenon in SERS rather than restricted to ultralow concentrations [83].

9.5.3.3 Plasmonic studies

Besides the locations, the energy distributions of hot spots can be investigated using single-molecule experiments based on the anti-Stokes to the Stokes intensity ratios (aS/S). In the single-molecule regime, some SERS spectra can exhibit unexpectedly high aS/S ratios that could not be explained by the thermal population of excited vibrational states. Santos and co-workers [84] investigated this parameter for a highly diluted mixture of crystal violet (CV) and brilliant green (BG) dyes using surface-enhanced resonant Raman scattering (SERRS). They were able to show experimentally that imbalances in the aS/S values could be explained through a resonance perspective. In this sense, the hot spot energies could eventually match the energy of either the anti-Stokes or the Stokes Raman scattering. This would lead to a preferential enhancement of the anti-Stokes over the Stokes scattering or vice-versa, which explains the observed deviations of the aS/S values from thermal equilibrium.

Since the hot spot energies vary with particle clustering, the aS/S ratios were further used to probe the aggregation states of metal colloids in a derived work [85]. The SERRS aS/S ratios for CV adsorbed on silver colloids were measured in the SM regime. The results were interpreted by comparing the estimated hot spot energies for different aggregate geometries using the generalized Mie theory. An important insight from this work is that intensity fluctuations in SM-SERS can be used to study the local structure of the hot spots in metal colloids without requiring sophisticated techniques.

9.6 Near-Field Imaging

As discussed earlier, super-resolution far-field imaging such as the localization-based method has limitations when it comes to analyzing plasmonic systems. Image reconstruction can be a difficult task due to the modified emission pattern of the molecule near the plasmonic particle. The gain in signal might come at a cost of significant resolution loss if the image fitting is not done appropriately.

The limited resolution of conventional microscopy comes from the loss of near-field information caused by diffraction in the far field. At optical frequencies, the near-field information is lost within only a few nanometers away from the emitting source due to the high contribution of evanescent components. Thus, an efficient way to overcome the diffraction barrier is by directly probing the near-field. Near-field information is assessed basically by exploring the features of evanescent waves [86].

Near-field optical imaging (NOM) is a branch of scanning probe microscopy (SPM). NOM relies on a short-range electromagnetic coupling between the sample and a point probe via the evanescent fields. The image is formed by collecting the optical response obtained from the raster scanning the probe at nanometric distances over the sample surface. There are two main types of raster probes for NOM. The aperture probe, usually made of aluminum-coated optical fiber, is used in scanning near-field optical microscopy (SNOM). The image resolution is determined by the characteristic diameter of the probe.

Smaller aperture diameters provide increased resolution but at the expense of low collection efficiencies. Unfortunately, this is critical for analyzing weak emitters or for intrinsically low signals, such as Raman scattering.

The other probe type is the metal tip, used in tip-enhanced near-field optical microscopy (TENOM). In TENOM, the metal tip is illuminated by a laser and generates enhanced local fields. TENOM tips of plasmonically active metals generate enhanced fields due to a combination of an electrostatic lightning rod effect and the excitation of LSPR.

TENOM is extremely relevant in plasmonics, being capable of enhancing fluorescence and Raman signals by the same mechanisms of SEF and SERS previously discussed. As two branches of TENOM, tip-enhanced fluorescence (TEF) and tip-enhanced Raman spectroscopy (TERS), have revolutionized imaging at the nanoscale. The techniques combine strong sensitivity, high-resolution chemical imaging, and topographic information in a single analysis. Despite the major advantages of tip-enhanced spectroscopies, the techniques provide only information on the surface of the sample. Additionally, they are restricted to the analysis of relatively flat substrates [87]. For conciseness, the term TENOM will be used along with the text when the discussed ideas refer to TERS and TEF simultaneously.

9.6.1 Near-Field and Sub-diffraction Resolution

The sub-diffraction resolution achieved with near-field techniques can be understood within the framework of the angular spectrum representation. Consider an illuminated object located at $z = 0$ and a detector at an arbitrary z-plane. The free-space propagation of the field emitted by this object in the z-direction is given by multiplying the field components in the Fourier space, $\widehat{\boldsymbol{E}}(k_x, k_y, z = 0)$, with the propagation term $e^{-ik_z z}$. The direction of the propagating wave is given by the wavevector $\boldsymbol{k}(k_x, k_y, k_z)$. The modulus of $\boldsymbol{k}$ depends upon the wavelength of the incoming light λ through the relation $|k| = \sqrt{k_x^2 + k_y^2 + k_z^2} = 2\pi n / \lambda$, where n is the refractive index of the propagating medium. Evanescent waves possess imaginary k_z components. And, from the previous relation, large lateral frequencies are possible for imaginary k_z values, since $k_x^2 + k_y^2 > (2\pi n / \lambda)^2$.

But the lateral components (k_x, k_y) are lost upon wave propagation and this information loss is what defines the diffraction limit. To enable high-resolution images beyond this limit, the spatial frequencies (k_x, k_y) must be retained at the image plane. This requires a broadening of the frequency range (k_x, k_y), which can only be achieved under strong coupling between the light source and the sample. In NOM, this coupling occurs by placing a probe that interacts very closely with the sample via the evanescent near fields. In TENOM, this coupling is particularly intense when an LSPR is involved [86].

9.6.2 Signal Enhancement in the Near-Field

As in surface-enhanced techniques, the field enhancement and hence the resolution in TENOM depends upon many parameters. High-quality, stable, and reproducible probes are crucial for successful TENOM measurements. But this research field faces severe limitations, due to the difficult fabrication and rapid degradation of TENOM tips. The tip geometry also plays a pivotal role in the near-field enhancements. To assure maximum field enhancement at the tip apex, the incident field should be oriented along the tip length. Various experimental setups can be used for this purpose. The most efficient illumination setup is the back-reflection geometry because it provides the best collection efficiency. But this configuration is restricted to transparent samples. Alternatively, side and top illumination schemes can be used for opaque samples but at a cost of elongated laser spots and shadowing effects, respectively. Radially or linearly polarized light can be used according to the illumination geometry (Fig. 9.8). All these different factors that influence TENOM measurements motivated strong research efforts for the optimization of probes and operation conditions [88–93].

Additionally, the enhancements also vary according to the nature of the tip-substrate interactions. The TENOM analysis of dielectric substrates is referred to as "non-gap mode". In this case, the resolution varies inversely to the diameter of the tip apex. On the other hand, when TENOM is performed on metal substrates, one operates on the so-called "gap mode". This nomenclature comes from the fact that gap plasmons are excited upon electromagnetic coupling between the probe and the substrate. As a result, strongly confined hot spots are formed at the tip apex. By moving the tip along the sample, it is possible to form the hot spots directly at the molecule location (Fig. 9.8), enabling high-resolution single-molecule detection.

As previously mentioned, the TERS and TEF enhancements will follow the same enhancement mechanisms of SERS and SEF. Consequently, the enhancement factors are expected to obey the same relations, that is, $F_{SERS} = F_{TERS}$ and $F_{SEF} = F_{TEF}$. As in SEF, the TEF enhancements will be higher for emitters with low quantum yield and at larger probe-sample distances than in TERS due to fluorescence quenching. TEF delivers enhancement factors comparable to the obtained in SEF, of around 10^3. TERS, on the other hand, usually leads to more modest enhancements compared to the huge SERS enhancements, delivering F_{TERS} values of $\sim 10^9$ [94]. In terms of imaging quality, an image resolution of ~10 nm has been reported for TEF [95]. Even better resolutions were achieved with TERS and an impressive sub-nanometer resolution has been reached [96].

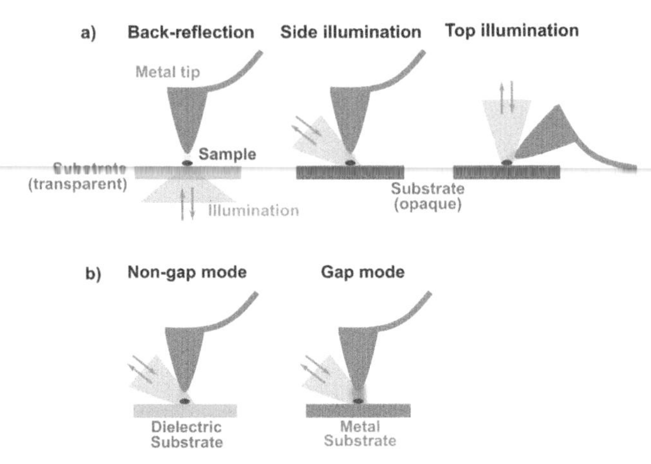

Figure 9.8 (a) Possible TERS measurement geometries by varying the relative orientation between the illumination incident and the metal tip. (b) Different TERS electromagnetic modes according to the nature of the substrate.

9.6.3 Applications on Single-Molecule Detection

When comparing the scientific production of tip-enhanced spectroscopies, it becomes evident that TERS has gained increasing research interest while TEF remains an unexplored technique.

Even with the issue of mislocalization effects, confocal microscopy has shown to be more attractive for single-molecule fluorescence imaging than TEF, due to the simpler working setup and high signals. Therefore, the most recent examples of single-molecule detection using TENOM found in the literature massively apply only TERS.

An important example of the power of combining TERS and TEF for spectral characterization of adsorbed molecules was reported by Van Duyne group [97]. They investigated a self-assembled monolayer of meso-tetrakis-(3,5-ditertiarybutylphenyl)-porphyrin (H2TBPP) adsorbed on an Ag(111) single crystal in ultra-high vacuum (UHV) conditions. The purpose of using UHV and low temperatures is to enhance molecule-tip stability and, consequently, to offer higher SNR. Usually, it is difficult to analyze both TERS and TEF spectra simultaneously. This obstacle arises because of the inverse dependencies of signal enhancements with molecule-probe distance caused by fluorescence quenching. But in this experiment, the TEF spectra were observed, which led to important conclusions on the geometry of the monolayer. The t-butyl phenyl terminations would induce a displacement of the porphyrin centers away from the surface. Then, the metal-molecule non-radiative coupling is reduced, increasing the contribution of radiative TEF enhancement.

TERS has pushed the resolution of optical imaging to unprecedented levels. Outstanding results from Zhang and co-workers [96] demonstrated the TERS capability to resolve the internal structure of a single molecule by mapping individual vibrational modes. This impressive resolution limit pointed to a field localization much beyond the effective volume of the gap plasmon mode. In fact, theoretical calculations later proved that the field localization in atomic dimensions is achievable because of the lightning rod effect at the atomic level that sums to the background plasmonic enhancement [98, 99]. Recently, Lee and co-workers [100] reported TERS measurements of single molecules with ångström-scale resolution. They studied a Co(II)–tetraphenyl porphyrin (CoTPP) molecule tightly anchored to a Cu(100) substrate. TERS measurements under UHV conditions at 6 K were performed with the aid of an Ag probe. The Raman signals were recorded by placing the tip at different regions above the CoTPP molecule. By constructing Raman maps for the different vibrational energies, they were able to image individual Raman modes for the very first time.

9.7 Final Remarks

As presented in this chapter, the unique electromagnetic features of plasmonic materials can be explored to improve the optical response of molecules. The main mechanisms of plasmon-modified secondary radiation were presented and thoroughly discussed. The focus was given to the electromagnetic pathways, due to their higher contribution to signal modification. Both radiative and non-radiative effects come into play and the relative contribution of each channel depends on the type of emission.

Additionally, it became clear that the plasmon-molecule coupling opens up a universe of applications. Particular attention was given to applications involving the single-molecule regime due to the intrinsic importance of increasing the sensitivity of optical devices. The future of plasmonics in sensing is promising, especially because the main analytical bottleneck of signal fluctuation has already been tackled. The proposed solutions pave the way for the future development of high-performance analytical devices. Moreover, single-molecule plasmon-enhanced spectroscopies offer a way to probe the properties of plasmonic hot spots in a straightforward manner. Ultimately, recent work showed that plasmonics helped transcend the diffraction barrier, which was thought to be an insuperable limitation of optical imaging. The possibility to confine light in single atoms under the action of a plasmonic near-field point that plasmonics may be the future of high-resolution optical imaging.

References

1. Bohren, C.F. (1983) How can a particle absorb more than the light incident on it? *Am. J. Phys.*, **51** (**4**), 323–327.
2. Kreibig, U. and Vollmer, M. (1995) *Optical Properties of Metal Clusters*, Springer, Berlin.
3. Maier, S.A. (2007) *Plasmonics: Fundamentals and Applications*, Springer, New York, NY.
4. Sanz, J.M., Ortiz, D., Alcaraz de la Osa, R., Saiz, J.M., González, F., Brown, A.S., Losurdo, M., Everitt, H.O., and Moreno, F. (2013) UV plasmonic behavior of various metal nanoparticles in the near- and far-field regimes: Geometry and substrate effects. *J. Phys. Chem. C*, **117** (**38**), 19606–19615.

5. Liu, X. and Swihart, M.T. (2014) Heavily-doped colloidal semiconductor and metal oxide nanocrystals: An emerging new class of plasmonic nanomaterials. *Chem Soc Rev*, **43** (**11**), 3908–3920.

6. Meng, X., Liu, L., Ouyang, S., Xu, H., Wang, D., Zhao, N., and Ye, J. (2016) Nanometals for solar-to-chemical energy conversion: From semiconductor-based photocatalysis to plasmon-mediated photocatalysis and photo-thermocatalysis. *Adv. Mater.*, **28** (**32**), 6781–6803.

7. Hartland, G.V. (2011) Optical studies of dynamics in noble metal nanostructures. *Chem. Rev.*, **111** (**6**), 3858–3887.

8. Bastús, N.G., Comenge, J., and Puntes, V. (2011) Kinetically controlled seeded growth synthesis of citrate-stabilized gold nanoparticles of up to 200 nm: Size focusing versus ostwald ripening. *Langmuir*, **27** (**17**), 11098–11105.

9. Ye, X., Jin, L., Caglayan, H., Chen, J., Xing, G., Zheng, C., Doan-Nguyen, V., Kang, Y., Engheta, N., Kagan, C.R., and Murray, C.B. (2012) Improved size-tunable synthesis of monodisperse gold nanorods through the use of aromatic additives. *ACS Nano*, **6** (**3**), 2804–2817.

10. Lee, J., Lee, P., Lee, H., Lee, D., Lee, S.S., and Ko, S.H. (2012) Very long Ag nanowire synthesis and its application in a highly transparent, conductive and flexible metal electrode touch panel. *Nanoscale*, **4** (**20**), 6408.

11. Yuan, H., Khoury, C.G., Hwang, H., Wilson, C.M., Grant, G.A., and Vo-Dinh, T. (2012) Gold nanostars: Surfactant-free synthesis, 3D modelling, and two-photon photoluminescence imaging. *Nanotechnology*, **23** (**7**), 075102.

12. Alves, R.S., Sigoli, F.A., and Mazali, I.O. (2020) Aptasensor based on a flower-shaped silver magnetic nanocomposite enables the sensitive and label-free detection of troponin I (cTnI) by SERS. *Nanotechnology*, **31** (**50**), 505505.

13. Skrabalak, S.E., Au, L., Li, X., and Xia, Y. (2007) Facile synthesis of Ag nanocubes and Au nanocages. *Nat. Protoc.*, **2** (**9**), 2182–2190.

14. Gallinet, B., Butet, J., and Martin, O.J.F. (2015) Numerical methods for nanophotonics: Standard problems and future challenges: Numerical methods for nanophotonics. *Laser Photonics Rev.*, **9** (**6**), 577–603.

15. Hohenester, U. and Trügler, A. (2012) MNPBEM – A Matlab toolbox for the simulation of plasmonic nanoparticles. *Comput. Phys. Commun.*, **183** (**2**), 370–381.

16. Ringler, M. (2008) Plasmonische Nahfeldresonatoren aus zwei biokonjugierten Goldnanopartikeln.
17. Johnson, P. and Christy, R. (1974) Optical constants of transition metals: Ti, V, Cr, Mn, Fe, Co, Ni, and Pd. *Phys. Rev. B*, **9** (**12**), 5056–5070.
18. Johnson, P.B. and Christy, R.W. (1972) Optical constants of the noble metals. *Phys. Rev. B*, **6** (**12**), 4370–4379.
19. Fraire, J.C., Pérez, L.A., and Coronado, E.A. (2012) Rational design of plasmonic nanostructures for biomolecular detection: Interplay between theory and experiments. *ACS Nano*, **6** (**4**), 3441–3452.
20. Paria, D., Zhang, C., and Barman, I. (2019) Towards rational design and optimization of near-field enhancement and spectral tunability of hybrid core-shell plasmonic nanoprobes. *Sci. Rep.*, **9** (**1**), 16071.
21. Rodrigues, T.S., da Silva, A.G.M., de Moura, A.B.L., Freitas, I.G., and Camargo, P.H.C. (2016) Rational design of plasmonic catalysts: Matching the surface plasmon resonance with lamp emission spectra for improved performance in Ag Au nanorings. *RSC Adv.*, **6** (**67**), 62286–62290.
22. Gersten, J. and Nitzan, A. (1980) Electromagnetic theory of enhanced Raman scattering by molecules adsorbed on rough surfaces. *J. Chem. Phys.*, **73** (**7**), 3023–3037.
23. Le Ru, E.C. and Etchegoin, P.G. (2009) *Principles of Surface-Enhanced Raman Spectroscopy and Related Plasmonic Effects*, Elsevier, Amsterdam, Boston.
24. Nordlander, P. and Prodan, E. (2004) Plasmon hybridization in nanoparticles near metallic surfaces. *Nano Lett.*, **4** (**11**), 2209–2213.
25. Gaponenko, S.V. (2010) *Introduction to Nanophotonics*, Cambridge University Press, Cambridge.
26. Purcell, E.M. (1946) Proceedings of the American physical society: Spontaneous emission probabilities at radio frequencies. *Phys. Rev.*, **69** (**11–12**), 674–674.
27. Blanco-Formoso, M., Pazos-Perez, N., and Alvarez-Puebla, R.A. (2020) Fabrication and SERS properties of complex and organized nanoparticle plasmonic clusters stable in solution. *Nanoscale*, **12** (**28**), 14948–14956.
28. Li, Z.-Y. (2018) Mesoscopic and microscopic strategies for engineering plasmon-enhanced raman scattering. *Adv. Opt. Mater.*, **6** (**16**), 1701097.
29. Förster, Th. (1948) Zwischenmolekulare Energiewanderung und Fluoreszenz. *Ann. Phys.*, **437** (**1–2**), 55–75.

30. Long, D.A. (2002) *The Raman Effect: A Unified Treatment of the Theory of Raman Scattering by Molecules*, Wiley.

31. Sala, O. (2008) Fundamentos de Espectroscopia no Raman and no Infravermelho, *Fundamentos da Espectroscopia Raman e no Infravermelho*, 2nd Ed., Unesp, São Paulo.

32. Szymanski, H.A. (1967) *Raman Spectroscopy: Theory and Practice*, Springer, Boston, MA.

33. Long, D.A. (2002) *The Raman Effect*, John Wiley & Sons, Ltd, Chichester, UK.

34. Le Ru, E.C., Blackie, E., Meyer, M., and Etchegoin, P.G. (2007) Surface enhanced Raman scattering enhancement factors: A comprehensive study. *J. Phys. Chem. C*, **111 (37)**, 13794–13803.

35. Langer, J., Jimenez de Aberasturi, D., Aizpurua, J., Alvarez-Puebla, R.A., Auguié, B., Baumberg, J.J., Bazan, G.C., Bell, S.E.J., Boisen, A., Brolo, A.G., Choo, J., Cialla-May, D., Deckert, V., Fabris, L., Faulds, K., García de Abajo, F.J., Goodacre, R., Graham, D., Haes, A.J., Haynes, C.L., Huck, C., Itoh, T., Käll, M., Kneipp, J., Kotov, N.A., Kuang, H., Le Ru, E.C., Lee, H.K., Li, J.-F., Ling, X.Y., Maier, S.A., Mayerhöfer, T., Moskovits, M., Murakoshi, K., Nam, J.-M., Nie, S., Ozaki, Y., Pastoriza-Santos, I., Perez-Juste, J., Popp, J., Pucci, A., Reich, S., Ren, B., Schatz, G.C., Shegai, T., Schlücker, S., Tay, L.-L., Thomas, K.G., Tian, Z.-Q., Van Duyne, R.P., Vo-Dinh, T., Wang, Y., Willets, K.A., Xu, C., Xu, H., Xu, Y., Yamamoto, Y.S., Zhao, B., and Liz-Marzán, L.M. (2020) Present and future of surface-enhanced Raman scattering. *ACS Nano*, **14 (1)**, 28–117.

36. Ding, S.-Y., You, E.-M., Tian, Z.-Q., and Moskovits, M. (2017) Electromagnetic theories of surface-enhanced Raman spectroscopy. *Chem. Soc. Rev.*, **46 (13)**, 4042–4076.

37. Dirac, P.A.M. (1927) The quantum theory of the emission and absorption of radiation. *Proc. R. Soc. Lond. Ser. Contain. Pap. Math. Phys. Character*, **114 (767)**, 243–265.

38. Fermi, E. (1974) *Nuclear physics: A course given by Enrico Fermi at the University of Chicago*, University of Chicago press, Chicago London.

39. Zheng, J., Yang, Z., Guo, Y., and Fang, Y. (2021) A detailed investigation in the enhancement factor of surface-enhanced Raman scattering in simulation. *Plasmonics*, **16 (6)**, 2207–2214.

40. Ausman, L.K. and Schatz, G.C. (2009) On the importance of incorporating dipole reradiation in the modeling of surface enhanced Raman scattering from spheres. *J. Chem. Phys.*, **131 (8)**, 084708.

41. Zhou, Y., Tian, Y., and Zou, S. (2015) Failure and reexamination of the Raman scattering enhancement factor predicted by the enhanced local electric field in a silver nanorod. *J. Phys. Chem. C*, **119 (49)**, 27683–27687.
42. Fleischmann, M., Hendra, P.J., and McQuillan, A.J. (1974) Raman spectra of pyridine adsorbed at a silver electrode. *Chem. Phys. Lett.*, **26 (2)**, 163–166.
43. Albrecht, M.G. and Creighton, J.A. (1977) Anomalously intense Raman spectra of pyridine at a silver electrode. *J. Am. Chem. Soc.*, **99 (15)**, 5215–5217.
44. Jeanmaire, D.L. and Van Duyne, R.P. (1977) Surface Raman spectroelectrochemistry. *J. Electroanal. Chem. Interfacial Electrochem.*, **84 (1)**, 1–20.
45. Moskovits, M. (1978) Surface roughness and the enhanced intensity of Raman scattering by molecules adsorbed on metals. *J. Chem. Phys.*, **69 (9)**, 4159–4161.
46. Chen, C.Y., Davoli, I., Ritchie, G., and Burstein, E. (1980) Giant Raman scattering and luminescence by molecules adsorbed on Ag and Au metal island films. *Surf. Sci.*, **101 (1–3)**, 363–366.
47. Gersten, J. and Nitzan, A. (1981) Spectroscopic properties of molecules interacting with small dielectric particles. *J. Chem. Phys.*, **75 (3)**, 1139–1152.
48. Glass, A.M., Liao, P.F., Bergman, J.G., and Olson, D.H. (1980) Interaction of metal particles with adsorbed dye molecules: Absorption and luminescence. *Opt. Lett.*, **5 (9)**, 368.
49. Weitz, D.A., Garoff, S., Hanson, C.D., Gramila, T.J., and Gersten, J.I. (1981) Fluorescent lifetimes and yields of molecules adsorbed on silver-island films. *J. Lumin.*, **24–25**, 83–86.
50. Wokaun, A., Lutz, H.-P., King, A.P., Wild, U.P., and Ernst, R.R. (1983) Energy transfer in surface enhanced luminescence. *J. Chem. Phys.*, **79 (1)**, 509–514.
51. Moskovits, M. (1985) Surface-enhanced spectroscopy. *Rev. Mod. Phys.*, **57 (3)**, 783–826.
52. Van Duyne, R.P. (1979) Laser excitation of Raman scattering from adsorbed molecules on electrode surfaces, In *Chemical and Biochemical Applications of Lasers, Vol. 4.* (ed. Moore, C.B.). Academic Press, New York, NY, 101–185.
53. Zhang, Y., Radjenovic, P.M., Zhou, X., Zhang, H., Yao, J., and Li, J. (2021) Plasmonic core–shell nanomaterials and their applications in spectroscopies. *Adv. Mater.*, 2005900.

54. Deckert-Gaudig, T., Taguchi, A., Kawata, S., and Deckert, V. (2017) Tip-enhanced Raman spectroscopy – From early developments to recent advances. *Chem. Soc. Rev.*, **46** (**13**), 4077–4110.

55. Fang, P.-P., Lu, X., Liu, H., and Tong, Y. (2015) Applications of shell-isolated nanoparticles in surface-enhanced Raman spectroscopy and fluorescence. *TrAC Trends Anal. Chem.*, **66**, 103–117.

56. Langer, J., Novikov, S.M., and Liz-Marzán, L.M. (2015) Sensing using plasmonic nanostructures and nanoparticles. *Nanotechnology*, **26** (**32**), 322001.

57. Liu, J., Jalali, M., Mahshid, S., and Wachsmann-Hogiu, S. (2020) Are plasmonic optical biosensors ready for use in point-of-need applications? *The Analyst*, **145** (**2**), 364–384.

58. Wang, M., Li, M., Jiang, S., Gao, J., and Xi, P. (2020) Plasmonics meets super-resolution microscopy in biology. *Micron*, **137**, 102916.

59. Willets, K.A., Wilson, A.J., Sundaresan, V., and Joshi, P.B. (2017) Super-resolution imaging and plasmonics. *Chem. Rev.*, **117** (**11**), 7538–7582.

60. Juan, M.L., Righini, M., and Quidant, R. (2011) Plasmon nano-optical tweezers. *Nat. Photonics*, **5** (**6**), 349–356.

61. Kitahama, Y., Funaoka, M., and Ozaki, Y. (2019) Plasmon-enhanced optical tweezers for single molecules on and near a colloidal silver nanoaggregate. *J. Phys. Chem. C*, **123** (**29**), 18001–18006.

62. Zhan, C., Wang, G., Yi, J., Wei, J.-Y., Li, Z.-H., Chen, Z.-B., Shi, J., Yang, Y., Hong, W., and Tian, Z.-Q. (2020) Single-molecule plasmonic optical trapping. *Matter*, **3** (**4**), 1350–1360.

63. Baumberg, J.J., Aizpurua, J., Mikkelsen, M.H., and Smith, D.R. (2019) Extreme nanophotonics from ultrathin metallic gaps. *Nat. Mater.*, **18** (**7**), 668–678.

64. Ciracì, C., Jurga, R., Khalid, M., and Della Sala, F. (2019) Plasmonic quantum effects on single-emitter strong coupling. *Nanophotonics*, **8** (**10**), 1821–1833.

65. Santos, D.P., Temperini, M.L.A., and Brolo, A.G. (2015) Surface-enhanced raman scattering: Principles and applications for single-molecule detection, In *Introduction to Plasmonics: Advances and Applications* (eds. Szunerits, S. and Boukherroub, R.). Jenny Stanford Publishing, Singapore, 275–317.

66. Moskovits, M. (1982) Surface selection rules. *J. Chem. Phys.*, **77** (**9**), 4408–4416.

67. Le Ru, E.C., Meyer, S.A., Artur, C., Etchegoin, P.G., Grand, J., Lang, P., and Maurel, F. (2011) Experimental demonstration of surface selection rules for SERS on flat metallic surfaces. *Chem. Commun.*, **47** (**13**), 3903.

68. Guerrero, A.R., Zhang, Y., and Aroca, R.F. (2012) Experimental confirmation of local field enhancement determining far-field measurements with shell-isolated silver nanoparticles. *Small*, **8** (**19**), 2964–2967.

69. Li, J.-F., Li, C.-Y., and Aroca, R.F. (2017) Plasmon-enhanced fluorescence spectroscopy. *Chem. Soc. Rev.*, **46** (**13**), 3962–3979.

70. Anger, P., Bharadwaj, P., and Novotny, L. (2006) Enhancement and quenching of single-molecule fluorescence. *Phys. Rev. Lett.*, **96** (**11**), 113002.

71. dos Santos, D.P., Temperini, M.L.A., and Brolo, A.G. (2019) Intensity fluctuations in single-molecule surface-enhanced Raman scattering. *Acc. Chem. Res.*, **52** (**2**), 456–464.

72. Moerner, W.E. and Fromm, D.P. (2003) Methods of single-molecule fluorescence spectroscopy and microscopy. *Rev. Sci. Instrum.*, **74** (**8**), 3597–3619.

73. Guerrero, A.R. and Aroca, R.F. (2011) Surface-enhanced fluorescence with shell-isolated nanoparticles (SHINEF). *Angew. Chem. Int. Ed.*, **50** (**3**), 665–668.

74. Acuna, G.P., Möller, F.M., Holzmeister, P., Beater, S., Lalkens, B., and Tinnefeld, P. (2012) Fluorescence enhancement at docking sites of DNA-directed self-assembled nanoantennas. *Science*, **338** (**6106**), 506–510.

75. Trofymchuk, K., Glembockyte, V., Grabenhorst, L., Steiner, F., Vietz, C., Close, C., Pfeiffer, M., Richter, L., Schütte, M.L., Selbach, F., Yaadav, R., Zähringer, J., Wei, Q., Ozcan, A., Lalkens, B., Acuna, G.P., and Tinnefeld, P. (2021) Addressable nanoantennas with cleared hotspots for single-molecule detection on a portable smartphone microscope. *Nat. Commun.*, **12** (**1**), 950.

76. Stranahan, S.M. and Willets, K.A. (2010) Super-resolution optical imaging of single-molecule SERS hot spots. *Nano Lett.*, **10** (**9**), 3777–3784.

77. Weber, M.L. and Willets, K.A. (2011) Correlated super-resolution optical and structural studies of surface-enhanced raman scattering hot spots in silver colloid aggregates. *J. Phys. Chem. Lett.*, **2** (**14**), 1766–1770.

78. Chattopadhyay, S. and Biteen, J.S. (2021) Super-resolution characterization of heterogeneous light–matter interactions between single dye molecules and plasmonic nanoparticles. *Anal. Chem.*, **93** (**1**), 430–444.

79. Fu, B., Isaacoff, B.P., and Biteen, J.S. (2017) Super-resolving the actual position of single fluorescent molecules coupled to a plasmonic nanoantenna. *ACS Nano*, **11** (**9**), 8978–8987.

80. Goldwyn, H.J., Smith, K.C., Busche, J.A., and Masiello, D.J. (2018) Mislocalization in plasmon-enhanced single-molecule fluorescence microscopy as a dynamical Young's interferometer. *ACS Photonics*, **5** (**8**), 3141–3151.

81. Lindquist, N.C., de Albuquerque, C.D.L., Sobral-Filho, R.G., Paci, I., and Brolo, A.G. (2019) High-speed imaging of surface-enhanced Raman scattering fluctuations from individual nanoparticles. *Nat. Nanotechnol.*, **14** (**10**), 981–987.

82. Bido, A.T., Nordberg, B.G., Engevik, M.A., Lindquist, N.C., and Brolo, A.G. (2020) High-speed fluctuations in surface-enhanced Raman scattering intensities from various nanostructures. *Appl. Spectrosc.*, **74** (**11**), 1398–1406.

83. Lindquist, N.C. and Brolo, A.G. (2021) Ultra-high-speed dynamics in surface-enhanced Raman scattering. *J. Phys. Chem. C*, **125** (**14**), 7523–7532.

84. dos Santos, D.P., Temperini, M.L.A., and Brolo, A.G. (2012) Mapping the energy distribution of SERRS hot spots from anti-stokes to stokes intensity ratios. *J. Am. Chem. Soc.*, **134** (**32**), 13492–13500.

85. dos Santos, D.P., Temperini, M.L.A., and Brolo, A.G. (2016) Single-molecule surface-enhanced (resonance) Raman scattering (SE(R)RS) as a probe for metal colloid aggregation state. *J. Phys. Chem. C*, **120** (**37**), 20877–20885.

86. Novotny, L. and Hecht, B. (2012) *Principles of Nano-Optics*, Cambridge University Press, Cambridge.

87. Mauser, N. and Hartschuh, A. (2014) Tip-enhanced near-field optical microscopy. *Chem Soc Rev*, **43** (**4**), 1248–1262.

88. Agapov, R.L., Sokolov, A.P., and Foster, M.D. (2013) Protecting TERS probes from degradation: Extending mechanical and chemical stability: Protecting TERS probes from degradation. *J. Raman Spectrosc.*, **44** (**5**), 710–716.

89. Bartolomeo, G.L., Goubert, G., and Zenobi, R. (2020) Tip recycling for atomic force microscopy-based tip-enhanced Raman spectroscopy. *Appl. Spectrosc.*, **74** (**11**), 1358–1364.

90. Bartolomeo, G.L., Zhang, Y., Kumar, N., and Zenobi, R. (2021) Molecular perturbation effects in AFM-based tip-enhanced Raman spectroscopy: Contact versus. *Anal. Chem.*, **93** (**46**), 15358–15364.

91. Gao, L., Zhao, H., Li, Y., Li, T., Chen, D., and Liu, B. (2018) Controllable fabrication of Au-coated AFM probes via a wet-chemistry procedure. *Nanoscale Res. Lett.*, **13** (**1**), 366.

92. Kumar, N., Spencer, S.J., Imbraguglio, D., Rossi, A.M., Wain, A.J., Weckhuysen, B.M., and Roy, D. (2016) Extending the plasmonic lifetime of tip-enhanced Raman spectroscopy probes. *Phys. Chem. Chem. Phys.*, **18** (**19**), 13710–13716.

93. Vasconcelos, T.L., Archanjo, B.S., Oliveira, B.S., Valaski, R., Cordeiro, R.C., Medeiros, H.G., Rabelo, C., Ribeiro, A., Ercius, P., Achete, C.A., Jorio, A., and Cançado, L.G. (2018) Plasmon-tunable tip pyramids: Monopole nanoantennas for near-field scanning optical microscopy. *Adv. Opt. Mater.*, **6** (**20**), 1800528.

94. Meng, L., Sun, M., Chen, J., and Yang, Z. (2016) A nanoplasmonic strategy for precision in-situ measurements of tip-enhanced Raman and fluorescence spectroscopy. *Sci. Rep.*, **6** (**1**), 19558.

95. Dong, Z., Guo, X.L., Trifonov, A., Dorozhkin, P., Miki, K., Kimura, K., Yokoyama, S., and Mashiko, S. (2004) Vibrationally resolved fluorescence from organic molecules near metal surfaces in a scanning tunneling microscope. *Phys. Rev. Lett.*, **92** (**8**), 086801.

96. Zhang, R., Zhang, Y., Dong, Z.C., Jiang, S., Zhang, C., Chen, L.G., Zhang, L., Liao, Y., Aizpurua, J., Luo, Y., Yang, J.L., and Hou, J.G. (2013) Chemical mapping of a single molecule by plasmon-enhanced Raman scattering. *Nature*, **498** (**7452**), 82–86.

97. Chiang, N., Jiang, N., Chulhai, D.V., Pozzi, E.A., Hersam, M.C., Jensen, L., Seideman, T., and Van Duyne, R.P. (2015) Molecular-resolution interrogation of a porphyrin monolayer by ultrahigh vacuum tip-enhanced Raman and fluorescence spectroscopy. *Nano Lett.*, **15** (**6**), 4114–4120.

98. Barbry, M., Koval, P., Marchesin, F., Esteban, R., Borisov, A.G., Aizpurua, J., and Sánchez-Portal, D. (2015) Atomistic near-field nanoplasmonics: Reaching atomic-scale resolution in nanooptics. *Nano Lett.*, **15** (**5**), 3410–3419.

99. Urbieta, M., Barbry, M., Zhang, Y., Koval, P., Sánchez-Portal, D., Zabala, N., and Aizpurua, J. (2018) Atomic-scale lightning rod effect in plasmonic picocavities: A classical view to a quantum effect. *ACS Nano*, **12** (**1**), 585–595.

100. Lee, J., Crampton, K.T., Tallarida, N., and Apkarian, V.A. (2019) Visualizing vibrational normal modes of a single molecule with atomically confined light. *Nature*, **568** (**7750**), 78–82.

Chapter 10

Intrinsic and Extrinsic Polymer Optical Fiber Schemes for Highly Sensitive Plasmonic Biosensors

Nunzio Cennamo,[a] Maria Pesavento,[b] and Luigi Zeni[a]
[a]*Department of Engineering, University of Campania Luigi Vanvitelli, Via Roma n. 29, 81031 Aversa (Italy)*
[b]*Department of Chemistry, University of Pavia, Via Taramelli n. 12, Pavia, 27100, Italy*
nunzio.cennamo@unicampania.it

This chapter presents several developed sensing approaches based on polymer optical fibers (POFs) used in different schemes to realize plasmonic platforms. More specifically, the POFs' capabilities are especially used to monitor or realize plasmonic sensors, useful in several bio/chemical application fields, exploiting specific receptor layers. The POFs' numerical aperture and their multimode characteristics have been extensively used to obtain highly sensitive extrinsic and intrinsic schemes of plasmonic probes, combined with different kinds of receptors (e.g., aptamers, antibodies, molecularly

Plasmonics-Based Optical Sensors and Detectors
Edited by Banshi D. Gupta, Anuj K. Sharma, and Jin Li

ISBN 978-981-4968-85-0 (Hardcover), 978-1-003-43830-4 (eBook)
www.jennystanford.com

imprinted polymers, etc.) to detect specific substances in the required range of detection. It is described in the following sections via an overview of the research work carried out by the authors in the last few years from about 2010.

10.1 Introduction

This chapter presents the POF-based plasmonic transmission-mode sensor configurations developed by our research group in the last 10 years. The plasmonic sensors presentation is organized in two sections, according to the employed schemes (intrinsic or extrinsic scheme); hence, plasmonic extrinsic and intrinsic POF sensor schemes are reported separately. This classification is based on two different schemes used to implement the POFs in the sensing devices. When the POF interacts with the analyzed medium directly, the sensor is defined as an intrinsic POF one; whereas, when the POF is used as an optical waveguide allowing the launch of the light into the sensing region, and/or its collection, the sensor is defined as an extrinsic POF. In the last years, both configurations have been implemented exploiting POFs' advantages, such as the excellent flexibility, the easy manipulation and modification features, the large diameter, the multimode characteristic, and the great numerical aperture. In particular, the numerical aperture is especially suitable to realize the extrinsic schemes, whereas the other characteristics can be exploited to implement the intrinsic schemes.

After the description of these POF-based plasmonic platforms, several bio/chemical applications based on these plasmonic POF platforms are reported. In particular, the first application part reports the detection of specific substances based on chemical recognition elements, such as molecularly imprinted polymers (MIPs) and receptors for metal ions. In the second part, applications based on aptamers and antibody recognition elements are presented. Actually, plasmonic probes and molecular recognition elements (MREs) must be efficiently combined to achieve highly sensitive and selective bio/chemical sensing.

All the plasmonic sensors here presented are used in transmission mode. They are monitored by simple equipment, such as a white light source, the plasmonic sensing head, and a spectrometer connected to a computer, as reported in the scheme shown in Fig. 10.1. This hardware can be combined with specific software with a user-friendly interface, useful in several application fields, and for the connection to the internet, as well [1].

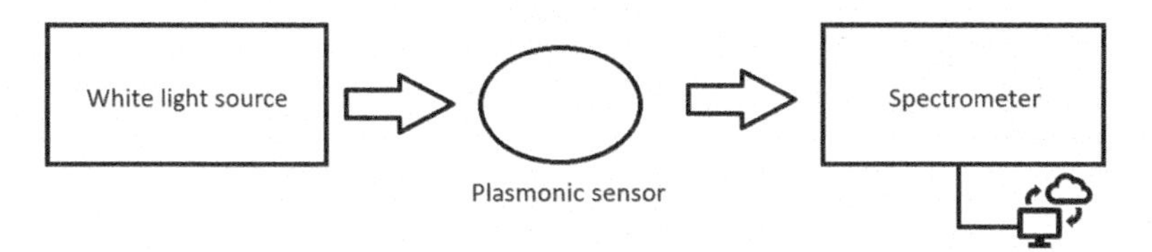

Figure 10.1 Outline of the spectral mode-based experimental setup used to implement all the proposed plasmonic sensors.

10.2 Plasmonic POF Sensors: Intrinsic Schemes

An extensively used intrinsic plasmonic POF sensing platform was proposed about 10 years ago by Cennamo et al., realized by modifying multimode POFs [2], which was simple to realize and at the same time highly sensitive. Figure 10.2 shows this type of surface plasmon resonance (SPR) D-shaped POF platform, along with an outline of the production steps [2]. The SPR sensor was obtained by removing the cladding and part of the core of the POF, spinning a thin layer of a photoresist (Microposit S1813) on the exposed core, with the refractive index greater than that of the POF's core, and finally sputtering a gold nanofilm. The used POF had a polymethyl methacrylate (PMMA) core of 980 μm (in diameter) and a fluorinated polymer cladding of 10 μm (1,000 μm in total diameter). The presence of the S1813 buffer layer under the gold nanofilm improves the performances of the SPR sensor and the gold film adhesion to the platform [2]. Thus, the sensing area, 10 mm long, presents a multilayer over the exposed POF core, consisting of a photoresist layer 1.5 μm thick and a thin gold film 60 nm thick. This plasmonic probe has been used to develop several bio/chemical applications exploiting different types of receptor layers, such as bioreceptors (aptamers and antibodies) and molecularly imprinted polymers (nanoMIPs and thin or thick MIP layers) [3].

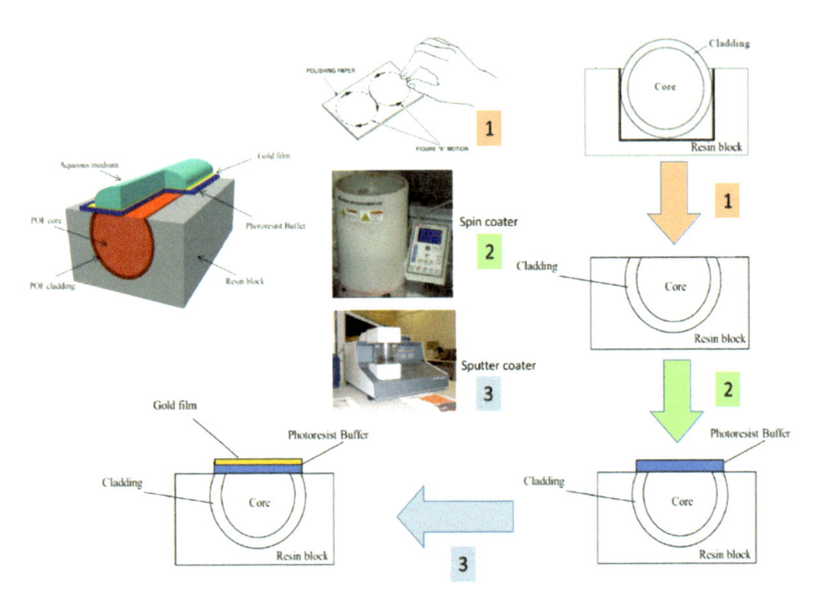

Figure 10.2 A simple to realize plasmonic platform: SPR in D-shaped POFs. Outline of the production steps [2].

Moreover, this plasmonic D-shaped POF sensor has been realized by exploiting tapered POFs to improve the sensor's performance [4]. In this tapered-POF-based sensor configuration, three steps are required to realize the device, as shown in Fig. 10.3, namely, heating and stretching (to realize the taper), polishing (to realize the D-shaped area), and sputtering (to cover the sensing area with the gold nanofilm) [4]. It is worth noting that, in these cases, the S1813 buffer layer is not required to improve the performance.

Cennamo et al. have demonstrated that, without the tapers, the SPR D-shaped POF sensors' performances worsen when the POF diameter decreases [5]. On the contrary, when the diameter decreases in the D-shaped tapered-POF sensing area, the performances improved because of the increase in taper ratio [4]. In fact, according to the theory, as confirmed by the experimental results, Cennamo et al. have verified that the sensitivity of the SPR sensor rises when the taper ratio increases [4].

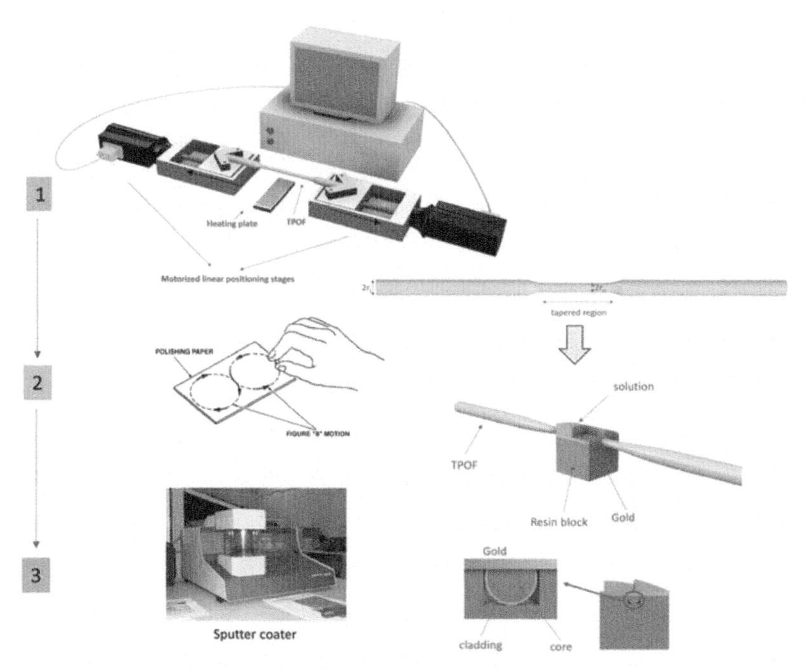

Figure 10.3 Outline of the production steps required to realize SPR D-shaped tapered-POFs sensors.

It is important to underline that plasmonic intrinsic optical fiber sensor configurations can be realized by exploiting fibers based on silica, specialty fibers, or polymer fibers, as well. In this regard, a novel type of specialty optical fibers, named light-diffusing fibers (LDFs), have been used to realize SPR sensors with high sensitivity by Cennamo et al. [6–9]. The particular LDF's characteristic, oriented to diffusing the propagated light on the external medium all along their length, can effectively excite the plasmonic phenomenon. These types of LDF-based SPR sensors can be implemented with silica [6, 7] or polymer [9] LDFs, covered by metal nanofilms, such as gold [6, 9] or silver nanofilm [8], and can be properly combined with bioreceptors to achieve the required selectivity [8]. Figure 10.4 shows an outline of an SPR LDF-based sensor connected to a white light source and a spectrometer. The experimental results reported in Refs. [6–9] reveal a very high sensitivity. For instance, the PMMA-based LDF covered by a gold nanofilm of 60 nm, as reported in

Fig. 10.4, shows a bulk sensitivity ranging from 1,000 to almost 3,000 (nm/RIU) for refractive indexes ranging from 1.332 to 1.392 [9]. These sensitivity values can be improved by using silica-based LDF (Fibrance® by Corning®, New York, NY, USA) covered by gold or silver nanofilms [6–8]. Moreover, Cennamo et al. have demonstrated in Ref. [10] that in plasmonic silica-based LDF sensors, the tapering process produces, instead of the enhancement of the sensitivity reported in Ref. [4], a significant loss of the bulk sensitivity, and a slight decrease in the full width at half maximum (FWHM) of the plasmonic spectra, the latter instrumental to increase the signal-to-noise ratio (SNR).

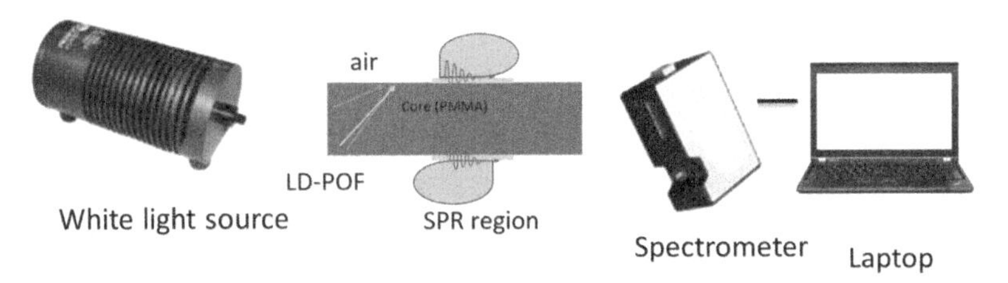

Figure 10.4 SPR sensor system based on a light-diffusing polymer optical fiber (LD-POF).

Usually, the plasmonic sensors based on intrinsic POF schemes exploit the multimode characteristics of the fibers, allowing to produce highly sensitive sensors in a wide refractive index range in which the resonance conditions are satisfied. In these cases, Cennamo et al. demonstrated that a modal filter, located after the plasmonic sensitive region, could create a trade-off between the sensitivity worsening and the desirable FWHM decrease, representing a good solution for several applications [9–11]. Furthermore, this modal filter can be realized by exploiting tapered optical fibers [10, 11] or covering the LDFs with an aqueous solution [9].

Figure 10.5 shows three possible schemes to realize the filtering of the higher modes after the plasmonic sensing region to reduce the FWHM of the SPR spectra, while keeping the sensitivity decrease at a very low value.

More specifically, Fig. 10.5a shows a silica-based SPR-LDF sensor combined with a tapered-LDF used as a modal filter [10], Fig. 10.5b shows an SPR LD-POF sensor with a modal filter obtained by covering the LD-POF with an aqueous solution [9], whereas Fig. 10.5c shows an SPR D-shaped POF sensor with a taper at the end used as a modal filter [11].

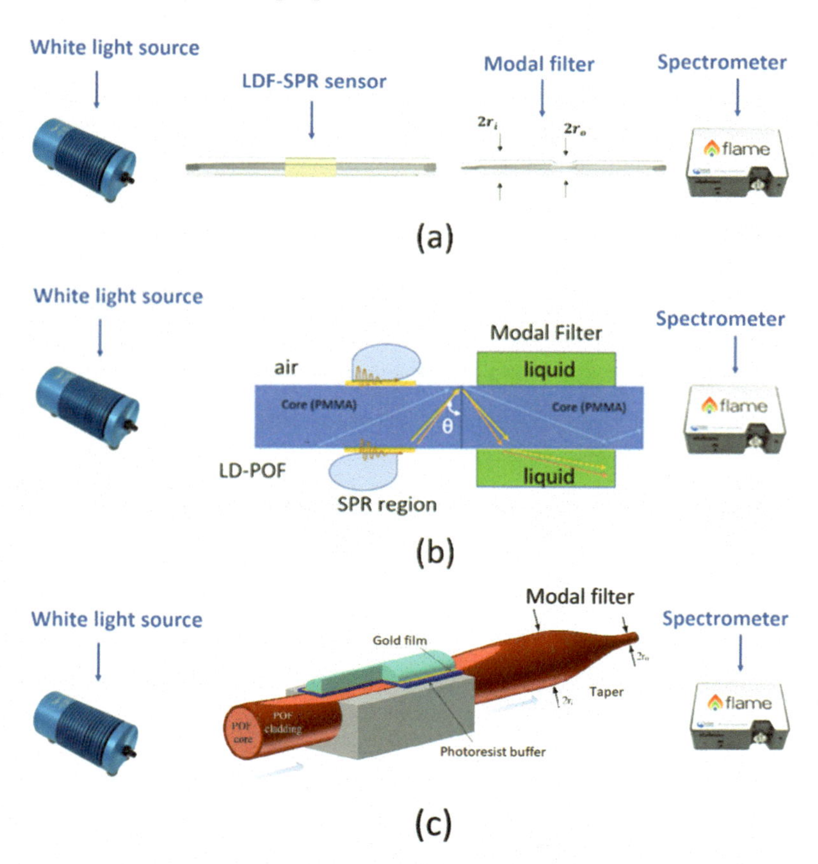

Figure 10.5 Several schemes of intrinsic SPR sensors combined with modal filters: (a) based on tapered silica LDFs, (b) based on LD-POFs, and (c) based on tapered POFs.

10.3 Plasmonic POF Sensors: Extrinsic Schemes

The great numerical aperture of the POFs is an effective characteristic to realize extrinsic optical fiber sensor configurations. Cennamo et al. have used POFs to monitor different plasmonic sensing devices. In particular, POFs have been combined with several sensor chips, such as those based on bacterial cellulose waveguides [12, 13], InkJet-Printed optical waveguides [14], nanostructured plasmonic sensor chips [15, 16], SPR sensors in slab waveguides [17, 18], and 3D-printed plasmonic sensors [19].

Figure 10.6 shows the plasmonic bacterial cellulose (BC) sensor configuration, based on BC waveguides, with and without gold. The BC without gold is used as a reference sensor, whereas the BC-based plasmonic sensor has been obtained by sputtering gold on the slab BC waveguide. In fact, the BC waveguide contains nanowires, covered by gold nanofilm, giving rise to a localized surface plasmon resonance (LSPR) phenomenon. Cennamo et al. have realized and tested some LSPR sensor configurations with different thicknesses of the BC waveguides, with and without the presence of ionic liquids inside the BC waveguides [12, 13].

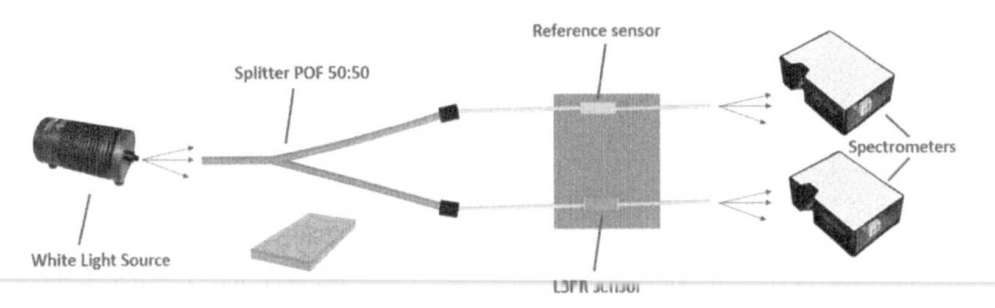

Figure 10.6 BC-based LSPR sensor monitored via POFs [12, 13].

Figure 10.7 shows an InkJet-Printed sensor chip, based on a PET (polyethene terephthalate) substrate with parallel lines realized by silver nanoparticles printing (via InkJet technology), covered by a thin molecularly imprinted polymer layer [14]. Two POFs connect this sensor chip with a white light source and a spectrometer. This optical-chemical sensor has been used in Ref. [14] to detect furfural (furan-2-carbaldehyde, 2-FAL) in an aqueous medium. The results obtained by exploiting this sensing approach are comparable to those of other 2-FAL sensors based on intrinsic POF configurations, but with several advantages, as reported in Ref. [14].

Furthermore, a 3D-printed custom holder, combined with POFs, has been used by Arcadio et al. to monitor PMMA-based nanoplasmonic sensor chips [15]. Figure 10.8 shows this kind of extrinsic POF sensor, based on gold nanograting, along with the interrogation equipment. The white light source is connected to two

POFs at the input of the holder via a POF optical coupler (50:50). One branch is used to illuminate the plasmonic sensor chip with the nanograting, while the other illuminates the reference sensor (a PMMA chip with the same gold film but without nanograting). At the output of the custom holder, exploiting two POFs, the transmitted light is collected and fed to two similar spectrometers. As reported in Fig. 10.8, in this sensor configuration, the PMMA substrate is considered a mere transparent substrate [15].

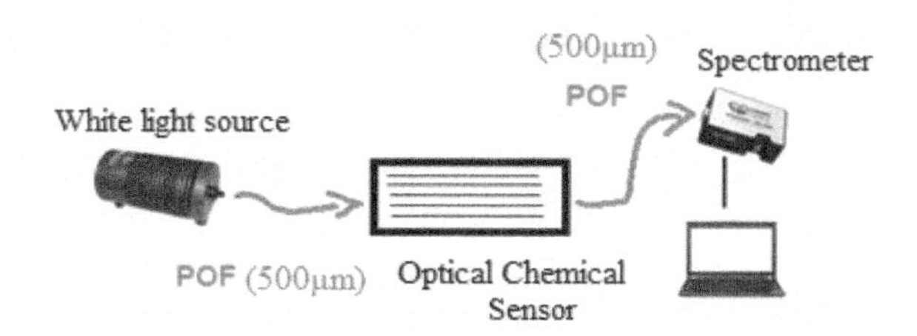

Figure 10.7 InkJet-Printed optical-chemical sensor chip is monitored via POFs [14].

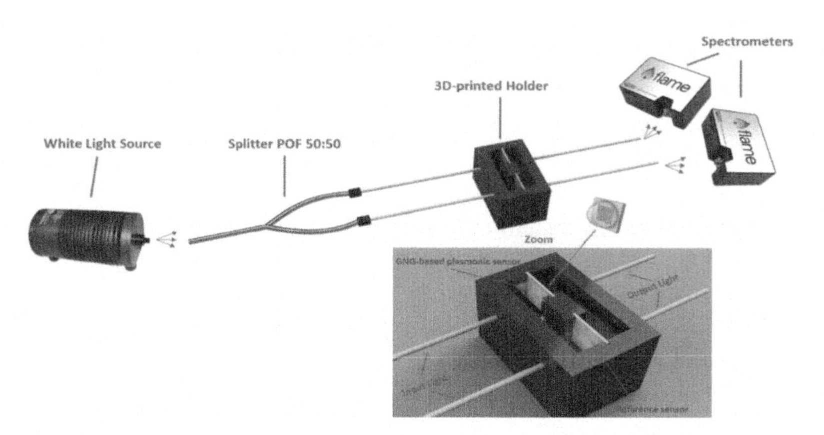

Figure 10.8 PMMA-based nanoplasmonic sensor chip monitored via a custom holder combined with POFs [15].

Figure 10.9 shows an alternative setup, proposed by Cennamo et al. [17], where the PMMA chip acts as a multimode slab waveguide [16–18]. As reported in Fig. 10.9a, the PMMA chip can be covered by a continuous gold film to realize an SPR sensor chip [17, 18]. Similarly, over the PMMA chip, a gold nanograting can be realized, via electron-beam lithography (EBL), to implement a nanoplasmonic sensor chip [16]. This nanoplasmonic sensor chip can be monitored via the same holder used for the uniform gold film chip, as reported in Fig. 10.9b.

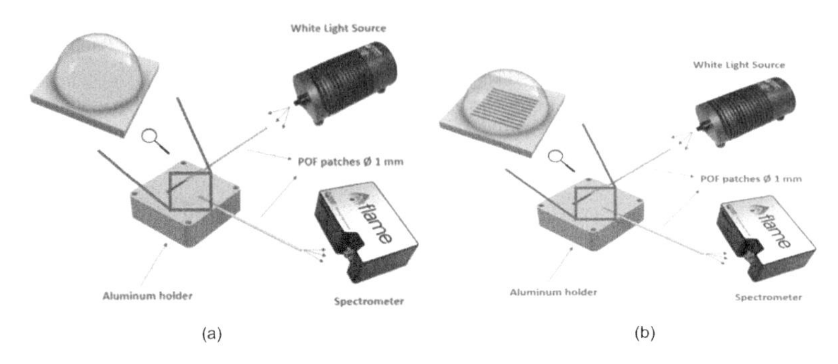

Figure 10.9 PMMA-based plasmonic sensor chip monitored via evanescent field. (a) SPR sensor chip is monitored via extrinsic POFs [17]. (b) Nanoplasmonic sensor chip is monitored via extrinsic POFs [16].

By using the sensor setup illustrated in Fig. 10.9, when a uniform gold nanofilm covers the PMMA chip, the sensor's performances are very similar to those obtained by an SPR D-shaped POF sensor [2], whereas, when a nanoplasmonic chip is used, the orientation of the nanostripes forming the grating pattern, with respect to the direction of the input light (longitudinal or orthogonal), influences the performances of the sensor [16].

Finally, Fig. 10.10 shows a low-cost and simple 3D-printed SPR sensor. This SPR sensing chip can be monitored by exploiting POFs, as illustrated in Fig. 10.10, and it has been realized by assembling different parts. The sensor chip was designed by using Autodesk® Fusion 360, then the STL files were generated [19] and sent to a PolyJet 3D printer Stratasys Objet260 Connex 1 (Stratasys, Los Angeles, CA, USA) for realization. The used material was a liquid photopolymer ink (VeroClear RGD810). This printed chip represents the cladding of the polymer-based waveguide, and its trench can be used the realize the core of the waveguide. To this end, a UV photopolymer adhesive (NOA88, Edmund Optics, Nether Poppleton York, UK) was microinjected into the trench (sensor channel) and cured for 10 min by means of a lamp bulb with UVA emission at

365 nm. Finally, on the cured core of the waveguide, a gold nanofilm was deposited by a sputtering process (Bal-Tec SCD 500, Schalksmühle, Germany) and a custom mask. The thickness of the sputtered gold film was about 60 nm [19].

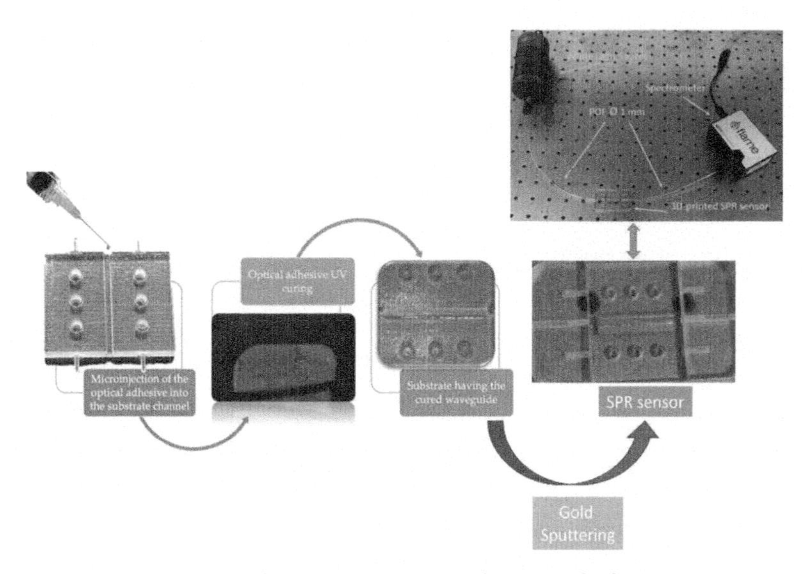

Figure 10.10 3D printed SPR sensor monitored via POFs [19].

10.4 Chemical Receptors on POF-Based Plasmonic Probes

An attractive class of chemical receptors is represented by molecularly imprinted polymers (MIPs). MIPs are porous solids or nanoparticles containing specific sites interacting with the molecule of interest (analyte), according to a "key and lock" model. In this type of artificial receptor, sites are functionally and dimensionally complementary to the analyte, similar to the receptor sites present in bioreceptors. The MIP's preparation is based on a template-assisted synthesis [20]. More specifically, the cavity in the porous solid is obtained by polymerization of the aggregate substrate-coordinating monomers after extraction of the template from the selective site, as schematically reported in Fig. 10.11.

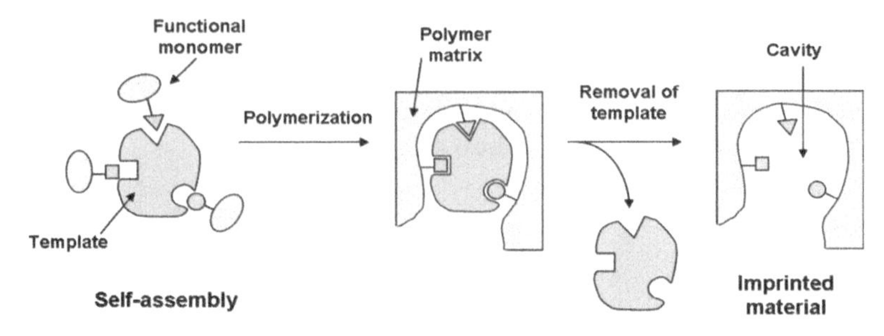

Figure 10.11 Outline of the MIP preparation procedure leading to the formation of selective cavities in the polymer.

MIPs often possess recognition properties analogous to natural receptors (aptamers or antibodies) but present the advantages of synthetic materials, such as stability, ease of preparation, micromachining, integrability, low-cost production, better stability out of the native environment, and reproducibility on an industrial scale. Moreover, MIP layers can be easily polymerized in tight contact with the sensing surface of the transducer both in intrinsic and extrinsic optical fiber sensing devices. While some extrinsic optical fiber sensors have been described in Refs. [14–16, 18], here only a few examples of intrinsic sensors developed by our research group are described to highlight some problems related to the combination of the receptor to the transducing surface.

Several approaches can be used for the formation of the MIP receptor layers, such as via self-assembled monolayers [15, 16, 18], exploiting photopolymerization [21], or dropping and spinning processes [22, 23]. When these last two deposition processes (photopolymerization and dropping and spinning) are considered, the MIP receptors can be viewed in a different way with respect to the usual bioreceptors, which are often self-assembled at the sensing surface as molecular monolayers (SAMs), since the thickness of the receptor layer can be tuned. This aspect is a favorable characteristic for several plasmonic-based sensing applications. For instance, when the detection range must be adapted to the specific application or when the considered real matrices present a very high refractive index and cannot be used in the resonance range of the transducer, the possibility to tune the MIP thickness comes in handy. In fact, by increasing the thickness of the MIP layer, the transducer can be isolated from the bulk solution; alternatively, by reducing the thickness of the MIP layer, the sensitivity of the sensors can be improved, etc. More specifically, depending on the MIP thickness, the plasmonic wave can penetrate only the MIP layer or reach the fluid above (the bulk solution), with the plasmonic phenomenon intensity exponentially decreasing.

For instance, in D-shaped POF-SPR platforms, an MIP layer can be easily deposited by a drop coating and spinning procedure, as previously described in several cases [3, 22, 24]. Figure 10.12 shows two SPR D-shaped POF platforms combined with MIP receptors to realize different kinds of plasmonic bio/chemical sensor configurations [25]. As demonstrated in Ref. [25] in the L-nicotine detection, using D-shaped tapered POFs instead of D-shaped POFs, the sensitivity of the optical-chemical sensor can be improved without changing the structure of the MIP layer, but taking advantage of the tapered structure of the POF as mentioned above.

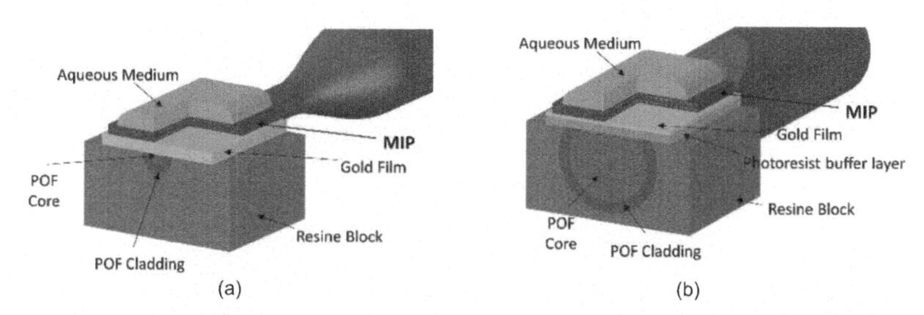

Figure 10.12 Outline of the MIP-based SPR-POF sensors. (a) SPR D-shaped tapered POF combined with MIPs. (b) SPR D-shaped POF combined with MIPs [25].

A sensing layer based on MIP mixed with gold nanostars (GNS), for the selective detection and analysis of 2,4,6-trinitrotoluene (TNT) in aqueous solution, has been spun over several POF platforms obtaining different responses, as demonstrated in Ref. [23].

Figure 10.13 summarizes the results reported in Ref. [23], showing several POF-based optical-chemical TNT sensors with real performances, in terms of limit of detections, sensitivity at low concentrations, etc. [23]. As shown in Fig. 10.13, Cennamo et al. [23] have presented a sensor configuration based on a tapered POF combined with a special sensing layer, composed of a specific MIP with gold nanostars inside. The gold nanostars are dispersed in the MIP's pre-polymeric mixture before spinning on the D-shaped tapered POF sensing area. There is no gold nanofilm in this sensor configuration, and we have an LSPR phenomenon caused by the gold nanostars dispersed in the polymer instead of the SPR produced by the gold nanofilm. In this GNS-based sensor configuration, the performances are improved with respect to the other configurations based on POFs and MIP; in fact, in this sensing layer, a larger number of specific sites that interact with the gold nanostars are present in the sensing layer deposited on the sensing POF area, as described in Fig. 10.13. So, we have a sort of 3D sensing with respect to a 2D sensing typical of a MIP layer deposited on a gold continuous film (see Fig. 10.13) [23].

As reported in the table and as shown in the dose-response curves of Fig. 10.13, the configuration based on MIP-GNS on D-shaped POF is better than the MIP layer on the gold film configuration, but the best one is the configuration based on MIP with GNS deposited on D-shaped tapered POFs. In particular, as reported in the table, the LOD is 10 times lower [23].

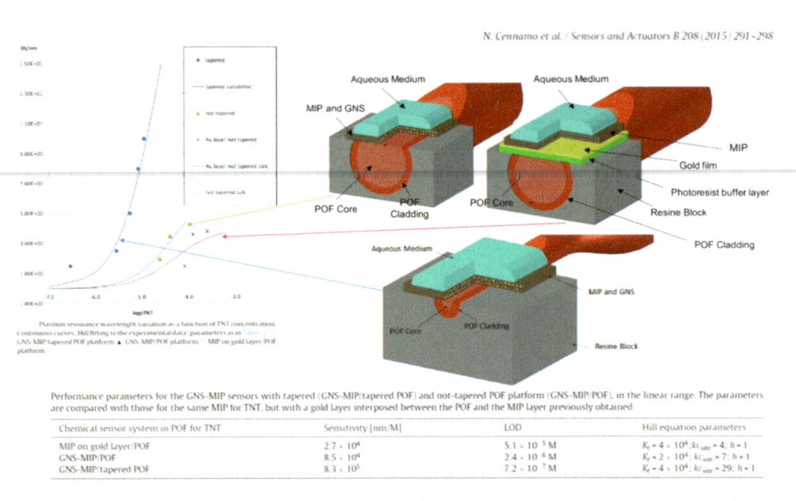

Performance parameters for the GNS-MIP sensors with tapered (GNS-MIP/tapered POF) and not-tapered POF platform (GNS-MIP/POF), in the linear range. The parameters are compared with those for the same MIP for TNT, but with a gold layer interposed between the POF and the MIP layer previously obtained

Chemical sensor system in POF for TNT	Sensitivity [nm/M]	LOD	Hill equation parameters
MIP on gold layer/POF	2.7×10^4	5.1×10^{-5} M	$K_f = 4 \times 10^4$; $\Delta\lambda_{sate} = 4$; $h = 1$
GNS-MIP/POF	8.5×10^4	2.4×10^{-6} M	$K_f = 2 \times 10^4$; $\Delta\lambda_{sate} = 7$; $h = 1$
GNS-MIP/tapered POF	8.3×10^5	7.2×10^{-7} M	$K_f = 4 \times 10^4$; $\Delta\lambda_{sate} = 29$; $h = 1$

Figure 10.13 TNT POF-MIP sensor configurations and the obtained dose-response curves. Outline of the MIP-POF sensors with a summary of the relative performances [23].

Another interesting aspect of the MIPs is that they can be used as receptors in sensing devices not only as a layer, but in other forms too, for example, nanoparticles. The peculiar deformable character of the nano-MIPs (molecularly imprinted polymeric nanoparticles) can be exploited to significantly enhance the sensor performances, allowing the detection of femtomolar concentrations of proteins, as demonstrated by Cennamo et al. [26, 27] by combining plasmonic POF probes with SAMs of nano-MIPs. In particular, in Ref. [26], an intrinsic SPR-POF platform has been combined with nanoMIPs, whereas Ref. [27] shows this kind of MIP receptor with an extrinsic POF sensor configuration.

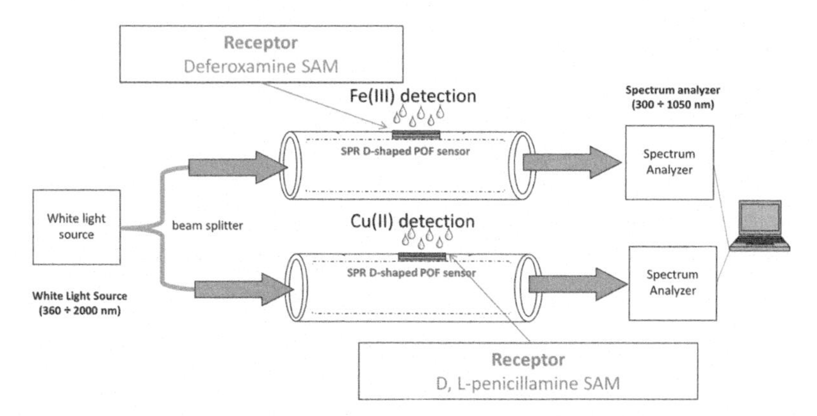

Figure 10.14 Simultaneous detection of iron (III) and copper (II) ions by chemical receptors on D-shaped POF sensors.

Another kind of synthetical receptor, which can be easily linked to a gold surface consists of metal ions complexing agents. Thus, they can be combined with plasmonic D-shaped POF platforms as self-assembled monolayers (SAMs) to detect metal ions, for example, iron (III) and copper (II). A SAM of deferoxamine has been used to detect iron (III), as described in Ref. [28]. Similarly, Cennamo et al. developed an SPR sensor for copper (II) detection in drinking water, but exploiting D,L-penicillamine as the receptor, deposited as a SAM on the gold nanofilm of the POF probe [29]. In Fig. 10.14, an experimental setup that can be used to test both these ions simultaneously is reported. It is based on a white light source, a POF beam splitter, two different optical-chemical sensors [28, 29], and two similar spectrometers. The sensors reported in Fig. 10.14 have been obtained by exploiting the same D-shaped POF probe in intrinsic sensor configuration combined with two different SAMs, one based on deferoxamine and the other on D,L-penicillamine [28, 29].

10.5 Bioreceptors (Aptamers and Antibodies) on POF-Based Plasmonic Probes

The bioreceptors have played a crucial role in sensor development, providing the necessary specificity and sensitivity to detect the substance of interest. Numerous SPR sensors based on bioreceptors have been reported in the literature, with different bio-macromolecules as MREs: proteins, nucleic acids, enzymes, aptamers, antibodies (Abs), etc. In this chapter, only aptamers and antibodies are considered, examining some examples of detection obtained by our research group via POF-based plasmonic biosensors, to show the capabilities of this sensing approach. In fact, Cennamo et al. have combined SPR D-shaped POF platforms with several SAMs of aptamers and antibodies to selectively detect different substances of interest.

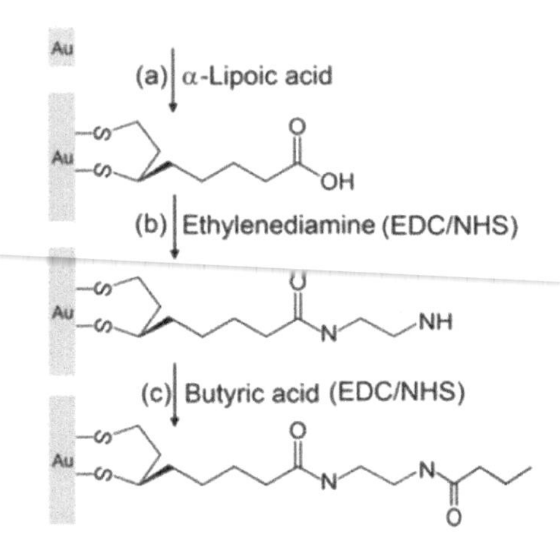

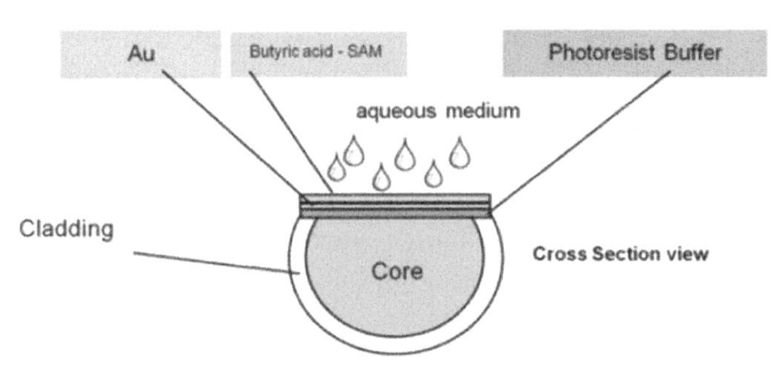

Figure 10.15 POF-based biosensor for detection of butanal [32].

In particular, antibodies have been exploited for the detection of PFAs [30] and naphthalene [31] in environmental monitoring applications; of butanal in food applications [32]; whereas, for biomedical applications, the same group has developed sensors based on antibodies as receptors for the detection of therapeutic antibodies, such as infliximab [33] and transglutaminase/anti-transglutaminase (useful in the diagnosis and/or follow-up of celiac disease) [34], pancreatic amylase in surgically-placed drain effluents [35], etc. Figure 10.15 shows the functionalization steps to develop the biosensor for butanal detection [32].

Similarly, different plasmonic POF aptasensors have been realized by depositing specific aptamers as SAMs over SPR D-shaped POF platforms. In particular, this type of POF aptasensor has been exploited in the detection of vascular endothelial growth factor (VEGF) (selected as a circulating protein potentially associated with cancer) [36], SARS-CoV-2 spike protein [37], and thrombin [38], as schematically reported in Fig. 10.16.

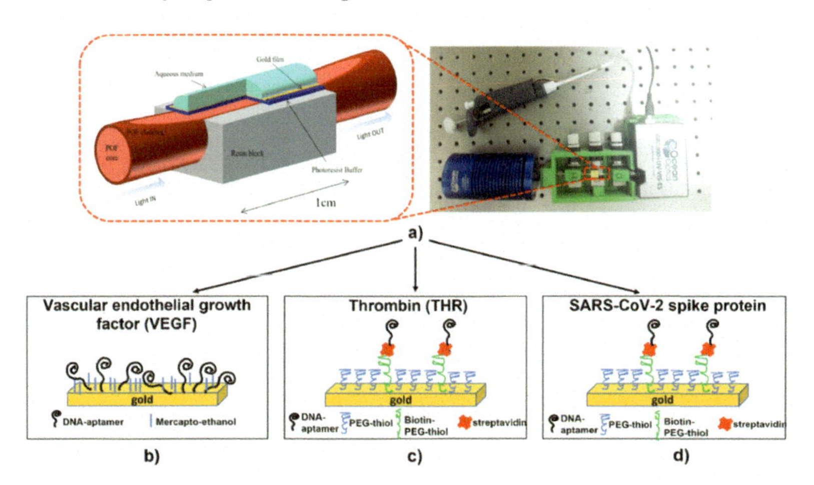

Figure 10.16 POF-based aptasensors: (a) SPR probe; (b) VEGF sensor; (c) Thrombin sensor; and (d) SARS-CoV-2 sensor.

10.6 Conclusions

To realize small-sized and low-cost sensing devices, highly sensitive plasmonic POF-based configurations exploiting intrinsic and extrinsic schemes can be combined with specific receptors (biological, biomimetic, or chemical MREs) and monitored via a universal setup based on a white light source and a spectrometer. More specifically, an overview of sensors developed by our research group in the last years in this category has been reported. Considering these examples, it has been demonstrated that the same general-purpose spectral mode-based setup can be used to monitor all the here described POF bio/chemical sensor chips, different for the type of POF-probe and/or for the kind of receptor layer, by showing detection capabilities from femtomolar to micromolar concentrations. Moreover, all the presented POF sensor configurations have been obtained by exploiting simple and low-cost technology and/or innovative materials. Finally, the POF-sensing technology's capabilities include biomedical, security, food and industrial applications, environmental monitoring, bioterrorism counteraction, and pandemic emergency control.

References

1. Cennamo, N., Arcadio, F., Capasso, F., Perri, C., D'Agostino, G., Porto, G., Biasiolo, A., and Zeni, L. (2020). Toward smart selective sensors exploiting a novel approach to connect optical fiber biosensors in internet, *IEEE T. Instrum. Meas.*, **69**, pp. 8009–8019.
2. Cennamo, N., Massarotti, D., Conte, L., and Zeni, L. (2011). Low cost sensors based on SPR in a plastic optical fiber for biosensor implementation, *Sensors*, **11**, pp. 11752–11760.
3. Cennamo, N., Pesavento, M., and Zeni, L. (2021). A review on simple and highly sensitive plastic optical fiber probes for bio-chemical sensing, *Sensor. Actuat. B: Chem.*, **331**, art. no. 129393.
4. Cennamo, N., Arcadio, F., Minardo, A., Montemurro, D., and Zeni, L. (2020). Experimental characterization of plasmonic sensors based on lab-built tapered plastic optical fibers, *Appl. Sci.*, **10**, art. no. 4389.
5. Cennamo, N., Massarotti, D., Galatus, R., Conte, L., and Zeni, L. (2013). Performance comparison of two sensors based on surface plasmon resonance in a plastic optical fiber, *Sensors*, **13**, pp. 721–735.
6. Cennamo, N., Zeni, L., Catalano, E., Arcadio, F., and Minardo, A. (2018). Refractive index sensing through surface plasmon resonance in light-diffusing fibers, *Appl. Sci.*, **8**, art. no. 1172.
7. Cennamo, N., Zeni, L., Arcadio, F., Catalano, E., and Minardo, A. (2019). A novel approach to realizing low-cost plasmonic optical fiber sensors: Light-diffusing fibers covered by thin metal films, *Fibers*, **7**, art. no. 34.

8. Cennamo, N., Trono, C., Giannetti, A., Baldini, F., Minardo, A., Zeni, L., and Tombelli, S. (2021). Biosensors exploiting unconventional platforms: The case of plasmonic light-diffusing fibers, *Sensor. Actuat. B: Chem.*, **337**, art. no. 129771.
9. Cennamo, N., Arcadio, F., Del Prete, D., Buonanno, G., Minardo, A., Pirozzi, S., and Zeni, L. (2021). A simple and efficient plasmonic sensor in light diffusive polymer fibers, *IEEE Sens. J.*, **21**, pp. 16054–16060.
10. Cennamo, N., Arcadio, F., Zeni, L., Catalano, E., Del Prete, D., Buonanno, G., and Minardo, A. (2021). The role of tapered light-diffusing fibers in plasmonic sensor configurations, *Sensors*, **21**, art. no. 6333.
11. Cennamo, N., Coelho, L., Santos, D. F., Baptista, J. M., Guerreiro, A., Jorge, P. A. S., and Zeni, L. (2015). Modal filtering for optimized surface plasmon resonance sensing in multimode plastic optical fibers, *IEEE Sens. J.*, **15**, pp. 6306–6312.
12. Cennamo, N., Trigona, C., Graziani, S., Zeni, L., Arcadio, F., Di Pasquale, G., and Pollicino, A. (2019). An eco-friendly disposable plasmonic sensor based on bacterial cellulose and gold, *Sensors*, **19**, art. no. 4894.
13. Cennamo, N., Trigona, C., Graziani, S., Zeni, L., Arcadio, F., Xiaoyan, L., Di Pasquale, G., and Pollicino, A. (2021). Green LSPR sensors based on thin bacterial cellulose waveguides for disposable biosensor implementation, *IEEE T. Instrum. Meas.*, **70**, art. no. 9507908.
14. Cennamo, N., Zeni, L., Pesavento, M., Marchetti, S., Marletta, V., Baglio, S., Graziani, S., Pistorio, A., and Andò, B. (2019). A novel sensing methodology to detect furfural in water, exploiting MIPs and InkJet-Printed optical waveguides, *IEEE T. Instrum. Meas.*, **68**, pp. 1582–1589.
15. Arcadio, F., Zeni, L., Montemurro, D., Eramo, C., Di Ronza, S., Perri, C., D'Agostino, G., Chiaretti, G., Porto, G., and Cennamo, N. (2021). Biochemical sensing exploiting plasmonic sensors based on gold nanogratings and polymer optical fibers, *Photonics Res.*, **9**, pp. 1397–1408.
16. Arcadio, F., Zeni, L., Minardo, A., Eramo, C., Di Ronza, S., Perri, C., D'Agostino, G., Chiaretti, G., Porto, G., and Cennamo, N. (2021). A nanoplasmonic-based biosensing approach for wide-range and highly sensitive detection of chemicals, *Nanomaterials*, **11**, art. no. 1961.
17. Cennamo, N., Mattiello, F., and Zeni, L. (2017). Slab waveguide and optical fibers for novel plasmonic sensor configurations, *Sensors*, **17**, art. no. 1488.
18. Arcadio, F., Zeni, L., Perri, C., D'Agostino, G., Chiaretti, G., Porto, G., Minardo, A., and Cennamo, N. (2021). Bovine serum albumin protein

detection by a removable SPR chip combined with a specific MIP receptor, *Chemosensors*, **9**, art. no. 218.

19. Cennamo, N., Saitta, L., Tosto, C., Arcadio, F., Zeni, L., Fragalá, M. E., and Cicala, G. (2021). Microstructured surface plasmon resonance sensor based on inkjet 3D printing using photocurable resins with tailored refractive index, *Polymers*, **13**, art. no. 2518.

20. Uzun, L. and Turner, A. P. F. (2016). Molecularly-imprinted polymer sensors: Realizing their potential, *Biosens. Bioelectron.*, **76**, pp. 131–144.

21. Paruli, E. III, Soppera, O., Haupt, K., and Gonzato, C. (2021). Photopolymerization and photostructuring of molecularly imprinted polymers, *ACS Appl. Polym. Mat.*, **3**, pp. 4769–4790.

22. Cennamo, N., D'Agostino, G., Galatus, R., Bibbò, L., Pesavento, M., and Zeni, L. (2013). Sensors based on surface plasmon resonance in a plastic optical fiber for the detection of trinitrotoluene, *Sensor. Actuat. B: Chem.*, **188**, pp. 221–226.

23. Cennamo, N., Donà, A., Pallavicini, P., D'Agostino, G., Dacarro, G., Zeni, L., and Pesavento, M. (2015). Sensitive detection of 2,4,6-trinitrotoluene by tridimensional monitoring of molecularly imprinted polymer with optical fiber and five-branched gold nanostars, *Sensor. Actuat. B: Chem.*, **208**, pp. 291–298.

24. Pesavento, M., Zeni, L., De Maria, L., Alberti, G., and Cennamo, N. (2021). SPR-optical fiber-molecularly imprinted polymer sensor for the detection of furfural in wine, *Biosensors*, **11**, 72.

25. Cennamo, N., D'Agostino, G., Pesavento, M., and Zeni, L. (2014). High selectivity and sensitivity sensor based on MIP and SPR in tapered plastic optical fibers for the detection of L-nicotine, *Sensor. Actuat. B: Chem.*, **191**, pp. 529–536.

26. Cennamo, N., Maniglio, D., Tatti, R., Zeni, L., and Bossi, A. M. (2020). Deformable molecularly imprinted nanogels permit sensitivity-gain in plasmonic sensing, *Biosens. Bioelectron.*, **156**, art. no. 112126.

27. Cennamo, N., Bossi, A. M., Arcadio, F., Maniglio, D., and Zeni, L. (2021). On the effect of soft molecularly imprinted nanoparticles receptors combined to nanoplasmonic probes for biomedical applications, *Front. Bioeng. Biotechnol.*, **9**, art. no. 801489.

28. Cennamo, N., Alberti, G., Pesavento, M., D'Agostino, G., Quattrini, F., Biesuz, R., and Zeni, L. (2014). A simple small size and low cost sensor based on surface plasmon resonance for selective detection of Fe(III), *Sensors*, **14**, pp. 4657–4661.

29. Pesavento, M., Profumo, A., Merli, D., Cucca, L., Zeni, L., and Cennamo, N. (2019). An optical fiber chemical sensor for the detection of copper(II) in drinking water, *Sensors*, **19**, art. no. 5246.

30. Cennamo, N., Zeni, L., Tortora, P., Regonesi, M. E., Giusti, A., Staiano, M., D'Auria, S., and Varriale, A. (2018). A high sensitivity biosensor to detect the presence of perfluorinated compounds in environment, *Talanta*, **178**, pp. 955–961.

31. Cennamo, N., Zeni, L., Ricca, E., Isticato, R., Marzullo, V. M., Capo, A., Staiano, M., D'Auria, S., and Varriale, A. (2019). Detection of naphthalene in sea-water by a label-free plasmonic optical fiber biosensor, *Talanta*, **194**, pp. 289–297.

32. Cennamo, N., Di Giovanni, S., Varriale, A., Staiano, M., Di Pietrantonio, F., Notargiacomo, A., Zeni, L., and D'Auria S. (2015). Easy to use plastic optical fiber-based biosensor for detection of butanal, *PLoS ONE*, **10**, art. no. e0116770.

33. Zeni, L., Perri, C., Cennamo, C., Arcadio, F., D'Agostino, G., Salmona, M., Beeg, M., and Gobbi, M. (2020). A portable optical-fibre-based surface plasmon resonance biosensor for the detection of therapeutic antibodies in human serum, *Sci. Rep.*, **10**, art. no. 11154.

34. Cennamo, N., Varriale, A., Pennacchio, A., Staiano, M., Massarotti, D., Zeni, L., and D'Auria, S. (2013). An innovative plastic optical fiber-based biosensor for new bio/applications. the case of celiac disease, *Sensor. Actuat. B: Chem.*, **176**, pp. 1008–1014.

35. Pasquardini, L., Cennamo, N., Malleo, G., Vanzetti, L. E., Zeni, L., Bonamini, D., Salvia, R., Bassi, C., and Bossi, A. M. (2021). A surface plasmon resonance plastic optical fiber biosensor for the detection of pancreatic amylase in surgically-placed drain effluent, *Sensors*, **21**, art. no. 3443.

36. Cennamo, N., Pesavento, M., Lunelli, L., Vanzetti, L. E., Pederzolli, C., Zeni, L., and Pasquardini, L. (2015). An easy way to realize SPR aptasensor: A multimode plastic optical fiber platform for cancer biomarkers detection, *Talanta*, **140**, pp. 88–95.

37. Cennamo, N., Pasquardini, L., Arcadio, F., Lunelli, L., Vanzetti, L. E., Carafa, V., Altucci, L., and Zeni, L. (2021). SARS-CoV-2 spike protein detection through a plasmonic D-shaped plastic optical fiber aptasensor, *Talanta*, **233**, art. no. 122532.

38. Cennamo, N., Pasquardini, L., Arcadio, F., Vanzetti, L. E., Bossi, A. M., and Zeni, L. (2019). D-shaped plastic optical fibre aptasensor for fast thrombin detection in nanomolar range, *Sci. Rep.*, **9**, art. no. 18740.

Chapter 11

2D Materials/Heterostructures/ Metasurfaces in Plasmonic Sensing and Biosensing

Maryam Ghodrati, Ali Mir, and Ali Farmani
School of Electrical Engineering, Lorestan University, Khorramabad, Iran
a.farmani@ieee.org

Materials behave notably distinct in reduced dimensions when compared with their bulk counterparts. 2D materials have exceptional electrical, optical, physical, and chemical properties that are well-suited to sensing and biosensing applications. Recently, the design and fabrication of biosensors based on the surface plasmon resonance (SPR) phenomenon have attracted great attention as a promising and accurate sensing technology. SPR phenomenon-based sensing is much more sensitive to small changes in biomolecules' refractive index than other optical sensing techniques. In this chapter, we present the characterization and optoelectronic properties of various 2D materials such as MXene,

Plasmonics-Based Optical Sensors and Detectors
Edited by Banshi D. Gupta, Anuj K. Sharma, and Jin Li

ISBN 978-981-4968-85-0 (Hardcover), 978-1-003-43830-4 (eBook)
www.jennystanford.com

graphene, transition metal dichalcogenides (TMDs), hexagonal boron nitride (h-BN), and black phosphorus (BP). Furthermore, we cover the usage of heterostructures and metasurfaces for plasmonic sensing and biosensing applications.

11.1 Introduction

In recent years, 2D materials and their heterostructures are much attractive as their extraordinary properties like generation, confinement, and transport of charge carriers can be controlled and manipulated [1–3]. Thanks to their unique physical, optical, and electronic properties, 2D-layered materials have been used in various fields such as electronics, biomedical, plasmonics, sensing, and optics. Layered 2D materials such as MXene, graphene, TMDs, h-BN, and BP, and their heterostructures have been explored by researchers to enhance the SPR sensor performance for sensing and biosensing applications [4–6]. SPR phenomenon is the resonant coupling of electromagnetic waves to the charge density oscillations at the interface of dielectrics and metals. Since there is an optical momentum mismatch between the SPR mode and light in free space, the optical excitations in the SPR are usually made possible by the attenuated total reflection (ATR) method proposed by Kretschmann [7] and Otto [8]. Furthermore, the optical gratings and waveguides with compact coupling structures are widely employed for SPR-integrated circuits. For surface plasmon waves (SPWs), only the TM-polarized electric field exists; this wave is decayed exponentially in the interface of dielectric and metal. The SPW is characterized by the propagation constant as [7–10]:

$$k_{sp} = \frac{\omega}{c}\sqrt{\frac{\varepsilon_d \varepsilon_m}{\varepsilon_d + \varepsilon_m}}, \tag{11.1}$$

where c is the velocity of light in a vacuum, ω is the angular frequency of the incident light, and ε_m and ε_d are the dielectric constants of metal and dielectric medium, respectively. The real part of ε_m must be negative, and its absolute value is smaller than ε_d to promise the metals could support SPW. According to Eq. (11.1), SPW is highly sensitive to changes in the properties of the metal and the dielectric media. To create the surface plasmon oscillation, it is needed to excite the electrons in the metal. So, imposing the light is necessary on the surface. The propagation constant of a light wave propagating with a frequency ω in free space is given by [7–10]:

$$k_d = \frac{\omega}{c}\sqrt{\varepsilon_d} \tag{11.2}$$

Based on the equations above, the permittivities of metal and dielectric being negative and positive, respectively, infer that the wave propagation constant of SPW should be always higher than that of the wave propagating in the dielectric [8, 9]. Therefore, surface plasmon cannot be excited with direct light; it needs light with extra momentum or energy with the same polarization state as the SPW. Furthermore, the propagation constant should be matched with the SPW. SPR structures are attracting platforms for the optical sensors, due to their strong confinement and enhancement of the electromagnetic field near the surface [8–10]. The SPR sensors are refractive index sensors, and these devices have potential advantages such as real-time and label-free detections, quick response, high sensitivity, cost-effectiveness, and high-resolution detection. SPR-based sensors are desirable for many different applications including environmental monitoring, and medical diagnostics such as the detection of human blood groups, DNA, glucose, virus, living cell analysis, and chemical and gas sensing [11–13].

The scientific community has also reported the usage of a metasurface-based sensor to achieve high sensitivity in biosensors. Metasurfaces are artificial photonic devices composed of subwavelength nanoresonators, and their performances can be engineered by rationally tailoring their geometric parameters. Metasurfaces, being 2D materials, can be easily integrated into other devices, which can make them a salient feature for nanophotonic circuits; this property will also allow them to be a part of lab-on-chip photonics. Metasurfaces can also be used in the development of a biosensor for applications such as medical diagnostic devices [12, 14]. The goal of this chapter is to (a) present the structure and characteristics of several 2D materials, (b) discuss their optoelectronic properties, and (c) introduce the use of 2D materials, heterostructures, and metasurfaces for plasmonic sensing and biosensing applications.

11.2 2D Materials Introduction

2D nanomaterials are one of the most attractive research topics, due to their outstanding potential applications in many fields, such as electronics, biomedicals, plasmonics, sensing, and optics [13, 14]. The 2D-layered nanomaterials have desirable physical and structural properties and are well-suited for applications in biosensing (Fig. 11.1). This is due to their large surface-to-volume ratio and direct bandgap compared to the indirect bandgap of bulk materials [12–16]. Up to now, at least 18 kinds of nanomaterials are included

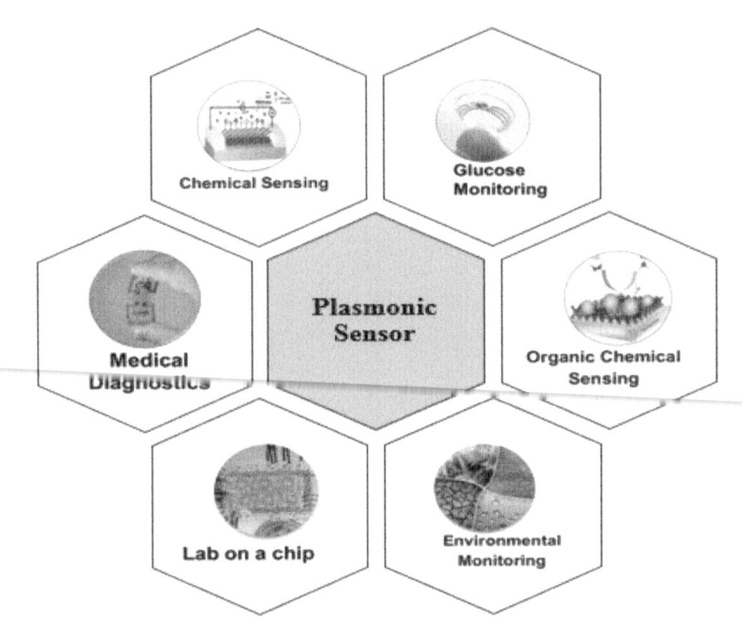

Figure 11.1 Various applications of SPR sensors.

in the family of 2D materials beyond graphene, h-BN, TMDs, BP, III–VI layered semiconductors, layered double hydroxides (LDHs), metal oxides, transition metal oxyhalides, metal halides, perovskites, silicates and hydroxides (clays), non-layer structured metal oxides, non-layer structured metal chalcogenides, and early transition metal carbides and/or nitrides (MXenes) [15–18]. Depending on their chemical compositions and structural configurations, atomically thin 2D materials can be categorized as metallic, semi-metallic, semiconducting, insulating, or superconducting. In this part, we introduce structure and optoelectronic properties of several 2D materials, including MXene, graphene, TMDs, h-BN, and BP (Fig. 11.2) [16–18].

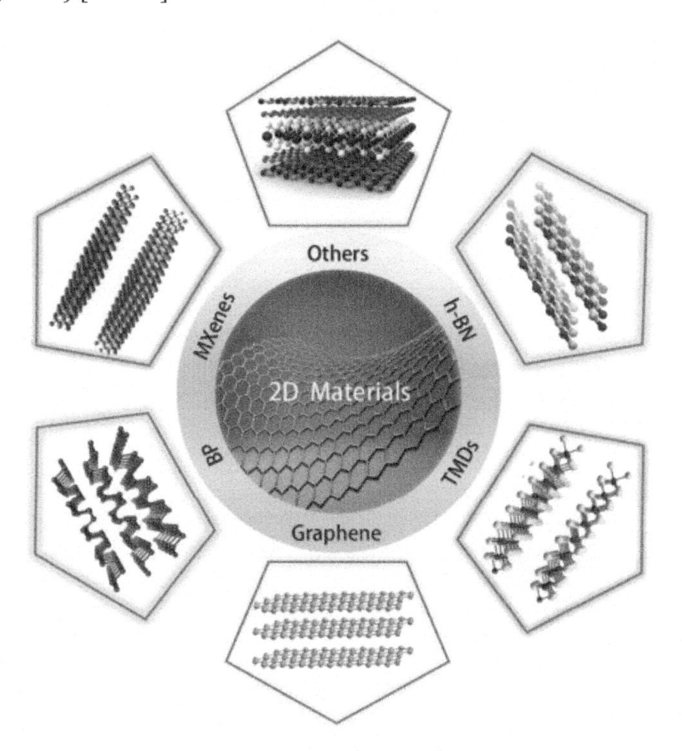

Figure 11.2 Schematic of 2D-layered materials [19].

11.2.1 Transition-Metal Carbides and Nitrides (MXene)

Transition-metal carbides and nitrides, known as MXenes, are 2D derivatives of ternary layered $M_{n+1}AX_n$ (MAX) phases, where M is an early transition metal, A is group IIIA–VIA element, X is carbon and/or nitrogen, and n = 1–3. MXenes are generally prepared by selective etching of the element of A from the parent MAX phase to form $M_{n+1}X_nT_x$ (n = 1–3), where M represents an early transition metal, X is carbon and/or nitrogen, and T_x is the surface termination groups ((–O), (–F), and (–OH)) (Fig. 11.3) [17, 20, 21]. The structural and electronic properties of MXenes are strongly affected by surface termination groups. The surface of MXene is terminated by O, F, and OH functional groups, depending on the synthesis method. For example, in HF-etched samples, F termination becomes dominant; whereas, in samples etched with LiF and HCl mixture, more O terminations are obtained. In both cases, there are significantly smaller OH terminations [21–23].

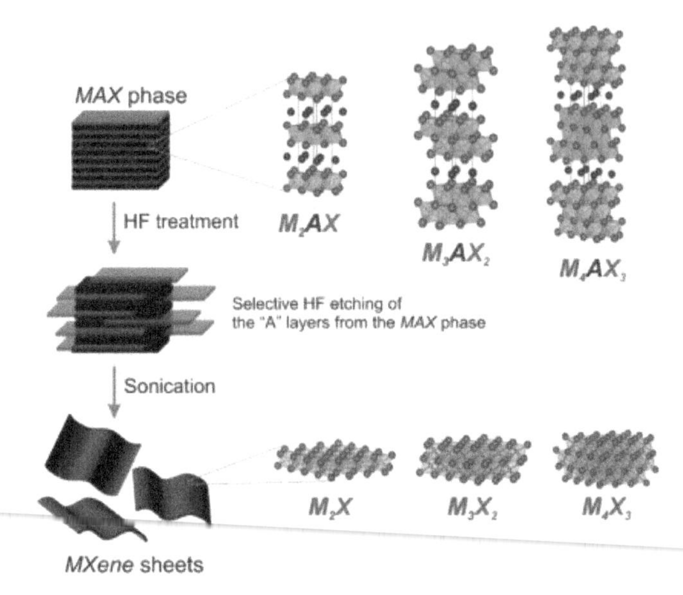

Figure 11.3 The schematic diagram is representing the process of synthesizing MXenes from MAX phases. Reproduced with permission from Ref. [17].

2D MXene materials, which offer many advantages such as electronic, optical, plasmonic, and thermoelectric properties, have attracted much interest recently. They are currently explored for a variety of applications, including energy, environment, catalysis, photocatalysis, optical devices, electronics, biomedicals, sensors, electromagnetic, others, etc. [17, 21–24]. Over 30 types of MXenes have already been synthesized via selective etching, chemical transformations, and bottom-up constructions, and many more have been theoretically proposed. $Ti_3C_2T_X$ has been the most studied MXene since its first discovery in 2011. $Ti_3C_2T_X$ (T_X = OH, F, and O) nanosheets with expanded layers were obtained by immersing the layer-structured in HF solution for 12 h to remove the Al atoms

[17, 20]. Based on theoretical calculations, many studies suggested that surface termination groups strongly influence the stability, electronic, optical, and transport properties of $Ti_3C_2T_x$. To date, $Ti_3C_2T_x$ MXene has shown potential in numerous applications, such as advanced supercapacitors, Li-batteries, electromagnetic shielding, biomedical, biosensing, and light emission [20, 21].

11.2.1.1 Optoelectronic properties

All the MXenes are metallic in the absence of surface functionalization. Due to the functionalization, some MXenes become semiconductors with energy gaps of about 0.25–2.0 eV [23, 24]. The density of free carriers is estimated to be $8 \pm 3 \times 10^{21}$ cm^{-3} while their mobility is estimated to be 0.7 ± 0.2 cm^2/Vs. The electronic band structure of $Ti_3C_2T_x$ MXene can be modified by the surface functionalization groups (T_x) added during the synthesis process [23, 24]. Figure 11.4 plots band structure diagrams for $Ti_3C_2(OH)_2$ and $Ti_3C_2O_2$, which show the metallic properties of the mentioned compounds.

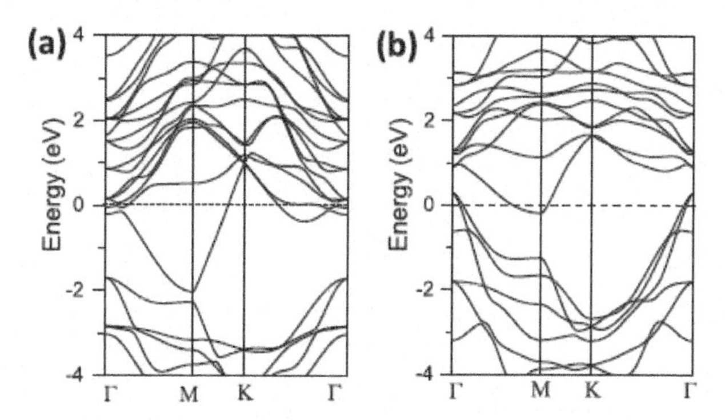

Figure 11.4 Band structure diagrams of $Ti_3C_2(OH)_2$ and $Ti_3C_2O_2$ (b) [24].

2D MXene materials with proper direct bandgap are especially desirable for optical applications. The optical properties of the $Ti_3C_2T_x$ MXene can be altered by the intercalation of cations between the negatively charged $Ti_3C_2T_x$ layers. This could, partly, be due to the increase/decrease of the lattice parameter and/or charge transfer between intercalants and the conducting $Ti_3C_2T_x$ layers [20, 24]. Figure 11.5 shows the crystal structure of Ti_3C_2, $Ti_3C_2F_2$, $Ti_3C_2O_2$, and $Ti_3C_2(OH)_2$. Pristine MXene Ti_3C_2 has a hexagonal lattice constant of

3.071 Å, and lattice parameters of $Ti_3C_2O_2$, $Ti_3C_2(OH)_2$, and $Ti_3C_2F_2$ are 3.12 Å, 3.071 Å, and 3.145 Å, respectively [21–23].

There is a significant optical tunability of MXenes by changing their chemistry that changes the absorption spectra. It should be noted that the absorbance is linearly dependent on the thickness of $Ti_3C_2T_x$ and the intercalated films. A ~1-nm-thick $Ti_3C_2T_x$ MXene absorbs about 3% of visible light at 550 nm wavelength [22–25]. Based on studies $Ti_3C_2T_x$ MXene is a plasmonic material at near-infrared wavelengths, a critical range for optical communication, biological imaging, chemical sensing, and thermal management. MXenes have the unique properties of metallic conductivity, hydrophilic surface, bandgap adjustment, strong light-matter interactions, and 2D-layered atomic structure; these properties make MXenes promising candidates in carriers, novel electronic materials, and parts of sensing materials for rapid, easy, and label-free detection [17, 21]. Besides, the high surface-to-volume ratio of 2D-layered MXene offers a larger contact area to attach the biomolecule/analyte and motivates researchers to explore its application in gas sensing, biochemical, and biosensing applications [22, 23].

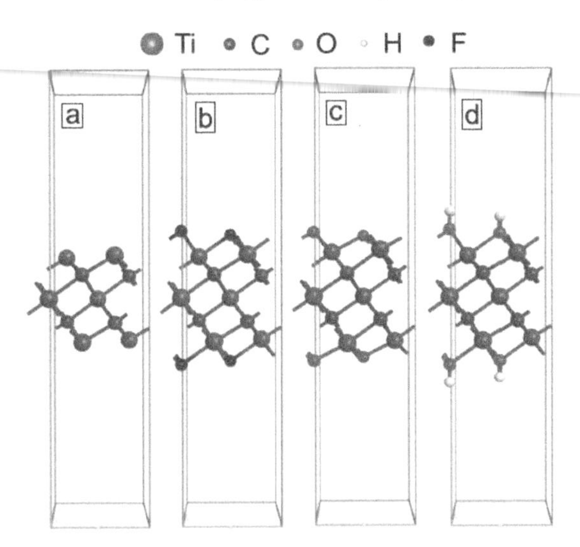

Figure 11.5 Top and side view of structures of Ti_3C_2 (a), $Ti_3C_2F_2$ (b), $Ti_3C_2O_2$ (c), and $Ti_3C_2(OH)_2$ (d) [23].

11.2.2 Transition Metal Dichalcogenides (TMDs)

TMDs are a class of 2D materials with the formula MX_2, where M represents a transition metal element from group IV (e.g., Ti, Zr, and Hf), group V (e.g., V, Nb, or Ta), or group VI (e.g., Mo and W), and X is a chalcogen (e.g., S, Se, or Te). Layered structures of these materials are formed in the X–M–X manner, where two chalcogen atoms hold one metal atom in the hexagonal plane [12, 13, 26]. Figure 11.6 shows the crystal structure of a hexagonal TMD monolayer.

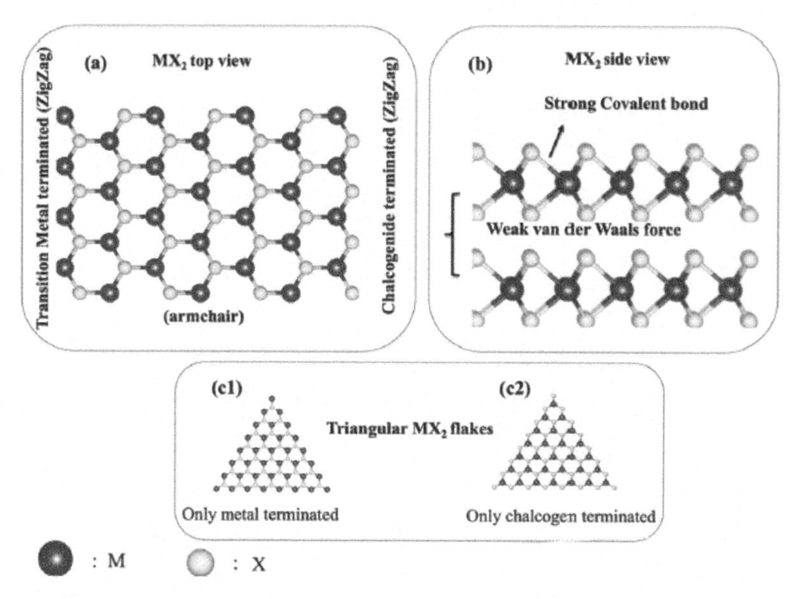

Figure 11.6 Structure of a hexagonal TMD monolayer (a) top view of 2D-MX_2, (b) side view of a bilayer 2D-MX_2, and (c) triangular MX_2 metal/chalcogen terminated zigzag edged structures [26].

11.2.2.1 Optoelectronic properties

2D TMDs exhibit unique electrical and optical properties that evolve from the quantum confinement and surface effects that arise during the transition of an indirect bandgap to a direct bandgap when bulk materials are scaled down to monolayers [26–28]. Among various 2D TMDs, molybdenum/tungsten sulfide/selenides (Mo/WS_2/Se_2) are known to be stable at ambient conditions and thus are known to be useful for energy-storage, sensing, electronic, and photonic device applications. Most semiconducting 2D TMDs reveal direct bandgap in monolayer; whereas, they are indirect bandgap in bulk [12, 13, 27]. At room temperature, the bandgaps of bulk MoS_2, $MoSe_2$, WS_2, and WSe_2 are, respectively, as follows: 0.75 eV, 0.80 eV, 0.89 eV, and 0.97 eV. For monolayer MoS_2, $MoSe_2$, WS_2, and WSe_2, the corresponding bandgaps, respectively, are 1.89 eV, 1.58 eV, 2.05 eV, and 1.61 eV [13, 27, 28].

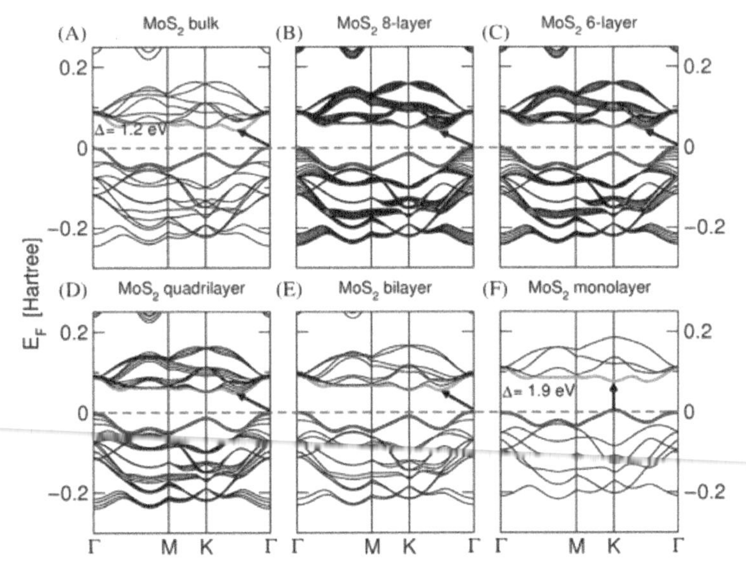

Figure 11.7 Band structures of (A) bulk MoS_2, (B) MoS_2 8-layer, (C) MoS_2 6-layer, (D) MoS_2 quardrilayer, (E) MoS_2 bilayer, and (F) MoS_2 monolayer [27].

Figure 11.7 shows the band structures of MoS_2 as a function of a number of layers; the horizontal dashed lines indicate the Fermi level, and the arrows indicate the approximate location of the bandgap. As can be seen, bulk MoS_2 has an indirect bandgap of 0.75 eV, and with the decreasing number of layers, the value of the indirect bandgap increases [27, 28]. Molybdenum disulfide (MoS_2) has unique properties including direct bandgap (1.89 eV), good mobility (~700 $cm^2\ V^{-1}\ s^{-1}$), high current on/off ratio of ~10^7–10^8, large optical absorption (~$10^7\ m^{-1}$) in the visible range, and it is widely used for optoelectronic devices [26, 27]. The monolayers of

TMDs have shown much higher absorbance at bandgap resonances, which usually fall in the visible and near-infrared range. The amount of adsorption depends on the thickness and number of layers; the monolayer thickness of MoS_2 and WS_2 are considered as 0.65 nm, while the monolayer thickness of $MoSe_2$ and WSe_2 are taken as 0.70 nm [26–28]. The strong absorption of light in semiconducting monolayer TMDs is due to the light-matter interactions of the material, which covers a broad bandwidth ranging from infrared to ultraviolet wavelengths. Besides, the high surface-to-volume ratio in 2D TMDs offers huge potential for the detection of large amounts of target analysts per unit area as well as rapid response and recovery with low power consumption. The 2D TMDs, such as MoS_2 and WS_2, are used for various sensing applications, including gas, chemical, and biosensors [26–28].

11.2.3 Graphene

Graphene can be considered as the parent for all other graphitic allotropes such as 0D fullerenes, 1D nanotubes, and 3D graphite [29–32]. It is made of an sp^2 hybridized hexagonal honeycomb carbon structure with a carbon-carbon distance of 0.142 nm, and the lattice constant is 2.460 Å [29–32]. Figure 11.8 shows the schematic of the graphene sheet and other graphitic allotropes made of graphene.

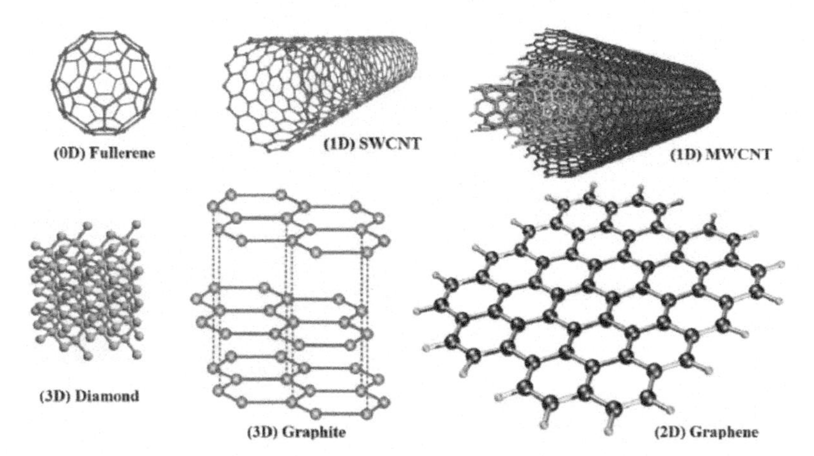

Figure 11.8 Schematic for the graphene sheet and other graphitic allotropes made of graphene [16].

Since its discovery in 2004, by virtue of its unique mechanisms and excellent thermal, electrical, optical, and other properties, it has received considerable scientific interest. Graphene has unique properties such as high strength, flexibility, high charge carrier mobility, high surface-to-volume ratio, and transparency. So, it is highly desired for many applications, including RF transistors, supercapacitors, gas sensors, and biosensors [30–33].

11.2.3.1 Optoelectronic properties

Graphene exhibits low-temperature mobility approaching 200000 cm^2/Vs for carrier densities below 5×10^9 cm^{-2}, which is not observable in semiconductors or non-suspended graphene sheets [14–16]. Furthermore, it shows a strong ambipolar electric field effect with the concentration of charge carriers up to 10^{13} cm^{-2} and room-temperature mobilities of ~10000 $cm^2V^{-1}s^{-1}$, when the gate voltage is applied [14–16]. Graphene's band structure is unique because it consists of two cones touching at the tips, or Dirac points. In these cones, the 2D energy dispersion relation is linear making electrons and holes degenerate. Since graphene does not have a bandgap, it acts like a semimetal, giving it a very poor on/off ratio in field-effect transistors (FETs) of up to ~5^5 [14–16]. Figure 11.9 shows the band structure near the Fermi level of graphene.

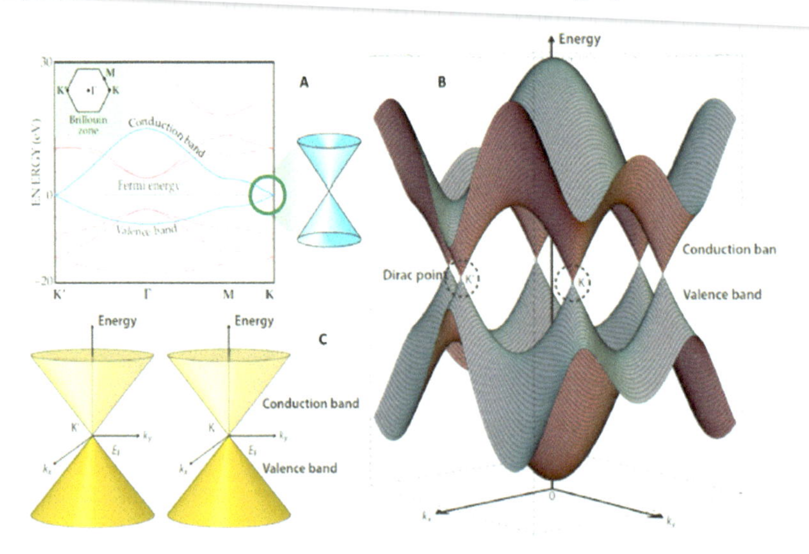

Figure 11.9 Band structure near Fermi level of graphene. (A) 2D schematic diagram, (B) 3D schematic diagram, and (C) Dirac cone of K and K′, which correspond to the Fermi level of (B) [32].

The optical properties of graphene do vary by the layer number and the absorbance increases linearly by increasing the graphene layers. So, the more graphene layers stacked on top of each other, the greater the light absorption and the lower the optical transparency [14, 15, 30]. For example, a pristine single layer of graphene only absorbs 2.3% of light, so 97.7% of light passes through a single layer, with around 0.1% reflected from its initial trajectory. It should be noted that the monolayer thickness of graphene is considered 0.34 nm [15, 16]. In particular, graphene can serve as a plasmonic material operating in the mid-IR to THz regions of the spectrum and its optical constants (such as permittivity and refractive index) can be adjusted by applying a gate voltage. The graphene SPRs occur in the mid-IR wavelength range and exhibit significant electromagnetic wave confinement. Graphene has also shown promise as a component of biological sensors, as it has a large surface area and is biocompatible with antibodies, enzymes, DNA, cells, and proteins. It can be introduced as a platform for nano-biosensor applications playing a crucial role in graphene plasmonic and supporting SPR modes at THz frequencies [14–16, 33].

11.2.4 Hexagonal Boron Nitride (h-BN)

The h-BN is a typical III-V group compound, which is composed of nitrogen atoms and boron atoms, and the two atoms are combined following different hybridization modes to form boron nitride with different phase structures [34–38]. The h-BN is the only phase structure in all phases of boron nitride, which is hexagonal, white, and similar to the layer structural features of graphene, also known as white graphene. Monolayer h-BN contains alternating boron and nitrogen atoms in arrangements of a hexagonal lattice with strong sp^2 bonds between them, which is similar to the lattice structure of graphene [9, 10, 35, 37]. Figure 11.10 shows the honeycomb structure of h-BN.

As can be seen, each of the six boron nitride atoms is an infinite extension of the hexagonal grid composed of B atoms and N atoms. The layers of h-BN are connected by a weak van der Waals force. The bond length is 0.6661 nm, the lattice constant is 0.2504 nm, and the density is 2.28 g/cm^3. The h-BN structure is a semiconductor such as graphene, while it is an insulator at room temperature [9, 10].

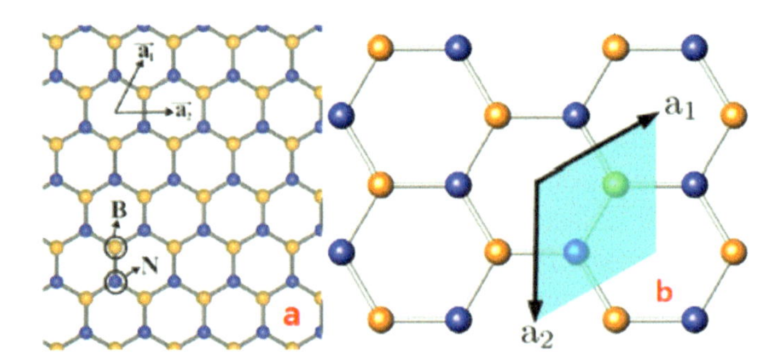

Figure 11.10 Honeycomb structure of 2D h-BN with Bravais lattice vectors and monolayer h-BN with highlighted unit cell [35, 37].

11.2.4.1 Optoelectronic properties

The h-BN is a wide bandgap semiconductor with high thermal stability and chemical stability. 2D h-BN is a direct bandgap, while the top of the valence band and the bottom of the belt are at the high symmetry point K. The band structure of 2D h-BN is not sensitive to the number of layers and is very close to that of the bulk crystal [9, 10, 35, 38]. Figure 11.11 plots the band structure of monolayer h-BN.

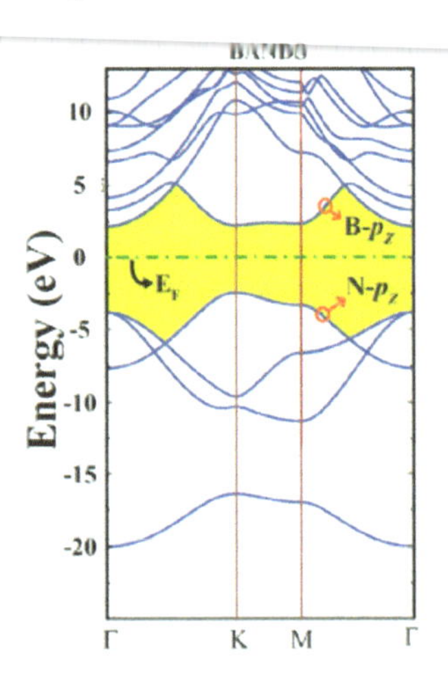

Figure 11.11 The band structure of monolayer h-BN [35].

The h-BN has a very wide bandgap; the theoretical results indicate that monolayer h-BN has an electronic bandgap of around

8.2 eV at the K point of the Brillouin zone and the related optical gap at 6.1 eV and optical absorption is as high as 7.5×10^5/cm. With its unique structural and physical advantages, h-BN is regarded as a promising assistant material to enhance the electronic and optoelectronic performance of other 2D materials [9, 10, 35, 37].

11.2.5 Black Phosphorus (BP)

Phosphorene or 2D BP has come into sight as a hopeful 2D material from its time of invention in 2014 through the flourishing exfoliation method. 2D BP has attracted enormous attention from researchers, due to its excellent structural, mechanical, electronic, magnetic, and vibrational properties [39–42]. In BP crystal structure, atoms are powerfully connected in-plane appearing as layers, although the layers interact via van der Waals forces. Phosphorus atoms in their layered structure have a valence shell configuration of $3s^2 3p^3$ with five valence electrons [39, 41–44]. Figure 11.12 shows the structures of BP for bulk BP and monolayer BP. As shown, phosphorene has a 2D puckered hexagonal honeycomb structure. The electronic band structure of BP is anisotropic, and its armchair nanoribbons appeared as semiconductors exhibiting indirect bandgap [39, 41, 42].

However, metallic behavior is revealed from the zigzag-oriented 2D BP nanoribbons. Most of the anisotropic properties of BP are due to the orthorhombic crystal structure. BP has unique physical properties, which help make BP a best-of-its-kind material related to the field of nanoelectronics and nanophotonics [39, 41, 42, 44].

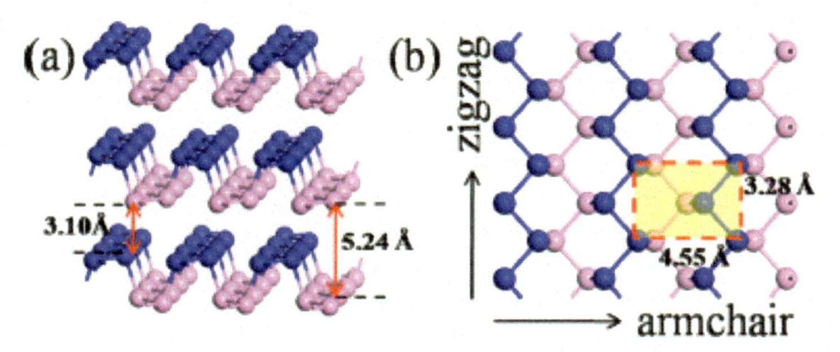

Figure 11.12 The structures of BP: (a) bulk BP and (b) top view of monolayer BP [44].

11.2.5.1 Optoelectronic properties

BP is one of the three main allotropes of phosphorus, and it is thermodynamically the most stable compared to its red and white counterparts. The BP has extraordinary properties, such as high mobility, ranging from 600 cm^2/Vs to 1000 cm^2/Vs at room temperature, and direct and tunable bandgap in both mono- and multi-layered forms [39, 41, 42]. Van der Waals interactions are present between the layers of BP, and the thickness of each layer of BP is about 0.6 nm. It is a semiconductor having unique properties such as in bulk form, its direct bandgap is about 0.3 eV and while it is considered a single-layered BP then its bandgap energy is about 2 eV [39, 42, 44].

Figure 11.13 shows the electronic band structures of multi-layer and single-layer BP materials, as the BP electronic band structure shows a direct bandgap at the Z point of the Brillouin zone, and ranges from ~0.3 eV (bulk) to ~2.0 eV (single-layer) [39, 42]. Since the BP bandgap depends on the number of layers, this property causes the BP to have a broadband optical response from visible to mid-infrared. The novel properties of BP, such as tunable bandgap, anisotropic electrical conductivity, high carrier mobility, large specific surface area, strong absorption in the ultraviolet (UV) and near-infrared regions, inherent in vivo biocompatibility, and biodegradability, make it a potential material for the construction of electronics and optoelectronics devices for biosensing [39, 41, 42].

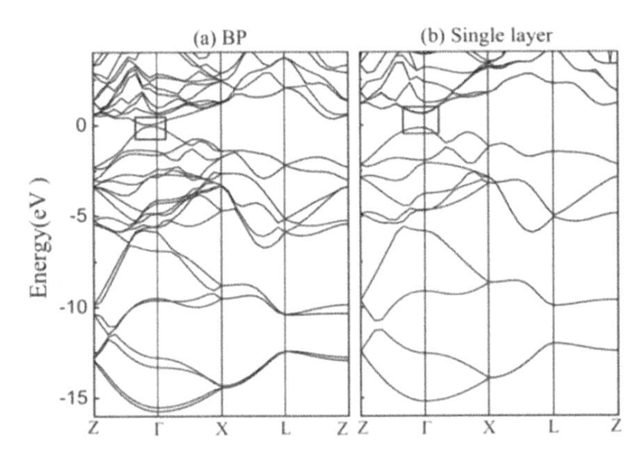

Figure 11.13 Electronic band structures of multi-layer and single-layer BP materials; the Fermi-energy has been set to zero [42].

11.3 Heterostructures

2D-layered materials can be used to build interesting heterostructures by mixing and matching for increased selectivity and sensitivity of the sensors and biosensors [43–45]. Heterostructures, although highly desirable, require careful considerations of lattice mismatch, misalignment of layers, and introduction of unforeseen defects during the deposition and the epitaxial growth [44, 46]. 2D-layered materials can be integrated into a monolayer (lateral 2D heterostructure) or a multilayer stack (vertical 2D heterostructure). Figure 11.14 shows a graphical representation of the heterostructures that can be made with the basic single layer of graphene, h-BN, and MoS_2 [43–45].

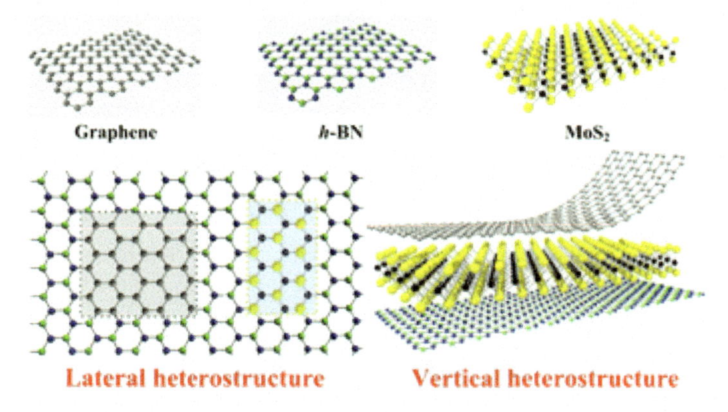

Figure 11.14 Schematic illustration of multilayer heterostructure formation [28].

TMDs materials, which have similar lattice structures, are ideal candidates for the construction of lateral heterostructures because of the similar structures and a lot of vibrations in the TMD 2D materials [44, 46]. For example, the MoS_2/WS_2, $MoS_2/MoSe_2$, or similar TMD heterostructures can be synthesized using the vapor-phase growth method. Layered crystals are characterized by strong intralayer covalent bonding and relatively weak interlayer van der Waals bonding [26, 35, 44]. A large number of vertical heterostructures, including G/h-BN (graphene on h-BN) stack, h-BN/G (h-BN on graphene) stack, G/TMD stack, TMD/TMD stack, and G/h-BN superlattice, have been investigated in experiment and theory and some of them show great potential for applications in electronics, optoelectronics, etc. 2D-layered materials have distinct properties and can be integrated into hybrid structures to realize novel functionalities and applications beyond their components of 2D atomic crystals [44–47].

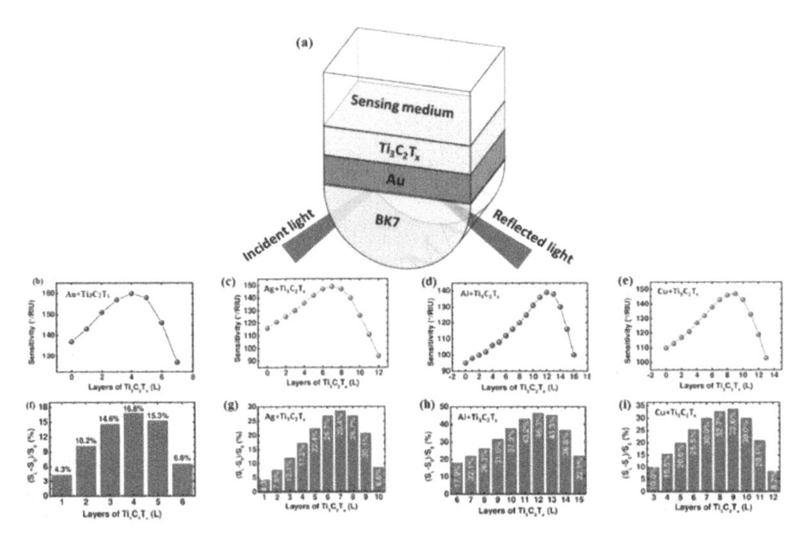

Figure 11.15 (a) Schematic diagram of the proposed SPR biosensor using $Ti_3C_2T_x$ MXene to enhance sensitivity [23]. (b–e) Diagrams show the change of sensitivity with the increase of $Ti_3C_2T_x$ MXene layers based on Au-$Ti_3C_2T_x$, Ag-$Ti_3C_2T_x$, Al-$Ti_3C_2T_x$, and Cu-$Ti_3C_2T_x$, respectively. (f–i) Diagrams show the variation of the sensitivity enhancement with respect to the layers of $Ti_3C_2T_x$ MXene at λ = 633 nm for the proposed biosensors based on Au-$Ti_3C_2T_x$, Ag-$Ti_3C_2T_x$, Al-$Ti_3C_2T_x$, and Cu-$Ti_3C_2T_x$, respectively [23].

11.3.1 MXene Heterostructures

Layer-by-layer van der Waals heterostructures (vdWh) with intriguing electronic properties can be designed on the base of MXenes and other 2D materials. Fabricating vdWh of MXenes, and their stacking with chalcogenides, graphene or other 2D materials can be an innovative route to develop materials with exotic properties [20, 21]. These heterostructures are suitable for optoelectronic device applications. By existing a large variety of MXenes and 2D materials, suitable heterostructures can be formed with a minor lattice mismatch. According to this, a Kretschmann configuration [23] prism-coupled SPR sensor [25] has been proposed as an efficient optical sensor for a biological or chemical analyte. It was found that the sensitivity highly depends on the thicknesses of MXene and the plasmonic metal film. Figure 11.15 shows that the sensitivity enhancements are 16.8%, 28.4%, 46.3%, and 33.6%, for the SPR biosensors based on Au with 4 layers $Ti_3C_2T_X$, Ag with 7 layers $Ti_3C_2T_X$, Al with 12 layers $Ti_3C_2T_X$, and Cu with 9 layers $Ti_3C_2T_X$, respectively. The highest sensitivity enhancement of 46% was obtained using an Al metal film with 12 atomic layers of $Ti_3C_2T_X$ [23].

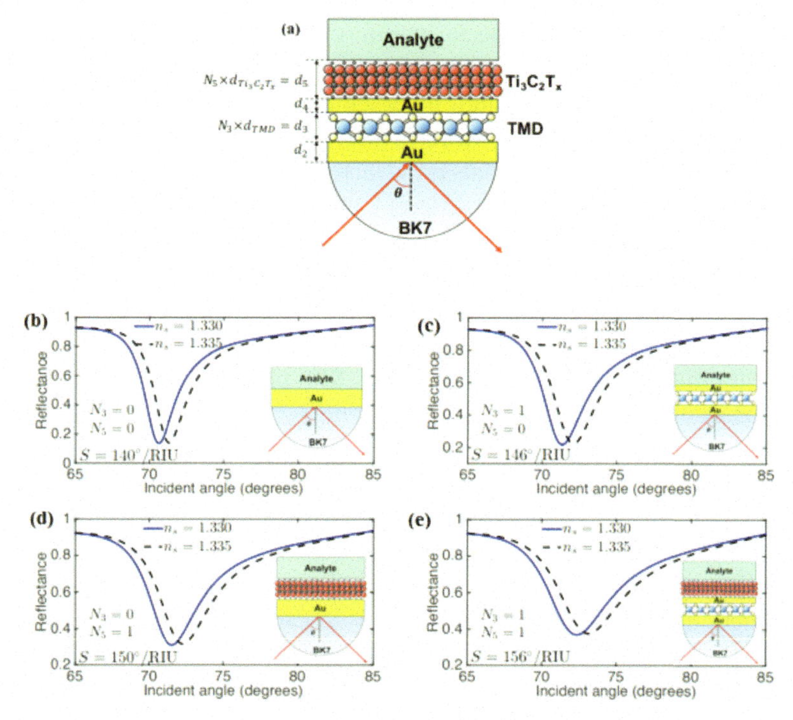

Figure 11.16 (a) Schematic diagram of the proposed SPR biosensor with $Ti_3C_2T_X$ and TMD layers [25]. (b–e) Diagrams show reflectance as a function of the incident angle for the $Ti_3C_2T_X$ -MoS_2-based SPR sensor with (b) $N_3 = 0$, $N_5 = 0$, (i.e., no 2D material); (c) $N_3 = 1$, $N_5 = 0$ (i.e., monolayer MoS_2); (d) $N_3 = 0$, $N_5 = 1$ (i.e., monolayer $Ti_3C_2T_X$), and (e) $N_3 = 1$, $N_5 = 1$ (i.e., monolayer MoS_2 and monolayer $Ti_3C_2T_X$) [25].

The metallic properties of 2D-layered MXenes offer an excellent platform for photon-electron coupling at their surface. So, in Ref.

[25], an SPR sensor based on a heterostructure of Au-$Ti_3C_2T_X$-Au-TMDs was also suggested with an enhancement of the sensitivity of 41% compared to the conventional Au film SPR sensors. Figure 11.16 shows reflectance as a function of the incident angle before and after the variation of the analyte refractive index for the $Ti_3C_2T_X$-MoS_2-based SPR sensor. As can be seen, using both $Ti_3C_2T_X$ MXene and MoS_2 layers, the sensitivity is increased to S = 156°/RIU.

11.3.2 TMDC Heterostructures

The availability of TMDC materials with different bandgaps and work functions allows the formation of heterostructures of many interesting electronic and optoelectronic properties [26, 27]. TMDC heterostructures can be built by assembling individual monolayers into functional multi-layer structures. The most important advantage of TMDC over other materials is the lack of dangling bonds that make it possible to re-stack with different TMDC materials in the vertical direction to produce heterostructures without the requirement of lattice matching [11, 12]. Heterostructure engineering of the TMDC materials offers an exciting opportunity to fabricate highly tunable optoelectronic devices such as photodetectors, photovoltaic devices, light-emitting diodes, and biosensors. SPR technique was used in a biosensor consisting of MoS_2/graphene hybrid structure with Au as a substrate was used to detect biomolecules using SPR [12]. Also in 2016, an SPR biosensor based on the Otto configuration made of MoS_2/Al film/MoS_2/graphene heterostructure was used to detect biomolecules [11]. Figure 11.17 shows a schematic diagram of the device and the change in resonance angle and sensitivity as the function of the number of MoS_2 layers. As can be seen, the maximum sensitivity ~ 190.36°/RIU is obtained with six layers of MoS_2 coating on both surfaces of Al thin film.

11.3.3 Graphene Heterostructures

Graphene heterostructures can be built by assembling 2D materials such as h-BN, TMDCs, MXene, and black phosphorous. For example, 2D h-BN is an appealing substrate dielectric for use in improved graphene-based devices [15, 16]. Graphene devices on h-BN substrates have mobilities and carrier inhomogeneities that are almost an order of magnitude better than devices on SiO_2. 2D h-BN has a similar lattice structure to graphene and has a lattice mismatch with graphene of less than 1.7% [32, 45]. The heterostructures of graphene and h-BN exhibit unique electrical properties and enable a broad prospect in the fabrication of optoelectronic devices. Besides, the WS_2/graphene vdWh are important because they gather the high carrier mobility and broadband absorption of graphene, as well as the direct bandgap and extremely strong light-matter interactions of a single layer WS_2 [45, 46].

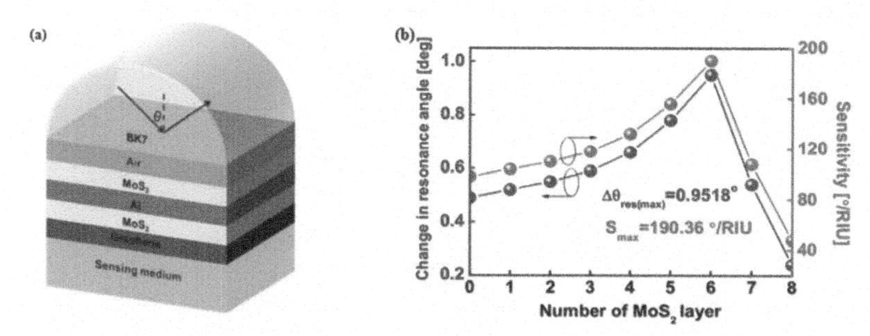

Figure 11.17 (a) Schematic diagram of the proposed SPR biosensor with the MoS_2 layer coated on both surfaces of Al film and monolayer graphene coated on the lower MoS_2 layer [11]. (b) The diagram shows the change in resonance angle and sensitivity as the function of the number of MoS_2 layers [11].

11.3.4 h-BN Heterostructures

In recent years, monolayer h-BN has been applied for the realization of van der Waals heterojunctions in combination with other 2D electronic systems such as graphene and TMDCs [44, 45]. Monolayer h-BN has been used as a dielectric substrate for graphene-based electronic devices, owing to its unique characteristics like thermal stability and as an insulator [44]. Among all the substrate materials, h-BN has an atomic flat surface, very-low roughness, and a surface without any hanging bonds [45, 46]. Due to their relatively small lattice mismatch (1.7%) and the same crystal structure, graphene and h-BN were considered to be integrated into a single atomic layer. A large number of experiments and research shows that graphene/h-BN heterostructures have broad prospects in the construction of nanoscale electronic and optoelectronic devices [44–46].

Graphene/h-BN-integrated devices have been used for DNA sequencing by current modulation and distinguishing nucleotides in DNA [45]. SPR-based biosensor consisting of graphene/h-BN hybrid structure for the detection of biomolecules was reported in 2019 [46]. Figure 11.18 shows schematics of the proposed device, the graphene/h-BN few-layer ribbon array on the top of CaF_2 ribbons is placed on a gold substrate separated by a CaF_2 dielectric spacer. Based on the results reported in Ref. [46], the interaction between the locally enhanced field around the graphene ribbon and its surrounding analyte leads to an ultra-high sensitivity (4.207 μm/RIU), with the FOM reaching approximately 58.

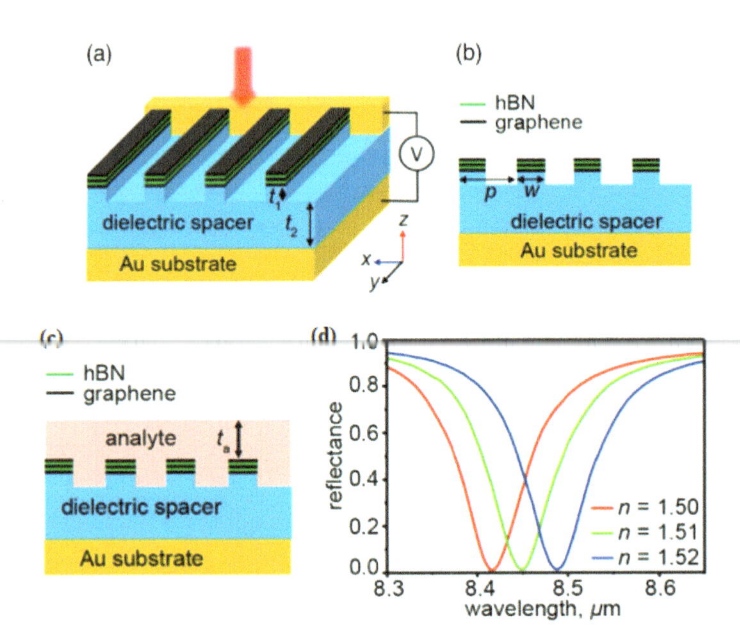

Figure 11.18 (a) Schematic diagram of the proposed SPR biosensor. (b) The cross-sectional view of the device in the x–z-plane, with a period p = 160 nm and width w = 80 nm. [46]. (c) The cross-sectional view of the proposed sensor with the analyte. (d) Reflectance spectra of different analytes with different RIs [46].

11.3.5 BP Heterostructures

BP-based vdWh have begun to attract great attention, due to their narrow bandgap and anisotropic lattice structure [41, 42]. Recently, vdWh based on BP and graphene have attracted a lot of attention, and the structural and electronic properties of graphene/BP heterostructure were studied [42, 44]. Since the lattice constants of BP are 3.34 and 4.57 Å along the zigzag and armchair direction and for graphene, the edge length of the basic hexagon is 2.46 Å. Supercells of both layers are created in commensuration with the common lattice, and the lattice constants of the graphene/BP heterostructure are chosen to be 9.92 and 4.42 Å along the zigzag and armchair direction, respectively, to reduce the strain in both layers [41, 42, 44]. A lattice mismatch of around 3% compressive (tensile) strain occurs along the armchair direction and less than 1% compressive (tensile) strain along the zigzag direction of BP (graphene). Based on studies, the graphene and BP heterostructures are found to be stabilized by weak van der Waals interactions and possess a small charge transfer across the interface. Furthermore, BP–TMDC heterostructures (BP–WSe_2 and BP–MoS_2) are promising candidates for mid-infrared light-emission applications [41, 42, 44]. In Table 11.1, we have summarized various sensors and biosensors made from 2D materials heterostructures.

Table 11.1 A brief report of various plasmonic sensors and biosensors

2D Materials heterostructures	Sensing mechanism	Ref.
MoS_2/Gr on Au	SPR	[12]
MoS_2/Gr-Al hybrid	SPR	[11]
Glass/Ag/Au/graphene	SPR	[1]
Ag/Si/BP/MXene hybrid	SPR	[47]
Gr/h-BN	SPR	[48]
Au/TMD/Au/MXene	SPR	[25]
Au/$Ti_3C_2T_x$ MXene	SPR	[23]

11.4 Metasurfaces

Metasurfaces have been used as wavefront control devices, nanoscale light sources, optical signal modulators, and optical quantum devices. For optical sensing, their capabilities to harvest light into nanoscale electromagnetic hotspots have been utilized to enhance the sensitivity of various detection techniques [48–50]. For instance, plasmonic metasurfaces have enabled the detection of monolayers and their real-time interactions with biomolecules in the mid-infrared (mid-IR) regime [49, 50]. The main advantages of metasurfaces with respect to the existing conventional technology include their low cost, low level of absorption in comparison with bulky metamaterials, and easy integration due to their thin profile. Due to these advantages, they are promising candidates for real-world solutions to overcome the challenges posed by the next generation of transmitters and receivers of future high-rate communication systems that require highly precise and efficient antennas, sensors, active components, filters, and integrated technologies [51–55].

In the design of sensing metasurfaces, several parameters should be carefully considered to achieve the targeted functionalities, including the working wavelength, resonance bandwidth, field enhancement, and resonance quality. Metasurfaces, the 2D assemblies of subwavelength optical antennas, have emerged as a new photonic platform for biosensing [49, 50]. Resonant excitation of optical antennas concentrates optical energy into volumes substantially smaller than diffraction-limited optical spots, which are suitable for detecting small quantities of biomolecules.

In Ref. [49], van der Waals's optical metasurface is suggested that the structure consisting of graphene and h-BN is placed on a silver grating. This structure has been investigated for the application of biosensing of multiple analytes in the mid-infrared region. Figure 11.19 shows a schematic diagram of the device and the absorption curve. The proposed device utilizes the strong and long interaction of hyperbolic phonon polaritons in h-BN. Furthermore, the use of graphene in the proposed structure imparts electrical tunability and turns the device into a multimolecular sensing device [49]. The proposed sensor enhances the absorption, making surface-enhanced infrared and absorption-based detection of analytes possible. Based on the numerical simulations, the biosensor shows high sensitivity and can be used to detect multiple analytes, i.e., 4,4'-bis(N-carbazolyl)–1,1'-biphenyl (CBP) and nitrobenzene in a small modal volume.

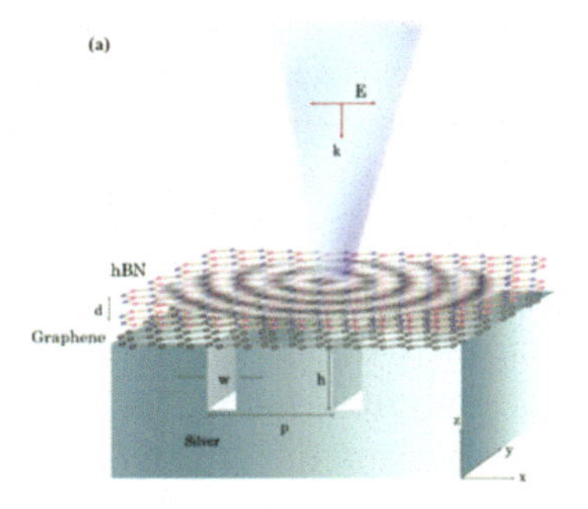

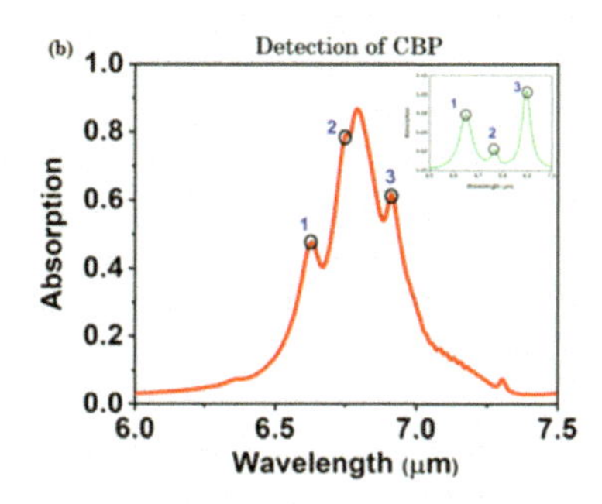

Figure 11.19 (a) Schematic of the h-BN/graphene van der Waals heterostructures on the silver grating. (b) Detection of 50 nm of CBP analyte placed on h-BN/silver grating structure, showing its three characteristic fingerprints marked by numbers at (1) 6.65 μm, (2) 6.77 μm, and (3) 6.9 μm [49].

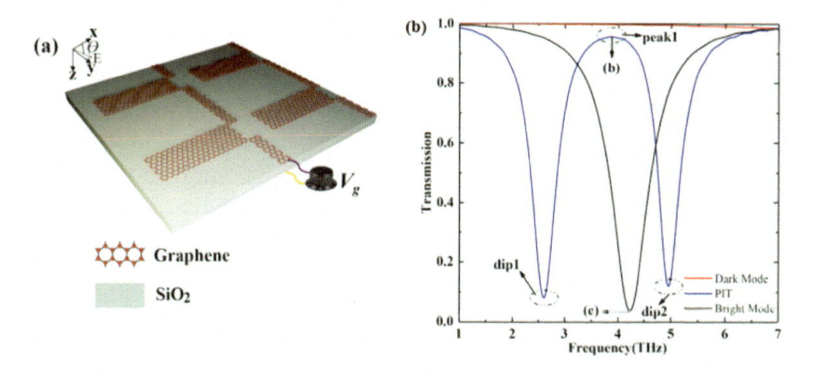

Figure 11.20 (a) Schematic illustration of monolayer graphene-silica metasurface. (b) Transmission spectra of proposed graphene metasurface structure [50].

In Ref. [39], researchers proposed a highly sensitive SPR biosensor composed of a few layered BP, monolayer graphene, and Au thin films. Light absorption and energy loss were well balanced by optimizing gold film thickness and the number of BP layers to generate the strongest SPR excitation [39]. In the device, the in-plane anisotropy of the BP layer can function as a polarizer, so the detection sensitivity of the proposed biosensor can be optically tuned by rotating the plasmonic biosensor. The BP-graphene hybrid layers significantly enhance the detection sensitivity of the proposed biosensor that produces the highest detection sensitivity (7.4914×10^4 deg/RIU) [39]. In Ref. [50], researchers presented a graphene-silica metasurface, an ultra-high sensitivity terahertz sensor based on the tunable PIT effect. Figure 11.20 depicts a schematic illustration of the monolayer graphene-silica metasurface and transmission spectra of the proposed graphene metasurface structure. A periodic structure is composed of wide horizontal graphene and two graphene sheets in bright mode and dark mode. The structure has excellent sensing properties and the sensitivity of the device can reach up to 1.7745 THz/RIU.

In Ref. [60], G-TMDC-G hybrid metasurfaces for ultra-sensitive plasmonic biosensing have been studied. The optimized G-TMDC-G/metal sensing substrates for four types of TMDC materials (i.e., WS_2, MoS_2, $MoSe_2$, and WSe_2) were respectively engineered. Our analysis demonstrated that the GTMDC-G hybrid nanostructure could enhance the SPR sensing performance significantly under optimum conditions [60]. For the G-WS_2-G hybrid metasurface, monolayer WS_2 is found to have the highest efficiency for the electron charge transferred to the metallic sensing surface to enhance the electric field. This G-TMDC-G hybrid plasmonic metasurface shows great potential in the detection of the analyte with small molecular weight or ultra-low concentration [60]. In Table 11.2, we have summarized various sensors and biosensors based on metasurfaces. In this area, several works are done by our research group [61–77].

Table 11.2 Summary of metasurface biosensors

Detection sample	Sample type	Nanostructure	Ref.
ssDNA	DNA	Graphene–gold	[54]
Fungi, penicillium chrysogenum, and bacteria	Bacteria	Gold square rings	[55]
Epidermal growth factor receptor	Breast cancer biomarker	Silicon nanopost array	[56]
Streptavidin-biotin binding	Small molecule	Metallic nanogroove array	[57]
HIV envelope glycoprotein	Protein	Nanograting	[58]
DNA aptamer-human ODAM binding	Protein	Elliptical zigzag array	[59]

11.5 Conclusion

This chapter presented the structure and optoelectronic properties of some 2D materials and investigated the use of 2D materials, heterostructures, and metasurfaces for plasmonic sensing and biosensing applications. The use of 2D-layered materials leads to significant improvement in sensing performance in terms of sensitivity, accuracy, response time, detection limit, etc., over a wide spectral range in the THz frequency range. Since SPR sensors have the potential to detect various types of biological and biochemical analytes. So, research on 2D materials/heterostructures/metasurfaces in plasmonic sensing and biosensing requires more theoretical and practical studies and has the potential for further development in the coming years.

References

1. Salah, N. H., Jenkins. D., Panina, L., Handy, R., Pan, G., and Awan, S. (2012). Self-sensing surface plasmon resonance for the detection of metallic nanoparticles. *Smart Nanosystems in Engineering and Medicine.*, **1**, (Global Institute of Nanotechnology in Engineering and Medicine).
2. Kravets, V., Schedin, F., Jalil, R., Britnell, L., Gorbachev, R.V., Ansell, D., Thackray, B., Novoselov, K.S., Geim, A.K., Kabashin, A.V., and Grigorenko, A.N. (2013). Singular phase nano-optics in plasmonic metamaterials for label-free single-molecule detection. *Nature materials.*, **12**, 304–309.
3. Verma, R., Gupta, B.D., and Jha, R. (2011). Sensitivity enhancement of a surface plasmon resonance based biomolecules sensor using graphene and silicon layers. *Sensors and Actuators B: Chemical.*, **160**, 623–631.
4. Vasic, B., Isic, G., and Gajic, R. (2013). Localized surface plasmon resonances in graphene ribbon arrays for sensing of dielectric environment at infrared frequencies. *Journal of Applied Physics.*, **113**, 013110-013110-013117.
5. Ye, Q., Wang, J., Liu, Z., Deng, Z. Ch., Kong, X.T., Xing, F., Chen, X.D., Zhou, W.Y., Zhang, C.P., and Tian, J.G. (2013). Polarization-dependent optical absorption of graphene under total internal reflection. *Applied Physics Letters.*, **102**, 021912 .

6. Dong, B., Wang, P., Liu. Z. B., Chen, X. D., Jiang, W. S., Xin, W., Xing, F., and Tian, J. G. (2014). Large tunable optical absorption of CVD graphene under total internal reflection by strain engineering. *Nanotechnology*, **25**, 455707.

7. Kretschmann, E. and Raether, H. (1965). Radiative decay of nonradiative surface plasmons excited by light. *Z. Naturforsch. A.*, **23**, 2135–2136.

8. Otto. A. (1968). Excitation of nonradiative surface plasma waves in silver by the method of frustrated total reflection. *Z. Phys.*, **216**, 398–410.

9. De Souza, F.A.L., Amorim, R.G., Scopel, W.L., and Scheicher, R.H. (2017). Electrical detection of nucleotides via nanopores in a hybrid graphene/h-BN sheet. *Nanoscale.*, **9**, 2207–2212.

10. Jiang, H., Choudhury, S., Kudyshev, Z. A., Wang, D., Prokopeva, L. J., Xiao, P., Jiang, Y., and Kildishev, A.V. (2019). Enhancing sensitivity to ambient refractive index with tunable few-layer graphene/hBN nanoribbons. *Photonics Res.*, **7**, 815–822.

11. Wu, L., Jia, Y., Jiang, L., Guo, J., Dai, X., Xiang, Y., and Fan, D. (2016). Sensitivity improved SPR biosensor based on the MoS_2/graphene-aluminum hybrid structure. *J. Lightwave Technol.*, **35**, 82–87.

12. Zeng, S., Hu, S., Xia, J., Anderson, T., Dinh, X.Q., Meng, X.M., Coqent, P., and Yong, K.T. (2015). Graphene-MoS_2 hybrid nanostructures enhanced surface plasmon resonance biosensors. *Sens. Actuators., B*, **207**, 801–810.

13. Naderi, A. and Ghodrati, M. (2017). Improving band-to-band tunneling in a tunneling carbon nanotube field effect transistor by multi-level development of impurities in the drain region. *Eur. Phys. J. Plus.*, **132**, 510.

14. Ando, T. (2009). The electronic properties of graphene and carbon nanotubes, *Npg Asia Mater.*, **1**, 17e21.

15. Geim, A.K. and Macdonald, A.H. (2007). Graphene: Exploring carbon flatland, *Phys. Today.*, **60**, 35e41.

16. Geim, A. K. and Novoselov, K. S. (2007). The rise of graphene. *Nature Materials.*, **6**, 183.

17. Naguib, M., Mochalin, V.N., Barsoum, M.W., and Gogotsi, Y. (2014). 25th anniversary article: MXenes: A new family of two-dimensional materials. *Adv. Mater.*, **26**, 992–1005.

18. Radisavljevic, B., Radenovic, A., Brivio, J., Giacometti, V., and Kis, A. (2011). Single-layer MoS_2 transistors. *Nature Nanotechnology.*, **6**, 147.

19. Dong, Z., Xu, H., Liang, F., Luo, C., Wang, C., Cao, Z.Y., Chen, X. J., Zhang, J., and Wu, X. (2019). Raman characterization on two-dimensional materials-based thermoelectricity. *Molecules.*, **24**, 88.

20. Xiao, Z., Ruan, S., Kong, L.B., Que, W., Zhou, K., Liu, W., and Zhang, T. (2020). *MXenes and MXenes-Based Composites. Springer.*

21. Chen, Y., Ge, Y., Huang, W., Li, Z., Wu, L., Zhang, H., and Li, X. (2020). Refractive index sensors based on $Ti_3C_2T_x$ MXene fibers. *ACS Appl. Nano Mater.*, **3**, 303–311.

22. Berdiyorova, G. R. (2016). Optical properties of functionalized $Ti_3C_2T_x$ (T = F, O, OH) MXene: First-principles calculations. *AIP Advances.*, **6**, 055105.

23. Wu, L., You, Q., Shan, Y., Gan, S., Zhao, Y., Dai, X., and Xiang, W. (2018). Few-layer $Ti_3C_2T_x$ MXene: A promising surface plasmon resonance biosensing material to enhance the sensitivity. *Sensors Actuators B.*, **277,** 210–215.

24. Zhang, L., Su, W., Shu, H., Lu, T., Fu, L., Song, K., Haung, X., Yu, J., Lin, C.T., and Tang, Y. (2019). Tuning the photoluminescence of large $Ti_3C_2T_x$ MXene flakes. *Ceramics International.*, **45**, 11468–11474.

25. Xu, Y., Ang, Y.S., Wu, L., and Ang, L.K. (2019). High sensitivity surface plasmon resonance sensor based on two-dimensional MXene and transition metal dichalcogenide: A theoretical study. *Nanomaterial.*, **9**, 165.

26. Ghatak, K., Kang, K. N., Yang, E, H., and Datta, D. (2020). Controlled edge dependent stacking of WS_2-WS_2 Homo- and WS_2-WSe_2 hetero-structures: A computational study. *Scientific reports.*, **10,** 1648.

27. Kuc, A., Zibouche, N., and Heine, T. (2011). Influence of quantum confinement on the electronic structure of the transition metal sulfide TS2. *Phys. Rev. B.*, **83**, 245213.

28. Wu, L. (2017). Sensitivity enhancement by using few-layer black phosphorus-graphene/TMDCs heterostructure in surface plasmon resonance biochemical sensor. *Sens. Actuators B.*, **249**, 542–548.

29. Ghodrati, M., Mir, A., and Naderi, A. (2021). Proposal of a doping-less tunneling carbon nanotube field-effect transistor. *Mater. Sci. Eng. A: B.*, **256**, 115016.

30. Baqir, M.A., Farmani, A., Fatima, T., Raza, M.R., Shaukat, S.F., and Mir, A. (2018). Nanoscale, tunable, and highly sensitive biosensor utilizing hyperbolic metamaterials in the near-infrared range. *Applied Optics.*, **57**, 9447–9454.

31. Ghodrati, M., Mir, A., and Naderi, A. (2020). New structure of tunneling carbon nanotube FET with electrical junction in part of drain region and step impurity distribution pattern. *AEU-Int. J. Electron. Commun.*, **117**, 153102.

32. Ando, T. (2009). The electronic properties of graphene and carbon nanotubes. *Npg Asia Mater.*, **1**, 17e21.

33. Farmani, A., Miri, M., and Sheikhi, M. H. (2018). Design of a high extinction ratio tunable graphene on white graphene polarizer. *IEEE Photon. Technol. Lett.*, **30**, 153–156.

34. Ghodrati, M., Farmani, A., and Mir, A. (2021). Non-destructive label-free biomaterials detection using tunneling carbon nanotube based biosensor. *IEEE Sensors J.*, **21**, 8847–8854.

35. Topsakal, M., Akturk, E., and Ciraci, S. (2009). First-principles study of two-and one-dimensional honeycomb structures of boron nitride. *Phys. Rev. B.*, **79**, 115442.

36. Naderi, A. and Ghodrati, M. (2018). An efficient structure for T-CNTFETs with intrinsic-n-doped impurity distribution pattern in drain region. *Turkish J. Electr. Eng. Comput. Sci.*, **26**, 2335–2346.

37. Ekuma, C.E., Dobrosavljevi, V., and Gunlycke, D. (2017). First-principles-based method for electron localization: Application to monolayer hexagonal boron nitride. *Phys. Rev. Lett.*, **118**, 106404.

38. Naderi, A. and Ghodrati, M. (2018). Cut off frequency variation by ambient heating in tunneling p-i-n CNTFETs. *ECS J. Solid State Sci. Technol.*, **7**, M6–M10.

39. Yuan, Y., Yu, X., Ouyang, Q., Shao, Y., Song, J., Qu, J., and Yong, K.T. (2018). Highly anisotropic black phosphorous-graphene hybrid architecture for ultrassensitive plasmonic biosensing: Theoretical insight. *2D Materials.*, **5**, 025015.

40. Ghodrati, M., Mir, A., and Farmani, A. (2020). Carbon nanotube field effect transistors-based gas sensors. *In Nanosensors for Smart Cities.* Amsterdam, The Netherlands: Elsevier, 171–183.

41. Zong, X., Hu, H., Ouyang, G., Wang, J., Shi, R., Zhang, L., Zeng, Q., Zhu, C., Chen, S., Cheng, C., Wang, B., Zhang, H., Liu, Z., Huang, W., Wang, T., Wang, L., and Chen, X. (2020). Black phosphorus-based van der Waals heterostructures for mid-infrared light-emission applications. *Light: Science & Applications.*, **9**, 114.

42. Yi, Y., Sun, Z., Li, J., Chu, P.K., and Yu, X.F. (2019). Optical and optoelectronic properties of black phosphorus and recent photonic and optoelectronic applications. *Small Methods.*, **3**, 1900165.

43. Naderi, A., Ghodrati, M., and Baniardalani, S. (2020). The use of a Gaussian doping distribution in the channel region to improve the performance of a tunneling carbon nanotube field-effect transistor. *J. Comput. Electron.*, **19**, 283–290.

44. Li, C., Xie, Z., Chen, Z., Cheng, N., Wang, J., and Zhu, G. (2018). Tunable bandgap and optical properties of black phosphorene nanotubes. *Materials.*, **11**, 304.

45. Souza, F.A.L., Amorim, R.G., Scopel, W.L., and Scheicher R.H. (2017). Electrical detection of nucleotides via nanopores in a hybrid graphene/h-BN sheet. *Nanoscale.*, **9**, 2207–2212.

46. Jiang, H., Choudhury, S., Kudyshev, Z.A., Wang, D., Prokopeva, L.J., Xiao, P., Jiang, Y., and Kildishev, A.V. (2019). Enhancing sensitivity to ambient refractive index with tunable few-layer graphene/hBN nanoribbons. *Photonics Res.*, **7**, 815–822.

47. Kumar, R., Pal, S., Verma, A., Prajapati, Y.K., and Saini, J. P. (2020). Effect of silicon on sensitivity of SPR biosensor using hybrid nanostructure of black phosphorus and MXene. *Superlattices Microstruct.*, **145**, 106591.

48. Zheng, Z., Lu, F., and Dai, X. (2019). Tunable reflected group delay from the graphene/hBN heterostructure at infrared frequencies. *Results Phys.*, **15**, 102681.

49. Kumari, R., Yadav, A., Sharma, S., Gupta, T. D., Varshney, S. K., and Lahiri, B. (2021). Tunable Van der Waal's optical metasurfaces (VOMs) for biosensing of multiple analytes, *Optics Express.*, **29**, 25800–25811.

50. He, Z., Li, L., Ma, H., Pu, L., Xu, H., Yi, Z., Cao, X., and Cui, W. (2021). Graphene-based metasurface sensing applications in terahertz band, *Results in Physics.*, **21**, 103795.

51. Briggs, N., Subramanian, S., Lin, Z., Li, X., Zhang, X., Zhang, K., Xiao, K., Geohegan, D., Wallace, R., and Chen, L.Q. (2019). A roadmap for electronic grade 2D materials. *2D Materials.*, **6**, 022001.

52. Ghodrati, M., Farmani, A., and Mir, A. (2019). Nanoscale sensor-based tunneling carbon nanotube transistor for toxic gases detection: A first-principle study. *IEEE Sensors J.*, **19**, 7373–7377.

53. Wu, J., Jiang, L., Guo, J., Dai, X., Xiang, Y., and Wen, S. (2016). Tunable perfect absorption at infrared frequencies by a graphene-hBN hyper crystal. *Opt. Express.*, **24**, 17103–17114.

54. Zeng, S., Sreekanth, K.V., Shang, J., Yu, T., Chen, C.K., and Yin, F. (2015). Graphene–gold metasurface architectures for ultrasensitive plasmonic biosensing. *Adv. Mater.*, **27**, 6163–6169.

55. Park. S. J., Hong, J. T., Choi, S. J., Kim, H. S., Park, W. K., Han, S. T., Park, J. Y., Lee, S., Kim, D. S., and Ahn, Y. H. (2015). Detection of microorganisms using terahertz metamaterials. *Sci. Rep.*, **4**, 4988.
56. Wang, Y., Ali, M.A., Chow, E.K.C., Dong, L., and Lu, M. (2018). An optofluidic metasurface for lateral flow-through detection of breast cancer biomarker. *Biosens. Bioelectron.*, **107**, 224–229.
57. Jiang, L., Zeng, S., Xu, Z., Quyang, Q., Zhang, D. H., Chong, P. H. J., Coquet, P., He, S., and Yong, K. T. (2017). Multifunctional hyperbolic nanogroove metasurface for submolecular detection. *Small.*, **13**, 1700600.
58. Ahmed, R., Ozen, M. O., Karaaslan, M. G., Prator, C.A., Thanh, C., Kumar, S., Torres, L., Iyer, N., Munter, S., Southern, S., Henrich, T. J., Inci, F., and Demirci, U. (2020). Tunable fano-resonant metasurfaces on a disposable plastic-template for multimodal and multiplex biosensing. *Adv. Mater.*, **32,** 1907160.
59. Leitis, A., Tittl, A., Liu, M., Lee, B. H., Gu, M. B., Kivshar, Y. S., and Altug, H. (2019). Angle-multiplexed all-dielectric metasurfaces for broadband molecular fingerprint retrieval. *Sci. Adv.*, **5**, eaaw2871.
60. Jiang, L., Zeng, S., Quyang, Q., Dinh, X. Q., Coquet, P., Qu, J., He, S., and Yong, K.T. (2017). Graphene-TMDC-graphene hybrid plasmonic metasurface for enhanced biosensing: A theoretical analysis. *Phys. Status Solidi A.*, **214**, 1700563
61. Farmani, A., Miri, M., and Sheikhi, M. H. (2017). Tunable resonant Goos–Hänchen and Imbert–Fedorov shifts in total reflection of terahertz beams from graphene plasmonic metasurfaces. *JOSA B*, **34**(6), 1097–1106.
62. Farmani, A., Mir, A., and Sharifpour, Z. (2018). Broadly tunable and bidirectional terahertz graphene plasmonic switch based on enhanced Goos-Hänchen effect. *Applied Surface Science*, **453**, 358–364.
63. Farmani, A. (2019). Three-dimensional FDTD analysis of a nanostructured plasmonic sensor in the near-infrared range. *JOSA B*, **36**(2), 401–407.
64. Farmani, A., Zarifkar, A., Sheikhi, M. H., and Miri, M. (2017). Design of a tunable graphene plasmonic-on-white graphene switch at infrared range. *Superlattices and Microstructures*, **112**, 404–414.
65. Farmani, A. and Mir, A. (2019). Graphene sensor based on surface plasmon resonance for optical scanning. *IEEE Photonics Technology Letters*, **31**(8), 643–646.
66. Farmani, H., Farmani, A., and Biglari, Z. (2020). A label-free graphene-based nanosensor using surface plasmon resonance for biomaterials detection. *Physica E: Low-dimensional Systems and Nanostructures*, **116**, 113730.

67. Hamzavi-Zarghani, Z., Yahaghi, A., Matekovits, L., and Farmani, A. (2019). Tunable mantle cloaking utilizing graphene metasurface for terahertz sensing applications. *Optics Express*, **27**(24), 34824–34837.

68. Amoosoltani, N., Zarifkar, A., and Farmani, A. (2019). Particle swarm optimization and finite-difference time-domain (PSO/FDTD) algorithms for a surface plasmon resonance-based gas sensor. *Journal of Computational Electronics*, **18**(4), 1354–1364.

69. Farmani, A., Soroosh, M., Mozaffari, M. H., and Daghooghi, T. (2020). Optical nanosensors for cancer and virus detections. *Nanosensors for Smart Cities*, Elsevier, 419–432.

70. Han, B., Tomer, V., Nguyen, T. A., Farmani, A., and Singh, P. K. (eds.) (2020). *Nanosensors for Smart Cities*, Elsevier.

71. Farmani, A., Farhang, M., and Sheikhi, M. H. (2017). High performance polarization-independent quantum dot semiconductor optical amplifier with 22 dB fiber to fiber gain using mode propagation tuning without additional polarization controller. *Optics & Laser Technology*, **93**, 127–132.

72. Khosravian, E., Mashayekhi, H. R., and Farmani, A. (2020). Tunable plasmonics photodetector in near-infrared wavelengths using graphene chemical doping method. *AEU-International Journal of Electronics and Communications*, **127**, 153472.

73. Farmani, H. and Farmani, A. (2020). Graphene sensing nanostructure for exact graphene layers identification at terahertz frequency. *Physica E: Low-dimensional Systems and Nanostructures*, **124**, 114375.

74. Omidniaee, A., Karimi, S., and Farmani, A. (2021). Surface plasmon resonance-based SiO2 kretschmann configuration biosensor for the detection of blood glucose. *Silicon*, 1–10.

75. Khajeh, A., Hamzavi-Zarghani, Z., Yahaghi, A., and Farmani, A. (2021). Tunable broadband polarization converters based on coded graphene metasurfaces. *Scientific Reports*, **11**(1), 1–11.

76. Khosravian, E., Mashayekhi, H. R., and Farmani, A. (2021). Highly polarization-sensitive, broadband, low dark current, high responsivity graphene-based photodetector utilizing a metal nano-grating at telecommunication wavelengths. *JOSA B*, **38**(4), 1192–1199.

77. Khani, S., Farmani, A., and Mir, A. (2021). Reconfigurable and scalable 2, 4-and 6-channel plasmonics demultiplexer utilizing symmetrical rectangular resonators containing silver nano-rod defects with FDTD method. *Scientific Reports*, **11**(1), 1–13.

Chapter 12

Design Considerations and Limitations for Miniature SPR Devices

Anand M. Shrivastav,[a] Marwan J. AbuLeil,[a,b] and Ibrahim Abdulhalim[a]

[a]*Department of Electrooptic and Photonics Engineering, Ilse Katz Institute of Nanoscale Science and Technology, ECE School, Ben Gurion University of the Negev, Beer Sheva 84105, Israel*

[b]*Photonicsys Ltd., House 54, Wahat Alsalam-Neveh Shalom 9976100, Israel*

anandmoh@post.bgu.ac.il

12.1 Introduction

Since the last three decades, surface plasmon resonance (SPR) has provided new horizons to the chemical/biomedical industry as an optical tool for the analysis of molecular interactions and sensing applications. In the current age of exponential growth of technologies, real-time and label-free detection of biomolecular interactions by optical means is unthinkable without SPR. The phenomenon basically refers to the electromagnetic oscillation of coherent collective-free electrons at a semi-infinite interface

Plasmonics-Based Optical Sensors and Detectors
Edited by Banshi D. Gupta, Anuj K. Sharma, and Jin Li

ISBN 978-981-4968-85-0 (Hardcover), 978-1-003-43830-4 (eBook)
www.jennystanford.com

associated with a metallic-dielectric interface when the incident photon momentum along the interface, with polarization in the plane of incidence, matches that of the plasmon. Due to the confinement of the field normal to the interface local enhancement appears. This is expressed as an integrated exponentially decaying electromagnetic field distribution along and normal to the interface and the associated wave is termed surface plasmon wave (SPW) or surface plasmon polariton (SPP) or simply extended plasmon. The reflectivity at the SPW has a large dependency upon the refractive index (RI) of the metal and dielectric materials. Hence, one can easily monitor the RI variation of the dielectric layer near the metal surface. Based on this methodology, SPR-based sensors are conceptualized [1–3].

Going back in time to 1902, the SPR concept was first visualized by observing metal-supported diffraction grating with an uneven distribution of the reflected spectrum [4]. Afterward, this light distribution was termed "surface plasmons (SPs)" by Ritchie, where he reported the SPs existence at a thin metal surface [5]. However, the main milestone in this field was given by Kretschmann in 1968, when he proposed a simple prism-based configuration to achieve SPR, where a thin metallic film was deposited on the base of a high index prism and the resonance is obtained at the other side of a metal-dielectric interface, named as Kretschmann configuration [6]. However, in terms of sensing applications, the first SPR-based biosensor work was pioneered by Liedberg and co-workers in 1983, where the detection of halothane gas and IgG antibodies was demonstrated [7]. Since then, the technology based on SPR has shown incredible growth, especially in the field of chemical/biosensing applications, due to the integration of nanotechnology, microfluidics, and biotechnology advancements. Seeking the quick, highly sensitive, and label-free detection of biological analytes, several commercial SPR-based devices are also available in the market. The first commercial SPR-based device was launched by Biacore Instruments in 1990, which was able to analyze biomolecular interactions with high sensitivity, and rapid response time in addition to accurate, reliable, and reproducible results [8]. After the first commercialized device and technological developments, SPR became one of the "gold standard" transducing mechanisms for real-time monitoring of biomolecular interactions. In continuation, IBIS technology launched the cuvette-based SPR system in 1995 and a modified version (IBIS II) along with two-channel SPR was launched in 1997 [9]. Since then, a number of companies launched their commercialized SPR systems ranging from high-cost, large with highly resolved devices to compact, low-cost with reasonable resolution. A comprehensive survey of several commercialized SPR products has been listed in Table 12.1.

Table 12.1 A list of commercial SPR systems available in the market

Model	Manufacturer	Website	Resolution claimed	Size (weight)
BI-4500 BI-2500 SPRm 200	Biosensing Instruments (USA)	www.biosensingusa.com	<0.06 RU <0.06 RU <0.6 RU	11.5 kg 8 kg 21 kg
Biacore T200 Biacore S200 Biacore X-100 Biacore 8K+	Biacore (USA)	www.biacore.com	<0.03 RU <0.015 RU <0.1 RU <0.02 RU	60 kg 60 kg - 141 kg
BIOSUPLAR-6	Biosuplar (Germany)	www.biosuplar.com	<3 mdeg	2.5 kg
Carterra LSA	Carterra (USA)	www.carterra-bio.com	–	–
IBIS MS96 IBIS iSPR6	IBIS Technologies B.V. (Netherlands)	www.ibis-spr.nl	–	–
MP-SPR-Navi	Bionavis (Finland)	www.bionavis.com	-	-
NanoSPR8 NanoSPR9 NanoSPR103 NanoSPR77	NanoSPR (USA)	www.nanospr.com	2×10^{-5} RIU 2×10^{-5} RIU 0.005 RIU 5×10^{-7} RIU	2.2 kg 2.2 kg 3 kg 3 kg
Open-SPR	Nicoya (USA)	www.nicoyalife.com	-	20 kg
P4SPR	Affinite Instruments (Canada)	www.affiniteinstruments.com	1×10^{-6} RIU	1.3 kg
Pioneer FE	SensiQ Technologies (USA)	www.sensiqtech.com	-	-
Plasmetrix CORGI	Plasmetrix (Canada)	www.plasmetrix.com	1×10^{-6} RIU	-

(Continued)

Table 12.1 (*Continued*)

Model	Manufacturer	Website	Resolution claimed	Size (weight)
Reichert 4SPR Reichert 2SPR	Reichert Technologies (USA)	www.reichertspr.com	1×10^{-8} RIU 1×10^{-7} RIU	- -
Sierra SPR® Pro	Bruker (USA)	www.bruker.com	< 0.02 RU	70 kg
SPR H-5 SPR MH-5 (Magnetic nanoparticles enabled)	Photonicsys (Israel)	www.photonicsys.com	1×10^{-5} RIU 1×10^{-6} RIU	1 kg 1 kg
White FOx system	Fox Biosystems (Belgium)	www.foxbiosystems.com	-	24 kg
Xel-Plex^{TM} SPRi-Arrayer OpenPlex	Horiba Scientific (Japan)	www.horiba.com	-	-

Although there are several companies listed providing SPR systems, there are several challenges while developing a commercial SPR device. These challenges include accurate signal measurements, affecting the signal noise such as light source, integration with microfluidic channels, binding analysis, etc. This chapter covers the design considerations and instrumental optics of a typical SPR device along with the interrogation schemes, various analytical methods to find the suitable SPR dip position, several parameters affecting the signal noise, and other basic units of the SPR system covering the flow channels and their properties along with sensogram of a generic SPR sensor at different sensing stages.

12.2 SPR Instrumental Optics

As mentioned in the introduction section, Kretschmann proposed the conventional prism-based configuration to realize SPR, also known as the Kretschmann configuration. The configuration consists of a high index prism coated with a few tens of nanometres thick plasmonic metals, followed by a dielectric layer over the other metallic interface, as shown in Fig. 12.1a. In this configuration, p-polarized is launched at the metal-prism interface facilitating total internal reflection (TIR). The evanescent wave generated due to TIR provides the photon momentum, necessary for the SPW to be generated at the metal-dielectric interface, and this coupling leads to a sharp dip in the reflectance spectra at the specific resonance parameters [10]. In addition to the worldwide broadly used prism-based configuration, optical fiber and grating-based SPR are the two other configurations, which are used to excite SPR as pictorially replicated in Fig. 12.1b,c, respectively. A more extensive understanding of these two SPR configurations can be found elsewhere [11–13].

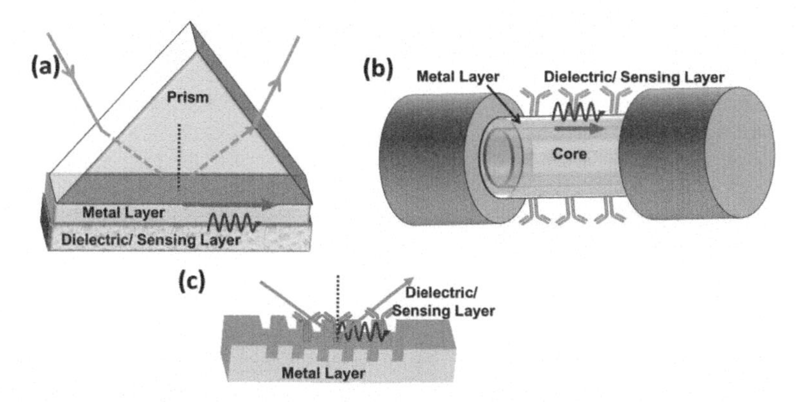

Figure 12.1 Optical configurations to realize SPR. (a) Conventional prism-based configuration. (b) Optical fiber configuration. (c) Grating coupled SPR configuration. Reprinted with permission from Springer Nature [13].

In general, prism-based SPR devices are configured in three different categories based on interrogation schemes: (i) intensity interrogation, (ii) angular interrogation, and (iii) spectral interrogation. We will discuss each interrogation scheme and its respective advantages and challenges one by one.

12.2.1 Intensity Interrogation Scheme

The intensity interrogation method corresponds to monitoring the intensity fluctuations in the reflectivity changes of SPR signal where SPR is excited at a fixed incident angle and wavelength of incident p-polarized light. The change in RI of the sensing layer, over the metallic thin film results in variation in the reflected intensity of reflected light, which can be measured by a detector. A basic understanding of the measurement process and the dynamic sensogram response is pictorially represented in Fig. 12.2.

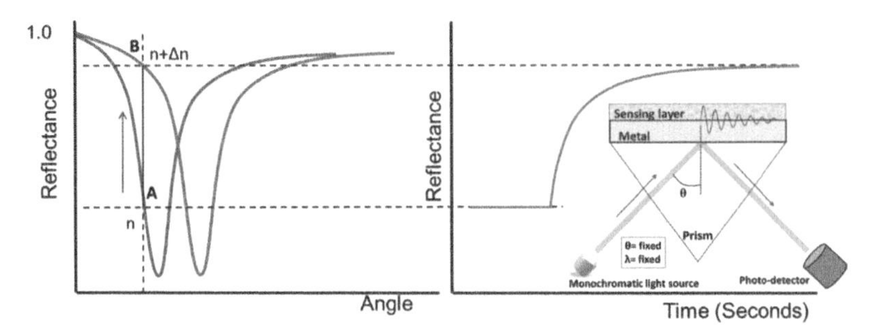

Figure 12.2 Reflectivity response and corresponding dynamic sensor response (sensogram) for intensity interrogation scheme. In this case, monochromatic light is detected at a fixed angle of the photodetector one line.

The SPR instruments based on intensity interrogation follow the change in the curve's reflected intensity, which is obtained due to a shift in SPR dip as shown through vertical lines A to B in the above figure. These types of SPR devices have a robotic stepper motor facility to find the required angle of incidence that provides maximum reflectivity change. A similar mechanism is also used as the basic principle for sensing devices based on SPR imaging (SPRi). SPRi adopts the Kretschmann configuration keeping a fixed angle and wavelength of the incident light, while the reflected light is recorded by a charge-coupled device (CCD) camera [14]. This setup allows the real-time visualization of the biochip and the sensor surface can be split into multiple sensing spots to achieve high throughput sensing. The first type of such sensor was reported by Rothenhausler and Knoll in 1988, and they call this innovative technique as SPR microscopy [15]. Jing et al. reported a combined protocol where SPR imaging and cell edge tracking were simultaneously used to analyze binding interactions. The first method provided the ligand mass binding kinetics with the surface, while the latter was used for small molecule binding detection capability [16]. The setup and corresponding results are represented in Fig. 12.3.

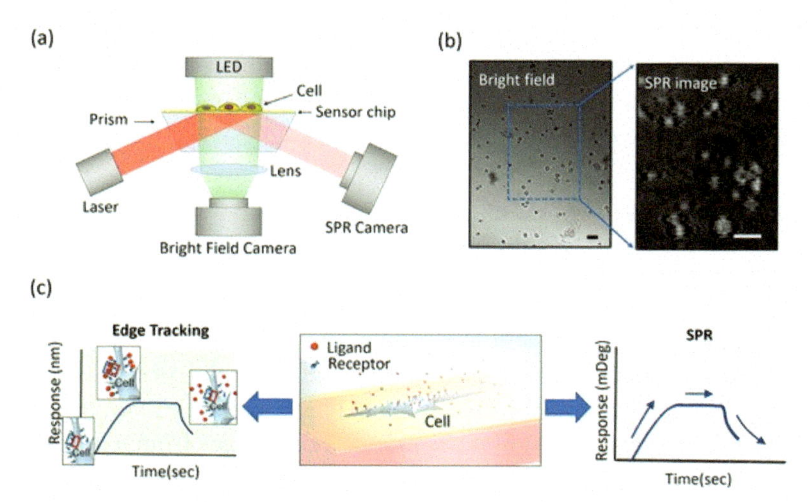

Figure 12.3 (a) Schematic of the setup for simultaneous operation of SPR imaging and bright field imaging. (b) A comparison of bright-field and SPR images at the scale bar of 20 μm. (c) A zoom-in schematic of one cell to illustrate binding affinity and kinetics analysis simultaneously by the edge tracking method and the SPR method. Reproduced with permission form [16].

Due to the intensity interrogation, the SPRi-based sensors suffer about one order worse resolution than the conventional reflectivity-based SPR sensors [11, 17, 18]. The reflected light intensity basically depends upon the coupling of incident p-polarized light with the plasmonic oscillations at the metal/dielectric interface and hence, it can be related to the RI of the medium over the metallic interface. Fu et al. introduce an SPR-based imaging sensor to increase the stability and SPR image contrast using a white light source along with a band pass filter [19]. The sensor was working at 853 nm operating wavelength with Ri resolution of 3×10^{-5} RIU and an upper limit of the spatial resolution with 50 μm. In addition, an SPRi configuration with increased resolution ($\sim5\times10^{-6}$) using the polarization contrast and a custom-designed multilayer structure was reported by Piliarik et al. [20]. In the study, the prism-based SPR device was kept between two crossed polarizers where the output polarizer blocked all the reflected light obtained from the inactive sensing regime. It would be important to mention that in such sensors, the factors which strongly affect the signal noise are the sensor surface roughness, fluctuations in light source intensity, wavelength deviations, photodiode resolution, CCD properties, etc. A brief discussion of these parameters will be done in the forthcoming sections of the chapter.

12.2.2 Angular Interrogation Scheme

Figure 12.4 shows a sensogram of the SPR device, operating in the angular interrogation mode, where the position of the SPR dip is monitored as a function of time using a monochromatic light source and varying the angles of incident light and corresponding photodetector. The resonance angle of the SPR shifts from A to B in the SPR curve as the RI of the sensing layer changes from n to $n+\Delta n$ (left side of the figure). Corresponding change with respect to time is tracked as a function of time as the dynamic behavior of the sensor in the sensogram as represented on the right side of Fig. 12.4. In such configuration, the SPR dip is fitted with a suitable analytical method corresponding to the resonance angle at minimum reflectance, to find the most reliable resonance angle. There are several methods are reported in the literature regarding the best ways to obtain suitable SPR dip quantitatively. These methods will be discussed in the next sections of the chapter.

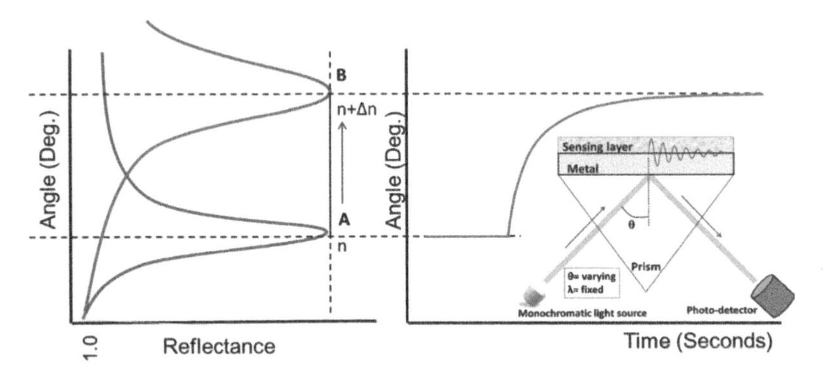

Figure 12.4 Sensogram formation of angular interrogation-based SPR sensor by tracking the resonance angle with respect to time.

The monitoring of SPR resonance angle change requires the angular movement of the photodiode with a range of angles at the output end, which makes a little complex instrumentation compared to the intensity interrogation method. The complexity can be reduced by using an array-based photodiode to record the reflected diverging beam replacing the requirement of photodiode rotation (Fig. 12.5). Another option, which is a property of photonicsys, is inserting a screen at the output and capturing the signal by the camera, which causes a dark black line at resonance angle and can be tracked by suitable image processing algorithm (Fig. 12.6).

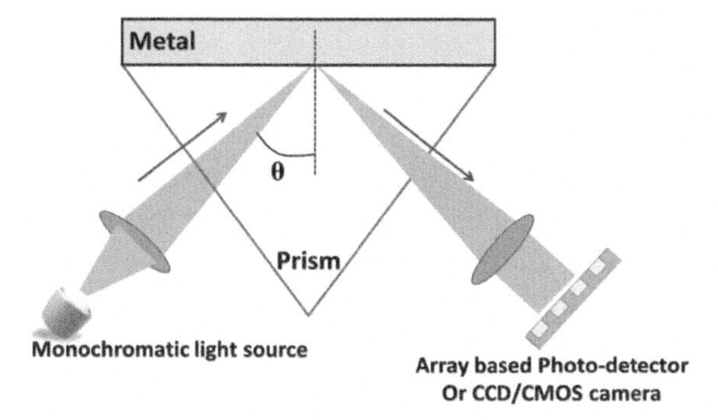

Figure 12.5 Converging beam SPR setup with various detectors to avoid any moving part.

The dark line in the screen can be divided into segments, which can later be used as the same amount of sensing channels. Several companies such as Biacore and Photonicsys use these configurations for simple and miniaturized device fabrication. This interrogation scheme was developed in the early 1990s with an impressive RI resolution of 2×10^{-6} RIU [21–23] and Biocore advanced this method with better optical design in a series of commercial SPR devices offering high performance (resolution down to the 10^{-7} RIU) along with high throughput sensing. It may also be noted that the incident light can also have two different types of configurations, which include the converging light beam and as well as the diverging light beam (see Fig. 12.5 and Fig. 12.6, respectively). The converging beam SPR setup allows the focusing of the SPR line down to a tiny narrow line at the sensor surface. However, in the case of diverging beam setup (Fig. 12.6), the SPR resonance dip walks along the sensor chips as any biomolecular interaction happens. This leads to a spatially undefined SPR minimum location on the sensor surface [24].

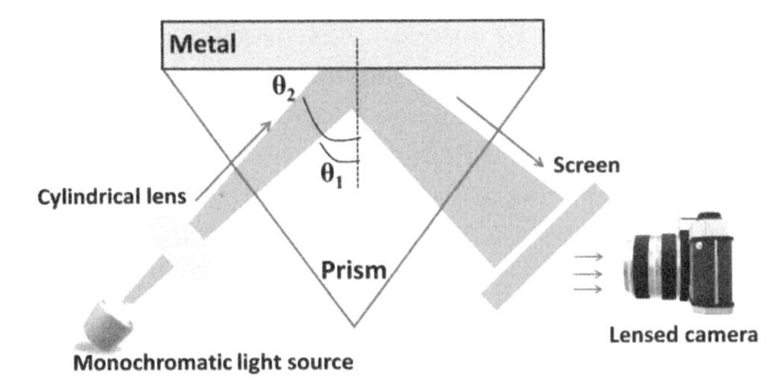

Figure 12.6 Diverging beam SPR setup based on angular interrogation.

Comparing both intensity and angular interrogation methods, reflectivity changes due to biomolecular interactions with respect to angular dip shift cannot be correlated unless one knows the information about the exact shape of SPR curves. Based on the simulations, one can say that the SPR dip position shows a linear relationship with respect to the RI of the sensing medium, hence binding measurements of guest analytes should be preferably done in angular interrogation mode [24].

12.2.3 Spectral Interrogation Scheme

This method basically refers to the monitoring of resonance wavelength change of the SPR curve where collimated white light is used as the source keeping the incident angle fixed. The reflected light through an optical fiber at the output is collected by a spectrometer, which will help find the wavelength corresponding to the minimum in reflection and can be tracked as a function of time. The main advantage of this configuration is that it can cover almost all RI ranges of operation without any additional mechanical movement devices. However, in this case, the penetration depth and propagation length of the SPW are not constant at all wavelengths of incident light. The SPR excitation at different wavelengths will produce different lateral resolutions (corresponding to propagation length) and penetration depths at each wavelength of operation, and hence the interaction volume of the evanescent wave with the sensing layer will be different. This is the reason, most commercial devices prefer single wavelength operation in their SPR devices. Figure 12.7 represents the setup configuration and corresponding dynamic response of the signal.

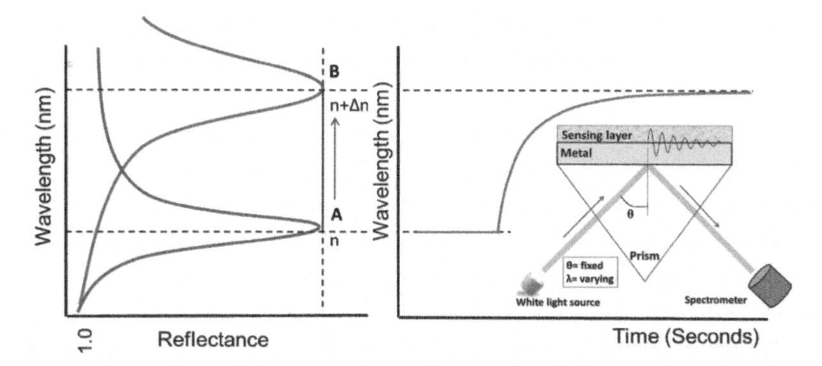

Figure 12.7 Sensogram based on spectral interrogation method, where the resonance wavelength is tracked with respect to time as it shifts from A to B by changing RI of the sensing layer from n to $n+\Delta n$.

The noise factors for the SPR sensing device based on the wavelength interrogation method are the light source properties such as tungsten halogen lamp (THL), white LEDs, light collimation, spectrometer resolution, etc. A brief discussion of these parameters will be given in the further sections of the chapter. It may be noted that the quality of the sensor surface is basically revealed by the shape of the SPR curve. As in the case of a non-uniform surface, particle agglomerations affect the SPR curve, which is used to get distorted in terms of broadening causing the spatial differences in resonance conditions within the active sensing regime. Hence, it is crucial to obtain a good SPR curve before starting the experiment. Another limitation of the spectral mode is the number of microchannels that can be monitored at the same time. Usually, either multiplexing or several light sources or spectrometers are used to monitor several channels simultaneously.

12.3 Signal Analysis in SPR-Based Sensors

As mentioned previously, minor changes in the analyte RI shift the resonance curve. Therefore, accurate and stable signal analysis has a crucial role in SPR instruments. The main obstacles are the SPR non-linearity and system noise. In addition to the minimum location in the SPR curve, the SPR signal has two main parameters: the reflectivity height at the minimum point and the dip width. While the analyte RI determines the minimum location, the thin metal layer determines the other two parameters. This section briefly presents a few methods of SPR signal analysis.

12.3.1 Fitting to a Theoretical Model

Fitting the measured signal to the theoretical model is a standard SPR processing method [25–27]. The fitting is usually based upon using Abeles matrices or Fresnel formulas for the reflection measurements targeted to the N-layer model as shown in Fig. 12.8. The best fitting through this model can be achieved by using accurate information about layers thickness and materials dispersion, in the experiments.

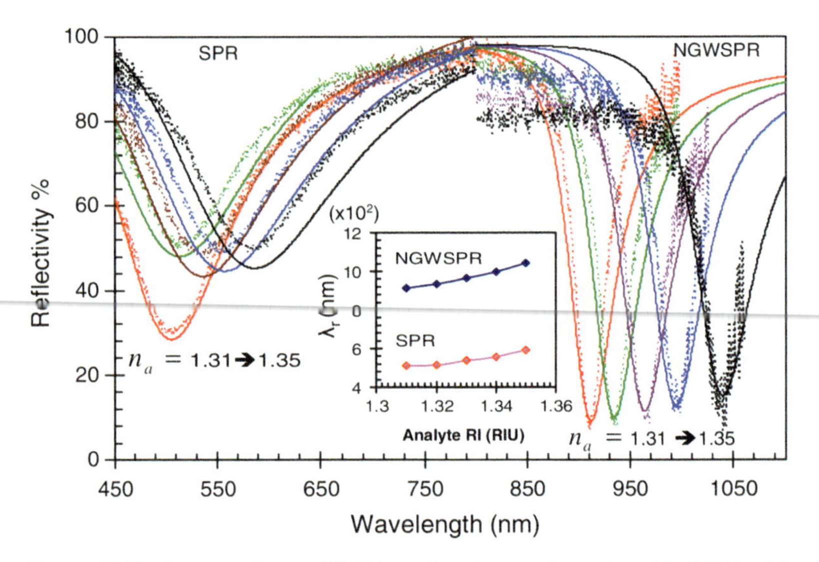

Figure 12.8 A comparison of SPR-based system and nearly guided SPR with varying analyte RI from 1.33 to 1.35 in the spectral mode. The dotted line corresponds to experimental results while solid lines are the SPR simulation results. Reprinted with permission from Ref. [26].

12.3.2 Minimum Hunt Method (MHM)

The MHM method is based on determining the minimum of the SPR curve in several ways, such as looking at the minimum point of the reflection curve or finding a location where the first derivative vanishes. However, the required minimum location can be found by fitting a parabola (or higher-order polynomial function) to the portion of the curve around the dip and finding the vertex [28], as Fig. 12.9 shows. Although this method has simplicity, the MHM suffers from different limitations such as sensitivity to noises, high-resolution data requirements, and different fits for each signal. In addition, automatic SPR analysis requires an additional step each time to choose the specific portion of the curve for fitting. Fitting the measured signal to the theoretical model is a standard SPR processing method.

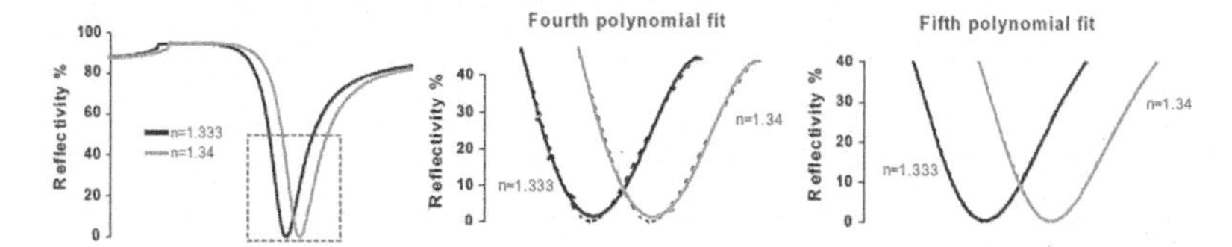

Figure 12.9 MHM method based on fitting the dip portion to different polynomial functions.

12.3.3 Centroid Method

Instead of analyzing the SPR curve by finding the accurate location of the minimum point, the centroid method [29–32] or center of mass technique is based on the dip width and reflectivity height at the minimum point parameters. It is based on determining the geometric center of the dip of the resulting resonant dip considering the curve asymmetry as depicted in Fig. 12.10a. The centroid/resonance location in pixels (pixel value can be translated to angle or wavelength) is

$$Pixel_{\text{res}} = \frac{\sum_{P_{th1}}^{P_{th2}} P.(R_{th} - R_P)}{\sum_{P_{th1}}^{P_{th2}} (R_{th} - R_P)}. \tag{12.1}$$

where P_{th1} and P_{th2} and P correspond to the start and end pixel numbers at the threshold and the pixel number, respectively. Additionally, R_{th} and R_p represent the reflectivity at the threshold and pixel P, respectively. The threshold here has a crucial role in this method. Choosing a higher threshold includes more pixels and reduces the noise level. The centroid and minimum location are equal when the curve below the threshold is symmetric (like MHM). Due to the fixed threshold, this method is highly sensitive to light source intensity fluctuations and background changes.

This noise level is reduced by the adjusted threshold method [32] by 6 orders of magnitude. This improvement was implemented in angular SPR. Figure 12.10b shows the main concept of the method. The suitable threshold is determined by the requirement of a constant ratio between the area below and above the threshold ($A1/A0$ = constant). Another constant that should be chosen is the integral range of the upper area. The fixed-boundary centroid method [32] proposed later is characterized by a higher signal-to-noise ratio (SNR), resolution, and speed. The concept is based on choosing fixed start and end points and calculating the centroid by:

$$Pixel_{res} = \frac{\sum_{P_{start}}^{P_{end}} P.R_P}{\sum_{P_{start}}^{P_{end}} R_P}. \tag{12.2}$$

Although the improved performance, the detection range of the fixed-boundary centroid is limited.

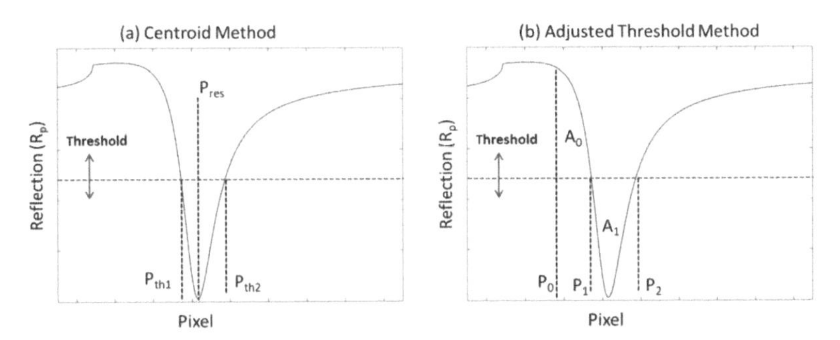

Figure 12.10 (a) Centroid method and (b) adjusted threshold method.

12.3.4 Dip Detection–Based Image Processing Techniques

The SPR signal appears as a dark line in imaging angular SPR and the analyte change appears as a position change of the dark line. In addition, these dark lines usually have the same direction (horizontal) on well-aligned systems. Therefore, image processing methods like Hough transform and Radon transform are suitable for candidates searching for the position. SPR line detecting Radon transform was proposed and implemented on diverging beam SPR [33].

In the first step, the Radon transform of the measured image is calculated in polar coordinates. In this case, the SPR line in the original image is represented by a minimum point in Radon space. To reduce the effect of the noise, the background is subtracted from the Radon transformation. Background can be estimated by a 2D median filter. Since the lines here are horizontal (angle independent), the method can be simplified to 1D transformation by integration along the angle. The noise level was minimized using this method, in spite of the noisy images, especially laser speckles. The background estimation filter type and parameters are crucial in this method and require optimization.

Recently, an image recognition-based deep learning method was proposed for SPR line detection [34]. The proposed method is based on convolutional neural network (CNN) architecture. First, the system was trained using simulated SPR images using Fresnel equations. To take noises into account, two types of noises are added to the training dataset: shot noise that originates from the low intensity at SPR dip and speckle. Then, the method has been tested on experimental SPR images showing a precision limit of 4.23×10^{-6} RIU. The CNN method is differentiated from other techniques by efficiently using the existing data by recognizing the relationship between the pixels instead of analyzing the SPR curve by finding the accurate location of the minimum.

12.4 Factors Producing Noise in SPR Signal

12.4.1 Plasmonic Surface Roughness

The surface roughness of plasmonic thin film is a very important factor that plays a crucial role in the sensor's performance, which has been exhaustively investigated in several studies [35–37]. The SPR characteristics, resonance condition, propagation length, and hence the reflected and transmitted spectra are the important parameters, which are affected by the plasmonic thin film roughness [38, 39]. For example, in an experimental study reported by Yang et al., it has been demonstrated that the plasmonic resonance angle and depth are highly dependent on the surface smoothness, which is varied by using different annealing temperatures [40]. Similarly, studies showed that in general, the bulk sensitivity of the sensor degrades due to the thin film roughness [41], but in some cases, due to the high surface area for the chemical/biological sensing along with the LSPR effect, an enhanced sensitivity has been observed [42].

The general methods for thin film deposition methods are sputtering, thermal evaporation, and e-beam deposition. However, the advancement in nanotechnology has provided us with several methods to obtain highly smooth metallic thin films after some surface treatments such as the stripping method [43], the thermal annealing method [44], the chemical polishing method [45], etc. Treebupachatsakul et al. performed a rigorous coupled wave analysis along with the Monte Carlo method to observe the SPR reflectance spectra based on the several surface roughness profile, generated by the low-pass frequency filtering method [46]. With the help of a comprehensive literature survey, the root means square (RMS) roughness of various coating methods and surface treatments was obtained and utilized to calculate several RI sensing parameters such as sensitivity (S), full width half maximum (FWHM), intensity contrast (ΔI), intensity at resonance angle (I_{sp}), and figure of merit (FOM). The schematic of the sensor design and flow chart of the calculation process are represented in Fig. 12.11a and Fig. 12.11b, respectively, while Table 12.2 represents the calculated RI sensing parameters based on RMS values of different deposition and surface treatment methods, which claims that surface roughness plays a huge role in the performance of an SPR sensor.

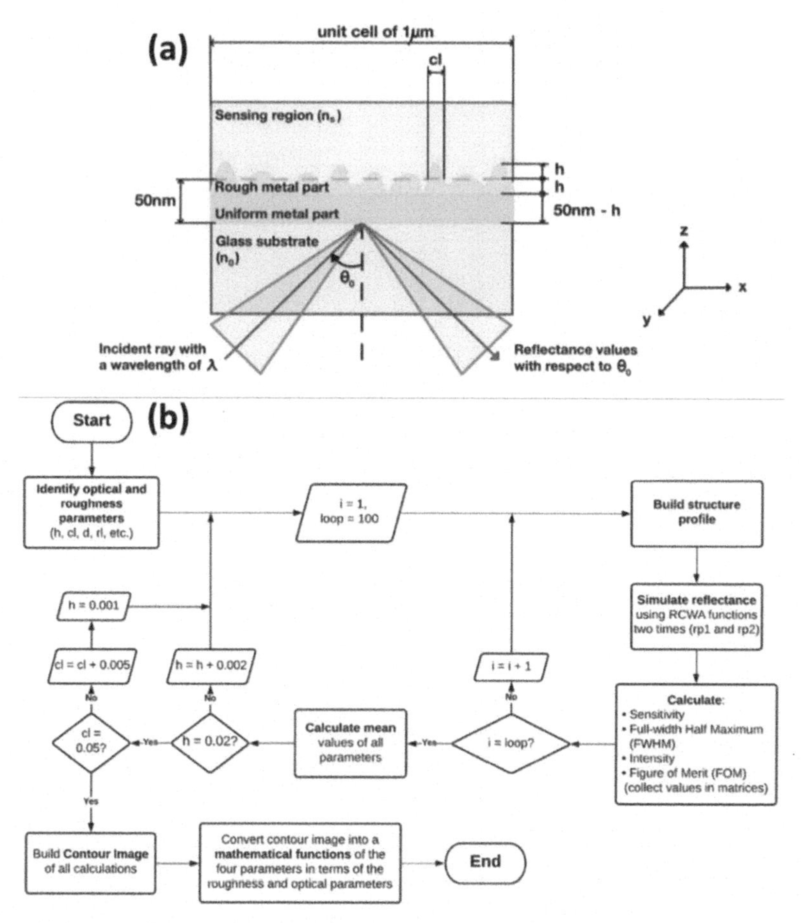

Figure 12.11 (a) Schematic design and (b) model formulated for calculating various parameters of SPR sensors upon varying the surface roughness for plasmonic thin film. Reprinted with permission from Ref. [46].

12.4.2 Light Source

Based on the type of interrogation used for the SPR device, the light source is decided. In the case of intensity or angular interrogation scheme, monochromatic light sources, such as LED and laser, are used. However, in the case of the spectral interrogation method, a broadband light source like a white LED and a tungsten halogen lamp (THL) can be used. There are several possibilities for the noise in the signal as follows:

Table 12.2 Effect of plasmonic surface roughness on various RI sensing performance parameters. Reprinted from Ref. [46]

Deposition method/ smoothing method	RMS roughness (nm)	Sensitivity of k-vector (rad/RIU. μm)	θ_{sp} (deg.)	FWHM in k-vector space (rad/μm)	ΔI_p	I_{sp}	FOM (RIU^{-1})
Ideal surface	0	7.46	71.40	0.039	0.64	0.007	139.16
Sputtering (without any modification)	1.2	7.55	71.51	0.040	0.60	0.024	90.09
Chemical polishing [45]	0.38	7.52	71.40	0.039	0.63	0.013	125.20
Mica substrate utilizing [43]	0.2	7.50	71.40	0.039	0.64	0.010	132.16
Chemically grown single-crystalline gold [47]	<1.0	>7.55	<71.51	<0.040	>0.61	<0.021	>98.77
Laser ablation [48]	0.17	7.50	71.40	0.039	0.64	0.010	132.77
Helium ion beam [49]	0.267	7.51	71.40	0.039	0.63	0.011	129.63
Thermal annealing [44]	<1.0	>7.55	<71.51	<0.040	>0.61	<0.021	>98.77

12.4.2.1 Spectral width

The angular interrogation method requires a light source with a single wavelength, but the light source used in the setup always has some spectral width, which causes some distorted SPR signal as for each wavelength of SPR light, there will be some different resonance angle. For example, considering the light source for 633 nm, the LED source with an FWHM value of 30 nm, and a laser source with a spectral width of 2 nm, the FWHM value of the corresponding SPR reflected beam is changed by a 0.05 degree along with a contrast degradation by 2% for the LED source. It can be simply observed in

Fig. 12.12, which corresponds to the SPR response for a conventional SPR chip designed by using SF11 glass as a substrate, 50 nm thick Au as plasmonic film, as the sensing medium with 1.33 RI.

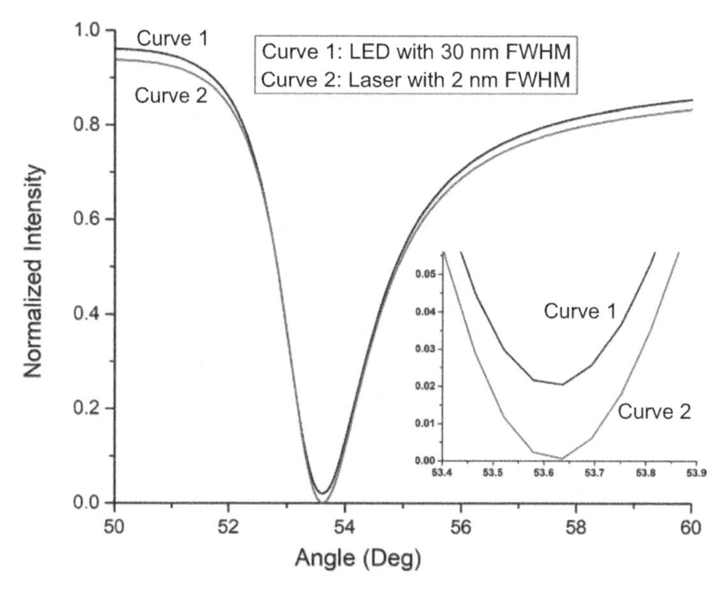

Figure 12.12 Effect in SPR reflectance curve for light sources with different spectral broadening ammeters.

12.4.2.2 Power fluctuations

Power fluctuations are the most important sources of noise in resonance-based sensing devices. These fluctuations are generally originated from the unstable power supply, thermal drift affecting the heat produced in the material used, spatial vibrations, etc. Several studies have been done in the literature to find a relationship between the intensity noise of the light source and the corresponding noise in the SPR signal. Ma et al. made an effort to find the effects of several spectral powers and corresponding noise levels on the SPR signal through simulations and experiments [50]. The study claimed that the optimal SPR dip wavelength is changed, and the RI resolution can even be nearly twice when the spectral SNR is increased [50]. Figure 12.13a shows the different light sources used in the study and corresponding RI resolution with the addition of noise level and its double is shown in Fig. 12.13b,c respectively.

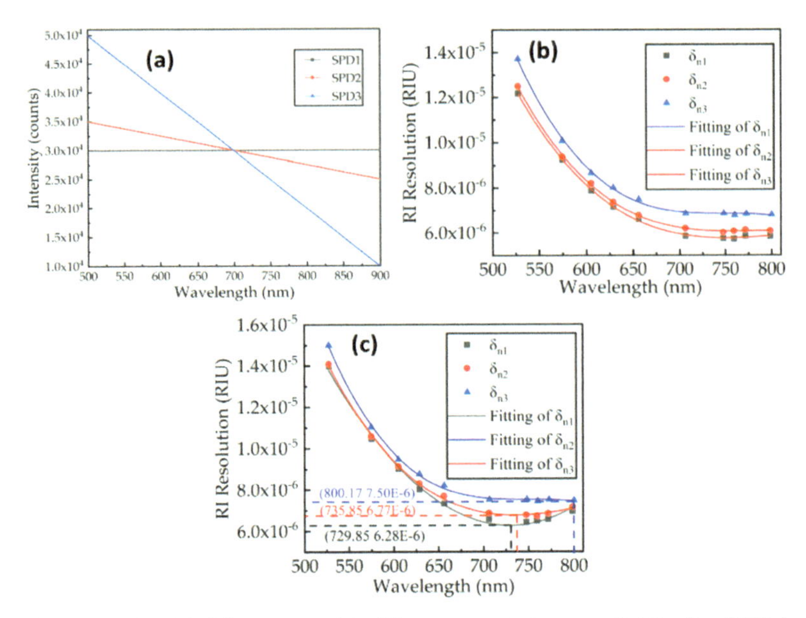

Figure 12.13 (a) Light source with different spectral power distribution (SPDs) and corresponding SPR RI resolution with the addition of (b) noise and (c) doubling the noise. Reprinted from Ref. [50].

Drayton et al. reported a study based on the various light sources (like Halogen, Fabry–Perot laser, and resonant cavity LED (RCLED) with wavelength filters and detectors (such as CCD and CMOS) to find the best combination based on low-cost sensing devices [51]. The study showed that the sensor based on lower Q-factors was more fault tolerant. The filtered RCLED is more advantageous to a laser source. The RCLED was preferred as compared to conventional LED, as it supports higher directionality and spectral density with the help of an additional filter. Although the filter added some extra cost, it helped to find a more defined central peak compared with that of the laser. The best-obtained resolution of the sensor was 1.7×10^{-5} using RCLED as a light source and a complementary metal oxide semiconductor (CMOS) camera as the detection. Figure 12.14 represents the power fluctuations obtained for the unfiltered laser diode, LED, and RCLED.

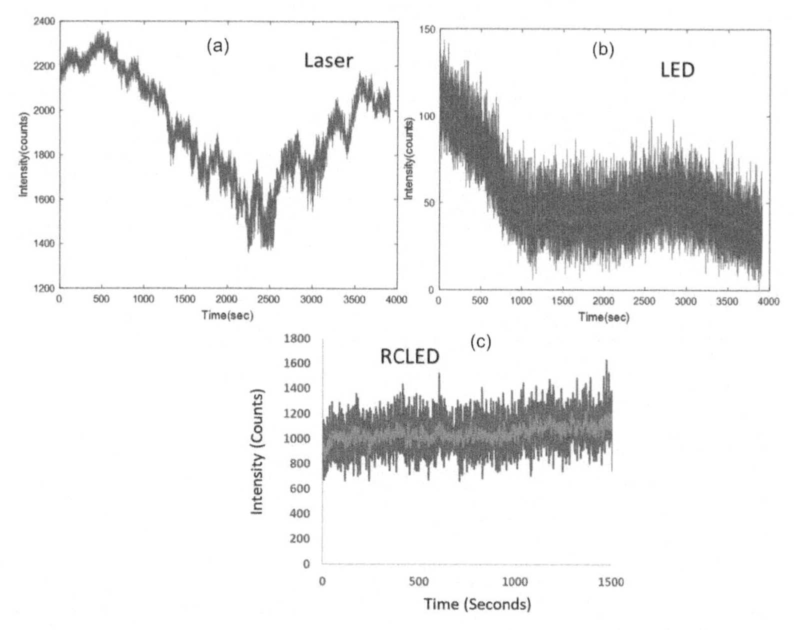

Figure 12.14 Power fluctuations in (a) unfiltered laser diode, (b) LED, and (c) RCLED.

12.4.3 SPR Signal Detector

There are several types of sources that are used as the detectors in SPR-based systems, these are mainly CCD and CMOS cameras, photodiodes, spectrometers, etc., based on the interrogation scheme used and obviously, the limitations of these detectors largely determine the resolution of the SPR system. The properties of these detectors, which affect the SPR signal, are the resolution, operating wavelength/wavelength range, temperature stability, long-term working ability, etc.

In the current era of smartphone devices, cameras are very simple and cost-effective solutions for the detector while working with intensity and angular-based interrogation schemes to avoid the use of mechanical equipment to cover a specific angle range. Broadly CCD and CMOS are the most used camera based on the technology's maturity of low-cost devices [52]. In the case of CCD, it uses a shift register for charge carrier transport and a signal amplifier at each row of a pixel. The repeatedly charge transportation during the signal readout produces temporal noise [53]. This noise cannot be easily noticed by the human eye, due to its time-dependent properties and the results based on this device impacts the sensor's performance. Additionally, due to excess charge leakage through nearby pixels, sometimes CCDs also suffer from a smear. While, in the case of CMOS cameras, due to an individual amplifier per pixel, these sensors do not get any temporal and bleeding curves, but each pixel can get its own sensitivity and offset leading to a time-independent pattern and gain noise. These noises can be eliminated by specially designed driving circuits depending on the manufacturing company. Another parameter is the dynamic range, here CCD wins the race as compared with CMOS cameras. Similar stays in terms of fill-factor, the CCDs get nearly 100% but CMOS cannot, since the driving electronic reduces their SNR. Experimentally, the CMOS camera showed a better response for guided mode resonance-based sensing devices [51]. Similarly, in the case of spectral interrogation, the resolution and signal stability play a strong role in the precise detection of resonance wavelength. In general, the cost-effective spectrometers in the market are available with a resolution of 1–2 nm, which corresponds to 2.5×10^{-4} RIU for an SPR sensor with a sensitivity of 4000 nm/RIU. To minimize the SPR resolution, the region of SPR curves is fitted with a suitable function to find the best dip, using one of the algorithms discussed in Section 3, and then the resolution of the sensor can be lowered down to the order of 10^{-7} RIU.

12.4.4 Ambient Temperature

The temperature surrounding the SPR active region also plays a significant role to increase the noise level in the system. The substrate RI, plasmonic thin-film RI, and solution refractive change as the ambient temperature changes. According to the work done by Shengming Zhang, for an SPR configuration with BK-9 glass/Au (50 nm)/water (RI: 1.33), the resonance angle gets shifted by 0.05° as the ambient temperate changed by 100 °C or simply we can say the conventional SPR has the temperature sensitivity of 5×10^{-4} deg/RIU [54]. Hence, for a conventional sensor with an RI sensitivity of approximately 70 deg/RIU, the changes in an ambient temperature of 10 °C will increase a drift of 5×10^{-5} RIU. This number does not include the effects on the performance of the light source and camera while increasing the temperature, which also plays a significant role in the SPR device.

Although 5×10^{-5} RIU is a small number when considering highly sensitive SPR sensors but to design an SPR-based device with high resolution (approx. 10^{-6} to 10^{-7}), it is very important to minimize the temperature drift or fluctuations in the devices. Hence, many devices used the temperature-controlled unit in their SPR systems, which led to minimum temperature drift in the signal. In addition to temperature-controlled units, a few works are reported to compensate for the noise level in the signal through some signal treatment. Filho et al. proposed a model for SPR-based biosensors, which includes roughly all possible disturbances and noise in the system. The possible variations in a monochromatic light source are the light source bias current, the temperature, and the exposure time. The temperature of the light source tunes the intensity and wavelength of the beam. The first affects the mean value of the signal providing an offset to the SPR curve, which can be minimized using a correction factor in the software with suitable feedback. The study reveals that the wavelength affected by temperature fluctuations has more pronounced effects. It varies the light beam aperture and materials used for SPR sensor design, along with photodetector sensitivity. Plasmonic thin film roughness, light source properties, detector configurations, and ambient temperature are broadly affecting parameters in the SPR signal along with beam uniformity, quality of optical components such as polarizers, lenses, and local vibrations. Hence, one needs to take care of all these factors while designing a stable and high-resolution SPR device.

12.5 A Commercial SPR Biosensor

Figure 12.15 represents a schematic of three main units of an SPR device, which includes SPR optics as discussed earlier, specially designed SPR chips, additional liquid handling systems such as flow cells, and microfluidic channels to minimize the sample requirement and leakage in the system. SPR chip is the backbone of a sensing device, which is designed based on the type of target molecule to be detected.

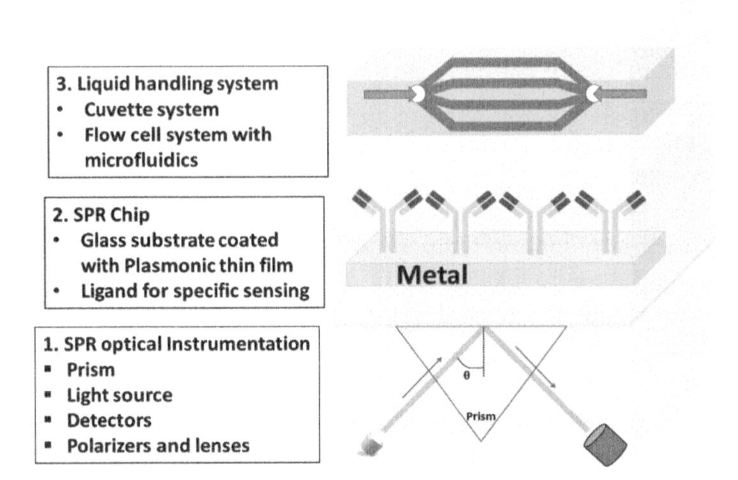

Figure 12.15 Main units of the optical head of a commercial SPR sensing instrument.

SPR chip with approximate 50 nm thick gold coating is mostly used for biosensing applications, due to high stability, chemical internees, and biocompatibility. Although silver (Ag) shows better SPR performance than Au, rapid oxidization limits its applications. The coating of a thin film of oxides of silver layer causes unwanted ionic effects and hence, drifts in the SPR signal. It may be noted that the Au-coated SPR chip possesses a penetration depth of around 200 nm, which limits its applications for large biomolecule applications. In this case, the insulator-metal-insulator (IMI) chip is used to realize a long-range SPR (LRSPR) phenomenon, where the penetration depth of the SPR signal is covering a few microns [55].

In addition to the SPR chip and the optical setup, the liquid handling system is also a very important unit of the SPR instrument. This system decides the way based on which the analyte will interact with the sensor surface, which directly correlates with the mass transportation limits, binding constants, surface depletion, diffusion gradient, etc. Cuvettes and flow cells are two main types of liquid handling systems, which are used in many SPR devices. In the past, cuvettes were the widely used technique in SPR instruments but due to the technology developments, microfluidics-based flow cells are in use. Although, a few devices such as Octet (ForteBio, Pall) still use the same technique.

Cuvette systems are equipped with an open vessel, which can be filled automatically or manually using a specific liquid-handling robotic motor. The system is mounted over the SPR chip without any leakage or any other disturbance in the optical setup. The sensing interactions of the biological analytes take place over the sensor surface and at the bottom of the cuvette system. Cuvette systems should contain a suitable mixing system, as uncontrolled mixing will cause uncontrolled mass transport resulting in a deformed sensogram. Compared with the flow systems, these are more compatible with aqueous solutions containing solid particles such as cell cultures, blood plasma, fermented media, etc., except the sensor surface is not affected by them. However, the main disadvantage of the cuvette system is its open architecture, which allows the uncontrolled evaporation of solvents in aqueous solutions, leading to increased concentration and RI.

Flow cells along with controlled microfluidics are currently used in many SPR-based sensing instruments, ranging from simple to highly automated cartridge systems. In this method, an aqueous solution is transported to the sensor surface for biomolecular interactions, through sample loops or pneumatic valves using syringes or flow pumps. The syringes or flow pumps are not providing sample transportation to the sensing surface but also, help control the hydrodynamic conditions of the fluid. These can mainly be fabricated using microfluid channels pressed with micromechanical devices. Biacore was the first who integrated the microfluidic channels for SPR technology using integrated microfluidic technology (IFC). This technology provides the analyte flow with a controlled and pulse-free flow with controlled constant analyte concentration and hydrodynamic conditions around the sensor surface. It has several other advantages such as minimal dispersion and dead volume, instantaneous sample transitions, and optimum thermo-stating of sample and surface chip [24].

Flow cell configurations are divided into three main categories: planer flow cells, hydrodynamic flow cells, and wall-jet flow cells. The first one contains simple inlet and outlet channels, which help occur sample interaction with the sensor surface. This type of flow cell can be modified up to the number of flow channels, as Biacore T-200 and S-200 are used for a four-channel system while Biacore 8K applied for an eight-channel flow cell-based device. High-end devices such as Biacore 4000 and MASS-1 (Sierra Sensors) are using hydrodynamic isolation flow cells in order to be working with a large surface area. The wall-jet flow cells are no longer used now as redial velocity is not uniform along the sensor surface. More detailed information on these can be obtained elsewhere [24].

12.6 A General SPR Sensogram

Till now, we have briefly covered the design of a general SPR instrument, using which one will be able to monitor the change in SPR signal with respect to time variation (sensogram). Sensogram is an important curve obtained from the sensor response, as it contains large information on binding analysis such as association constant, decay constant, and concentration of the biomolecule, while its interaction with the sensor surface occurs. Figure 12.16 shows a typical sensogram for an SPR sensor, which covers different phases of the molecular interactions with the sensor surface.

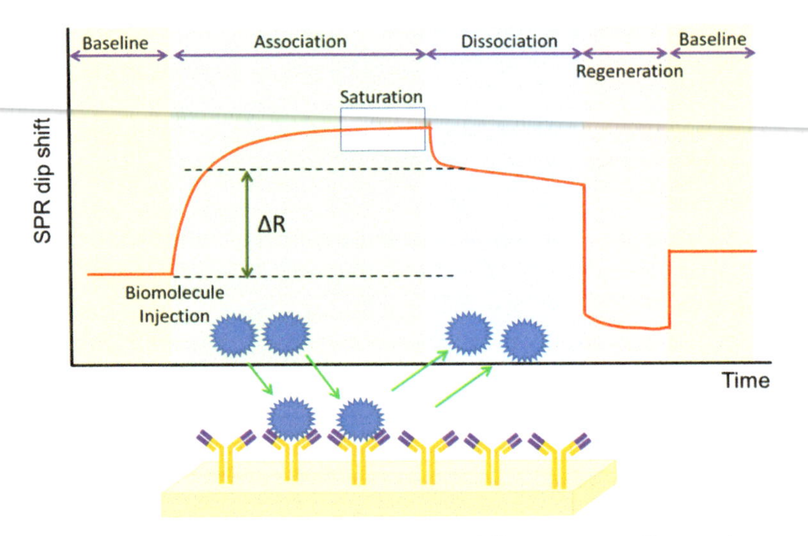

Figure 12.16 Typical sensogram response at different stages of biosensing.

Each sensing measurement with a specific sensor surface contains the following phases:

I. *Baseline:* This corresponds to having a stable baseline of the sensor signal before binding starts. It is done by flowing the analyte solvent such as buffer and DI, over the sensor. This is used to get a stable signal of device covering to minimize the effect of analyte solvent RI with sensor surface, temperature effects, hydrogel swelling, etc. In this case, the sensor surface is ready to capture the guest target molecule.

II. *Association:* During the sample injection, the biomolecule starts binding with the sensor surface and the RI of the sensor surface begins to change with time. This phase is used to analyze the absorption kinetics of the sensor surface. In this case, the molecules other than the analyte can also stick with the surface in the absence of a selective sensing layer, which corresponds to the non-specific binding.

III. *Dissociation:* After reaching the association curve to saturation, the buffer is flowed over the sensor surface to wipe out the non-specifically bound components, called as dissociation phase. From the figure, ΔR is the accumulated mass loaded of the target analyte of the sensing surfaces. Additionally, in some cases, the dissociation of the analyte also starts and the kinetic helps analyze the dissociation constant.

IV. *Regeneration:* Finally, the sensor surface is cleaned by flowing a regeneration solution to remove the specific binding and drain out all the biomolecules, keeping the recognition medium at the sensing surface. To reuse the sensor chip again, it is important to use the regeneration solution, which can remove the guest target molecule without altering the activities of ligands.

A more detailed understanding of binding kinetics analysis using SPR along with its applications can be found elsewhere [24, 56–58].

12.7 Summary

To summarize, we have represented the technical aspects of designing a commercial SPR sensing device. As discussed earlier, the SPR technique has shown great potential for the detection of biomolecular interactions along with label-free, dynamic sensing, and high sensitivity, and this led to the development of a broad range of commercial SPR devices. SPR-based devices have demonstrated huge potential in the field of biomolecule sensing industry, the future of SPR will be dedicated to developing cost-effective SPR devices having more miniaturization along with high resolution especially applicable for on-site detection. A certain area is also focused on even smartphone-based plasmonic sensors [59, 60] but is still limited to research articles. A few recent studies are also using deep learning methods to decrease the noise level of an SPR sensogram [61, 62]. Thus, along with the integration of advancement in the technology, SPR-based devices will not be limited only to the research labs, these will be used more often in hospitals and clinical labs. Miniature SPR devices are becoming more and more available with high resolution, in particular, the ultraminiature SPR of photonicsys with multichannel, remote control, and magnetic field to control magnetic nanoparticles is remarkable. Thus, we tried to provide a brief knowledge of SPR-based devices. At the beginning of the chapter, we discussed a brief survey of companies, which are producing several SPR devices with their respective design and innovations. The next few sections of the chapter have been devoted to SPR measurement schemes, various approaches to finding SPR dip positions along with several factors, which introduce noise in the SPR systems. The chapter also provided a brief of commercial SPR sensors and their flow channels and a typical SPR sensogram. This chapter is targeted to new graduate students and research professionals, which are willing to do research in the plasmonic biosensor techniques but not only limited to research publications but also planning the fabrication of new SPR devices.

Acknowledgments

This work was supported by the EU MSCA-RISE H2020 project IPANEMA grant agreement N° 872662 and the Cordial-S programme under grant agreement No. 101016038.

References

1. Sharma, A. K., Jha, R., and Gupta, B. D. (2007). Fiber-optic sensors based on surface plasmon resonance: A comprehensive review, *IEEE Sens. J.*, **7(8)**, pp. 1118–1129.
2. Homola, J. (2008). Surface plasmon resonance sensors for detection of chemical and biological species, *Chem. Rev.*, **108(2)**, pp. 462–493.
3. Abdulhalim, I., Zourob, M., and Lakhtakia, A. (2008). Surface plasmon resonance for biosensing: A mini-review, *Electromagnetics*, **28(3)**, pp. 214–242.
4. Wood, R. W. (1902). XLII. On a remarkable case of uneven distribution of light in a diffraction grating spectrum, *Lond. Edinb. Dublin philos.*, **4(21)**, pp. 396–402.
5. Ritchie, R. H. (1957). Plasma losses by fast electrons in thin films, *Phys. Rev.*, **106(5)**, pp. 874–881.
6. Kretschmann, E. and Raether, H. (1968). Radiative decay of non-radiative surface plasmons excited by light, *Z. Naturforschung A*, **23(12)**, pp. 2135–2136.
7. Liedberg, B., Nylander, C., and Lunström, I. (1983). Surface plasmon resonance for gas detection and biosensing, *Sens. Act.*, **4**, pp. 299–304.
8. Jönsson, U., Fägerstam, L., Ivarsson, B., Johnsson, B., Karlsson, R., Lundh, K., Löfås, S., Persson, B., Roos, H., and Rönnberg, I. (1991). Real-time biospecific interaction analysis using surface plasmon resonance and a sensor chip technology, *Biotechniques*, **11(5)**, pp. 620–627.
9. Wink, T., de Beer, J., Hennink, W. E., Bult, A., and van Bennekom, W. P. (1999). Interaction between plasmid DNA and cationic polymers studied by surface plasmon resonance spectrometry, *Anal. Chem.*, **71(4)**, pp. 801–805.
10. Shalabney, A. and Abdulhalim, I. (2011). Sensitivity-enhancement methods for surface plasmon sensors, *Laser Photon. Rev.*, **5(4)**, pp. 571–606.
11. Homola, J., Yee, S. S., and Gauglitz, G. (1999). Surface plasmon resonance sensors, *Sens. Act. B*, **54(1–2)**, pp. 3–15.
12. Gupta, B. D., Shrivastav, A. M., and Usha, S. P. (2016). Surface plasmon resonance-based fiber optic sensors utilizing molecular imprinting, *Sensors*, **16(9)**, pp. 1381.
13. Shrivastav, A. M., Cvelbar, U., and Abdulhalim, I. (2021). A comprehensive review on plasmonic-based biosensors used in viral diagnostics, *Comm. Bio.*, **4(1)**, pp. 1–12.

14. Peng, W., Liu, Y., Fang, P., Liu, X., Gong, Z., Wang, H., and Cheng, F. (2014). Compact surface plasmon resonance imaging sensing system based on general optoelectronic components, *Opt. Exp.*, **22(5)**, pp. 6174–6185.
15. Rothenhäusler, B. and Knoll, W. (1988). Surface–plasmon microscopy, *Nature*, **332(6165)**, pp. 615–617.
16. Jing, W., Hunt, A., Tao, N., Zhang, F., and Wang, S. (2020). Simultaneous quantification of protein binding kinetics in whole cells with surface plasmon resonance imaging and edge deformation tracking, *Membranes*, **10(9)**, pp. 247.
17. Spoto, G. and Minunni, M. (2012). Surface plasmon resonance imaging: What next?, *J. Phy. Chem. Lett.*, **3(18)**, pp. 2682–2691.
18. Piliarik, M. and Homola, J. (2009). Surface plasmon resonance (SPR) sensors: Approaching their limits, *Opt. Exp.*, **17(19)**, pp. 16505–16517.
19. Fu, E., Foley, J., and Yager, P. (2003). Wavelength-tunable surface plasmon resonance microscope, *Rev. Sci. Inst.*, **74(6)**, pp. 3182–3184.
20. Piliarik, M., Vaisocherová, H., and Homola, J. (2005). A new surface plasmon resonance sensor for high-throughput screening applications, *Biosens. Bioelectron.*, **20(10)**, pp. 2104–2110.
21. Liedberg, B., Lundström, I., and Stenberg, E. (1993). Principles of biosensing with an extended coupling matrix and surface plasmon resonance, *Sens. Act. B.*, **11(1–3)**, pp. 63–72.
22. Sjoelander, S. and Urbaniczky, C. (1991). Integrated fluid handling system for biomolecular interaction analysis, *Anal. Chem.*, **63(20)**, pp. 2338–2345.
23. Löfås, S., Malmqvist, M., Rönnberg, I., Stenberg, E., Liedberg, B., and Lundström, I. (1991). Bioanalysis with surface plasmon resonance, *Sens. Act. B*, **5(1–4)**, pp. 79–84.
24. Schasfoort, R. B. (ed.). (2017). *Handbook of Surface Plasmon Resonance*, Royal Society of Chemistry.
25. Yuan, X. C., Ong, B. H., Tan, Y. G., Zhang, D. W., Irawan, R., and Tjin, S. C. (2006). Sensitivity–stability-optimized surface plasmon resonance sensing with double metal layers, *J. Opt. A*, **8(11)**, pp. 959.
26. Shalabney, A. and Abdulhalim, I. (2012). Figure-of-merit enhancement of surface plasmon resonance sensors in the spectral interrogation, *Opt. Lett.*, **37(7)**, pp. 1175–1177.
27. Shalabney, A. and Abdulhalim, I. (2010). Electromagnetic fields distribution in multilayer thin film structures and the origin of sensitivity enhancement in surface plasmon resonance sensors, *Sens. Act. A*, **159(1)**, pp. 24–32.

28. Gentleman, D. J., Obando, L. A., Masson, J. F., Holloway, J. R., and Booksh, K. S. (2004). Calibration of fiber optic based surface plasmon resonance sensors in aqueous systems, *Anal. Chim. Act.*, **515(2)**, pp. 291–302.

29. Karabchevsky, A. and Abdulhalim, I. (2015). Techniques for signal analysis in surface plasmon resonance sensors, *Nanomaterials for Water Management—Signal Amplification for Biosensing from Nanostructures*, pp. 163–186.

30. Jacobus, C. J. and Chien, R. T. (1981). Two new edge detectors, *IEEE PAMI*, **5**, pp. 581–592.

31. Zhan, S., Wang, X., and Liu, Y. (2010). Fast centroid algorithm for determining the surface plasmon resonance angle using the fixed-boundary method, *Meas. Sci. Tech.*, **22(2)**, pp. 025201.

32. Thirstrup, C. and Zong, W. (2005). Data analysis for surface plasmon resonance sensors using dynamic baseline algorithm, *Sens. Act. B*, **106(2)**, pp. 796–802.

33. Karabchevsky, A., Karabchevsky, S., and Abdulhalim, I. (2011). Fast surface plasmon resonance imaging sensor using Radon transform, *Sens. Act. B*, **155(1)**, pp. 361–365.

34. Thadson, K., Sasivimolkul, S., Suvarnaphaet, P., Visitsattapongse, S., and Pechprasarn, S. (2022). Measurement precision enhancement of surface plasmon resonance based angular scanning detection using deep learning, *Sci. Rep.*, **12(1)**, pp. 1–14.

35. Braundmeier Jr, A. J. and Arakawa, E. T. (1974). Effect of surface roughness on surface plasmon resonance absorption, *J. Phy. Chem. Sol.*, **35(4)**, pp. 517–520.

36. Rahman, T. S. and Maradudin, A. A. (1980). Surface-plasmon dispersion relation in the presence of surface roughness, *Phy. Rev. B*, **21(6)**, pp. 2137.

37. Crowell, J. and Ritchie, R. H. (1970). Surface-plasmon effect in the reflectance of a metal, *JOSA*, **60(6)**, pp. 794–799.

38. Kolomenski, A., Kolomenskii, A., Noel, J., Peng, S., and Schuessler, H. (2009). Propagation length of surface plasmons in a metal film with roughness, *App. Opt.*, **48(30)**, pp. 5683–5691.

39. Parmigiani, F., Scagliotti, M., Samoggia, G., and Ferraris, G. P. (1985). Influence of the growth conditions on the optical properties of thin gold films, *Thin Solid Films*, **125(3–4)**, pp. 229–234.

40. Yang, Z., Liu, C., Gao, Y., Wang, J., and Yang, W. (2016). Influence of surface roughness on surface plasmon resonance phenomenon of gold film, *Chin. Opt. Lett.*, **14(4)**, pp. 042401.

41. Agarwal, S., Giri, P., Prajapati, Y. K., and Chakrabarti, P. (2016). Effect of surface roughness on the performance of optical SPR sensor for sucrose detection: Fabrication, characterization, and simulation study, *IEEE Sens. J.*, **16(24),** pp. 8865–8873.
42. Byun, K. M., Yoon, S. J., and Kim, D. (2008). Effect of surface roughness on the extinction-based localized surface plasmon resonance biosensors, *App. Opt.*, **47(31),** pp. 5886–5892.
43. Diebel, J., Löwe, H., Samori, P., and Rabe, J. P. (2001). Fabrication of large-scale ultra-smooth metal surfaces by a replica technique, *App. Phy. A*, **73(3),** pp. 273–279.
44. Zhang, J., Irannejad, M., Yavuz, M., and Cui, B. (2015). Gold nanohole array with sub-1 nm roughness by annealing for sensitivity enhancement of extraordinary optical transmission biosensor, *Nano. Res. Lett.*, **10(1)**, pp. 1–8.
45. Miller, M. S., Ferrato, M. A., Niec, A., Biesinger, M. C., and Carmichael, T. B. (2014). Ultrasmooth gold surfaces prepared by chemical mechanical polishing for applications in nanoscience, *Langmuir*, **30(47),** pp. 14171–14178.
46. Treebupachatsakul, T., Shinnakerdchoke, S., and Pechprasarn, S. (2021). Analysis of effects of surface roughness on sensing performance of surface plasmon resonance detection for refractive index sensing application, *Sensors*, **21(18),** pp. 6164.
47. Wieduwilt, T., Kirsch, K., Dellith, J., Willsch, R., and Bartelt, H. (2013). Optical fiber micro-taper with circular symmetric gold coating for sensor applications based on surface plasmon resonance, *Plasmonics*, **8(2),** pp. 545–554.
48. Ng, D. K., Bhola, B. S., Bakker, R. M., and Ho, S. T. (2011). Ultrasmooth gold films via pulsed laser deposition, *Adv. Func. Mat.*, **21(13),** pp. 2587–2592.
49. Zhang, C., Li, J., Belianinov, A., Ma, Z., Renshaw, C. K., and Gelfand, R. M. (2020). Nanoaperture fabrication in ultra-smooth single-grain gold films with helium ion beam lithography, *Nanotechnology*, **31(46),** pp. 465302.
50. Ma, L., Xia, G., Jin, S., Bai, L., Wang, J., Chen, Q., and Cai, X. (2021). Effect of spectral signal-to-noise ratio on resolution enhancement at surface plasmon resonance, *Sensors*, **21(2),** pp. 641.
51. Drayton, A., Li, K., Simmons, M., Reardon, C., and Krauss, T. F. (2020). Performance limitations of resonant refractive index sensors with low-cost components, *Opt. Exp.*, **28(22),** pp. 32239–32248.

52. Bigas, M., Cabruja, E., Forest, J., and Salvi, J. (2006). Review of CMOS image sensors, *Microelectron. J.*, **37(5)**, pp. 433–451.

53. Tian, H., Fowler, B., and Gamal, A. E. (2001). Analysis of temporal noise in CMOS photodiode active pixel sensor, *IEEE J. Solid-State Circuits*, **36(1)**, pp. 92–101.

54. Zhang, S. (2011). *The Effect of Temperature and Electric Current on the SPR Sensors*, Indiana University of Pennsylvania.

55. Isaacs, S. and Abdulhalim, I. (2015). Long range surface plasmon resonance with ultra-high penetration depth for self-referenced sensing and ultra-low detection limit using diverging beam approach, *App. Phy. Lett.*, **106(19)**, pp. 193701.

56. Zhang, P., Ma, G., Dong, W., Wan, Z., Wang, S., and Tao, N. (2020). Plasmonic scattering imaging of single proteins and binding kinetics, *Nat. Meth.*, **17(10)**, pp. 1010–1017.

57. Wolf, L. K., Fullenkamp, D. E., and Georgiadis, R. M. (2005). Quantitative angle-resolved SPR Imaging of DNA–DNA and DNA–drug kinetics, *J. Am. Chem. Soc.*, **127(49)**, pp. 17453–17459.

58. Im, H., Sutherland, J. N., Maynard, J. A., and Oh, S. H. (2012). Nanohole-based surface plasmon resonance instruments with improved spectral resolution quantify a broad range of antibody-ligand binding kinetics, *Anal. Chem.*, **84(4)**, pp. 1941–1947.

59. Guner, H., Ozgur, E., Kokturk, G., Celik, M., Esen, E., Topal, A. E., Ayas, S., Uludag, Y., Elbuken, C., and Dana, A. (2017). A smartphone based surface plasmon resonance imaging (SPRi) platform for on-site biodetection, *Sens. Act.*, **239**, pp. 571–577.

60. Lertvachirapaiboon, C., Baba, A., Shinbo, K., and Kato, K. (2018). A smartphone-based surface plasmon resonance platform, *Anal. Meth.*, **10(39)**, pp. 4732–4740.

61. Thadson, K., Sasivimolkul, S., Suvarnaphaet, P., Visitsattapongse, S., and Pechprasarn, S. (2022). Measurement precision enhancement of surface plasmon resonance based angular scanning detection using deep learning, *Sci. Rep.*, **12(1)**, pp. 1–14.

62. Gomes, J. C. M., Souza, L. C., and Oliveira, L. C. (2021). SmartSPR sensor: Machine learning approaches to create intelligent surface plasmon based sensors, *Biosens. Bioelectron.*, **172**, pp. 112760.

Chapter 13

Bottom-Up Fabrication of Plasmonic Sensors and Biosensors

Nabarun Polley and Claudia Pacholski
Institute of Chemistry, University of Potsdam,
Karl-Liebknecht-Str. 24-25 Potsdam, 14476, Germany
nabapolley@uni-potsdam.de, cpachols@uni-potsdam.de

13.1 Introduction

The projected market value of plasmonic sensors for 2027 will be USD 1.24 billion [1]. In comparison, the projected market value of all sensor types combined for 2026 will be USD 411.2 billion [2]. From these numbers, it is clear that plasmonic sensors have a negligible share of the overall sensor market, i.e., not many companies are working with plasmonic sensors. One of the possible reasons for the low interest of the industry is the large gap between academia and industry [3]. Despite the wealth of knowledge about plasmonics, a few commercial applications have been carried out in the last decades.

Plasmonics-Based Optical Sensors and Detectors
Edited by Banshi D. Gupta, Anuj K. Sharma, and Jin Li

ISBN 978-981-4968-85-0 (Hardcover), 978-1-003-43830-4 (eBook)
www.jennystanford.com

As a probable solution, it has been recommended by experts in this field to take a more long-term view and approach to this topic, which means that more trained and knowledgeable personnel in this field is needed. This brings us to the point of defining the motivation and expected audience for this book chapter.

The performance and potential for the commercialization of plasmonic sensors very much depend on the adaptation of specific fabrication strategies. In general, there are two basic fabrication strategies available: top-down and bottom-up. The essence of bottom-up manufacturing strategies and their corresponding applications are presented in this book chapter. You will get a sense of how the field has developed in the past, where we are today, and what the opportunities will be in the future. This chapter is intended for new researchers and for students who want to work in the field. It is expected to add value to those already working in the field, but not to experts. Before we get to the heart of bottom-up fabrication, we will answer some basic questions that might arise.

Why do we need plasmonic sensors?

Basic or contemporary optical detection methods are based on the direct interaction of light with the analyte of interest. They require that the analyte molecules absorb light directly or subsequently produce fluorescence. In both cases, the absorbance or fluorescence signal is further correlated with the analyte concentration. These detection strategies work fine when the analyte molecules have either a high extinction coefficient or a high quantum yield. However, there are two major problems with these strategies. First, not every molecule has at least one of the two properties (high extinction coefficient or quantum yield). The reverse is true, most of the molecules, especially smaller molecules (enzyme, protein, and DNA), do not show significant absorbance or fluorescence [4]. Second, the relevant detection range for a specific molecule (often in the ppm or ppb range) is often far below the concentration, which is necessary to generate a measurable signal using one of these detection strategies. As a solution, the capture probe (or another molecule involved in the detection) is attached to some known molecules with either a high extinction coefficient or a high quantum yield. This process is called labeling. The gold standard of analytical assays, enzyme-linked immunosorbent assay (ELISA), is based on this strategy. It adds up a lot of preparation steps (hence higher cost) and requires expensive equipment to perform any sensing [4]. There is no other way around this problem with contemporary optical detection. This is where plasmonic sensing comes in, allowing us to detect small molecules down to the single-molecule level without labeling steps [5]. In addition, plasmonic sensing enables real-time monitoring of binding events and thus access to binding or reaction kinetics, which is difficult to achieve with conventional strategies. Another advantage of plasmonic sensing is the small volume of the sensing region, which is limited to a small space near the surface of the nanostructured plasmonic material.

What is a plasmonic sensor?

By definition, the term 'plasmonic sensor' refers to a special category of sensors in which the interaction of light with a plasmonic, usually metallic, nanostructure is explored as a transduction strategy. Plasmonic sensors come in a variety of designs, and depending on their geometric arrangement, different types of plasmonic properties can be explored for sensing. In the case of plasmonic nanoparticles, the excitation is called localized surface plasmon resonance (LSPR), which can directly be excited by light and can detect molecules that are in close proximity to the plasmonic surface (a few nanometers). On the contrary, in thin gold films propagating surface plasmon resonance (PSPR) can only be excited indirectly using prism coupling. In this case, the detection volume above the plasmonic surface is significantly increased. Metallic films with periodic "voids" can show both: LSPR and PSPR. Furthermore, plasmonic nanostructures can be used for sensing applications in the form of surface-enhanced Raman scattering (SERS), surface-enhanced infrared absorption (SEIRA) spectroscopy, and surface-enhanced fluorescence (SEF) [6].

What are plasmonic materials?

When we talk about plasmonics, the material of choice automatically narrows down to elements with a lot of free conduction electrons to spare. Gold (Au), silver (Ag), aluminum (Al), and copper (Cu) are most often considered for fabricating plasmonic sensors. It is worth mentioning that other novel materials like graphene also show plasmonic properties with promising applications [7–9]. However, given the intent of this chapter to guide researchers new to the field of bottom-up fabrication of plasmonic sensors, we continue our discussion with the conventional approach of using metals as plasmonic materials. Among the mentioned metals, silver has the most prominent plasmonic responses. But the reactivity of silver with atmospheric oxygen to form silver oxide restricts the real-life usability of the material as a plasmonic sensor. This stability issue can simply be bypassed by proper sensor storing protocols [10, 11]. Unlike silver, gold (Au) is known for its chemical inertness and excellent biocompatibility. The plasmonic response of Au can be tuned from the visible to the mid-IR range and is based on the selection of the geometric arrangement of the nanostructure. This behavior provides us the freedom to choose the structure based on the requirements. All these advantages make Au the material of choice for plasmonic applications.

Combinations of materials like Au-Ag either in form of core-shell particles or alloys, are also viable options for plasmonic sensing applications. It facilitates fine-tuning of the optical response and the chemical stability of the plasmonic sensors by varying the material composition [12]. For example, the sensitivity of silver and the biocompatibility of gold can be combined in alloyed Ag/Au nanospheres [13]. In addition, the enhancement of plasmon-driven catalytic selective hydrogenation activity by utilizing a hybrid Au–Ag core-shell structure has also been reported [14].

What are the fabrication strategies?

By now, we are aware of the importance of plasmonic sensors in the development of improved sensing strategies. The next obvious question might be, how they are fabricated? The straight answer would be that there are two strategies: top-down and bottom-up. The top-down approach starts with a solid material that is cut to the desired structure. In contrast, in the bottom-up method, we need to build the structure starting from the “unit” (atomic/molecular/structural) level piece by piece. A good analogy to distinguish between the two methods could be obtained by comparing the construction of buildings (Fig. 13.1). The construction of building brick by brick (conventional) can be compared with the bottom-up technique. In principle, we can also start with a solid structure like a rock and then work on it as needed, which is a top-down approach. Kailasa Temple, Ellora, India is a classic example of such a construction.

Why bottom-up?

In this book chapter, we have focused on presenting mainly bottom-up fabrication. The obvious question that might arise is: Why bottom-up? The top-down approach is the current standard for nanostructures production due to its associated accuracy, geometric design reliability, and high nanometer-scale precision. The most used top-down methods are optical lithography and electron-beam lithography (EBL). Optical lithography has a limited resolution, while EBL is expensive to operate and not suitable for mass production. In contrast, bottom-up fabrication is extremely cost-effective and promising [15] because it does not generate waste or unused materials and does not require the disposal of parts of the final system. With an increasing understanding of atomic and molecular interactions at the nanoscale, it is possible to fabricate periodic nanostructures using low-cost processes while maintaining high order over large areas with excellent reproducibility [16].

Figure 13.1 Malbork Castle (Malbork, Poland), the largest brick castle in the world, is an example of bottom-up construction (left). Kailasa Temple (Ellora, India), the largest of the rock-cut Hindu temples, is an example of top-down construction (right) [17, 18].

It looks like the top-down approach is quickly reaching its physical and economic limits and the bottom-up approach is the future of nanofabrication. Although these are two conceptually different manufacturing strategies, the combination of the two methods offers potential opportunities [19] (Fig. 13.2). This might be the true future of nanofabrication.

Let us now discuss what we expect to learn from this book chapter.

Key learnings:

- *Basic building block of knowledge about plasmonics*
- *Bottom-up fabrication techniques for plasmonic sensors*
- *Structural and functional characterization techniques for fabricated plasmonic sensors*
- *Popular applications of plasmonic nanostructures in sensors and biosensors*

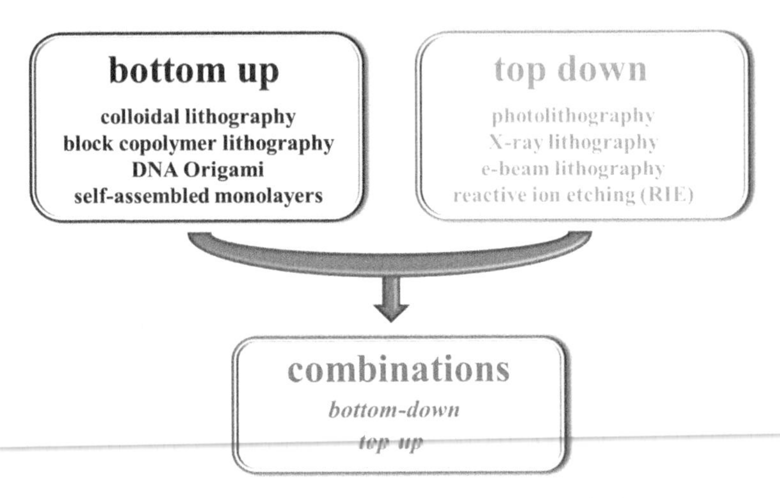

Figure 13.2 Fabrication routes to plasmonic nanomaterials: bottom-up, top-down, and combinations of both.

13.2 Building Blocks of Structures and 'Knowledge'

13.2.1 Building Block of Knowledge

In order to gain a comprehensive understanding of the subject, it is worthwhile to take a look at the physical fundamentals of plasmonics and the corresponding characterization techniques even before learning the basic fabrication procedure. This "building block" of knowledge will help us to understand the subject "from the bottom up".

13.2.1.1 Concept of plasmonics

The operation of plasmonics is based on the coherent oscillation of free-conduction electrons in response to an external electromagnetic field. In bulk metals, the free electrons respond to the time-dependent "local" change in the amplitude of the electric field and oscillate 180° out of phase. Each metal has its characteristic frequency, which is called plasma frequency [20]. The plasma frequency of a metal depends on the electron density and the effective mass of the electrons. If an electromagnetic wave (light) has a lower frequency than the plasma frequency, the light is reflected due to the response of the electrons. However, if the frequency is higher than the plasma frequency, then the electrons are not fast enough to react to the external electromagnetic field, and it is rather transmitted or absorbed in the interband transitions of the metal. This simply means that in order to create any kind of plasmon response in metallic nanostructures, we must operate in a frequency range lower than the plasma frequency. Since we are more comfortable with the concept of wavelength than frequency, let's rethink the matter from the perspective of wavelength. To excite the free conduction electrons, we need to work with a wavelength that is longer than the "plasma wavelength". Normally, the "plasma wavelength" is in the UV to deep UV range. Therefore, it is necessary to choose a wavelength in the visible to the infrared range for the electrons to respond to the electromagnetic wave (light). Please keep in mind that the plasma frequency depends on several parameters, including the material. It is worth noting that the combination of interband transitions and the plasmonic response results in the characteristic colors of the metal under normal ambient illumination (Fig. 13.3).

PSPR:

For thin films (thickness in the subwavelength range), electronic oscillation in response to an external electromagnetic field (light) can occur only at the surface. The generated charge wave propagates and is called propagating SPPs. Due to the limited dimensions and the additional constraints imposed by the interface between the film and the surrounding medium, the electronic oscillations are selective for a certain angle (compared to any angle in the bulk) and a certain frequency (compared to any frequency below the plasma frequency in the bulk). However, there is a catch to matching the momentum of the oscillating electrons to the incident light, the light must be introduced via a prism (Kretschmann geometry) [21] or a grating structure [20]. The electron density oscillations at the interface between the metal and the surrounding medium are the key to any plasmonic sensing. The local electric field generated by the displacement of electrons extends up to 200 nm into the dielectric. Any minute change in the immediate environment of the film, such as the attachment of molecules, disturbs the motion of the electrons and changes the SPP frequency and the angle. Both can be measured and used to monitor molecular interactions in real-time.

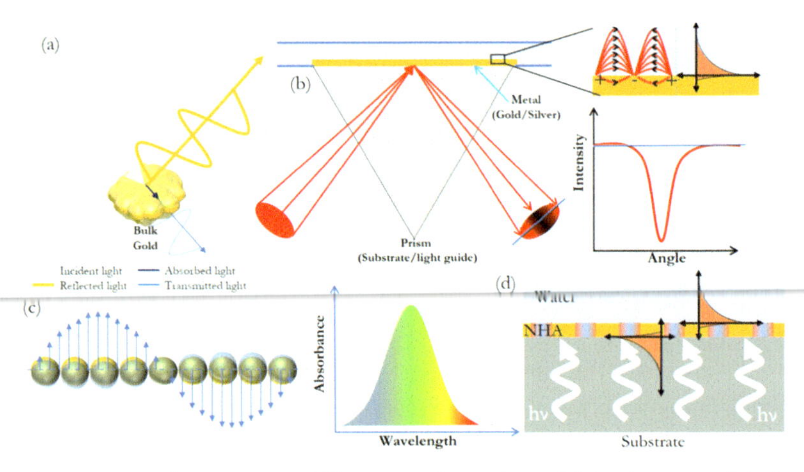

Figure 13.3 Plasmonics in different configurations. (a) In the case of bulk metal (gold), the light above the plasma wavelength is reflected. The rest is either absorbed in the transition levels between the bands or transmitted. (b) In a metal film, SPPs are generated upon momentum matching. The amplitude decreases exponentially with distance from the interface (top inset). The intensity profile that appears due to this energy transfer shifts to the change in the refractive index of the dielectric medium (lower inset). (c) Metallic nanoparticle with free conduction electrons responds to the local electric field amplitude with a 180º phase shift. The collective oscillations generate an amplified local EM field known as localized surface plasmons and the resonance condition known as LSPR. The LSPR response can be determined by colorimetric information (see box). (d) Light can induce both LSPR and SPP in nanohole arrays due to lattice-induced momentum matching.

LSPR:

When the film is broken down into smaller "pieces" with nanometer dimensions (much smaller than the wavelength), the "propagating surface plasmons" are localized only to the nanosized "pieces", hence it is called LSPR. The electromagnetic field generated by the displacement of electrons is on the order of 10–30 nm. A conceptual difference between SPPs and LSPR is that LSPR does not require momentum matching between the oscillating electrons and the incident light, since the additional momentum is provided by the limited geometry of the nanoparticles themselves. The LSPR phenomenon results in a characteristic oscillation of the electrons at a specific frequency/wavelength known as the LSPR peak/band (depending on size and shape as well as material and environment). Like SPPs, the LSPR is very sensitive to interface changes and is the key to any LSPR-based sensing mechanism.

Plasmon coupling:

Both forms of plasmons can be coupled together, leading to intriguing new plasmonic responses. When two nanoparticles are physically close to each other (within the LSP region), both plasmons can be coupled, leading to a change in the LSPR peak or to a completely new plasmonic peak in the optical spectrum. Similarly, the LSP of nanoparticles can be coupled to a thin metal film [22].

13.2.1.2 Characterization techniques

The dimensions of the plasmonic material are usually in the sub-wavelength range, and the physical properties differ significantly from those of their bulk counterpart. The larger surface-to-volume ratio increases reactivity at the molecular level [23]. Broadly speaking, characterization can be divided into physical characterization (shape, size) and functional characterization (SPR, LSPR). A comprehensive review on this topic can be found in an article by Mourdikoudis et al. [23].

Physical characterization:

Optical microscopy is usually the first choice for determining the physical properties of structures with small dimensions. However, since the dimension of plasmonic nanomaterials (PNMs) are on the nanoscale, the resolution of the optical microscope is not sufficient to achieve the required dimensions. However, due to their high scattering cross-sections, it is possible to detect PNMs using special microscopy techniques such as dark-field microscopy and confocal microscopy. In practice, the shape and size of PNMs are determined by electron microscopy [24]. Both scanning electron microscopy (SEM) and transmission electron microscopy (TEM) are regularly performed to determine the size and shape of the nanomaterial. SEM is able to produce surface images from secondary electrons scattered from the sample, while TEM is able to provide detailed information about the structure of the material. With high-resolution TEM, it is possible to achieve sub-nanometer resolution. Both TEM and SEM produce a 2D image of the PNMs, so the depth information obtained is not accurate. The depth information can be accurately obtained by atomic force microscopy (AFM). Since the images are generated by raster scanning with an atomically sharp tip, the surface and depth information can be determined with nanometer accuracy. Strategies similar to AFM have been developed, but in the optical domain, called scanning near-field optical microscopy (SNOM) [25]. Another powerful optical technique based on light scattering is quite widely used in determining the hydrodynamic diameter of PNMs in suspensions and is called dynamic light scattering (DLS).

Functional characterizations:

The determination of functional parameters such as plasmonic response, particle concentration, agglomeration state, and surface charge are important for the characterization of PNMs [23]. The optical response of PNMs can be determined by absorption, transmission, or reflection spectroscopy (more generally: optical spectroscopy). We now know that the optical response of a PNM depends on its size, shape, and composition. Another important parameter, the information about the interface between the PNMs and the environment, can be obtained by UV-Vis spectroscopy. This is an important step toward the use of PNMs as a sensing element, as UV-Vis spectroscopy allows us to directly study the transduction element in the sensor. Additional information such as nanoparticle concentration of gold and silver nanoparticles can also be obtained by UV-Vis spectroscopy [26, 27].

13.2.2 Building Blocks for Bottom-Up Fabrication

Bottom-up fabrication of nanostructures can be divided broadly into two categories: direct assembly and template-assisted fabrication. In any case, the fabrication always starts with a 'unit' or building block. In this section, we will focus on the preparation of building blocks.

13.2.2.1 Gold and silver nanoparticles

Chemical reduction of a precursor by a reducing agent is the most common method for the synthesis of gold nanoparticles. Nucleation followed by a "growth" step and subsequent stabilization are the main steps in the formation of nanoparticles (NPs). When nucleation and growth occur simultaneously, it is referred to as in situ synthesis. This form of synthesis is ideal for the production of spherical and quasi-spherical particles with comparatively small sizes (<30 nm) [28]. If another growth step is performed separately after the nucleation step, it is called the seed growth method. It is possible to produce Au NPs in different sizes (Fig. 13.4) and shapes with comparatively large physical dimensions [29].

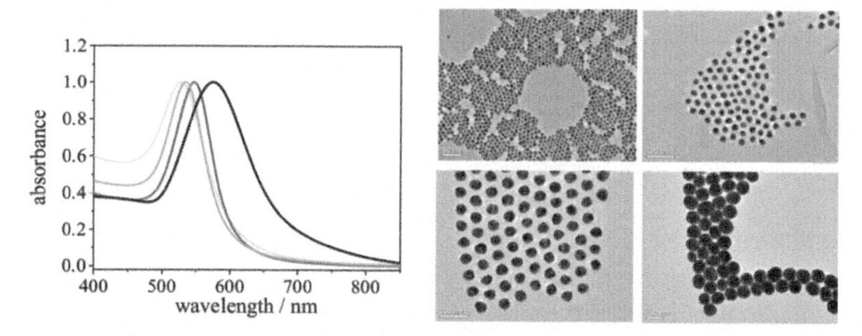

Figure 13.4 Wet-chemically synthesized gold nanoparticles with different diameters: absorbance spectra (left) and corresponding transmission electron micrographs (right).

The most popular method of gold particle synthesis is the chemical reduction and subsequent stabilization of $HAuCl_4$ by trisodium citrate dihydrate. It was proposed by Turkevich and further modified by Frens to obtain better control over the size of the nanoparticles. Briefly, an aqueous solution of $HAuCl_4$ is heated to a boiling temperature of 100 °C, and then trisodium citrate dihydrate is added rapidly with vigorous stirring. The resulting wine-red solution is cooled to room temperature after a predefined time, usually 15–30 minutes [30]. The size of the nanoparticles can be controlled between 15 and 100 nm depending on the Au and trisodium citrate ratio. A higher concentration of citrate leads to smaller monodisperse nanoparticles. Conversely, a lower concentration leads to larger nanoparticles, but stability and monodispersity decrease. In general, better synthetic control over colloidal synthesis can be achieved when the required reaction temperature is lower [24]. The need for high temperatures can be avoided by using reducing agents with high reducing power. The use of tannic acid as a reducing agent and citrate as a stabilizer reduces the required temperature to 60 °C [31]. It is possible to synthesize gold nanoparticles at room temperature by using a reducing agent such as $NABH_4$ with high reducing power and citrate as a stabilizer [32]. The reduction can be achieved by initiation with UV light at room temperature with citrate as a stabilizer [30].

The seed-growth technique is popular for the synthesis of large-diameter nanoparticles, nanorods, nanocubes, nanotriangles, and other geometries [24, 33]. Silver nanoparticles can be synthesized based on similar strategies to the synthesis of gold nanoparticles, but in this case silver nitrate salt ($AgNO_3$) is used as the metal source [24].

Role of ligands in stability, size, and shape:

Stabilizing ligands are required in order to prevent NPs from agglomerating. Here, the stability of the NPs depends on the ligand of choice. Furthermore, the concentration of the ligands directly affects the final particle size. The Au NPs synthesized by the Brust-Schiffrin method are quite robust due to the use of thiol ligands (stability). At the same time, the synthesized particles have a diameter of less than 10 nm (size) [33] due to the high affinity of the thiol ligand for the gold surface. The final shape of the NPs synthesized by the seed-growth method is also influenced by the chosen ligand. Nanorods are synthesized in the presence of the ligand and the shaping agent cetyltrimethylammonium bromide (CTAB) [24]. Depending on the end-use of the NPs (e.g., biocompatibility [34], solubility in solvents), the ligands can be modified by ligand exchange processes.

Plasmonic properties:

The plasmonic properties of NPs depend on their size, shape, composition, and environment. The plasmonic response of Au NPs with a size of about 10 nm has an LSPR peak at ~520 nm (Fig. 13.5). In contrast, Ag NPs of similar size have a peak at ~400 nm. In both cases, a redshift in the LSPR peak is observed as the size increases because the resonance of the electrons on the surface also changes with size. When the structure is asymmetric, multiple plasmonic responses can be observed. As shown in Fig. 13.5, Au nanorods exhibit two plasmonic bands, one at ~520 nm and another at ~850 nm, corresponding to the plasmonic response along the width and length of the NRs, respectively. For a given shape, size, and composition, the plasmonic response of the NPs depends on the local RI of the surrounding medium. As the local RI changes, the plasmonic band also changes its spectral position to either the red (redshift) or blue (blueshift) end of the spectrum. The amplitude of the shift depends on the refractive index (RI) difference, the thickness of the medium around the NPs, and the electromagnetic field generated around the NPs. NPs with sharper edges cause a high density of the electromagnetic field compared to symmetric NPs. Asymmetric NPs, e.g., nanotriangles, are therefore more sensitive to changes in refractive index, which is desirable from a sensing point of view.

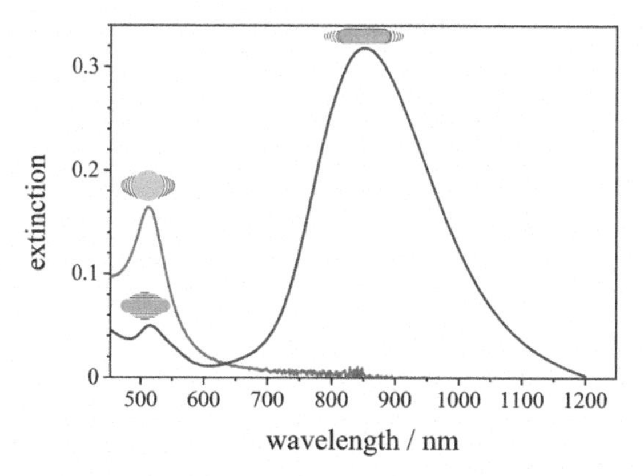

Figure 13.5 Asymmetric gold nanostructures: comparison of the optical properties of spherical and rod-shaped gold nanoparticles.

13.2.2.2 Polystyrene latices, silica colloids, polyNIPAM microgel

Building blocks for the bottom-up fabrication of 2D and 3D nanomaterials can also be made of materials not showing plasmonic behavior such as polymers. These building blocks are used as masks or templates to fabricate plasmonic nanostructures. For this purpose, colloids showing so-called hard sphere behavior are most often utilized. They are composed of polystyrene or silica and their synthesis is well-established. Moreover, colloidal dispersion of polystyrene (polystyrene latices) nanospheres and of silica nanospheres are commercially available supporting their wide-spread application in the bottom-up production of nanomaterials.

Monodisperse polystyrene latices can be obtained using different synthetic methods such as emulsion polymerization (with and without surfactants), seed swelling polymerization, and dispersion polymerization [35]. A well of preparation protocols has been reported leading to monodisperse polystyrene colloids that can present different chemical groups/molecules at their surface. These groups/molecules foster the exploitation of electrostatic interactions in self-assembly and offer substantial potential for surface modification. Moreover, the long-lasting experience in emulsion polymerization allows for easy implementation in every chemistry lab and even undergraduate teaching [36]. The same is true for the preparation of silica colloids, which is often based on the hydrolysis of tetraethylorthosilicate (TEOS) and subsequent condensation reactions between generated silanol groups in solutions containing water, short-chain alcohols, and ammonia as a catalyst. This synthesis strategy was published by Stöber, Fink, and Bohn in 1968 and is commonly referred to as the Stöber synthesis [37]. Until today, it is utilized and further developed to obtain monodisperse silica colloids with tailor-made properties [38].

In addition to colloids showing hard sphere behavior, other building blocks with soft sphere properties are investigated to fabricate PNMs. Hydrogel microgels composed of poly-N-isopropylacrylamide (polyNIPAM) are the most prominent colloids to exploit the advantages and challenges of soft spheres in colloidal lithography and compare their performance with that of hard spheres made of polystyrene or silica. Pelton and Chibante reported in 1986 the first synthesis of monodisperse aqueous latex dispersions based on N-isopropylacrylamide (NIPAM) in combination with acrylamide and N,N′-methylenebisacrylamide as cross-linkers [39]. They observed changes in the diameter of the prepared polyNIPAM microgels upon temperature changes. In Fig. 13.6, a graph is displayed that demonstrates this switch between the swollen and collapsed states of polyNIPAM microgels in response to temperature changes (observed by DLS). Since then, the stimuli-responsive behavior of polyNIPAM colloids has been studied intensively and by incorporation of co-monomers, their volume can nowadays be abruptly changed by several stimuli including pH or light.

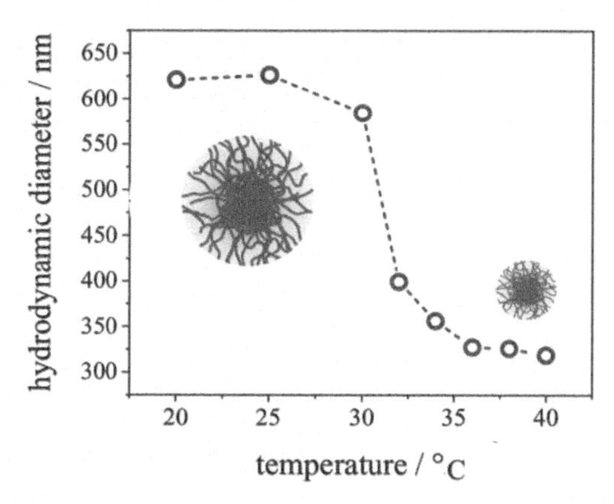

Figure 13.6 Changes in the hydrodynamic diameter of polyNIPAM microgels in response to temperature changes determined by DLS.

It is important to mention that polyNIPAM colloids can show both hard-sphere and soft-sphere behavior [40], making them ideal candidates for the bottom-up fabrication of nanomaterials.

13.2.2.3 Biological templates

Self-assembly of plasmonic nanoparticles into defined nanostructures can also be accomplished using biological templates such as DNA, viruses, S-layer proteins, and amyloid fibers [41]. DNA is especially appealing to prepare dimers or trimers of plasmonic nanoparticles for SERS. Here, the distance between the plasmonic nanoparticles is small providing so-called 'hot spots' for highly sensitive detection of molecules. How is this possible? Short DNA molecules, referred to as oligonucleotides, with defined chemical structures (sequences) can be synthesized in an automated process for decades. The reaction of a long DNA single strand with several smaller DNA strands (staplers) allows oligonucleotides to be folded into nanoscale objects with a previously determined shape [42]. This DNA Origami process is not only suitable for the production of differently shaped DNA nanoparticles but also allows the incorporation of chemical groups/DNA strands to hold, e.g., plasmonic nanoparticles at a predefined location on a DNA scaffold. Hence, a careful design of DNA Origami structures paves the way to create plasmonic nanoparticle arrangements with a precision of a few nanometers.

Another route to highly ordered 2D structures of plasmonic nanoparticles is based on using proteins as biological templates. A prominent example is S-layer proteins. They are 2D crystalline membranes, which can be found on the outer surface of bacterial cell envelopes. Their symmetry (trigonal, square, and hexagonal) on the nanoscale can be directly employed for the in situ nucleation of inorganic nanoparticles or the self-assembly of gold nanoparticles into hexagonal arrays [43].

13.2.2.4 Other templates for assembly

Top-down strategies have often been used to create arrays of pillars or cavities that facilitate the formation of highly ordered structures by template-assisted assembly—often on relatively small areas due to the capabilities of EBL and focused ion beam (FIB) milling. Nevertheless, polydimethylsiloxane (PDMS) stamps have been fabricated using masters, i.e., microstructured or nanostructured silicon. These stamps have been successfully utilized to stamp molecules onto suitable surfaces, resulting in a defined chemical topography that is perfect for building nanostructures [44].

PDMS can further be exploited for generating wrinkled surfaces [45, 46]. For this purpose, an elastomeric PDMS substrate is stretched and subsequently exposed to oxygen plasma to form a thin and hard SiO_x layer on top by cross-linking reactions. Upon release of the stress, wrinkled surfaces are formed due to buckling instability. The resulting pattern can be controlled by the applied strain field. For example, a uniaxial strain field will lead to parallel lines of waves. By increasing the thickness of the SiO_x layer, the wavelength and amplitude of the wrinkles can be determined. Other patterns of wrinkles such as chevrons topology or radial patterns were obtained by generating corresponding strain fields [47]. These wrinkled surfaces are versatile platforms for self-assembly and can provide macroscopic templates.

The same goal can be achieved by phase separation of immiscible polymer blends that can produce micro-/nanoporous structures at the wafer scale [48]. More regular and highly ordered templates can be obtained by using block copolymers instead of polymer blends. Block copolymers are polymers composed of at least two different homopolymer subunits and linked by covalent bonds. The length and the material of the subunits can easily be controlled during synthesis opening up a simple chemical route to polymers with tailor-made properties. It has been shown that in thin films of block copolymers arrays with periodicities on the nanoscale can be realized [49]. For example, the morphology of solvent-cast polystyrene-polybutadiene-polystyrene triblock copolymer thin films is influenced by the solvent evaporation rate and post-evaporation annealing [50]. How these templates can be used to generate plasmonic nanostructures will be discussed later in this chapter. The strategies presented for fabricating templates for the generation of plasmonic nanostructures provide only a small glimpse of the large variety of methods reported but should whet the appetite for applying these techniques to the production of plasmonic sensors using self-assembly.

13.3 Self-Assembly

The beauty of material and structure design using bottom-up approaches is its simplicity and modular character, neither extensive procedures nor expensive equipment is required, and yet well-ordered and complex structures can be created. Self-assembly of colloids impressively demonstrates the potential and colloidal lithography is an established technique for creating plasmonic nanostructures on large areas. Different strategies have been developed for obtaining the desired design with the required geometry and precision. Thereby, it can be distinguished between template-free and template-assisted self-assembly procedures. When only the building blocks interact to form plasmonic 2D and 3D materials in solution or on substrate surfaces, these are referred to as template-free methods. Another route to highly defined PNMs is based on the interaction between building blocks and pre-structured templates. In this case, the template has a topographical or/and chemical micro-/nanostructure that controls the self-assembly of the plasmonic building blocks. In the following part, we provide an overview of established bottom-up methods to fabricate 2D and 3D PNMs for sensing applications.

13.3.1 Template-Free Assembly

Colloidal lithography is a very active research field and highly interdisciplinary profiting from contributions made by polymer physics as well as inorganic and macromolecular chemistry, fundamental physics, and materials science. Its attractiveness is based on its simplicity and cost-efficiency for fabricating highly ordered 3D and 2D colloidal crystals with large dimensions, which are obtained by self-assembly [51, 52]. The colloids can be made of different materials or possess different sizes providing easy access to complex and hierarchical structures.

Self-assembly of colloids is influenced by a variety of parameters such as the properties of the colloid (material, size, surface functionalization, and dispersion medium), the choice of the deposition procedure (convective assembly, controlled evaporation, electrostatic deposition, template-assisted assembly, etc.), and the supporting substrate (for example, hydrophilicity/hydrophobicity, charge, etc.). Only by finding the right balance between attractive and repulsive forces as well as external forces acting on the colloids during the assembly process, the desired structure can be obtained. Figure 13.7 provides an overview of the various forces that can be involved in the self-organization process [53]. Please refer to the relevant literature on this topic for more details [51]. It has to be emphasized that it is highly favorable to work in a dust-free environment in order to achieve 2D and 3D colloidal crystals with long-range order.

To create plasmonic nanostructures using template-free colloidal self-assembly, the building blocks themselves must have plasmonic properties. Hence, gold nanoparticles are most often employed. They are well-studied and their surface can be easily modified with different ligands (molecules) such as amine ligands with different lengths or thiol-terminated polymers [54]. For example, Yang and Hallinan have developed a simple method to prepare ordered gold nanoparticle monolayers covering areas of a square centimeter [55]. For this purpose, a droplet of aqueous gold nanoparticle dispersion was placed in a petri dish filled with an organic solvent containing amine-terminated ligands. A three-phase system of air/aqueous gold nanoparticle dispersion/organic solvent formed. Upon the addition of ethanol, gold nanoparticle islands were generated at the aqueous gold nanoparticle dispersion/organic solvent interface. These gold nanoparticle islands transfer from the aqueous/organic solvent interface to the water/air interface and then form the desired highly ordered gold nanoparticle monolayer. The inter-particle distances in these gold monolayers depend on the chain length of the ligand and are quite small, making the monolayers perfect candidates for SERS.

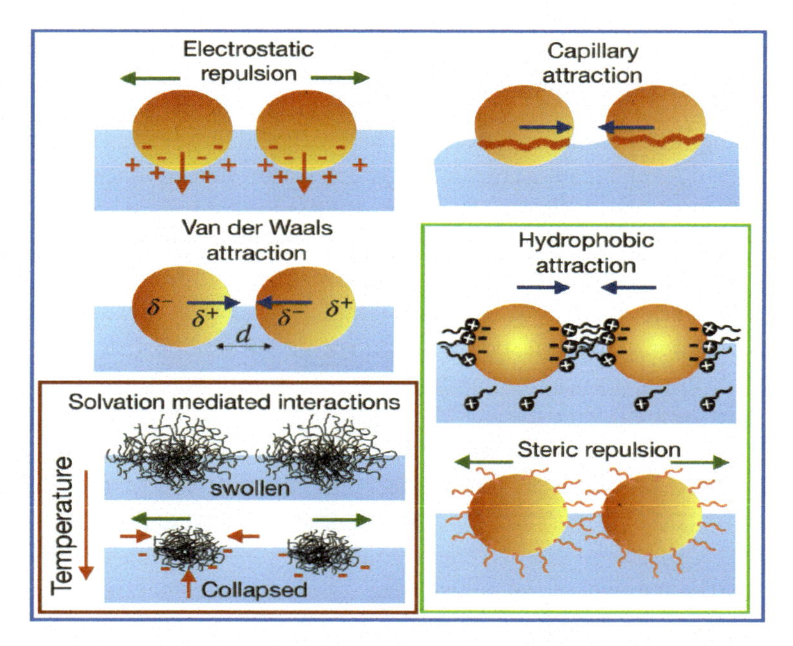

Figure 13.7 Basic forces involved in colloidal self-assembly: electrostatic interaction, van der Waals attraction, capillary forces, solvation-mediated interactions, hydrophobic attraction, and steric repulsion. Reproduced from Ref. [53], with permission from Elsevier.

Larger inter-particle distances can be achieved by using thiol-terminated polystyrenes as ligands. In this case, the polystyrene acts as a soft shell supporting the formation of highly ordered 2D arrays. Again, here, the spacing between the particles can easily be controlled by the molecular weight of the polystyrene used [56]. In addition, more sophisticated plasmonic nanostructures such as gold nanoparticle double layers or honeycomb structures can be obtained by sequential assembly [57]. This method paves the way to prepare a large variety of structures, where the particle sizes, shapes, distances, materials, and mixtures can be controlled. Since gold nanoparticles are deposited on substrate surfaces by spin-coating, large homogenous nanostructures of several square centimeters can be prepared in a few minutes. The performance of the resulting plasmonic structures in surface-enhanced spectroscopy has already been investigated [58].

13.3.2 Template-Assisted Assembly

13.3.2.1 Colloidal masks

2D colloidal monolayers have often been utilized for creating surface patterns, which can be exploited as masks for subsequent etching or deposition of materials. Most often so-called hard spheres, colloids typically composed of silica or polystyrene, are employed for the formation of close-packed hexagonally ordered colloidal monolayers. They can serve as templates for the arrangement of plasmonic nanoparticles into arrays of nanoparticle rings, representing a pure bottom-up strategy. For example, Lerond et al. self-assembled polystyrene microspheres into quasi-hexagonally ordered close-packed arrays on glass substrates. Smaller gold nanoparticles were arranged into rings upon drying of the gold nanoparticle dispersion, due to capillary forces acting at the interface between polystyrene colloids and substrate interface [59]. After the removal of the polystyrene colloids by thermal treatment, plasmonic nanoring arrays remained on the substrate ready to use for sensing applications. It must be emphasized that a variety of plasmonic nanoparticle arrays was realized by using 2D and 3D colloidal crystals as masks for subsequent deposit of plasmonic material using physical vapor deposition or sputtering [60, 61].

In addition, colloidal masks have been employed for the fabrication of plasmonic nanohole arrays. Loosely packed, highly ordered colloidal arrays are required for this purpose. When the arrays are self-assembled from hard spheres, the result is closed-packed arrays that cannot directly be used as a mask for the deposition of a gold film. The necessary size reduction to convert densely packed arrays to loosely packed arrays is often achieved by reactive ion etching, a typical top-down method [62]. An elegant way to directly access loosely packed colloidal arrays is the utilization of soft colloids made of hydrogels or core-shell particles having a hard sphere core and a soft shell of polyNIPAM. The self-assembly of polyNIPAM colloids into loosely packed hexagonally ordered arrays was already observed in the first article published on their wet-chemical synthesis by Pelton and Chibante [39]. Just by placing a droplet of diluted polyNIPAM microgel dispersion onto a TEM grid and drying at room temperature 2D colloidal crystals with defined lattice constant were obtained. Since then an impressive amount of articles on the assembly of soft colloids at interfaces and their properties on solid substrates including swelling behavior, softness, and geometrical arrangement has been published [63]. Highly ordered 2D arrays were mainly achieved by assembling polyNIPAM microgels at an air/liquid or liquid/liquid interface and subsequently transferring the formed monolayer to a solid substrate. Here, a Langmuir trough was often used, as it also allows for studying compression isotherms of the polyNIPAM microgel monolayers, in order to gain a deeper understanding of the self-assembly process. The benefits of exploiting hydrogel microgels to fabricate highly ordered plasmonic nanostructures are presented in Section 13.4.3.3: Nanohole array.

13.3.2.2 Biological templates

Differently shaped DNA scaffolds are commercially available today. The size, shape, and position of functional groups/molecules of these DNA scaffolds can theoretically be chosen at will by careful design of the DNA single strand and the DNA stapler strands used in the DNA Origami process. Figure 13.8 shows one example of preparing gold nanoparticle dimers using triangular DNA scaffolds [64]. In this case, the DNA scaffolds present several single-stranded DNA extensions at a specific location on both sides of the DNA triangle. Gold nanoparticles coated with complementary DNA strands bind to the exposed single-strand DNA extensions on the DNA scaffold and form the desired gold nanoparticle dimers. This strategy has not only successfully been exploited for the fabrication of gold nanoparticle dimers, but also for the assembly of nanoparticle trimers (plasmonic lens) [65]. These plasmonic nanostructures are perfectly suited for surface-enhanced spectroscopy.

Similar strategies are reported for the use of proteins as biological templates for the formation of plasmonic nanostructures. For example, S-layers have a spatially defined physical and chemical topography that can be further tailored for capturing plasmonic nanoparticles by genetic modification or by conjugation of exposed carboxylic or amino groups with functional molecules such as amino acids [66]. Highly ordered arrays of gold nanoparticles have successfully been prepared using this strategy [67].

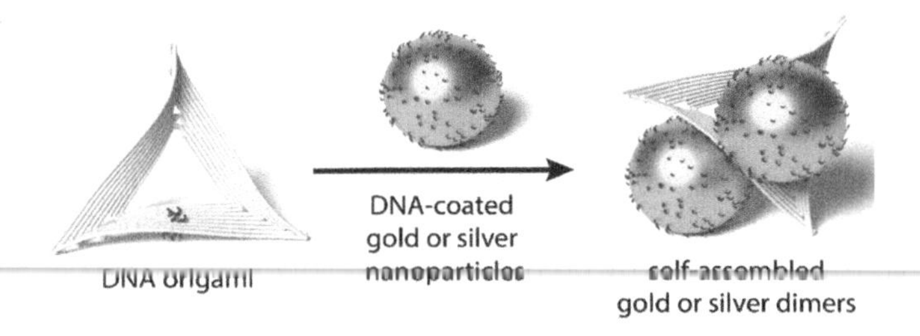

Figure 13.8 Assembly of gold nanoparticle dimers using DNA Origami: DNA triangles carrying groups of several single-stranded DNA extensions are mixed with gold nanoparticles whose surface is covered with DNA of a complementary sequence. Gold nanoparticle dimers are formed with a gap of a few nanometers between them. Reproduced from Ref. [64].

13.3.2.3 Other templates

There is a large variety of templates, which cannot be all covered in this short chapter. To give a taste of the possibilities already published, templates consisting of block copolymers and wrinkled polymer surfaces are presented below.

As already discussed, templates can be prepared by exploiting the microphase separation of block copolymers. For example, hexagonally ordered nanopore templates were obtained by depositing a monolayer of polystyrene-block-poly(2-vinylpyridine) micelles on a substrate and subsequent treatment with acetic acid for morphology reconstruction. These nanopores can serve as a mask for the growth of plasmonic nanostructures by galvanic displacement reactions [68, 69]. In addition, block copolymer templates have successfully been employed for the self-assembly of plasmonic nanoparticles [70–72]. Moreover, polystyrene-block-poly(vinylpyridine) inverse micelles can be filled with an inorganic precursor before depositing a monomicellar layer on a substrate by dip-coating. A hexagonally ordered array of gold nanoparticles can be obtained from these samples by plasma treatment. This method is called block copolymer micelle nanolithography (BCML) and is very versatile in terms of nanoparticle composition, size, and spacing [73–75].

Wrinkled polymer surfaces have also the potential for the production of homogenous PNMs on large areas. Two different routes can be followed leading to plasmonic nanostructures with sizes on either centimeter scale or larger. On the one hand, wrinkles can be used to assist plasmonic nanostructure formation on flat surfaces by stamping or by providing confinement. Also, the plasmonic nanoparticle can directly be assembled out of suspensions in the wrinkles due to capillary forces upon drying [76, 77].

13.3.2.4 Substrate of choice

Bottom-up fabrication of plasmonic sensors often involves the deposition of building blocks on substrate surfaces. Almost all kinds of materials have been investigated as substrates, including thin films of gold, glass slides, polymer layers, and optical fibers. The selection of a substrate has a significant influence not only on the sensor performance but also on the usability of the sensor. The presence of a substrate itself has an influence on the distribution of the electromagnetic field (surface plasmon resonance) on plasmonic nanostructures [78]. The substrate leads to a change in the dielectric function at one side of the plasmonic nanostructure, i.e., the plasmonic nanostructure is exposed to at least two different media having two different refractive indices. This can have a great impact on the performance of plasmonic sensors [79]. Furthermore, only one side of the plasmonic nanostructure is accessible for refractive index change detection, so the sensitivity of plasmonic sensors composed of plasmonic building blocks immobilized on substrates may be lower in comparison to plasmonic nanoparticles dispersed in a solution. However, the amount of accessible surface area on the immobilized plasmonic nanostructure can be considerably increased by using porous substrates such as hydrogel layers or porous silicon [80]. In addition, flexible or stretchable substrates are highly attractive for the fabrication of plasmonic sensors because they allow the control of the distance between the plasmonic building blocks [81–83]. As a result, the sensitivity of these sensors can be significantly increased by exploiting plasmon coupling [84]. The incorporation of optical fibers as substrates empowers us to perform sensing applications with minimal sample volume due to their miniature dimensions, and even in vivo sensing due to their excellent biocompatibility.

13.4 Characterization and Applications

13.4.1 Quantitative Analysis of Nanostructures

The performance of a plasmonic sensor depends on the quality of the fabricated structure. In Section 13.2, we discussed various instrumental techniques for characterizing the nanostructures. A qualitative understanding of all these parameters can be obtained by the methods described in Section 13.2, but further quantitative studies are needed to address issues that might otherwise arise. To ensure reproducibility and applicability, we need additional quality control parameters. Size distribution, surface quality, periodicity, and homogeneity are just a few of the many parameters that need to be defined for the successful application of nanostructures. In the following section, we have discussed various quantitative analysis techniques.

13.4.1.1 Size distribution

The size distribution in terms of the diameter (hydrodynamic diameter) of nanoparticles in suspension can be easily measured by DLS. However, there are some limitations in DLS measurements. The hydrodynamic diameter is larger than the actual diameter of the particles, and also the measurement is not valid for asymmetric particles such as nanorods. In practice, the size distribution is usually determined using images obtained from TEM and SEM examinations.

To define the size distribution independent of the number of visible particles, a histogram of the diameter of NPs vs. the number of such particles is created and normalized based on the total number of particles. The data used to generate the histogram can be collected manually or easily automated using readily available software such as ImageJ [85]. The distribution of particle size is indicated by the width of the Gaussian fit over the histogram data. In practice, the size distribution is expressed as Gaussian peak ± (width of Gaussian fit/2), in other words, particle size ± standard deviation.

13.4.1.2 Degree of order

The determination of the degree of order is crucial for the specific characterization of periodic nanostructures, e.g., an array of nanoparticles [86], an array of nanodisks [87], or a nanohole array [88]. The quality of an ordered nanostructure can be expressed in terms of short-range and long-range orders. The short-range order indicates the distance and geometric arrangements between the nearest nanostructures. The long-range order, on the other hand, indicates the consistency between repeating nanostructures. The quality of short-range geometries can be illustrated with a Voronoi diagram [89]. This is usually expressed by the Voronoi parameter, which is the ratio between the number of saucer-like nanogroups formed and the total number of nanogroups detected. For a perfectly ordered structure, the value of the Voronoi parameter should be equal to one. For example, in the case of self-assembled gold nanoparticles in a hexagonal arrangement, six particles are expected to surround a central particle. If the nanoparticles are not arranged in perfect order, the number will vary across the surface. Therefore, the Voronoi parameter will be less than one, due to the quality of the short-range consistency. Another, more robust method to express the arrangement over short and long distances in a periodic nanostructure can be expressed by the radial distribution function [90] (RDF, Fig. 13.9). The RDF characterizes the density distribution of particles on a surface and does not consider the geometric arrangement of particles. The RDF($g(r)$) is expressed as:

$$g(r) = \frac{\rho(r)}{\rho_0}. \tag{13.1}$$

where $\rho(r)$ is the local number density of nanofeatures at a given distance r with thickness dr, and ρ_0 is the overall number density.

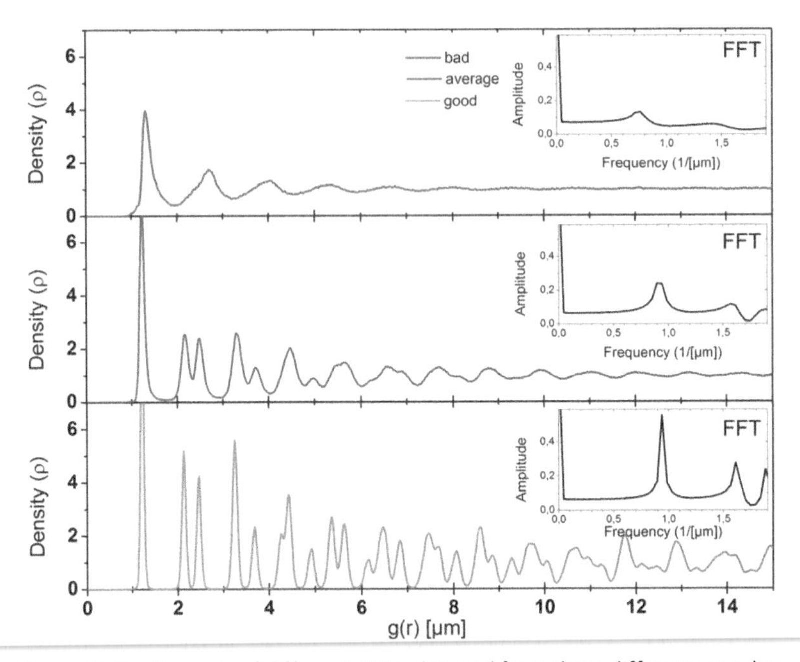

Figure 13.9 Example of different RDFs obtained from three different samples. The degree of order of a nanostructure over long distances is indicated by the prominent peak in the RDF and the sharper FFT of the RDF profile. Examples of poor, average, and good long-range periodicity of nanostructures with respect to RDF and the corresponding FFT are shown in the top, middle, and bottom plots, respectively. Image courtesy: Dr. Markus Weiler.

13.4.2 Performance Evaluation

The basic performance of a plasmonic sensor is determined by the refractive index sensitivity of the sensor. The plasmonic peak changes linearly with the change in the refractive index of the surrounding medium. The sensitivity of a plasmonic sensor is expressed as the unit of change in plasmonic peak (degrees (°) for SPR and nm for LSPR sensors) per refractive index unit (RIU). The value of the bulk sensitivity may differ from that of the local sensitivity at the surface of the sensor. In any case, it always indicates the change in the environment of the sensor. Another important parameter for determining the detection performance of a plasmonic sensor is the FOM value. It is determined by dividing the sensitivity of the plasmonic sensor by the width of the plasmonic peak. Here, the "width" indicates the full width of the plasmonic peak, which is calculated from the values of the half maxima (FWHM). In practice, the LSPR response of a sensor is recorded with a collinear optical setup in transmission or reflection mode, depending on the transparency and requirement of the substrate. It is worth noting that the information obtained in transmission and reflection modes is not comparable. This is because the scattering cross-section and the absorption cross-section vary independently of the wavelength. The transmitted signal carries the contribution of the scattering and absorption of the sensor and, in addition, some contribution of the analyte being measured [91] (the contribution is significant if the analyte has its own absorption characteristics). It is reported in the literature that reflectance measurement offers higher sensitivity compared to transmission measurement [92, 93]. However, the FOM is higher in the case of transmission measurement. Depending on the application, other performance parameters, such as limit of detection (LOD), selectivity, response time, and linearity, become relevant [94, 95].

13.4.3 SPR Sensors

13.4.3.1 Single particles

Although they are a building block for bottom-up manufacturing, NPs themselves are excellent for sensing and biosensing applications. Based on the use of nanoparticles, they can be divided into two categories: (a) in suspension and (b) on substrates.

(a) In suspension: Since the NPs are synthesized by wet chemical methods, resulting in a stable colloidal suspension, it is easier to use them for further sensing applications. NPs can be studied individually, with the sensing information coming from all of these particles. Instead of a singular response of NPs, when multiple NPs are brought closer together, the coupling of the plasmonic field of the individual particles results in a shift in the LSPR response and thus a change in color [20]. This detection strategy is ideal for visually representing the presence of the analyte for a variety of

applications such as environmental monitoring and biosensing. However, quantitative analysis of the analyte can also be performed by colorimetric detection [96]. A representative example of such a detection strategy is a recent article on the detection of SARS-CoV-2 nucleocapsid proteins [97].

(b) On substrates: Using NPs in suspension may be easier from a manufacturing point of view, but not from a practical point of view. In suspension, too many open parameters, such as temperature, pH change, and specific and non-specific bindings, can lead to changes in the plasmonic response. It is difficult to distinguish the contribution due only to the change in local RI. NPs bound to the substrate minimize the complexity and solidify the possibility of using NPs in real sensing applications [98]. Based on the use of LSPR "on the substrate," sensing can be further divided into two categories: (a) single particles [99–101] and (b) particle ensemble [102]. In the case of single particles, the NPs are usually examined microscopically in dark-field mode. Due to their high scattering cross-section, NPs are usually visible. This type of measurement is highly appreciated in the literature to understand the mechanistic behavior of plasmonically mediated light-matter interactions. However, due to the requirement of expensive devices, it is not suitable for point of need applications. Whereas, the instrumentation required for particle ensemble-based detection is extremely cheap and miniature, hence ideal for point-of-care applications.

13.4.3.2 Plasmon coupling

In the previous section, we briefly discussed plasmon coupling, which used nanoparticle aggregation for sensor application. Plasmon coupling itself belongs to an important class of sensors. The idea behind plasmon coupling is that a coupled plasmon is generated when two plasmonic entities are brought close to each other. The newly generated plasmon coupled reaction can be detected by the spectroscopic signature. There are two possible types of plasmon coupling: nanoparticle-nanoparticle (NP-NP) and nanoparticle-thin film (NP-film).

Plasmon coupling between particles produces very narrow plasmon lines [103] and exceptionally large electromagnetic enhancements ($|E|^2 > 10^7$) [86]. Plasmon coupling of NP-NP has been utilized as a plasmonic molecular ruler (PMR) [104]. The concept of PMR is analogous to Förster resonance energy transfer (FRET) [105]. PMR can be used to measure the size of biomolecules in a label-free manner [20]. However, PMR has an advantage over FRET such as the measurement range is longer at 70 nm than at ~10 nm [104], and there are no problems with photobleaching and blinking. In addition to NP-NP conjugation, an NP fluorophore-based sensing mechanism based on the principle of nano-surface energy transfer (NSET) [106–108] has also been reported. The interplay between the localized plasmon resonance of NPs and the SPPs of the gold film can be observed in the spectroscopic study of NP-film configurations [109, 110]. In this case, the spectroscopic information can be viewed from two different perspectives. From the perspective of the nanoparticle, the spectroscopic information obtained under dark-field illumination provides information about the plasmonic coupling between the film and the NP. From the perspective of the film in the Kretschmann configuration, the same information can be obtained, but in this case, the sensitivity of the refractive index can be improved by coupling the NP to the film surface, which is known as "enhanced SPR" [102].

13.4.3.3 Nanohole array

In contrast to NPs arranged in an ensemble, repeating holes in a metal film with dimensions in the nanometer range are called nanohole arrays (NHA). The repetitions of holes in the film can be periodic [111] or quasi-periodic [112]. The resulting plasmonic response of NHA is a combination of different optical mechanisms, including Bloch wave SPP (BW-SPP), LSPR, and Fano resonance [113]. From a sensing perspective, the positions of the plasmonic peaks are determined with a change in the local refractive index, similar to NPs. The plasmonic peak(s), as revealed by spectroscopic data, was first described by Ebbesen et al. based on the phenomenon of "extraordinary optical transmission (EOT)" [114]. The EOT response of the NHA depends on many physical parameters such as the thickness of the film, hole size, periodicity, and lattice constant. The techniques described in Sections 13.4.1 and 13.2.1.2 are useful to define the quality of the sensor and to evaluate the reliability and sensor control. The main advantage of NHA over other structures is its large surface area and a higher FOM [113].

Despite the tremendous progress in colloid lithography, top-down techniques such as FIB and EBL have been used to fabricate

NHA. These techniques have some limitations, such as a time-consuming and expensive process, small nanostructured areas, and low throughput. As an alternative, the use of a self-assembling colloidal mask for NHA fabrication has emerged as a potential bottom-up fabrication strategy (Section 13.3.2.1). The bottom-up approach has major advantages over the top-down approach because it is inexpensive, fast, and can be used to fabricate NHA with areas on the order of square centimeters. These advantages make the bottom-up strategy ideal for the mass production of NHA. An exclusively chemical approach to NHA fabrication was proposed by Quint and Pacholski [115]. In this study, self-assembled, and loosely packed hydrogel microspheres were used as a mask for the fabrication of NHA. The use of close-packed polystyrene microsphere arrays has also been reported [116] but requires size reduction of the polystyrene spheres by reactive ion etching. However, the use of "soft" hydrogel has advantages over "hard" polystyrene in the fabrication of NHA because it is loosely packed and has self-healing properties [117]. In addition, the incorporation of additional plasmonic or reactive materials into the holes is easy and results in hybrid plasmonic structures with improved sensor performance [118, 119].

The use of NHA in sensing has generally emerged as chip-based sensing for point-of-care applications using NHA on a flat substrate [120–122]. The use of optical fibers as a substrate for sensing applications provides an additional opportunity for miniaturization with potential opportunities for in vivo sensing [88, 112, 123]. Regardless of the choice of substrate, NHA-based sensing can be broadly divided into two categories: flow over and flow through. The majority of applications using NHA as sensing elements operate in flow-over mode. The flow-through mode, on the other hand, offers the added advantages of fast response time and improved sensitivity [124, 125].

13.4.4 Surface-Enhanced Sensors

13.4.4.1 SERS

Raman spectroscopy has attracted considerable attention, due to its ability to identify structural fingerprints by which individual molecules can be recognized. The application of Raman spectroscopy spans a variety of fields, including sensing, medical diagnostics [126], chemistry [127], and biology [128]. The way Raman spectroscopy works is based on the inelastic scattering of photons by molecules. When light from a source (usually a monochromatic light source such as a laser) is scattered from molecules, the scattered light contains information about the vibrational state of the molecule. Unfortunately, due to the weak signal (< 1 photon per million is scattered inelastically) [96, 129], it is often difficult to perform Raman measurements in real applications. This is where surface enhancement comes into play. According to our previous understanding of plasmonics, light induced in a plasmonic structure can produce coherent oscillations of conduction electrons, increasing the density of the electromagnetic field near the plasmonic surface. As a result, the Raman signal from the surface-bound molecule is significantly enhanced. In addition, charge transfer can further enhance the Raman signal. Literature reports a local sensitivity gain that is several orders of magnitude higher than the conventional strategy [96]. Because of this significant amplification of the Raman signal, it is possible to record SERS in very dilute samples, and even spectra recorded from a single molecule are also possible [130]. Currently, SERS has established itself as a vibrational spectroscopic technique that has applications in various disciplines of chemistry, materials, and especially life sciences [131–134].

13.4.4.2 Surface-enhanced infrared absorption (SEIRA)

Infrared absorption spectroscopy (IRA) measures the amount of IR light absorbed by the molecular vibrational modes of a molecule. Raman spectroscopy measures the scattering of light as opposed to the amount of light absorbed; the information obtained with IRA spectroscopy complements the Raman measurement. Like Raman spectroscopy, IRA has played an increasingly important role in material identification [96, 135–137]. However, a major limitation of IRA is its sensitivity. Due to the small molecular absorption cross-section in the IR region, it is not as effective for detecting molecules at low concentrations [135]. This is where surface enhancement by the collective oscillations of electrons in the infrared region can contribute to its applicability. With SEIRA, signal amplification of up to 10^7 is possible compared to conventional IRA [138].

13.4.4.3 Surface-enhanced fluorescence (SEF)

Fluorescence is known to be one of the most suitable optical methods for the detection of biological and chemical substances [139]. However, the signal-to-noise ratio (SNR) of conventional fluorescence techniques is low. In addition, photostability and detectability of fluorescence are common problems of conventional systems. High sensitivity, better detection limit, and shorter analysis time can be achieved with the SEF strategy [140]. Fluorophores located in close proximity to a plasmonic nanostructure surface can interact with the surface plasmons, which, in turn, enhances the absorption and emission cross-sections of the fluorophores. As a result, the emitted fluorescence is dramatically enhanced, and the field enhancement factor can reach more than 10^3 orders of magnitude [141]. Due to the involvement of the plasmonic effect of metallic nanostructures in fluorescence enhancement, it is also referred to as plasmon-enhanced fluorescence (PEF) or metal-enhanced fluorescence (MEF) [142, 143].

13.5 Outlook

Bottom-up manufacturing has gained significant attention in nanofabrication, due to its extreme cost-efficiency. The need for instrumentation is minimal compared to top-down methods. Therefore, it is easy to start nanofabrication using the bottom-up strategy. Because of the lower initial investment, the cost of the fabricated sensors also remains low. Over time, and with a better understanding of the critical factors at the nanoscale, it is now possible to fabricate nanostructures with physical dimensions in the millimeter range while maintaining harmony at the nanoscale. The usefulness of the fabricated sensors has changed our perception of the applicability of plasmonic sensing in everyday life. However, the adaptation of plasmonics to commercial sensing applications is quite slow. We agree with the experts who recommend patience and a long-term perspective in this regard.

Despite the unbeatable track record of bottom-up fabrication, there are still some puzzles to be solved. These include reproducibility, surface preparation and conditioning for controlled deposition of atoms, control of impurities and site uniformity, reactant quality, etc. As the top-down approach reaches its theoretical and economic limits, adapting the bottom-up approach in commercial nanofabrication is the possible way forward. We often end up having to justify which strategy is better than the other. However, we believe that it might be beneficial to combine both strategies to harvest the advantages of bottom-up and top-down methods at the same time.

References

1. https://www.coherentmarketinsights.com, C.M.I. Surface Plasmon Resonance Market to Surpass US$ 1,244.7 Million by 2027.
2. Sensors Market Size, Share & Growth Analysis Report.
3. (2015) Commercializing plasmonics. *Nature Photon*, **9** (8), 477–477.
4. Kumar, S. and Singh, R. (2021) Recent optical sensing technologies for the detection of various biomolecules: Review. *Optics & Laser Technology*, **134**, 106620.
5. Zijlstra, P., Paulo, P.M.R., and Orrit, M. (2012) Optical detection of single non-absorbing molecules using the surface plasmon of a gold nanorod. *Nature Nanotech*, **7** (6), 379–382.
6. Shrivastav, A.M., Cvelbar, U., and Abdulhalim, I. (2021) A comprehensive review on plasmonic-based biosensors used in viral diagnostics. *Commun Biol*, **4** (1), 70.
7. Liu, M., Yin, X., Ulin-Avila, E., Geng, B., Zentgraf, T., Ju, L., Wang, F., and Zhang, X. (2011) A graphene-based broadband optical modulator. *Nature*, **474** (7349), 64–67.
8. Wang, R., Ren, X.-G., Yan, Z., Jiang, L.-J., Sha, W.E.I., and Shan, G.-C. (2019) Graphene based functional devices: A short review. *Front. Phys.*, **14** (1), 13603.
9. Ogawa, S., Fukushima, S., and Shimatani, M. (2020) Graphene plasmonics in sensor applications: A review. *Sensors*, **20** (12), 3563.
10. Pinto, V.V., Ferreira, M.J., Silva, R., Santos, H.A., Silva, F., and Pereira, C.M. (2010) Long time effect on the stability of silver nanoparticles in aqueous medium: Effect of the synthesis and storage conditions. *Colloids and Surfaces A: Physicochemical and Engineering Aspects*, **364** (1), 19–25.
11. Rebe Raz, S., Leontaridou, M., Bremer, M.G.E.G., Peters, R., and Weigel, S. (2012) Development of surface plasmon resonance-based sensor for detection of silver nanoparticles in food and the environment. *Anal Bioanal Chem*, **403** (10), 2843–2850.

12. Borah, R. and Verbruggen, S.W. (2020) Silver–gold bimetallic alloy versus core–shell nanoparticles: Implications for plasmonic enhancement and photothermal applications. *J. Phys. Chem. C*, **124** (22), 12081–12094.
13. Gao, C., Hu, Y., Wang, M., Chi, M., and Yin, Y. (2014) Fully alloyed Ag/Au nanospheres: Combining the plasmonic property of Ag with the stability of Au. *J. Am. Chem. Soc.*, **136** (20), 7474–7479.
14. Yin, Z., Wang, Y., Song, C., Zheng, L., Ma, N., Liu, X., Li, S., Lin, L., Li, M., Xu, Y., Li, W., Hu, G., Fang, Z., and Ma, D. (2018) Hybrid Au–Ag nanostructures for enhanced plasmon-driven catalytic selective hydrogenation through visible light irradiation and surface-enhanced Raman scattering. *J. Am. Chem. Soc.*, **140** (3), 864–867.
15. Chen, Z., Zhan, P., Dong, W., Li, Y., Tang, C., Min, N., and Wang, Z. (2010) Bottom-up fabrication approaches to novel plasmonic materials. *Chin. Sci. Bull.*, **55** (24), 2600–2607.
16. Iqbal, P., Preece, J.A., and Mendes, P.M. (2012) Nanotechnology: The "top-down" and "bottom-up" approaches, in *Supramolecular Chemistry*, Gale, P.A. and Steed, J.W. (eds.), John Wiley & Sons, Ltd, Chichester, UK, smc195.
17. Gregy, CC BY-SA 3.0 PL <https://creativecommons.org/licenses/by-sa/3.0/pl/deed.en>, via Wikimedia Commons.
18. By Pratheepps Own work, CC BY-SA 2.5, https://commons.wikimedia.org/w/index.php?curid=794937.
19. Cheng, J.Y., Ross, C.A., Smith, H.I., and Thomas, E.L. (2006) Templated Self-assembly of block copolymers: Top-down helps bottom-up. *Adv. Mater.*, **18** (19), 2505–2521.
20. Li, M., Cushing, S.K., and Wu, N. (2015) Plasmon-enhanced optical sensors: A review. *Analyst*, **140** (2), 386–406.
21. Vinogradov, A.P., Dorofeenko, A.V., Pukhov, A.A., and Lisyansky, A.A. (2018) Exciting surface plasmon polaritons in the Kretschmann configuration by a light beam. *Phys. Rev. B*, **97** (23), 235407.
22. Geddes, C.D. (ed.) (2012) *Reviews in Plasmonics 2010*, Springer New York.
23. Mourdikoudis, S., Pallares, R.M., and Thanh, N.T.K. (2018) Characterization techniques for nanoparticles: comparison and complementarity upon studying nanoparticle properties. *Nanoscale*, **10** (27), 12871–12934.
24. Newhouse, R.J. and Zhang, J.Z. (2012) Optical properties and applications of shape-controlled metal nanostructures, in *Reviews in Plasmonics 2010*, Geddes, C.D. (eds.), Springer, New York, 205–238.

25. Fleischer, M. (2012) Near-field scanning optical microscopy nanoprobes. *Nanotechnology Reviews*, **1** (4), 313–338.
26. Haiss, W., Thanh, N.T.K., Aveyard, J., and Fernig, D.G. (2007) Determination of size and concentration of gold nanoparticles from UV–Vis spectra. *Anal. Chem.*, **79** (11), 4215–4221.
27. Paramelle, D., Sadovoy, A., Gorelik, S., Free, P., Hobley, J., and Fernig, D.G. (2014) A rapid method to estimate the concentration of citrate capped silver nanoparticles from UV-visible light spectra. *Analyst*, **139** (19), 4855.
28. Zhao, P., Li, N., and Astruc, D. (2013) State of the art in gold nanoparticle synthesis. *Coordination Chemistry Reviews*, **257** (3–4), 638–665.
29. Stanglmair, C., Scheeler, S.P., and Pacholski, C. (2014) Seeding growth approach to gold nanoparticles with diameters ranging from 10 to 80 nanometers in organic solvent: Seeding growth approach to gold nanoparticles. *European Journal of Inorganic Chemistry*, **2014** (23), 3633–3637.
30. Kimling, J., Maier, M., Okenve, B., Kotaidis, V., Ballot, H., and Plech, A. (2006) Turkevich method for gold nanoparticle synthesis revisited. *J. Phys. Chem. B*, **110** (32), 15700–15707.
31. Slot, J.W. and Geuze, H.J. (1985) A new method of preparing gold probes for multiple-labeling cytochemistry. *Eur J Cell Biol*, **38** (1), 87–93.
32. Brown, K.R., Fox, A.P., and Natan, M.J. (1996) Morphology-dependent electrochemistry of cytochrome c at Au colloid-modified SnO2 electrodes. *J. Am. Chem. Soc.*, **118** (5), 1154–1157.
33. Daruich De Souza, C., Ribeiro Nogueira, B., and Rostelato, M.E.C.M. (2019) Review of the methodologies used in the synthesis gold nanoparticles by chemical reduction. *Journal of Alloys and Compounds*, **798**, 714–740.
34. Nicol, J.R., Dixon, D., and Coulter, J.A. (2015) Gold nanoparticle surface functionalization: A necessary requirement in the development of novel nanotherapeutics. *Nanomedicine*, **10** (8), 1315–1326.
35. Piaopiao, W. and Zihui, M. (2019) Progress in polystyrene microspheres. *IOP Conf. Ser.: Mater. Sci. Eng.*, **563** (2), 022001.
36. Murshid, N., Cathcart, N., and Kitaev, V. (2019) Room-temperature synthesis of size-uniform polystyrene latex and characterization of its properties: Third-year undergraduate teaching lab. *J. Chem. Educ.*, **96** (7), 1479–1485.
37. Stöber, W., Fink, A., and Bohn, E. (1968) Controlled growth of monodisperse silica spheres in the micron size range. *Journal of Colloid and Interface Science*, **26** (1), 62–69.

38. Ghimire, P.P. and Jaroniec, M. (2021) Renaissance of Stöber method for synthesis of colloidal particles: New developments and opportunities. *Journal of Colloid and Interface Science*, **584**, 838–865.
39. Pelton, R.H. and Chibante, P. (1986) Preparation of aqueous latices with N-isopropylacrylamide. *Colloids and Surfaces*, **20** (3), 247–256.
40. Eckert, T. and Richtering, W. (2008) Thermodynamic and hydrodynamic interaction in concentrated microgel suspensions: Hard or soft sphere behavior? *The Journal of Chemical Physics*, **129** (12), 124902.
41. Deschaume, O., De Roo, B., Van Bael, M.J., Locquet, J.-P., Van Haesendonck, C., and Bartic, C. (2014) Synthesis and properties of gold nanoparticle arrays self-organized on surface-deposited lysozyme amyloid scaffolds. *Chem. Mater.*, **26** (18), 5383–5393.
42. Rothemund, P.W.K. (2006) Folding DNA to create nanoscale shapes and patterns. *Nature*, **440** (7082), 297–302.
43. Hall, S.R., Shenton, W., Engelhardt, H., and Mann, S. (2001) Site-specific organization of gold nanoparticles by biomolecular templating. *ChemPhysChem*, **2** (3), 184–186.
44. Qin, D., Xia, Y., Xu, B., Yang, H., Zhu, C., and Whitesides, G.M. (1999) Fabrication of ordered two-dimensional arrays of micro- and nanoparticles using patterned self-assembled monolayers as templates. *Adv. Mater.*, **11** (17), 1433–1437.
45. Bowden, N., Brittain, S., Evans, A.G., Hutchinson, J.W., and Whitesides, G.M. (1998) Spontaneous formation of ordered structures in thin films of metals supported on an elastomeric polymer. *Nature*, **393** (6681), 146–149.
46. Huck, W.T.S., Bowden, N., Onck, P., Pardoen, T., Hutchinson, J.W., and Whitesides, G.M. (2000) Ordering of spontaneously formed buckles on planar surfaces. *Langmuir*, **16** (7), 3497–3501.
47. Schweikart, A., Pazos-Pérez, N., Alvarez-Puebla, R.A., and Fery, A. (2011) Controlling inter-nanoparticle coupling by wrinkle-assisted assembly. *Soft Matter*, **7** (9), 4093.
48. Guo, X., Liu, L., Zhuang, Z., Chen, X., Ni, M., Li, Y., Cui, Y., Zhan, P., Yuan, C., Ge, H., Wang, Z., and Chen, Y. (2015) A new strategy of lithography based on phase separation of polymer blends. *Sci Rep*, **5** (1), 15947.
49. Spatz, J.P., Sheiko, S., and Möller, M. (1996) Substrate-induced lateral micro-phase separation of a diblock copolymer. *Advanced Materials*, **8** (6), 513–517.
50. Kim, G. and Libera, M. (1998) Morphological development in solvent-cast polystyrene–polybutadiene–polystyrene (SBS) triblock copolymer thin films. *Macromolecules*, **31** (8), 2569–2577.

51. Vogel, N., Retsch, M., Fustin, C.-A., del Campo, A., and Jonas, U. (2015) Advances in colloidal assembly: The design of structure and hierarchy in two and three dimensions. *Chemical Reviews*, **115** (13), 6265–6311.
52. van Dommelen, R., Fanzio, P., and Sasso, L. (2018) Surface self-assembly of colloidal crystals for micro- and nano-patterning. *Advances in Colloid and Interface Science*, **251**, 97–114.
53. Maestro, A. (2019) Tailoring the interfacial assembly of colloidal particles by engineering the mechanical properties of the interface. *Current Opinion in Colloid & Interface Science*, **39**, 232–250.
54. Schulz, F., Pavelka, O., Lehmkühler, F., Westermeier, F., Okamura, Y., Mueller, N.S., Reich, S., and Lange, H. (2020) Structural order in plasmonic superlattices. *Nat Commun*, **11** (1), 3821.
55. Yang, G. and Hallinan, D.T. (2016) Gold nanoparticle monolayers from sequential interfacial ligand exchange and migration in a three-phase system. *Sci Rep*, **6** (1), 35339.
56. Ullrich, S., Scheeler, S.P., Pacholski, C., Spatz, J.P., and Kudera, S. (2013) Formation of large 2D arrays of shape-controlled colloidal nanoparticles at variable interparticle distances. *Particle & Particle Systems Characterization*, **30** (1), 102–108.
57. Scheeler, S.P., Mühlig, S., Rockstuhl, C., Hasan, S.B., Ullrich, S., Neubrech, F., Kudera, S., and Pacholski, C. (2013) Plasmon coupling in self-assembled gold nanoparticle-based honeycomb islands. *The Journal of Physical Chemistry C*, **117** (36), 18634–18641.
58. Stanglmair, C., Neubrech, F., and Pacholski, C. (2018) Chemical routes to surface enhanced infrared absorption (SEIRA) substrates. *Zeitschrift für Physikalische Chemie*, **232** (9–11), 1527–1539.
59. Lerond, T., Proust, J., Yockell-Lelièvre, H., Gérard, D., and Plain, J. (2011) Self-assembly of metallic nanoparticles into plasmonic rings. *Appl. Phys. Lett.*, **99** (12), 123110.
60. Wang, Z., Ai, B., Möhwald, H., and Zhang, G. (2018) Colloidal lithography meets plasmonic nanochemistry. *Advanced Optical Materials*, **6** (18), 1800402.
61. Ai, B., Yu, Y., Möhwald, H., Zhang, G., and Yang, B. (2014) Plasmonic films based on colloidal lithography. *Advances in Colloid and Interface Science*, **206**, 5–16.
62. Vogel, N., Goerres, S., Landfester, K., and Weiss, C.K. (2011) A convenient method to produce close- and non-close-packed monolayers using direct assembly at the air-water interface and subsequent plasma-induced size reduction. *Macromolecular Chemistry and Physics*, **212** (16), 1719–1734.

63. Rey, M., Hou, X., Tang, J.S.J., and Vogel, N. (2017) Interfacial arrangement and phase transitions of PNiPAm microgels with different crosslinking densities. *Soft Matter*, **13** (46), 8717–8727.

64. Heck, C., Kanehira, Y., Kneipp, J., and Bald, I. (2019) Amorphous carbon generation as a photocatalytic reaction on DNA-assembled gold and silver nanostructures. *Molecules*, **24** (12), 2324.

65. Heck, C., Kanehira, Y., Kneipp, J., and Bald, I. (2018) Placement of single proteins within the SERS hot spots of self-assembled silver nanolenses. *Angew. Chem. Int. Ed.*, **57** (25), 7444–7447.

66. Weinert, U., Pollmann, K., Barkleit, A., Vogel, M., Günther, T., and Raff, J. (2015) Synthesis of S-layer conjugates and evaluation of their modifiability as a tool for the functionalization and patterning of technical surfaces. *Molecules*, **20** (6), 9847–9861.

67. Tang, J., Badelt-Lichtblau, H., Ebner, A., Preiner, J., Kraxberger, B., Gruber, H.J., Sleytr, U.B., Ilk, N., and Hinterdorfer, P. (2008) Fabrication of highly ordered gold nanoparticle arrays templated by crystalline lattices of bacterial S-layer protein. *ChemPhysChem*, **9** (16), 2317–2320.

68. Wang, Y., Becker, M., Wang, L., Liu, J., Scholz, R., Peng, J., Gösele, U., Christiansen, S., Kim, D.H., and Steinhart, M. (2009) Nanostructured gold films for SERS by block copolymer-templated galvanic displacement reactions. *Nano Lett.*, **9** (6), 2384–2389.

69. Aizawa, M. and Buriak, J.M. (2006) Nanoscale patterning of two metals on silicon surfaces using an ABC triblock copolymer template. *J. Am. Chem. Soc.*, **128** (17), 5877–5886.

70. Zhu, H., Masson, J.-F., and Bazuin, C.G. (2020) Templating gold nanoparticles on nanofibers coated with a block copolymer brush for nanosensor applications. *ACS Appl. Nano Mater.*, **3** (1), 516–529.

71. Yan, N., Liu, X., Zhu, J., Zhu, Y., and Jiang, W. (2019) Well-ordered inorganic nanoparticle arrays directed by block copolymer nanosheets. *ACS Nano*, **13** (6), 6638–6646.

72. Bigall, N.C., Nandan, B., Gowd, E.B., Horechyy, A., and Eychmüller, A. (2015) High-resolution metal nanopatterning by means of switchable block copolymer templates. *ACS Appl. Mater. Interfaces*, **7** (23), 12559–12569.

73. Spatz, J.P., Herzog, T., Mößmer, S., Ziemann, P., and Möller, M. (1999) Micellar inorganic-polymer hybrid systems—A tool for nanolithography. *Advanced Materials*, **11** (2), 149–153.

74. Glass, R., M ller, M., and Spatz, J.P. (2003) Block copolymer micelle nanolithography. *Nanotechnology*, **14** (10), 1153–1160.

75. Möller, M., Spatz, J.P., and Roescher, A. (1996) Gold nanoparticles in micellar poly(styrene)-b-poly(ethylene oxide) films—Size and interparticle distance control in monoparticulate films. *Advanced Materials*, **8** (4), 337–340.
76. Yu, Y., Ng, C., König, T.A.F., and Fery, A. (2019) Tackling the scalability challenge in plasmonics by wrinkle-assisted colloidal self-assembly. *Langmuir*, **35** (26), 8629–8645.
77. Hanske, C., Tebbe, M., Kuttner, C., Bieber, V., Tsukruk, V.V., Chanana, M., König, T.A.F., and Fery, A. (2014) Strongly coupled plasmonic modes on macroscopic areas via template-assisted colloidal self-assembly. *Nano Lett.*, **14** (12), 6863–6871.
78. Vernon, K.C., Funston, A.M., Novo, C., Gómez, D.E., Mulvaney, P., and Davis, T.J. (2010) Influence of particle–substrate interaction on localized plasmon resonances. *Nano Lett.*, **10** (6), 2080–2086.
79. Mahmoud, M.A. and El-Sayed, M.A. (2013) Substrate effect on the plasmonic sensing ability of hollow nanoparticles of different shapes. *J. Phys. Chem. B*, **117** (16), 4468–4477.
80. Balderas-Valadez, R.F., Schürmann, R., and Pacholski, C. (2019) One spot—two sensors: Porous silicon interferometers in combination with gold nanostructures showing localized surface plasmon resonance. *Frontiers in Chemistry*, **7**, 593.
81. Mizuno, A. and Ono, A. (2021) Dynamic control of the interparticle distance in a self-assembled Ag nanocube monolayer for plasmonic color modulation. *ACS Appl. Nano Mater.*, **4** (9), 9721–9728.
82. Shir, D., Ballard, Z.S., and Ozcan, A. (2016) Flexible plasmonic sensors. *IEEE J. Select. Topics Quantum Electron.*, **22** (4), 12–20.
83. Choe, A., Yeom, J., Shanker, R., Kim, M.P., Kang, S., and Ko, H. (2018) Stretchable and wearable colorimetric patches based on thermoresponsive plasmonic microgels embedded in a hydrogel film. *NPG Asia Mater*, **10** (9), 912–922.
84. Lee, J.-E., Park, C., Chung, K., Lim, J.W., Marques Mota, F., Jeong, U., and Kim, D.H. (2018) Viable stretchable plasmonics based on unidirectional nanoprisms. *Nanoscale*, **10** (8), 4105–4112.
85. https://imagej.nih.gov/ij/.
86. Chu, Y., Schonbrun, E., Yang, T., and Crozier, K.B. (2008) Experimental observation of narrow surface plasmon resonances in gold nanoparticle arrays. *Appl. Phys. Lett.*, **93** (18), 181108.
87. Kasani, S., Curtin, K., and Wu, N. (2019) A review of 2D and 3D plasmonic nanostructure array patterns: Fabrication, light management and sensing applications. *Nanophotonics*, **8** (12), 2065–2089.

88. Polley, N., Basak, S., Hass, R., and Pacholski, C. (2019) Fiber optic plasmonic sensors: Providing sensitive biosensor platforms with minimal lab equipment. *Biosensors and Bioelectronics*, **132**, 368–374.

89. Zámbó, D., Suzuno, K., Pothorszky, S., Bárdfalvy, D., Holló, G., Nakanishi, H., Wang, D., Ueyama, D., Deák, A., and Lagzi, I. (2016) Self-assembly of like-charged nanoparticles into Voronoi diagrams. *Phys. Chem. Chem. Phys.*, **18** (36), 25735–25740.

90. Kaatz, F.H., Bultheel, A., and Egami, T. (2009) Real and reciprocal space order parameters for porous arrays from image analysis. *Journal of Materials Science*, **44** (1), 40–46.

91. Kedem, O., Vaskevich, A., and Rubinstein, I. (2014) Critical issues in localized plasmon sensing. *J. Phys. Chem. C*, **118** (16), 8227–8244.

92. Bendikov, T.A., Rabinkov, A., Karakouz, T., Vaskevich, A., and Rubinstein, I. (2008) Biological sensing and interface design in gold island film based localized plasmon transducers. *Anal. Chem.*, **80** (19), 7487–7498.

93. Tesler, A.B., Chuntonov, L., Karakouz, T., Bendikov, T.A., Haran, G., Vaskevich, A., and Rubinstein, I. (2011) Tunable localized plasmon transducers prepared by thermal dewetting of percolated evaporated gold films. *J. Phys. Chem. C*, **115** (50), 24642–24652.

94. Spackova, B., Wrobel, P., Bockova, M., and Homola, J. (2016) Optical biosensors based on plasmonic nanostructures: A review. *Proc. IEEE*, **104** (12), 2380–2408.

95. Gupta, B.D., Shrivastav, A.M., and Usha, S.P. (2017) *Optical Sensors for Biomedical Diagnostics and Environmental Monitoring*, CRC Press, Boca Raton, FL: CRC Press, Taylor & Francis Group.

96. Kołątaj, K., Krajczewski, J., and Kudelski, A. (2020) Plasmonic nanoparticles for environmental analysis. *Environ Chem Lett*, **18** (3), 529–542.

97. Behrouzi, K. and Lin, L. (2022) Gold nanoparticle based plasmonic sensing for the detection of SARS-CoV-2 nucleocapsid proteins. *Biosensors and Bioelectronics*, **195**, 113669.

98. Martinsson, E., Otte, M.A., Shahjamali, M.M., Sepulveda, B., and Aili, D. (2014) Substrate effect on the refractive index sensitivity of silver nanoparticles. *J. Phys. Chem. C*, **118** (42), 24680–24687.

99. Mock, J.J., Smith, D.R., and Schultz, S. (2003) Local refractive index dependence of plasmon resonance spectra from individual nanoparticles. *Nano Lett.*, **3** (4), 485–491.

100. Nusz, G.J., Marinakos, S.M., Curry, A.C., Dahlin, A., Höök, F., Wax, A., and Chilkoti, A. (2008) Label-free plasmonic detection of biomolecular binding by a single gold nanorod. *Anal. Chem.*, **80** (4), 984–989.

101. Truong, P.L., Cao, C., Park, S., Kim, M., and Sim, S.J. (2011) A new method for non-labeling attomolar detection of diseases based on an individual gold nanorod immunosensor. *Lab Chip*, **11** (15), 2591.

102. Hill, R.T. (2015) Plasmonic biosensors. *WIREs Nanomed Nanobiotechnol*, **7** (2), 152–168.

103. Kinnan, M.K. and Chumanov, G. (2010) Plasmon coupling in two-dimensional arrays of silver nanoparticles: II. Effect of the particle size and interparticle distance. *J. Phys. Chem. C*, **114** (16), 7496–7501.

104. Sönnichsen, C., Reinhard, B.M., Liphardt, J., and Alivisatos, A.P. (2005) A molecular ruler based on plasmon coupling of single gold and silver nanoparticles. *Nat Biotechnol*, **23** (6), 741–745.

105. Polley, N., Singh, S., Giri, A., Mondal, P.K., Lemmens, P., and Pal, S.K. (2015) Ultrafast FRET at fiber tips: Potential applications in sensitive remote sensing of molecular interaction. *Sensors and Actuators B: Chemical*, **210**, 381–388.

106. Chen, C., Midelet, C., Bhuckory, S., Hildebrandt, N., and Werts, M.H.V. (2018) Nanosurface energy transfer from long-lifetime terbium donors to gold nanoparticles. *J. Phys. Chem. C*, **122** (30), 17566–17574.

107. Sarkar, P.K., Polley, N., Chakrabarti, S., Lemmens, P., and Pal, S.K. (2016) Nanosurface energy transfer based highly selective and ultrasensitive "turn on" fluorescence mercury sensor. *ACS Sens.*, **1** (6), 789–797.

108. Polley, N., Sarkar, P.K., Chakrabarti, S., Lemmens, P., and Pal, S.K. (2016) DNA biomaterial based fiber optic sensor: Characterization and application for monitoring *in situ* mercury pollution. *ChemistrySelect*, **1** (11), 2916–2922.

109. Mock, J.J., Hill, R.T., Degiron, A., Zauscher, S., Chilkoti, A., and Smith, D.R. (2008) Distance-dependent plasmon resonant coupling between a gold nanoparticle and gold film. *Nano Lett.*, **8** (8), 2245–2252.

110. Hill, R.T., Mock, J.J., Hucknall, A., Wolter, S.D., Jokerst, N.M., Smith, D.R., and Chilkoti, A. (2012) Plasmon ruler with angstrom length resolution. *ACS Nano*, **6** (10), 9237–9246.

111. Monteiro, J.P., Carneiro, L.B., Rahman, M.M., Brolo, A.G., Santos, M.J.L., Ferreira, J., and Girotto, E.M. (2013) Effect of periodicity on the performance of surface plasmon resonance sensors based on subwavelength nanohole arrays. *Sensors and Actuators B: Chemical*, **178**, 366–370.

112. Jia, P., Yang, Z., Yang, J., and Ebendorff-Heidepriem, H. (2016) Quasiperiodic nanohole arrays on optical fibers as plasmonic sensors: Fabrication and sensitivity determination. *ACS Sens.*, **1** (8), 1078–1083.

113. Xu, Y., Bai, P., Zhou, X., Akimov, Y., Png, C.E., Ang, L., Knoll, W., and Wu, L. (2019) Optical refractive index sensors with plasmonic and photonic structures: Promising and inconvenient truth. *Advanced Optical Materials*, **7** (9), 1801433.

114. Ebbesen, T.W., Lezec, H.J., Ghaemi, H.F., Thio, T., and Wolff, P.A. (1998) Extraordinary optical transmission through sub-wavelength hole arrays. *Nature*, **391** (6668), 667–669.

115. Quint, S.B. and Pacholski, C. (2009) A chemical route to sub-wavelength hole arrays in metallic films. *J. Mater. Chem.*, **19** (33), 5906–5908.

116. Lee, S.H., Bantz, K.C., Lindquist, N.C., Oh, S.-H., and Haynes, C.L. (2009) Self-assembled plasmonic nanohole arrays. *Langmuir*, **25** (23), 13685–13693.

117. Quint, S.B. and Pacholski, C. (2011) Extraordinary long range order in self-healing non-close packed 2D arrays. *Soft Matter*, **7** (8), 3735–3738.

118. Weiler, M., Menzel, C., Pertsch, T., Alaee, R., Rockstuhl, C., and Pacholski, C. (2016) Bottom-up fabrication of hybrid plasmonic sensors: Gold-capped hydrogel microspheres embedded in periodic metal hole arrays. *ACS Applied Materials & Interfaces*, **8** (39), 26392–26399.

119. Weiler, M., Quint, S.B., Klenk, S., and Pacholski, C. (2014) Bottom-up fabrication of nanohole arrays loaded with gold nanoparticles: Extraordinary plasmonic sensors. *Chem. Commun.*, **50** (97), 15419–15422.

120. Najiminaini, M., Vasefi, F., Kaminska, B., and Carson, J.J.L. (2012) Nano-hole array structure with improved surface plasmon energy matching characteristics. *Appl. Phys. Lett.*, **100** (4), 043105.

121. Escobedo, C. (2013) On-chip nanohole array based sensing: A review. *Critical Review*, 19.

122. Prasad, A., Choi, J., Jia, Z., Park, S., and Gartia, M.R. (2019) Nanohole array plasmonic biosensors: Emerging point-of-care applications. *Biosensors and Bioelectronics*, **130**, 185–203.

123. Wang, Q. and Wang, L. (2020) Lab-on-fiber: plasmonic nano-arrays for sensing. *Nanoscale*, **12** (14), 7485–7499.

124. Eftekhari, F., Escobedo, C., Ferreira, J., Duan, X., Girotto, E.M., Brolo, A.G., Gordon, R., and Sinton, D. (2009) Nanoholes as nanochannels: Flow-through plasmonic sensing. *Anal. Chem.*, **81** (11), 4308–4311.

125. Yanik, A.A., Huang, M., Artar, A., Chang, T.-Y., and Altug, H. (2010) Integrated nanoplasmonic-nanofluidic biosensors with targeted delivery of analytes. *Appl. Phys. Lett.*, **96** (2), 021101.

126. Ozaki, Y. (1988) Medical application of Raman spectroscopy. *Applied Spectroscopy Reviews*, **24** (3–4), 259–312.

127. Das, R.S. and Agrawal, Y.K. (2011) Raman spectroscopy: Recent advancements, techniques and applications. *Vibrational Spectroscopy*, **57** (2), 163–176.

128. Stöckel, S., Kirchhoff, J., Neugebauer, U., Rösch, P., and Popp, J. (2016) The application of Raman spectroscopy for the detection and identification of microorganisms. *Journal of Raman Spectroscopy*, **47** (1), 89–109.

129. Jin, H.M., Kim, J.Y., Heo, M., Jeong, S.-J., Kim, B.H., Cha, S.K., Han, K.H., Kim, J.H., Yang, G.G., Shin, J., and Kim, S.O. (2018) Ultralarge area sub-10 nm plasmonic nanogap array by block copolymer self-assembly for reliable high-sensitivity SERS. *ACS Appl. Mater. Interfaces*, **10** (51), 44660–44667.

130. Almehmadi, L.M., Curley, S.M., Tokranova, N.A., Tenenbaum, S.A., and Lednev, I.K. (2019) Surface enhanced Raman spectroscopy for single molecule protein detection. *Sci Rep*, **9** (1), 12356.

131. Campion, A. and Kambhampati, P. (1998) Surface-enhanced Raman scattering. *Chem. Soc. Rev.*, **27** (4), 241–250.

132. Stewart, M.E., Anderton, C.R., Thompson, L.B., Maria, J., Gray, S.K., Rogers, J.A., and Nuzzo, R.G. (2008) Nanostructured plasmonic sensors. *Chem. Rev.*, **108** (2), 494–521.

133. Schlücker, S. (2014) Surface-enhanced Raman spectroscopy: Concepts and chemical applications. *Angewandte Chemie International Edition*, **53** (19), 4756–4795.

134. Sharma, B., Frontiera, R.R., Henry, A.-I., Ringe, E., and Van Duyne, R.P. (2012) SERS: Materials, applications, and the future. *Materials Today*, **15** (1), 16–25.

135. Zhou, H., Li, D., Hui, X., and Mu, X. (2021) Infrared metamaterial for surface-enhanced infrared absorption spectroscopy: Pushing the frontier of ultrasensitive on-chip sensing. *International Journal of Optomechatronics*, **15** (1), 97–119.

136. Yang, X., Sun, Z., Low, T., Hu, H., Guo, X., García de Abajo, F.J., Avouris, P., and Dai, Q. (2018) Nanomaterial-based plasmon-enhanced infrared spectroscopy. *Advanced Materials*, **30** (20), 1704896.

137. Beć, K.B., Grabska, J., and Huck, C.W. (2020) Biomolecular and bioanalytical applications of infrared spectroscopy – A review. *Analytica Chimica Acta*, **1133**, 150–177.

138. Dong, L., Yang, X., Zhang, C., Cerjan, B., Zhou, L., Tseng, M.L., Zhang, Y., Alabastri, A., Nordlander, P., and Halas, N.J. (2017) Nanogapped Au antennas for ultrasensitive surface-enhanced infrared absorption spectroscopy. *Nano Lett.*, **17** (9), 5768–5774.

139. Dong, J., Zhang, Z., Zheng, H., and Sun, M. (2015) Recent progress on plasmon-enhanced fluorescence. *Nanophotonics*, **4** (4), 472–490.

140. Bauch, M., Toma, K., Toma, M., Zhang, Q., and Dostalek, J. (2014) Plasmon-enhanced fluorescence biosensors: A review. *Plasmonics*, **9** (4), 781–799.

141. Li, J.-F., Li, C.-Y., and Aroca, R.F. (2017) Plasmon-enhanced fluorescence spectroscopy. *Chem. Soc. Rev.*, **46** (13), 3962–3979.

142. Jeong, Y., Kook, Y.-M., Lee, K., and Koh, W.-G. (2018) Metal enhanced fluorescence (MEF) for biosensors: General approaches and a review of recent developments. *Biosens. Bioelectron.*, **111**, 102–116.

143. Hang, Y., Boryczka, J., and Wu, N. (2022) Visible-light and near-infrared fluorescence and surface-enhanced Raman scattering point-of-care sensing and bio-imaging: A review. *Chem. Soc. Rev.*, **1**.

Section II
Plasmonics-Based Photodetectors

Chapter 14

Metallic Nanostructures/Gratings for Plasmon-Enhanced Photodetection

Ankit Kumar Pandey
Department of ECE, Bennett University, Greater Noida, India
ankit.pandey@bennett.edu.in

The light absorption enhancement in various plasmonic devices is greatly achieved by engineered metallic nanostructures. This enhancement plays a crucial role in the improvement of the responsivity of conventional photodetectors. Light harvesting in plasmonic nanostructures for photodetection is enabled by extraordinary optical transmission (EOT) and extraordinary optical absorption (EOA). In addition, the responsivity tuning is facilitated by these structures by tailoring the grating dimensions, selection of incident angle, and wavelength. This chapter provides insight into the application of grating nanostructures for the photodetection utilizing the plasmonic phenomenon.

Plasmonics-Based Optical Sensors and Detectors
Edited by Banshi D. Gupta, Anuj K. Sharma, and Jin Li

ISBN 978-981-4968-85-0 (Hardcover), 978-1-003-43830-4 (eBook)
www.jennystanford.com

14.1 Introduction

Surface plasmon resonance (SPR) is a phenomenon of free-electron oscillation in a coherent manner. In general, radiative and non-radiative decays are the two ways by the which decay of surface plasmons (SPs) occurs [1]. Radiative decays are observed as re-emitted photons and hot electrons are non-radiative decays of SPs. It is evident that the enhancement in hot electron generation can be achieved by utilizing the strong absorption characteristics of metallic nanostructures [2]. The extensive applications of grating-based metallic/dielectric nanostructures can be observed in photovoltaic devices [3], plasmonic sensors [4], polarizers [5], and plasmonic absorbers [6] for field enhancement. These grating nanostructures are referred to as plasmonic gratings, as these support propagation of plasmonic modes alongside their surface. The focused ion beam (FIB) milling method is usually preferred for nano-gratings fabrication than traditional photolithography processes in view of the precise grating structure formation [7]. Furthermore, the hot electrons are responsible for the thermal losses, which, in turn, affects the performance of plasmonic devices. However, a pre-thermalized collection of hot electrons can be utilized for application in photodetectors in specific nanostructures [8].

Metallic element in plasmonic photodetectors has two main roles to play: (i) act as an absorber in hot-carrier devices and (ii) provide field enhancement field inside an absorber. Figure 14.1 represents the classification of plasmonic photodetectors and the field enhancement approach.

Plasmonic enhancement in each of the detector types can be accomplished by three basic forms: (a) grating type plasmon, (b) localized plasmon polariton in a metallic nanoparticle, and (c) surface plasmonic polariton (SPP) in a waveguide. High responsivities and external quantum efficiencies are experienced by plasmonic photodetectors as compared with conventional photodetectors because of the hot electrons generation ability and enhanced light absorption [9]. For this reason, they are referred to as hot electron photodetectors. In these photodetectors, hot electrons possess energies greater than the barrier height so that they cross the barrier before thermalization [10]. Further, these hot electrons are injected into the conduction band of the semiconductor, which results in photocurrent. External circuitry is employed that can detect the photocurrent. There are three widely used hot electron photodetector configurations: (a) metal-insulator-metal (MIM) configuration, (b) metal-semiconductor-metal (MSM) configuration, and (c) metal-semiconductor (MS) configuration.

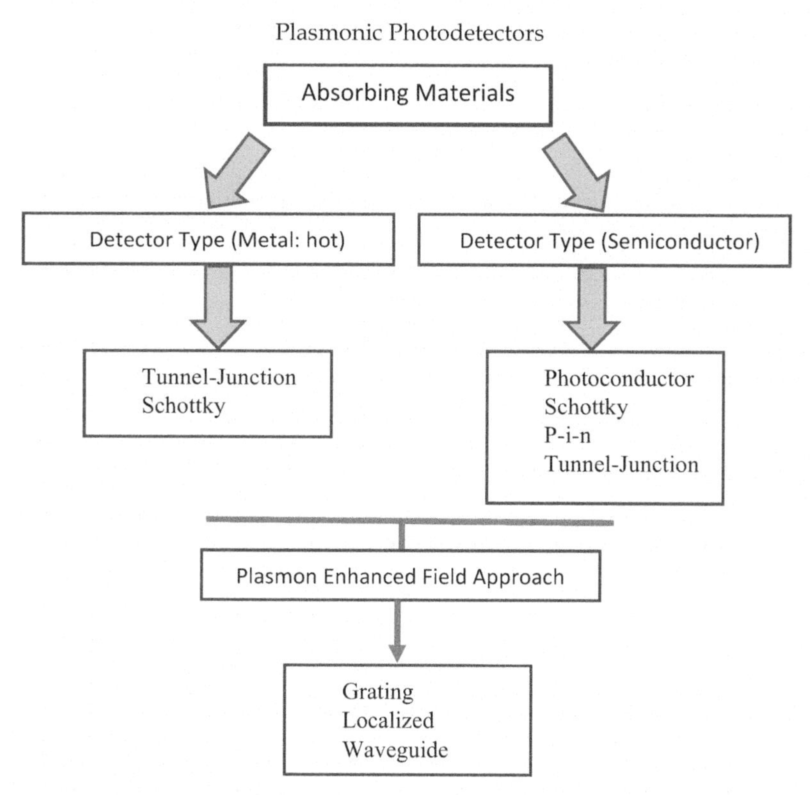

Figure 14.1 Plasmonic photodetector classifications (a schematic approach).

The biggest drawback associated with the MIM configuration is the less efficient hot electron generation and propagation across the high metal-insulator (M/I) barrier [11]. The photodetection in the ultraviolet (UV) and visible ranges is limited by the high M/I barrier. The M/I barrier lowers the probability barrier crossing by hot electrons excited by low-frequency photons [11, 12]. The MSM and MS configurations have resolved the high barrier problem. In hot electron photodetectors with MS configurations, such as Au gratings [13], antennas [11, 14], nanowires [15], absorbers [16], and waveguides [17], a unidirectional photocurrent due to hot electrons in a single metallic layer is generated. In MSM configurations, the semiconductor layer is usually placed between two opposite metallic layers (top/bottom or left/right). These hot electrons are generated in both metallic layers. Asymmetrical absorption is responsible for the generation of net photocurrent in the MSM configuration. Different metals or structures are used to create absorption in the two metallic layers. In this direction, a plasmonic grating structure in MSM photodetector (at wavelength 980 nm) for enhanced absorption was reported by Tan et al. [18].

In another work, Tan et al. [19] numerically demonstrated the applications of nano-gratings and metallic nano-particles in absorbance enhancement in plasmonic MSM photodetectors. Recently, grating-enhanced plasmonic MSM photodetector on Si substrate based on GeSn has been experimentally reported by Zhou et al. for the photodetection at 2 μm [20]. The Schottky barrier (SB) height between Au and GeSn was also demonstrated. In the last few decades, Si-based SB photodetectors have shown rapid growth pertaining to numerous advantages such as low capacitance, large bandwidth, low contact resistance, and affluence of fabrication for integration [9, 21]. In this context, the next section is focused on the grating nanostructure-based plasmon-enhanced Schottky photodetectors.

14.2 Plasmon-Enhanced SB Photodetector

SPs excitation relation (bulk material) for grating structure is given as [22]:

$$\frac{2\pi}{\lambda} \times n_s \sin\theta_{\text{inc}} + m\frac{2\pi}{\Lambda} = \pm Re\left(\frac{\omega}{c}\sqrt{\frac{\epsilon_m \epsilon_i}{\epsilon_m + \epsilon_i}}\right). \tag{14.1}$$

In the above expression, the incident light wavelength is λ, the angle of incidence is θ_{inc}, ω, and m are the angular frequency and diffraction order, respectively. c is the velocity of light in vacuum. ε_m and ε_i are the dielectric constant of the metal layer and incident medium, respectively. The photodetection process can be explained as: initially, non-radiative decay of plasmons, i.e., the hot electrons are generated by the incident photons on optical absorption. This absorption is greatly enhanced by plasmonic nanostructures. In continuation, generated hot electrons are diffused toward the Schottky interface. Further, these are transmitted to the conduction band of the semiconductor through internal photoemission.

In the context of grating-based Schottky photodetector, Mahdieh et al. [9] reported absorption enhancement in an SB photodetector structure by utilizing the Au gratings. However, the information regarding internal quantum efficiency and $\mathcal{R}$ values were not addressed. In this direction, a narrowband (54 meV) Au grating-based Schottky photodetector with zero bias responsivity of 0.6 mA/W was reported by Sohbani et al. [13]. Further, Chen et al., theoretically and experimentally, demonstrated Au gratings on *n*-type Si SB photodetector for the photodetection in the NIR region [23]. The grating-based structure and band diagram are shown in Fig. 14.2. The effect of incident angle on responsivity was also analyzed. An Au-Si dual-wavelength SB photodetector was reported by Yang et al. [24] with $\mathcal{R}$ values of 2.4 mA/W and 4.5 mA/W at wavelengths of 1.55 μm and 1.31 μm, respectively. The calculation of internal photoemission (IPE) was based on Fowler's model. In continuation, Yang et al. [25] demonstrated the silica nanocore array-based SB (Au/TiO_2) photodetector with a responsivity of 0.360 mA/W at a wavelength of 620 nm. Recently, a triple Schottky junction based-photodetector has been reported by Shao et al. [26]. A high absorption efficiency (~100%) leading toward a $\mathcal{R}$ value of ~2 mA/W has been numerically analyzed. Further, Vahdani et al. [27] reported a platinum silicide (PtSi)-Si junction SB photodetector or sensing of CO_2 gas. This photodetector at a wavelength of 4.3 μm provided nearly 84% incident light absorption, however, the responsivity description was missing. In another work [28], free carrier absorption and hot carrier photocurrent generation schemes were integrated in order to achieve a narrowband high responsivity response. Recently, Pandey and Sharma theoretically demonstrated a stacked tungsten grating-based SB photodetector with a peak responsivity of 6.16 mA/W (λ 559.38 nm, without bias) reported in the case of normal incidence [29].

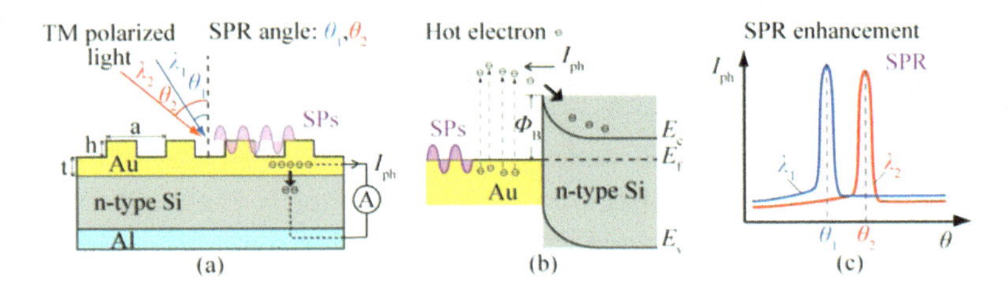

Figure 14.2 Schematic representation of the (a) structure, (b) band diagram, and (c) SPR enhancement of Schottky photodetector [23]. © 2016 Optica Publishing Group.

14.2.1 Performance Evaluation

Initiating the performance evaluation of the SB photodetectors, the structure absorbance needs to be calculated first. For grating-based SB photodetectors, the rigorous coupled wave analysis (RCWA) method is used in general for absorbance (A) computation [30]. It should be noted that the wavelength-dependent dielectric properties of materials must be used for precise calculation. Apart from RCWA, finite difference time domain (FDTD) and finite element methods can also be employed for absorbance computation. Considering the IPE process, the internal quantum efficiency (η) is obtained based on Fowler's model [24, 31, 32] with the inclusion of hot holes [33] as follows:

$$\eta = \frac{(h\nu - \Phi_b)^2}{8E_F h\nu}, \tag{14.2}$$

where E_F is the Fermi energy of metal. Φ_b and q are the barrier height and elementary charge, respectively. Further, the responsivity ($\mathcal{R}$) is the ratio of generated photocurrent to the incident optical power of the radiation of frequency ν (h is Planck's constant) can be represented as:

$$\mathcal{R} = A\eta q/h\nu. \tag{14.3}$$

This relationship between responsivity and absorbance is depicted in Fig. 14.3 with varying incident wavelengths. The mathematical calculation is performed in the visible region for the structure reported in Ref. [34].

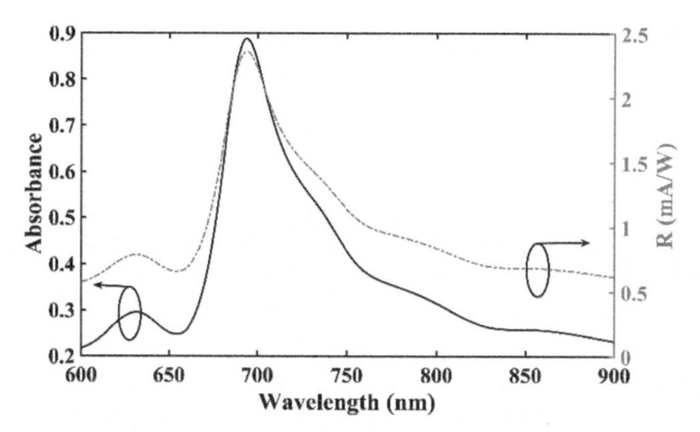

Figure 14.3 Simulated absorbance and responsivity variations with wavelength for Au grating-based SB photodetector structure [34].

Table 14.1 provides insight into the responsivity performance of some of the recently reported SB photodetectors. In addition to the responsivity value, the zero bias dark current density (J_{dark}) for an SB photodetector can be given as follows [37]:

$$J_{\text{dark}} = A^{*}T^{2}e^{-q\Phi_{b}/k_{B}T}. \tag{14.4}$$

Table 14.1 Responsivity values of some of the recently reported SB photodetectors

Ref.	SB structure	$\mathcal{R}$ (mA/W)
Gao et al. [8]	Porous Ag/TiO_2/Ti	3.3 (λ = 450 nm)
Sobhani et al. [13]	Au/Ti/Si	0.6 (λ = 1460 nm)
Chen et al. [23]	Au/n-Si/Al	0.013 (λ = 1550 nm)
Yang et al. [25]	Au/TiO_2/Al	0.36 (λ = 620 nm)
Shao et al. [26]	Au/TiO_2 cavities	2.0 (λ = 950 nm)
Sharma and Pandey [29]	W/Ni/Aln	6.164 (λ = 559.4 nm)
Matsui et al. [35]	Au/$SrTiO_3$/Si	0.23 (λ = 1350 nm)
Yang et al. [36]	Au/Si nanopillar	2.5 (λ = 1310 nm)
Sharma and Pandey [34]	Au/Ti/6H-SiC	4.396 (λ = 694.5 nm) 2.496 (λ = 1035 nm)

Here, A^* is the effective Richardson constant, k_B is the Boltzmann constant, the absolute temperature is T (usually 300 K), and the electronic charge is q. Moreover, another parameter, i.e., the detectivity (D^*) [8] is defined as

$$D^* = \frac{1}{\sqrt{2qJ_{\text{dark}}}}\mathcal{R}. \tag{14.5}$$

The experimental setup responsivity measurement is shown in Fig. 14.4 considering room temperature and dark environment [23]. For demonstration purposes, a wavelength-tunable laser source is utilized. TM-polarized light is generated by passing the laser light through a polarizer. A rotation stage is used generally for placing the photodetector with the grating grooves perpendicular to the plane of incidence. A stepper motor is used to control the rotation stage. The photocurrent generated is measured by a voltage-current monitor. A power meter is required at this stage for the measurement of laser power.

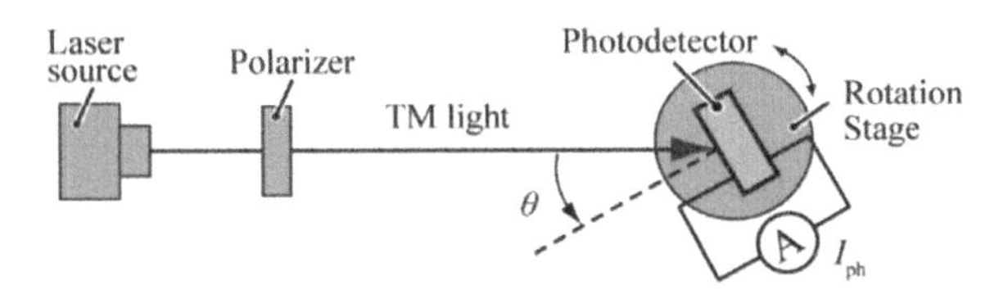

Figure 14.4 Experimental setup for measuring the responsivity matrix [23]. © 2016 Optica Publishing Group.

14.2.2 Au Gratings on SiC Substrate-Based SB Photodetector Description

Si is the most preferred substrate material for SB photodetector. Other substrate materials need to be explored for their application in optoelectronic devices. In this continuation, silicon carbide (SiC) substrate-based SB diodes have been demonstrated earlier [38–40]. SiC is a wide bandgap material (2.3–3.2 eV) [41] that possesses high-temperature applicability, saturation velocity, thermal conductivity, and fabrication compatibility [42]. In 2015, for the detection of UV radiation, a 4H-SiC-based SB photodetector with nickel silicide (Ni_2Si)-interdigitated contacts was reported [43]. In continuation, Ni/4H-SiC SB photodetectors (with sensitive areas of 1–4 cm^2) were experimentally demonstrated for pulsed X-rays/UV-light and alpha particles [44].

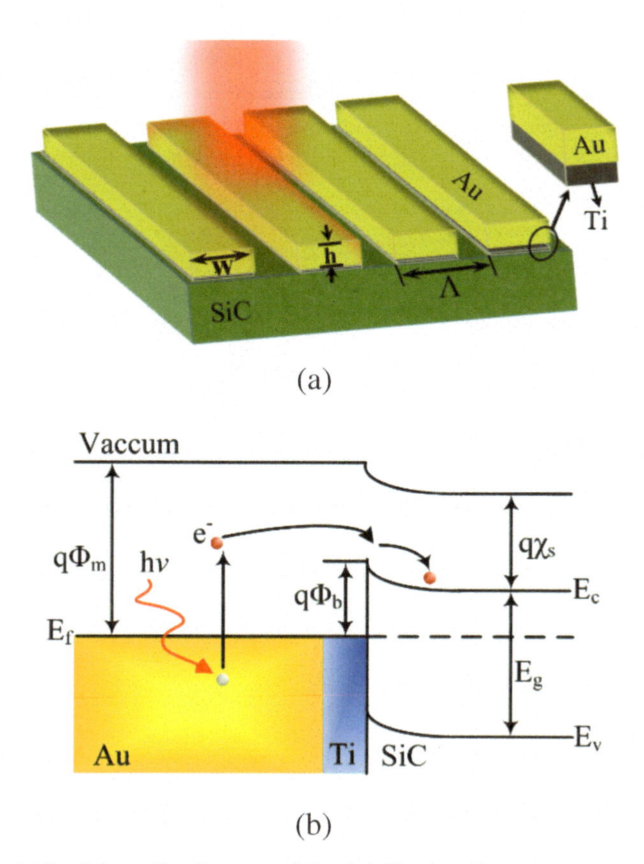

Figure 14.5 Schematic diagram of the (a) SB photodetector structure with 2 nm Ti layer as an adhesive interlayer and (b) energy band diagram. Reprinted from Ref. [34] with permission of IOP Science.

Recently, metallic grating (Au) enhanced SB photodetector with 6H-SiC substrate has been reported for photodetection application in visible and near-infrared (NIR) spectral regions [34]. The schematic of Au gratings over SiC substrate is illustrated in Fig. 14.5a with consideration of incident angle $\theta = 0°$. Air is the incident medium with a refractive index (RI) of 1.00. Here, the width, height, and periodicity variables of the grating structure are indicated by '*w*', '*h*', and '*Λ*', respectively. The variable values are w = 360 nm, h = 100 nm, and Λ = 400 nm. Figure 14.5b depicts the energy band diagram with the transfer process of the hot electron. Here the energy of incident photons is $h\nu$. Initially, non-radiative decay of plasmons, i.e., the hot electrons are generated by the incident photons (on optical absorption). These generated hot electrons are further diffused toward the Schottky interface. Further, these are transmitted to the conduction band of the semiconductor through IPE.

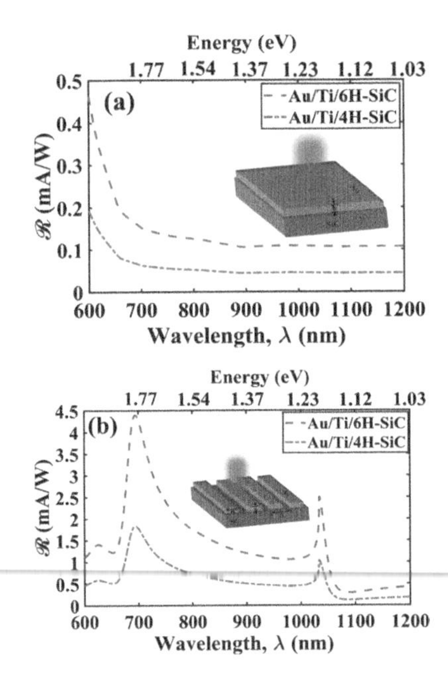

Figure 14.6 Variation of responsivity (in mA/W) with varying incident wavelength for Au/Ti/6H-SiC and Au/Ti/4H-SiC structures (a) without Au gratings (conventional), and (b) with Au gratings. Reprinted from Ref. [34] with permission of IOP Science.

Consider the case of a conventional SB photodetector structure with bulk metal film (without any grating or interdigital structure as shown in Fig. 14.6a), the $\mathcal{R}$ variation with wavelength is shown in Fig. 14.6a. In addition, the variation of $\mathcal{R}$ with λ for a grating-based structure is shown in Fig. 14.6b. Figure 14.6 signifies that a larger $\mathcal{R}$ (~10 times) is obtained in the case of grating-based photodetector structure as compared with that of conventional photodetector structure. Localized absorption losses [13] are responsible for absorbance enhancement in the case of grating-based structures. This enhancement contributes toward a significant increase in $\mathcal{R}$ value. For both the device structures, narrowband responses are obtained as shown in Fig. 14.6b, giving two prominent values of $\mathcal{R}$. The definition of a narrowband response is related to the full width at half maximum (FWHM < 100 nm) of the $\mathcal{R}$ curve variation with wavelength [45]. The narrowband response of a photodetector is responsible for the streamlined wavelength selective detection. The criteria for efficient detection of the photons (with energy $h\nu$) by SB photodetector: $\phi_b < h\nu < E_g$, E_g is the bandgap of semiconductor material (E_g= 3.2 eV for 4H-SiC and 3.0 eV for 6H-SiC) [41]. The highest $\mathcal{R}$ value of 4.396 mA/W is obtained for Au/Ti/6H-SiC at λ = 694.5 nm. In the case of Au/Ti/4H-SiC structure, a relatively smaller peak value of $\mathcal{R}$ (1.824 mA/W) is obtained (at λ = 694.5 nm) mainly due to a larger ϕ_b value. In addition, the $\mathcal{R}$ values of 2.496 mA/W (at λ = 1035 nm) and 1.039 mA/W (at λ = 1034 nm) are obtained in the case of 6H-SiC and 4H-SiC-based SB photodetectors, respectively, which correspond to the second peak in absorbance spectrum. Thus, an efficient photodetection can be achieved in the vicinity of two wavelengths (694.5 nm and 1034 nm).

The effective Richardson constant for 6H-SiC is ≅ 156 A cm^{-2} K^{-2} [46]. Considering room temperature (T = 300 K) and only the absorption area of unit cell structure (with Au grating length, l_g of 1 μm), $S = l_g \times w = 0.36$ μm^2, the dark current is 1.6×10^{-15} A. At a peak wavelength of 694.50 nm, the detectable power ($I_{dark}/\mathcal{R}$) and value of D^* are 0.363×10^{12} W and 1.162×10^{10} cm Hz$^{1/2}$/W, respectively, for Au/Ti/6H-SiC SB photodetector. This D^* value is in a comparable range when compared with the previously reported value of 9.80×10^{10} cm Hz$^{1/2}$/W at λ = 450 nm for Ag/TiO_2 SB photodetector [8] and as 3.311×10^{10} cm Hz$^{1/2}$/W at λ = 559.38 nm for W/Ni/AlN SB photodetector [29].

14.2.3 Effect of Grating Parameters on Absorbance (A)

As the responsivity behavior of the photodetector is directly related to the structure's absorbance, in this section, the grating parameter optimization is focused on the structure referred to in Ref. [34]. During the numerical analysis, a normal incident of TM-polarized light is taken into account. Figure 14.7a shows the absorbance variation with λ (600–1200 nm) for different values of h (80–120 nm). The variables w and Λ are fixed at 360 nm and 400 nm, respectively, with a Ti layer thickness of 2 nm. It is evident from Fig. 14.7a that peak absorbance shifted toward a longer wavelength (from λ = 690 nm to 710 nm) on increasing *h*. It can be marked that the maximum *A* value of 0.852 (at λ = 694.50 nm and *h* = 100 nm) is obtained for *h* values in the range of 90–110 nm. Thus, the preferable *h* value in view of both light absorption and thus electron collection capability can be opted in the vicinity of 100 nm with a fabrication tolerance of $\pm$10 nm.

In addition, the effect of *w* (from 300 nm to 380 nm) on *A* with λ with fixed values of Λ (400 nm) and *h* (100 nm) is presented in Fig. 14.7b. The highest normalized *A* value of 0.8729 is obtained for *w* = 360 nm. However, for the second wavelength peak in the NIR region, the peak absorbance shifts toward larger wavelengths, and the highest absorbance is attained at *w* = 380 nm. Thus, *w* value in the range of 360–380 nm can be the preferable choice.

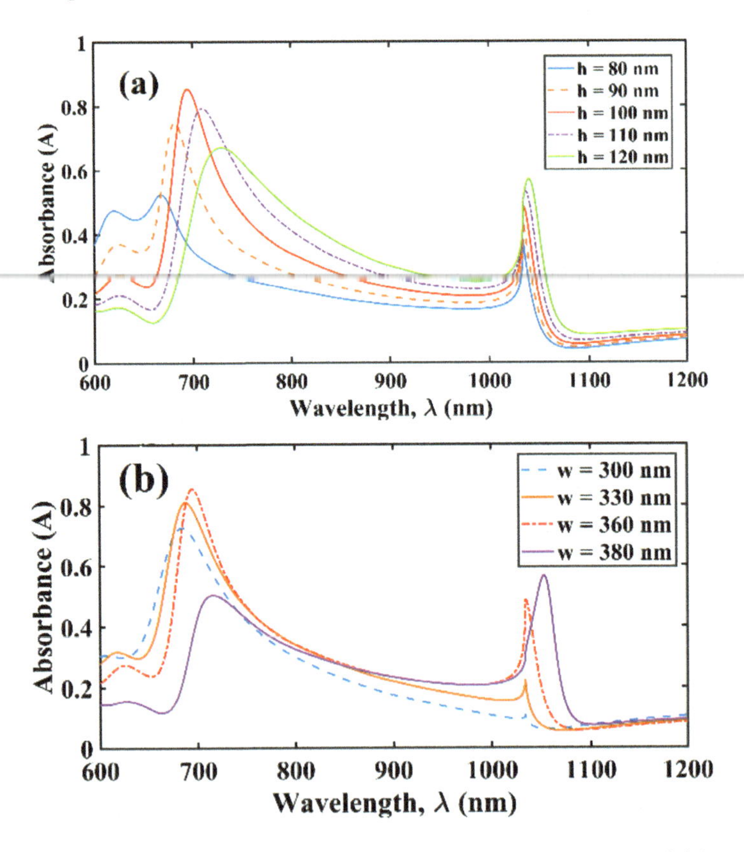

Figure 14.7 Variation of absorbance with λ for grating parameters (a) *h* and (b) *w*. Here, Λ = 400 nm and θ = 0°.

In continuation, the variation of A with λ for different Λ values is shown in Fig. 14.8 with fixed h (100 nm) and w (360 nm) values. The fill factor (FF) is defined as the ratio of w to Λ. Initially at Λ = 380 nm, the FF is high (~95%), which causes a weak plasmonic coupling [2]. This weak plasmonic coupling is an element responsible for a decrease in A value on increasing the fill factor. Moreover, no significant shift in peak values of A is observed (first peak) on increasing Λ. However, the second peaks in the NIR region are shifted toward longer λ. The highest A value of 0.852 (at λ = 694.5 nm) is obtained for Λ = 410 nm. Thus, Λ value of 410 nm can be opted with a fabrication tolerance value of ±10 nm.

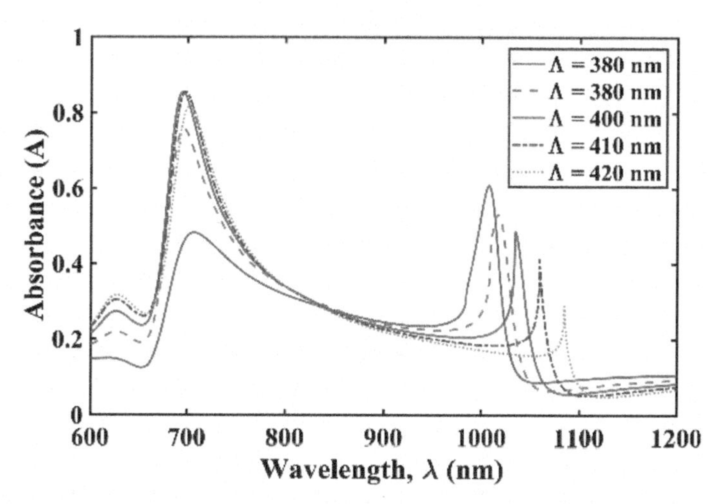

Figure 14.8 Variation of absorbance with λ for different Λ values, h = 100 nm, and w = 360 nm.

14.2.4 Fabrication Steps for Grating-Based SB Photodetector

The fabrication process for a grating-based SB photodetector is shown in Fig. 14.9 [23]. Initially, to remove native silicon dioxide, an n-type Si wafer was placed in a 1% hydrofluoric acid solution. Next, the wafer was coated with a photoresist using spin coating. The linear grating design was exposed to the photoresist. Further, a grating pattern was formed by developing the photoresist. Then the deposition of 100-nm-thick Au using an electron beam

(EB) evaporator was made on the photoresist grating design. In continuation, a stripper solution was used to remove the photoresist after the deposition of Au on the grating. Now, Au was lifted off to form Au grating over the n-Si. In order to completely cover the *n*-Si surface, the deposition of thick Au was done. For the cathode electrode, a thick aluminum (Al) film was deposited.

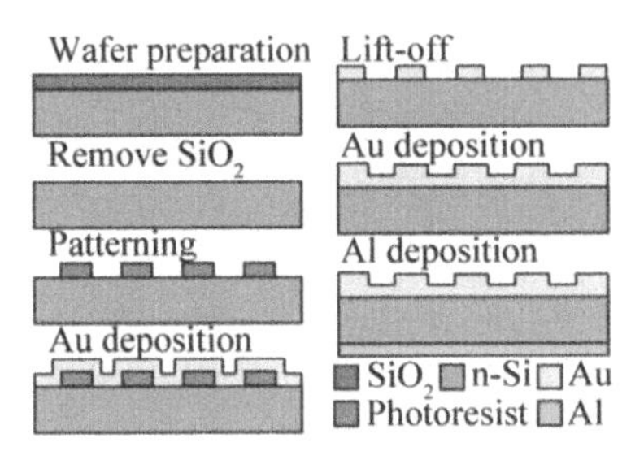

Figure 14.9 Fabrication process [23]. © 2016 Optica Publishing Group.

14.3 Conclusion and Future Aspects

Grating enhanced photodetection is focused on in this chapter emphasizing SB structure. The grating structure succeeded in achieving enhanced absorbance when compared with photodetector structures without grating. This absorbance enhancement leads toward responsivity increment. However, the dark current reduction is still a challenge. In SB photodetector, the dark current reduction is achieved by proper selection of work function (materials accordingly). Nowadays, 2D materials have witnessed a burgeoning application in optoelectronic devices. Graphene Schottky diodes have been experimentally demonstrated earlier [47]. More recently, graphene/Si SB-based photodetector has been reported for application at 633 nm [48]. In another work, MoS_2/Au SB-based photodetector has been presented in Ref. [49]. In view of these discussions, the future aspects will be the utilization of 2D materials and nanocomposites in grating-based photodetector for responsivity enhancement and dark current reduction. In addition, the photodetection can be extended toward a range of wavelengths as per the requirement by tuning various physical and chemical parameters (e.g., chemical potential in the case of graphene).

References

1. Maier S A (2004). *Plasmonics : Fundamentals and Applications*. Vol 677. doi:10.1016/j.aca.2010.06.020
2. Wu K, Zhan Y, Zhang C, Wu S, and Li X (2015). Strong and highly asymmetrical optical absorption in conformal metal-semiconductor-metal grating system for plasmonic hot-electron photodetection application. *Sci Rep*. 5(1):14304. doi:10.1038/srep14304
3. Watjen JI, Liu XL, Zhao B, and Zhang ZM (2017). A computational simulation of using tungsten gratings in near-field thermophotovoltaic devices. *J Heat Transfer*. 139(5):1–8. doi:10.1115/1.4035356
4. Sharma AK and Pandey AK (2020). Metal oxide grating based plasmonic refractive index sensor with Si layer in optical communication band. *IEEE Sens J*. 20(3):1275–1282. doi:10.1109/JSEN.2019.2947627
5. Tetz KA, Pang L, and Fainman Y (2006). High-resolution surface plasmon resonance sensor based on linewidth-optimized nanohole array transmittance. *Opt Lett*. 31(10):1528–1530. doi:10.1364/OL.31.001528
6. Shi L, Shang J, Liu Z, Li Y, Fu G, Liu X, Pan P, Luo H, and Liu G (2020). Ultra-narrow multi-band polarization-insensitive plasmonic perfect absorber for sensing. *Nanotechnology*. 31(46):465501. doi:10.1088/1361-6528/abad60
7. Tong J, Suo F, Ma J, Y. M. Tobing L, Qian L, and Hua Zhang D (2019). Surface plasmon enhanced infrared photodetection. *Opto-Electronic Adv*. 2(1):18002601–18002610. doi:10.29026/oea.2019.180026
8. Gao XD, Fei GT, Xu SH, Zhong BN, Ouyang HM, Li XH, and Zhang LD (2019). Porous Ag/TiO_2-Schottky-diode based plasmonic hot-electron photodetector with high detectivity and fast response. *Nanophotonics*. 8(7):1247–1254. doi:10.1515/nanoph-2019-0094
9. Hashemi M, Hosseini Farzad M, Mortensen NA, and Xiao S (2013). Enhanced plasmonic light absorption for silicon Schottky-barrier photodetectors. *Plasmonics*. 8(2):1059–1064. doi:10.1007/s11468-013-9509-y
10. Xiao H, Lo SC, Tai YH, Clark JK, Ho YL, Deng CZ, Wei PK, and Delaunay J-J (2020). Hot electron photodetection with spectral selectivity in the C-band using a silicon channel-separated gold grating structure. *Nano Express*. 1(1):010015. doi:10.1088/2632-959x/ab82e4
11. Chalabi H, Schoen D, and Brongersma ML (2014). Hot-electron photodetection with a plasmonic nanostripe antenna. *Nano Lett*. 14(3):1374–1380. doi:10.1021/nl4044373

12. Wang F and Melosh NA (2011). Plasmonic energy collection through hot carrier extraction. *Nano Lett.* 11(12):5426–5430. doi:10.1021/nl203196z

13. Sobhani A, Knight MW, Wang Y, Zheng B, King NS, Brown LV, Fang Z, Nordlander P, and Halas NJ (2013). Narrowband photodetection in the near-infrared with a plasmon-induced hot electron device. *Nat Commun.* 4:1643. doi:10.1038/ncomms2642

14. Lin K Te, Chen HL, Lai YS, and Yu CC (2014). Silicon-based broadband antenna for high responsivity and polarization-insensitive photodetection at telecommunication wavelengths. *Nat Commun.* 5:1–10. doi:10.1038/ncomms4288

15. Kim K, Yoon SJ, and Kim D (2006). Nanowire-based enhancement of localized surface plasmon resonance for highly sensitive detection : A theoretical study. *Opt Express.* 14(25):3737–3742.

16. Chou Chau YF, Chou Chao CT, Huang HJ, Rahimi Kooh MR, Kumara NTRN, Lim CM, and Chiang HP (2020). Perfect dual-band absorber based on plasmonic effect with the cross-hair/nanorod combination. *Nanomaterials.* 10(3):493. doi:10.3390/nano10030493

17. Mizutani A, Shinji U, and Kikuta H (2015). Highly sensitive refractive index sensor using a resonant grating waveguide on a metal substrate. *Appl Opt.* 54(13):4161. doi:10.1364/AO.54.004161

18. Tan CL, Lysak V V., Alameh K, and Lee YT (2010). Absorption enhancement of 980 nm MSM photodetector with a plasmonic grating structure. *Opt Commun.* 283(9):1763–1767. doi:10.1016/j.optcom.2009.11.044

19. Tan CL, Karar A, Alameh K, and Lee YT (2013). Optical absorption enhancement of hybrid-plasmonic-based metal-semiconductor-metal photodetector incorporating metal nanogratings and embedded metal nanoparticles. *Opt Express.* 21(2):1713. doi:10.1364/oe.21.001713

20. Zhou H, Zhang L, Tong J, Wu S, Son B, Chen Q, Zhang DH, and Tan CS (2021). Surface plasmon enhanced GeSn photodetectors operating at 2 μm. *Opt Express.* 29(6):8498. doi:10.1364/oe.420543

21. Dorodnyy A, Salamin Y, Ma P, Plestina JV, Lassaline N, Mikulik M, Gomez PR, Morrol AF, and Leuthold J (2018). Plasmonic photodetectors. *IEEE J Sel Top Quantum Electron.* 24(6):1–13. doi:10.1109/JSTQE.2018.2840339

22. Roh S, Chung T, and Lee B (2011). Overview of the characteristics of micro- and nano-structured surface plasmon resonance sensors. *Sensors.* 11(2):1565–1588. doi:10.3390/s110201565

23. Chen W, Kan T, Ajiki Y, Matsumoto K, and Shimoyama I (2016). NIR spectrometer using a Schottky photodetector enhanced by grating-based SPR. *Opt Express*. 24(22):25797. doi:10.1364/oe.24.025797

24. Yang Z, Liu M, Liang S, Zhang W, Mei T, Zhang D, and Chua SJ (2017). Hybrid modes in plasmonic cavity array for enhanced hot-electron photodetection. *Opt Express*. 25(17):20268. doi:10.1364/oe.25.020268

25. Yang Z, Du K, Lu F, Pang Y, Hua S, Gan X, Zhang W, Chua SJ, and Mei T (2019). Silica nanocone array as a template for fabricating a plasmon induced hot electron photodetector. *Photonics Res*. 7(3):294. doi:10.1364/prj.7.000294

26. Shao W, Yang Q, Zhang C, Wu S, and Li X (2019). Planar dual-cavity hot-electron photodetectors. *Nanoscale*. 11(3):1396–1402. doi:10.1039/c8nr05369c

27. Vahdani M, Yaraghi S, Neshasteh H, and Shahabadi M (2019). Narrow-band 4.3 μm plasmonic schottky-barrier photodetector for CO_2 sensing. *IEEE Sensors Lett*. 3(3):1–4. doi:10.1109/lsens.2019.2895968

28. Tanzid M, Ahmadivand A, Zhang R, Cerjan B, Sobhani A, Yazdi S, Norlander P, and Halas NJ (2018). Combining plasmonic hot carrier generation with free carrier absorption for high-performance near-infrared silicon-based photodetection. *ACS Photonics*. 5(9):3472–3477. doi:10.1021/acsphotonics.8b00623

29. Pandey AK and Sharma AK (2020). Schottky barrier photodetector utilizing tungsten grating nanostructure. *J Nanophotonics*. 14(04):035001–035007. doi:10.1117/1.JNP.14.046002

30. Sharma AK and Pandey AK (2019). Design and analysis of plasmonic sensor in communication band with gold grating on nitride substrate. *Superlattices Microstruct*. 130:369–376. doi:10.1016/j.spmi.2019.05.006

31. Fowler RH (1931). The analysis of photoelectric sensitivity curves for clean metals at various temperatures. *Phys Rev*. 38:107.

32. Scales C and Berini P (2010). Thin-film schottky barrier photodetector models. *IEEE J Quantum Electron*. 46(5):633–643. doi:10.1109/JQE.2010.2046720

33. Tagliabue G, Jermyn AS, Sundararaman R, Welch AJ, DuChene JS, Pala R, Davoyan AR, Narang P, and Atwater HA (2018). Quantifying the role of surface plasmon excitation and hot carrier transport in plasmonic devices. *Nat Commun*. 9(1). doi:10.1038/s41467-018-05968-x

34. Sharma AK and Pandey AK (2020). Au grating on SiC substrate: Simulation of high performance plasmonic Schottky barrier photodetector in visible and NIR regions. *J Phys D Appl Phys.* 53(17):175103. doi:10.1088/1361-6463/ab6fcd

35. Matsui T, Li Y, Hsu MHM, Merckling C, Oulton RF, Cohen LF, and Maier SA (2018). Highly stable plasmon induced hot hole transfer into silicon via a SrTiO3 passivation interface. *Adv Funct Mater.* 28(17):1–6. doi:10.1002/adfm.201705829

36. Yang Z, Du K, Wang H, Lu F, Pang Y, Wang J, Gan X, Zhang W, Mei T, and Chua SJ (2019). Near-infrared photodetection with plasmon-induced hot electrons using silicon nanopillar array structure. *Nanotechnology.* 30(7):075204. doi:10.1088/1361-6528/aaf4a6

37. Zhou Q, Wu H, Li H, Tang X, Qin Z, Dong D, Lin Y, Lu C, Qiu R, Zheng R, Wang J, and Li B (2019). Barrier inhomogeneity of schottky diode on nonpolar aln grown by physical vapor transport. *IEEE J Electron Devices Soc.* 7(April):662–667. doi:10.1109/JEDS.2019.2923204

38. Urushidani T, Kobayashi S, and Matsunami H (1993). High-voltage (> 1 kV) SiC schottky barrier diodes with low on-resistances. *IEEE Electron Device Lett.* 14(12):548–550. doi:10.1109/55.260785

39. Wang YH, Zhang YM, Zhang YM, Song QW, and Jia RX (2011). Al/Ti/4H-SiC Schottky barrier diodes with inhomogeneous barrier heights. *Chinese Phys B.* 20(8):087305. doi:10.1088/1674-1056/20/8/087305

40. Lee SK, Zetterling CM, and Östling M (2001). Schottky barrier height dependence on the metal work function for p-type 4H-silicon carbide. *J Electron Mater.* 30(3):242–246. doi:10.1007/s11664-001-0023-1

41. Chelnokov VE and Syrkin AL (1997). High temperature electronics using SiC: Actual situation and unsolved problems. *Mater Sci Eng B.* 46(1–3):248–253. doi:10.1016/S0921-5107(96)01990-3

42. Roccaforte F, La Via F, Di Franco S, and Raineri V (2003). Dual metal SiC Schottky rectifiers with low power dissipation. *Microelectron Eng.* 70(2–4):524–528. doi:10.1016/S0167-9317(03)00465-9

43. Lioliou G, Mazzillo MC, Sciuto A, and Barnett AM (2015). Electrical and ultraviolet characterization of 4H-SiC Schottky photodiodes. *Opt Express.* 23(17):21657. doi:10.1364/oe.23.021657

44. Liu LY, Wang L, Jin P, Liu J, Zhang XP, Chen L, Zhang JF, Ouyang XP, Liu A, Huang RH, and Bai S (2017). The fabrication and characterization of Ni/4H-SiC Schottky diode radiation detectors with a sensitive area of up to 4 cm^2. *Sensors (Switzerland).* 17(10). doi:10.3390/s17102334

45. Armin A, Jansen-van Vuuren RD, Kopidakis N, Burn PL, and Meredith P (2015). Narrowband light detection via internal quantum efficiency manipulation of organic photodiodes. *Nat Commun.* 6(1):6343. doi:10.1038/ncomms7343

46. Im HJ, Ding Y, Pelz JP, and Choyke WJ (2001). Nanometer-scale test of the Tung model of Schottky-barrier height inhomogeneity. *Phys Rev B - Condens Matter Mater Phys.* 64(7):753101–753109. doi:10.1103/physrevb.64.075310

47. Di Bartolomeo A (2016). Graphene Schottky diodes: An experimental review of the rectifying graphene/semiconductor heterojunction. *Phys Rep.* 606:1–58. doi:https://doi.org/10.1016/j.physrep.2015.10.003

48. Ji P, Yang S, Wang Y, Li K, Wang Y, Suo H, Woldu YT, Wang X, Wang F, Zhang L, and Jiang Z (2022). High-performance photodetector based on an interface engineering-assisted graphene/silicon Schottky junction. *Microsystems Nanoeng.* 8(1). doi:10.1038/s41378-021-00332-4

49. Li Z, Hu S, Zhang Q, Tian R, Gu L, Zhu Y, Yuan Q, Yi R, Li C, Liu Y, Hao Y, Gan X, and Zhao J (2022). Telecom-band waveguide-integrated MoS_2 photodetector assisted by hot electrons. *ACS Photonics.* Published online. doi:10.1021/acsphotonics.1c01622

Chapter 15

Graphene Plasmonic Mid-Infrared Photodetector

Junxiong Guo,[a] Lin Lin,[b] Jianbo Chen,[a] Shangdong Li,[b] and Yuhao He[b]

[a]*School of Electronic Information and Electrical Engineering, Chengdu University, No. 2025, Chengluo Avenue, Chengdu, 610106, China*

[b]*School of Electronic Science and Engineering, University of Electronic Science and Technology of China, No. 4, Section 2, Jianshe North Road, Chengdu 610054, China*

guojunxiong@cdu.edu.cn, linlin@std.uestc.edu.cn

Graphene plasmons can resonantly enhance the incident light absorption and offer an attractive feature of potential for tunable spectral selectivity for mid-infrared detection. It can be tuned by a passive method such as chemical doping and dimension of graphene, and it can be also tuned by an active method such as changing the applied voltages. So far, it has remained challenging to efficiently excite graphene plasmons and couple them to incident wavelengths in mid-infrared photodetectors. In the past few years, we have designed several types of graphene plasmonic mid-infrared

Plasmonics-Based Optical Sensors and Detectors
Edited by Banshi D. Gupta, Anuj K. Sharma, and Jin Li

ISBN 978-981-4968-85-0 (Hardcover), 978-1-003-43830-4 (eBook)
www.jennystanford.com

photodetectors with tunable selectivity and ultrahigh responsivity. In this book chapter, more details for developing those graphene plasmonic devices and the discussion of the tunable graphene plasmonic mechanism are performed, which can be very useful for the next-generation photodetectors.

15.1 Introduction

Photodetectors that can convert optical signals into electrical signals for postprocessing play an important part in photonic chips for various practical applications in information communication, astronomy observation, imaging, and macromolecule detection [1–3]. People have discovered and exploited several semiconductors for photodetection from the ultraviolet to far-infrared regions such as GaN, silicon, InGaAs, and HgCdTe [3–5]. For the mid-infrared region, these photodetectors are dissatisfactory in performance, resolution, and fabrication. Hence, it is highly significant to scale down the photodetector to the nanoscale with superior responsivity and detectivity for the next generation of mid-infrared photonic chips.

2D materials such as the graphene family and layered transition metal dichalcogenides (TMDs) have aroused great attention, due to their intriguing physical properties among photoelectricity and condensed matter physics [6–9]. Especially, graphene has unique characteristics, including semimetallic properties and graphene surface plasmon polariton (SPP) [10–14], contributing to the potential for tuning carrier density and ultrahigh absorption. However, the tuning approach for graphene carriers and the induction of graphene SPP are still complex and ineffective for practical applications.

Graphene SPP naturally depends on the physical properties of graphene itself. For the current problems, we introduce and conclude two tuning approaches of graphene SPP, containing periodical patterned graphene and doping graphene by periodical polarized ferroelectric domains, and its applications in the mid-infrared detector. We also introduce three kinds of mid-infrared photodetectors applying two tuning methods. (i) The photodetector is based on graphene etched to periodical strips one and partially covered by MoS_2 [15]. (ii) The photodetector is formed by graphene transferred on a strip-like polarized ferroelectric domains array [16]. (iii) The photodetector consists of graphene, which is transferred on periodical ring-shaped polarized ferroelectric superdomains [17]. These detectors all feature great responsivity and detectivity for mid-infrared detection up to 1×10^7 A/W and 6.2×10^{13} Jones (1 Jones = 1 cm $Hz^{1/2}$ W^{-1}), respectively. The tunable resonance wavelengths of three detectors cover from 4 to 52 μm. The mechanism of graphene SPP induction and the details of preparation and discussion are illustrated in this chapter.

15.2 Tuning of Graphene Plasmons

15.2.1 Patterned Graphene

Graphene [18, 19] is a flat sheet arranged by carbon atoms in a honeycomb lattice with an atom thick, and a patterned one is a graphene with a specific shape fabricated by etching or self-assembly. Benefiting from the specific shape of graphene, graphene SPP is effectively confined and induced under a free-space beam [20–22].

A normal and separate patterned graphene is not usually working to excite graphene SPP. In general, periodically patterned graphene is popular, such as periodical graphene ribbon grating. The excitation of graphene SPP in patterned graphene is much more relevant to the direction of polarized light when the shape of graphene is anisotropy. For periodical graphene ribbon grating, different performances are shown in Fig. 15.1a,b when the electric field is parallel and perpendicular to the graphene ribbon lines, respectively.

The dispersion relationship of graphene plasmon waves can be calculated by referring to the model of noble plasmon metals since the behavior of the electrons in graphene is similar to that in noble metals. Applying the Drude model, the dispersion relationship is shown in Eq. (15.1):

$$\beta(\omega)=\frac{\pi\hbar^2\varepsilon_0\left(\varepsilon_{r1}+\varepsilon_{r2}\right)}{e^2E_f}\left(1+\frac{i}{\omega\tau}\right)\omega^2, \tag{15.1}$$

where $\beta(\omega)$, $\hbar$, ε_0, ε_{r1}, ε_{r2}, e, E_f, ω, and τ represent the wavevector of the graphene SPP, reduced Planck constant, the dielectric constant of vacuum, the dielectric constant of the material above graphene, the dielectric constant of below the graphene, the elemental charge, the Fermi level of graphene, the radian frequency, and the carrier relaxation time, respectively [23]. To compensate for the wavevector difference by the period graphene ribbons, there is a condition need to be satisfied as Eq. (15.2):

$$\mathrm{Re}\big(\beta(\omega)\big)-\frac{\omega}{c}\sin\theta=\frac{2\pi}{P}, \tag{15.2}$$

where c and P represent the speed of light and the period of the graphene ribbons array, respectively [15]. Considering the incident wave is at a normal angle (θ = 0) from the top, the resonance frequency (ω_0) is obtained as Eq. (15.3) [15]:

$$\omega_0=\sqrt{\frac{2e^2E_\mathrm{f}}{\hbar^2\varepsilon_0\left(\varepsilon_{\mathrm{r}1}+\varepsilon_{\mathrm{r}2}\right)P}}. \tag{15.3}$$

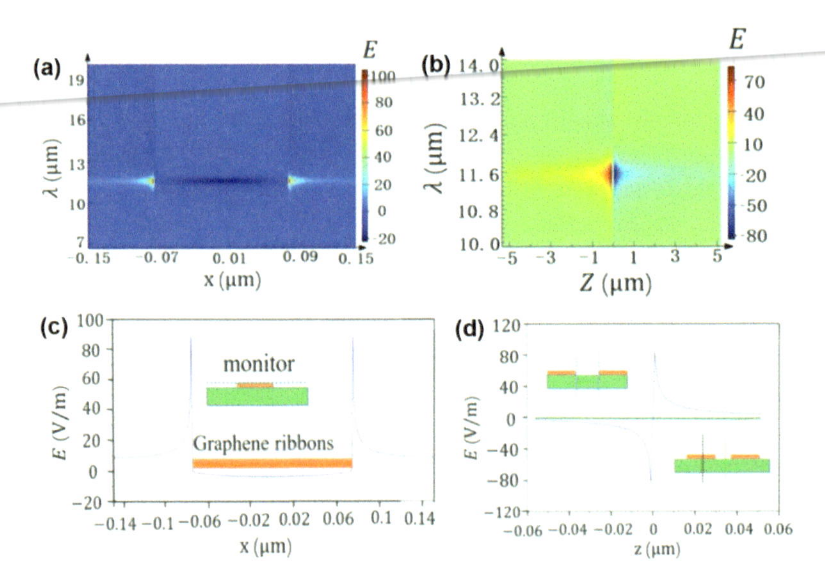

Figure 15.1 Plasmon excitation in graphene nanoribbon/MoS_2 heterostructure [15]. Electric field (a) parallel and (b) perpendicular to the graphene nanoribbons. (c, d) The position of the monitor and the corresponding electric field value in panels (a, b), respectively.

As a result, the Fermi level of graphene and the period of the graphene ribbons array are two changeable parameters, which approve the feasibility of tuning patterned graphene.

Here, we discuss the situation that the width of the graphene ribbon is 150 nm and the gap between the two ribbons is 150 nm as well. It can be observed that the electric field is intensively enhanced around the edge of graphene and SiO_2, which proves that the SPP is generated by patterned graphene instead of a large flat area of the graphene surface. In Fig. 15.1b, there is a flipped sign of the electric field appearing at the edge of graphene and SiO_2, which is typical behavior of localized plasmon resonance. As shown in Fig. 15.1c,d, the curves represent the intensities of the electric field at the position of dotted lines. In Fig. 15.1d, the intensity of the electric field flips at the edge of graphene and SiO_2 from –80 to 80 V/m with a short decay length of ~10 nm, which further demonstrates the path of graphene SPP propagation.

15.2.2 Ferroelectric Doping of Graphene

Concluding from the patterned graphene part, the position of graphene SPP is generally located at the edge of two different materials. Indeed, one material with different complex permittivity from that of intrinsic graphene and specific shape and period of patterned graphene matter a lot. Therefore, doping a pure flat graphene film with periodical specific areas is an inventive approach.

In the past, graphene doping methods are various, such as chemical vapor deposition, ion implantation, and oxidized graphene [24]. However, these methods can hardly achieve doping in the specific area and tunable doping. The ferroelectric material can dope graphene precisely and controllably, resulting from the tunable directions of ferroelectric polarization and the ultrathin domain boundary [25–27]. The chemical potential of graphene (μ_c) is almost equal to the Fermi level of graphene, which is directly influenced by the geometric equivalent capacitance (C_d) and the remanent polarization (P_r) of ferroelectric domains, as shown in Eq. (15.4) and Fig. 15.2a:

$$\frac{P_r}{C_d} = \frac{E_f}{e} + \phi, \tag{15.4}$$

where ϕ represents electrostatic potential difference [16, 17]. The remanent polarization of the ferroelectric domain depends on the intensity and direction of the applied external electric field. As a result, the chemical potential of graphene, which can decide the carrier concentration as shown in Eq. (15.5), can be tuned flexibly and influence the optical conductivity of graphene (σ_g) [28].

$$\mu_c \approx E_f \approx \hbar v_f \left(\pi N_0\right)^{1/2}, \tag{15.5}$$

where v_f and N_0 represent Fermi velocity and carrier concentration of graphene, respectively. For graphene, its optical conductivity can be divided into two parts, the interband (σ_{inter}) and the intraband (σ_{intra}), resulting from two kinds of carrier transitions, from the valance band to the conduction band and within the valance band or conduction band, as shown in Eq. (15.6):

$$\sigma_g = \sigma_{\text{intra}} + \sigma_{\text{inter}}. \tag{15.6}$$

The intraband and interband optical conductivities are calculated by the Falkovsky formula, as shown in Eq. (15.7) and Eq. (15.8):

$$\sigma_{\text{intra}} = \frac{2ie^2 k_B T}{\pi\hbar^2\left(\omega + i\tau^{-1}\right)} \ln\left[2\cosh\left(\frac{\mu_c}{2k_B T}\right)\right] \tag{15.7}$$

$$\sigma_{\text{inter}} = \frac{e^2}{4\hbar}\left[G\left(\frac{\hbar\omega}{2}\right) + i\frac{4\hbar\omega}{\pi}\int_0^\infty d\gamma \frac{G(\gamma) - G\left(\frac{\hbar\omega}{2}\right)}{(\hbar\omega)^2 - 4\gamma^2}\right], \tag{15.8}$$

where k_B and T represent the Boltzmann constant and temperature, respectively. $G(\gamma)$ in Eq. (15.8) is another function shown in Eq. (15.9) [16, 17]. The calculated imaginary part of graphene optical conductivity is shown in Fig. 15.2b.

$$G(\gamma) = \frac{\sinh\left(\frac{\gamma}{k_B T}\right)}{\cosh\left(\frac{\mu_c}{k_B T}\right) + \cosh\left(\frac{\gamma}{k_B T}\right)}. \tag{15.9}$$

The optical conductivity of graphene can be estimated in relationship to permittivity as shown in Eq. (15.10):

$$\varepsilon = \varepsilon_r + \frac{i\sigma}{\omega}, \tag{15.10}$$

where ε, ε_r, and σ represent the complex permittivity, the real part of permittivity, and the conductivity of the material, respectively [16, 17]. Therefore, doping graphene by ferroelectric domains is feasible and promising.

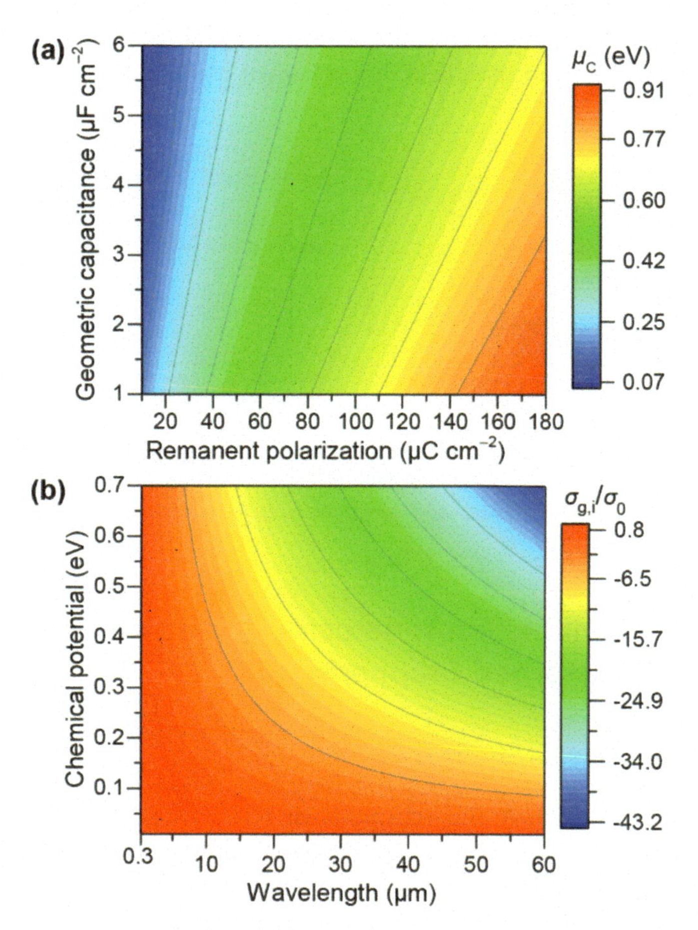

Figure 15.2 Doping of graphene by ferroelectric polarization [17]. (a) Ferroelectric polarization doped graphene as a function of the geometric capacitance and remanent polarization. (b) The imaginary part of the graphene conductivity is a function of the chemical potential and frequency at room temperature.

Alternately arranging up- and down-polarized ferroelectric strip domains is similar to the periodical patterned graphene ribbons. It is similar to periodical patterned graphene ribbons in the mechanism of graphene SPP inducing, but much more flexible and controllable. As shown in Fig. 15.3a,b, the graphene SPP is induced at the interface between graphene and down-polarized ferroelectric domains. Graphene on up-polarized ferroelectric domains is unable to propagate graphene plasmon, contributing to the highly confined graphene SPP in graphene upon down-polarized ferroelectric domains.

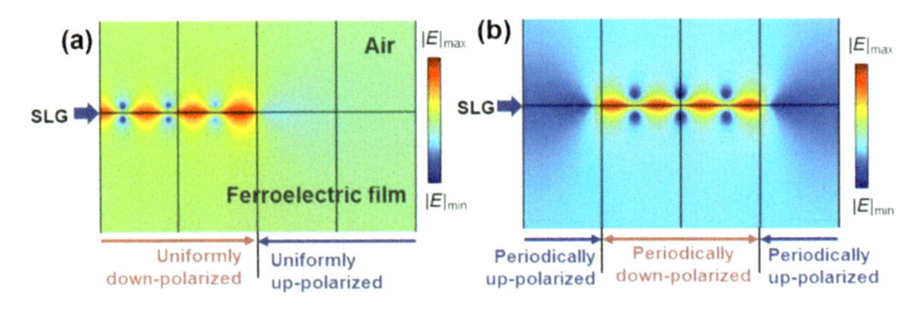

Figure 15.3 Graphene plasmon transformation in graphene/ferroelectric heterostructure [10]. (a) Uniformly up- and down-polarized ferroelectric domains. (b) Periodically up- and down-polarized ferroelectric domains.

15.3 Tunable Photodetectors Based on Patterned Graphene

15.3.1 Device Structure

Figure 15.4a shows the schematic of MoS_2/graphene plasmonic mid-infrared photodetectors by patterning graphene into periodical strips. A flat graphene film can be transferred on the Si/SiO_2 substrate through a wet process and then etched into graphene ribbons after the lithography mask. The MoS_2 film is exfoliated and transferred to graphene. Figure 15.4b shows the side view of this detector and the silicon at the bottom is prepared to apply voltage for tuning photoresponse. The MoS_2 and graphene are connected to the Pt and Ag electrodes, respectively, as shown in Fig. 15.4c.

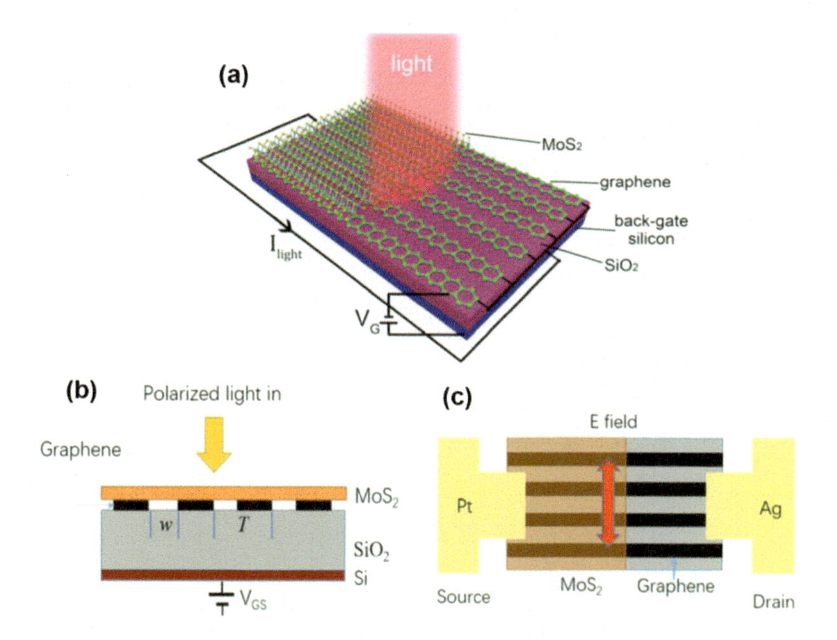

Figure 15.4 Schematic of the graphene ribbons/MoS_2 heterostructure photodetector [15]. (a) 3D view, (b) side view, and (c) top view of the device.

15.3.2 Tunable Photoresponse

According to the previous illustration, tuning the Fermi level of graphene and the width and period of the graphene ribbons array are two significant approaches.

In the initial situation, the period and the width of graphene ribbons are 300 and 150 nm, respectively. The chemical potential of graphene is ~0.01eV when the back-gate voltage is set to zero, and able to reach ~0.46 eV by applying back-gate voltage. The scatting rate is set to 0.001 eV. As a result, controlling the Fermi level of graphene from 0.46 to 0.3 eV with -0.02 eV step by changing the back-gate voltage, the wavelength of the absorption peak is changed from ~11.6 to ~14.5 μm, as shown in Fig. 15.5a,b. The absorptivity is up to 80%.

Further, to study the effect of the width of graphene ribbons on the photoresponse, the period of the graphene ribbons array is fixed as 300 nm and the scatting rate and chemical potential are set to 0.001 and 0.46 eV, respectively. Finally, the wavelength of the absorption peak is changed from ~9.4 to ~13.6 μm, according to the width of graphene ribbons from 0.1 to 0.17 μm, as shown in Fig. 15.5c,d.

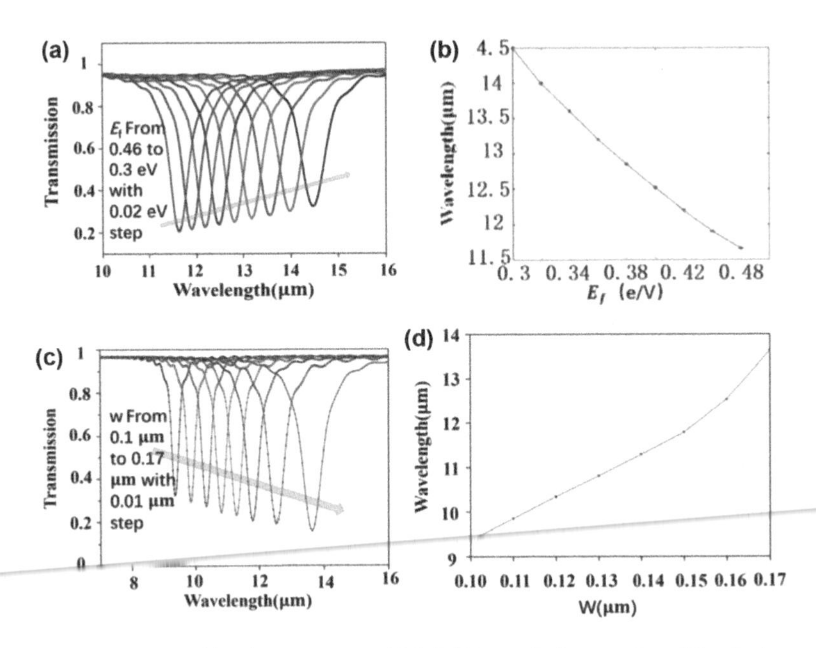

Figure 15.5 Tunable photoresponse [15]. (a) Transmission spectra of the device with different Fermi energy levels in graphene. (b) The resonant wavelength is a function of Fermi energy. (c) Transmission spectra with different ribbon widths *w*. (d) The resonant wavelength is a function of graphene nanoribbon widths.

15.3.3 Photodetection Performances

The patterned-graphene-based tunable photodetector is studied at room temperature, and applied with a free-space beam from the top at a normal angle. The absorption is vastly enhanced due to the induced SPP and the absorbed photons are transformed into electron-hole pairs, which are separated by the built-in electric field among patterned graphene and MoS_2. Enhanced by the strongly induced graphene SPP, the built-in electric field, and the high carrier mobility of both graphene and MoS_2, this photodetector displays superior performance. As the back-gate voltage changed from 0 to 50 V, the responsivity which is defined as photocurrent divided by the power of the incident beam, increased slowly up to 1×10^7 A/W, and the absorption shifts from ~16 to ~6.8 μm, as shown in Fig. 15.6a. It also features a stable photoresponse (1×10^7 to 5×10^6 A/W) ranging from 6 to 16 μm in Fig. 15.6b.

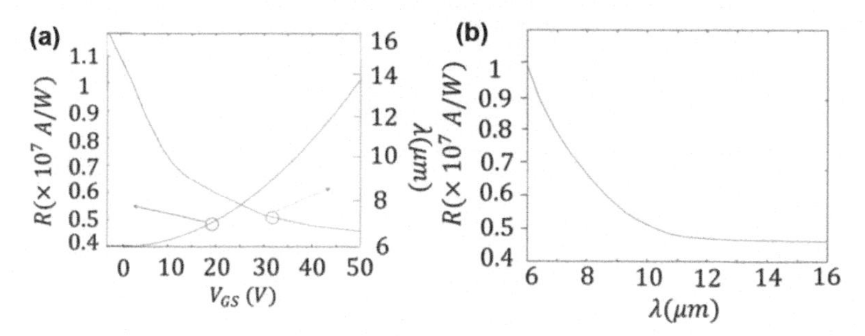

Figure 15.6 Photoresponsivity *R* of a device [15]. (a) The relationships among responsivity *R* (left), resonant wavelength (right), and gate voltage V_{GS}. (b) The relationships between responsivity *R* and wavelength.

15.4 Tunable Photodetectors Based on Ferroelectric-Doping Graphene

15.4.1 Device Structure

Figure 15.7 shows a schematic of graphene plasmonic mid-infrared photodetectors based on strip-like ferroelectric doping graphene. Compared to patterned graphene, graphene doping by ferroelectric film avoids etching graphene into a specific shape, which maintains the advanced performance and integrity of graphene. As shown in Fig. 15.7, the ferroelectric film ($BiFeO_3$) is epitaxial growth on the bottom electrode layer ($La_{1/3}Sr_{2/3}MnO_3$), which is prepared on the substrate ($SrTiO_3$). The periodical strip-like ferroelectric domain is switched into the direction perpendicular to the surface of the ferroelectric film, containing up-polarized domains and down-polarized domains by the piezoresponse force microscopy (PFM) technique. Graphene can be transferred to the ferroelectric film by a dry method or wet method.

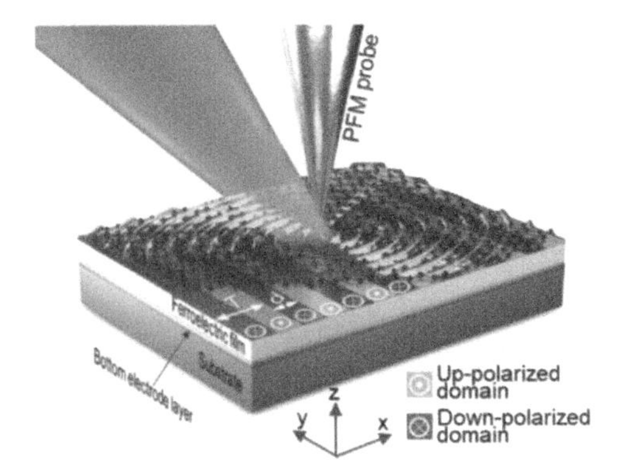

Figure 15.7 Schematic of a device based on graphene/periodic polarized ferroelectric domains with nanoribbon patterns [16].

15.4.2 Tunable Photoresponse

The tuning methods of ferroelectric doping graphene are mainly concluded as two aspects: the remanent polarization and the parameters according to the designed shapes.

Initially, we set the width of the down-polarized strip domain and the period as 25 and 50 nm, respectively. The chemical potential of graphene on up-polarized domains and down-polarized domains is 0.001 eV behaving as intrinsic graphene and 0.36 eV behaving as p-doped graphene, respectively. The chemical potential of graphene on down-polarized domains ranges from 0.12 to 0.64 eV by changing the remanent polarization, contributing to the absorption shifts from ~18 to ~5 μm, as shown in Fig. 15.8a.

Furthermore, in the condition that the widths of down-polarized strip domains and up-polarized stripe domains are the same and equal to half of the period, the various periods match different absorption bands. Figure 15.8b shows the wavelength of the absorption peak, which is shifted from ~6 to ~20.5 μm as the period changes from 20 to 500 nm.

15.4.3 Photodetection Performances

This tunable photodetector is based on strip-like ferroelectric doping graphene work at a temperature motivated by an incident beam at a normal angle. The graphene SPP is induced at the interface of graphene and down-polarized domains, which enormously enhances the absorption and performance of the device. In detail, a set the period as 50 nm, the responsivity increased rapidly up to ~8×10^6 A/W when the chemical potential of graphene raises to 0.64 eV, and the detectivity, which describes the ability to detect tiny signal increases to ~6.2×10^{13} Jones. As shown in Fig. 15.9a,b, this detector also features excellent responsivity and detectivity from ~1×10^6 to ~5.5×10^6A/W and from ~2.5×10^{13} to ~5.7×10^{13} Jones, respectively, according to the chemical potential of graphene setting as 0.36 eV and the decrease of the period from 500 to 20 nm.

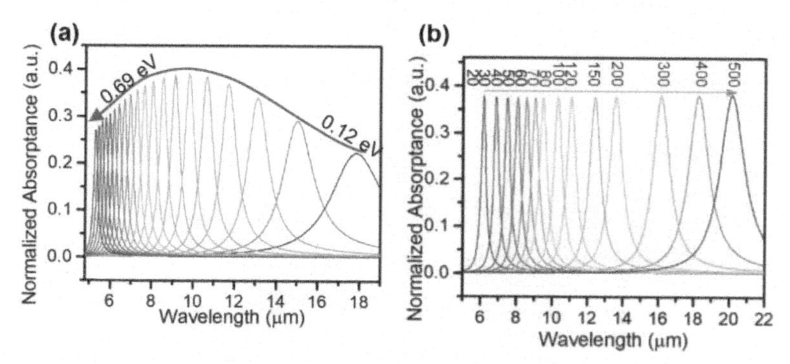

Figure 15.8 Tunable photoresponse of a device based on graphene/periodic polarized ferroelectric domains with nanoribbon patterns [16]. (a) Absorptance spectra as functions of ferroelectric-doped graphene Fermi levels and (b) period of ferroelectric polarization nanoribbons.

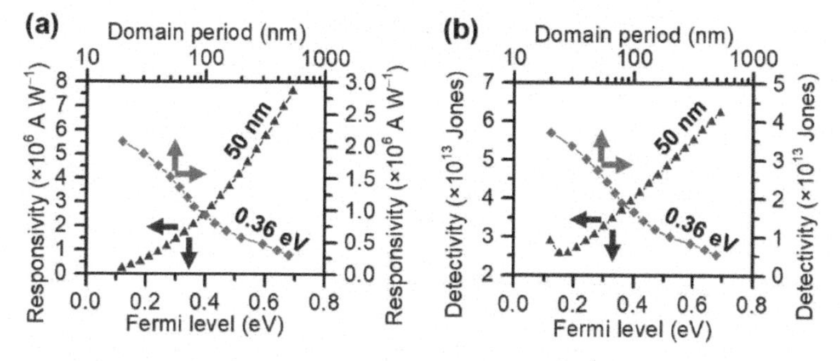

Figure 15.9 Photodetection performance of a device based on graphene/periodic polarized ferroelectric domains with nanoribbon patterns [16]. (a) Photoresponsivity and (b) detectivity as functions of ferroelectric-doped graphene Fermi levels and period of ferroelectric polarization nanoribbons.

15.5 Dual-Band Photodetectors Based on Ferroelectric-Superdomain-Doping Graphene

15.5.1 Device Structure

Figure 15.10 shows another mid-infrared photodetector based on graphene SPP induced by ferroelectric domains. With a similar structure containing substrate, bottom electrode, ferroelectric film, single layer graphene, and Au electrode, this mid-infrared photodetector features a dual-band absorption resulting from the periodical ring-shaped ferroelectric superdomains. There are two radiuses to determine the area of ring-shaped ferroelectric domains, which provide the possibility to match different resonance frequencies of graphene SPP.

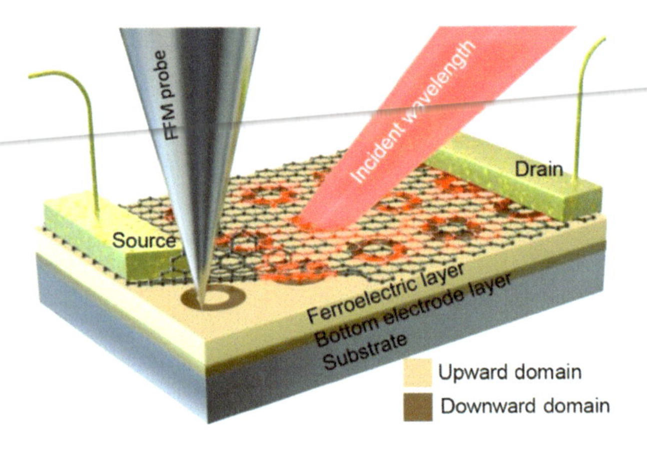

Figure 15.10 Schematic of a device based on graphene/periodic polarized ferroelectric domains with nanoring patterns [17].

15.5.2 Dual-Band Photoresponse

The radius of the inner and the outer circle of ring-shaped ferroelectric domains, the period of polarized domains, and the chemical potential of graphene are initially set as 20 nm, 30 nm, 200 nm, and 0.54 eV, respectively.

Figure 15.11 shows the transmission extinction spectra vary with the chemical potential of graphene on down-polarized ferroelectric domains ranging from 0.10 to 0.70 eV. The two peaks exhibit a similar blue shift from ~14 to ~6 μm and from ~45 to ~25 μm as the chemical potential of graphene increases, as shown in Fig. 15.11 and Fig. 15.12a. When increasing the unit of the ferroelectric domain proportionally, the two bands display the same redshifts, as shown in Fig. 15.12b. These two bands shift from ~4 to ~13 μm and from ~15 to ~52 μm as the period increases from 50 to 800 nm and two radiuses scale up.

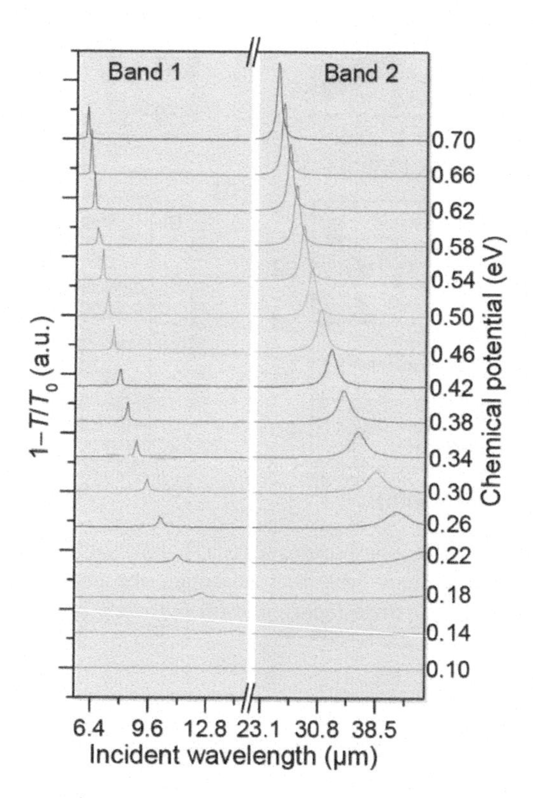

Figure 15.11 Tunable photoresponse with dual-band selectivity features [17].

To study the effect of two radiuses on resonance peaks, we fixed the chemical potential of graphene and the period as 0.54 eV and 200 nm, respectively. As the inner radius is set as two-thirds of the outer radius, the resonance wavelengths increase from ~10 to

~16 μm and ~21 to ~36 μm, according to the outer radius ranging from 15 to 90 nm, as shown in Fig. 15.12c. In addition, when the outer radius is fixed at 60 nm and the inner radius increases from 10 to 50 nm, the two resonance wavelengths feature opposite shifts from ~9 to ~5.5 μm and ~16 to ~40 μm, as shown in Fig. 15.12d.

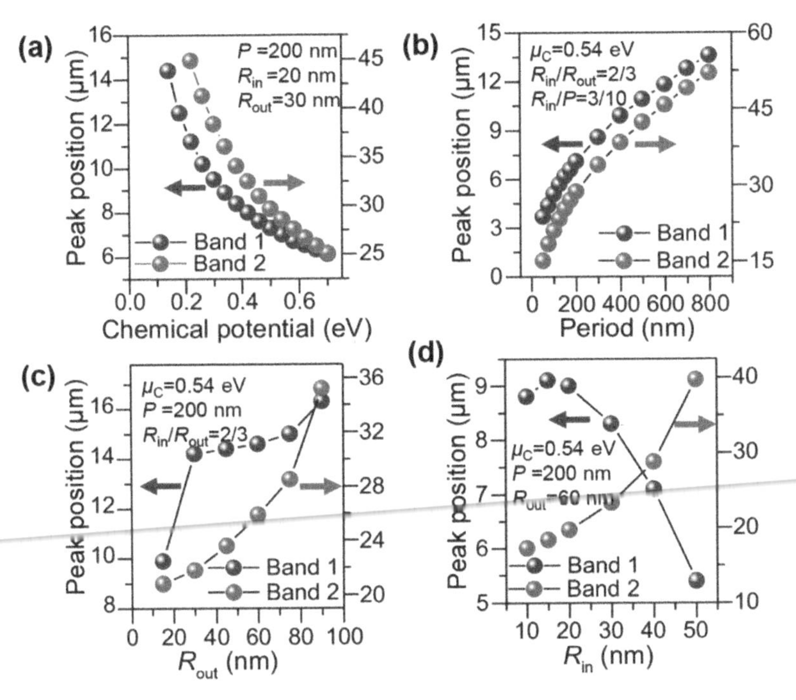

Figure 15.12 Tunable peak resonances [17]. (a) Peak resonance positions as functions of graphene chemical potential, (b) period of ferroelectric superdomain, (c) inner circle radius, and (d) out circle radius of ferroelectric superdomain.

15.5.3 Photodetection Performances

This tunable dual-band photodetector is based on superdomain ferroelectric doping graphene-inducing graphene SPP. The graphene SPP is induced around the edge of graphene upon ring-shaped down-polarized domains, which enormously improves absorption and photodetection. As shown in Fig. 15.13a, when the chemical potential, the period, the inner radius, and the outer radius are set as 0.54 eV, 200 nm, 40 nm, and 60 nm, respectively, the photocurrent density is sharply boosted to 667–1080 mA/cm^2, which exhibits great enhancement compared to that of single down-polarized ferroelectric domains (2.2 to 5.4 mA/cm^2). Figure 15.13b features the responsivity increased rapidly from 0.67×10^3 to 1.08×10^3 A/W, according to the resonance wavelength from 50 to 5 μm.

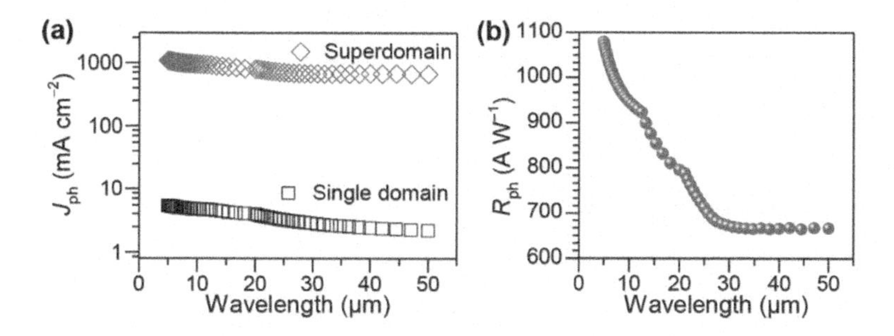

Figure 15.13 Photodetection performance of device based on graphene/periodic polarized ferroelectric domains with nanoring patterns [17]. (a) Photocurrent density (J_{ph}) as functions of incident wavelengths in devices based on integrating graphene with ferroelectric superdomain and single ferroelectric domain, respectively. (b) Photoresponsivity (R_{ph}) as a function of incident wavelengths in devices based on integrating graphene with ferroelectric superdomain.

15.6 Applications

Mid-infrared photodetectors can convert the optical signal into an electric one, which is much more convenient for postprocessing. These photodetectors are expected for integration to satisfy practical applications. The key parameters of our proposed photodetector are listed in Table 15.1.

These mid-infrared photodetectors are useful for infrared imaging and dynamic detection. As shown in Fig. 15.14a, each photodetector based on strip-like graphene plasmonic works as a pixel for infrared detection. Benefiting from the nanoscale of each photodetector and the tunable absorption from 5 to 20 μm, the application is the potential for infrared imaging with high resolution. Additionally, the tunable absorption and high detectivity provide the possibility for specific materials detection, as shown in Fig. 15.14b.

Table 15.1 Key parameters of graphene plasmonic mid-infrared photodetectors

Structure/ concept	Maximum photoresponsivity [A W^{-1}]	Maximum detectivity [Jones]	Wavelength tunability	Ref.
Graphene nanoribbons/MoS_2	1×10^7	N.A.	6~16 μm	[15]
Graphene/ ferroelectric stripe superdomain	8×10^6	6.2×10^{13}	5~20 μm	[16]
Graphene/ ferroelectric ring superdomain	1.08×10^3	NA	5.5~9 μm 16~40 μm	[17]

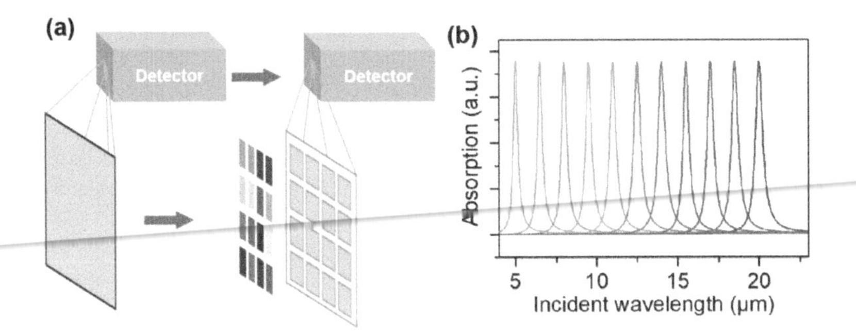

Figure 15.14 Proposed application of micro-spectrometer [16]. (a) The schematic of the graphene plasmonic photodetector array is driven by ferroelectric domains. (b) The target spectrum ranges from 5 to 20 μm of the proposed micro-spectrometer.

The dual-band photodetectors are also promising for infrared human body imaging and warning systems, resulting from one of the absorption bands located at ~10 μm, as shown in Fig. 15.15a,b. Another absorption band can reach up to ~52 μm, which is promising for terahertz applications. The dual-band photodetector is a natural multifunctional device suitable for different application situations. Furthermore, the dual-band photodetectors have the potential for parallel detection due to the obvious different photo responses according to two resonance wavelengths.

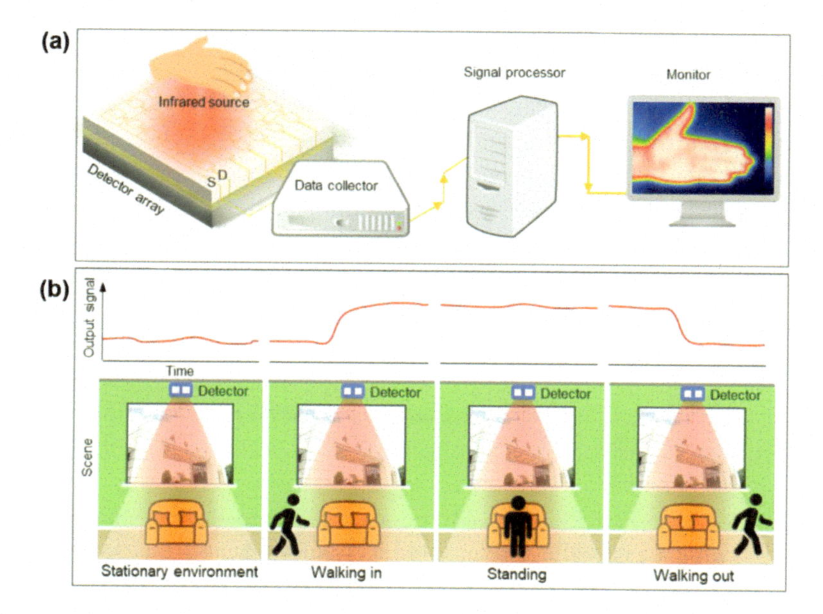

Figure 15.15 Proposed applications of dual-band mid-infrared photodetector [17]. (a) A thermal imaging system and (b) motional detection.

15.7 Summary

Tunable mid-infrared photodetectors with ultrahigh performance are significant for practical applications. We have successfully designed three mid-infrared photodetectors based on graphene plasmon with different tuning approaches. (i) The detector is based on graphene etched to periodical strips one and partially covered by MoS_2, which features ultrahigh absorption (80%) from 6 to 16 μm, and responsivity (1×10^7 A/W) due to the induced graphene SPP. (ii) A photodetector is formed by graphene transferred on a strip-like polarized ferroelectric domains array. The absorption band ranges from 5 to 20 μm and the responsivity and detectivity reach up to ~8×10^6 A/W and ~6.2×10^{13} Jones, respectively. (iii) Graphene is transferred on periodical ring-shaped polarized ferroelectric superdomains. The special polarized shape can match two different resonance wavelengths, which makes the detector possesses a wide absorption range (~4 to ~16 μm and ~15 to ~52 μm) and high responsivity (1.08×10^3 A/W). These photodetectors exhibit excellent responsivity and a wide absorption range with nanoscale structure, which indicates the potential in practical applications such as infrared imaging and warning system. Further, these tuning approaches not only promote the performance of mid-infrared detection but also provide a simple method for the smart fabrication of mid-infrared photodetectors.

References

1. Koppens, F. H. L., Mueller, T., Avouris, P., Ferrari, A. C., Vitiello, M. S., and Polini, M. (2014). Photodetectors based on graphene, other two-dimensional materials and hybrid systems, *Nat Nanotechnol*, **9**(10), pp. 780–793.
2. Xie, C., Mak, C., Tao, X., and Yan, F. (2017). Photodetectors based on two-dimensional layered materials beyond graphene, *Adv Funct Mater*, **27**, pp. 1603886.
3. Long, M., Wang, P., Fang, H., and Hu, W. (2019). Progress, challenges, and opportunities for 2D material based photodetectors, *Adv Funct Mater*, **29**(19), pp. 1803807.
4. Kinch, M. and Borrello, S. (1975). 0.1 ev HgCdTe photodetectors, *Infrared Phys*, **15**(2), pp. 111–124.
5. Sakaki, H., Noda, T., Hirakawa, K., Tanaka, M., and Matsusue, T. (1987). Interface roughness scattering in GaAs/AlAs quantum wells, *Appl Phys Lett*, **51**(23), pp. 1934–1936.
6. Novoselov, K., Mishchenko, A., Carvalho, A., and Neto, A. C. (2016). 2D materials and van der Waals heterostructures, *Science*, **353**(6298), pp. aac9439.
7. Mas-Balleste, R., Gomez-Navarro, C., Gomez-Herrero, J., and Zamora, F. (2011). 2D materials: to graphene and beyond, *Nanoscale*, **3**(1), pp. 20–30.
8. Manzeli, S., Ovchinnikov, D., Pasquier, D., Yazyev, O. V., and Kis, A. (2017). 2D transition metal dichalcogenides, *Nat Rev Mater*, **2**(8), pp. 17033.
9. Neto, A. C., Guinea, F., Peres, N. M., Novoselov, K. S., and Geim, A. K. (2009). The electronic properties of graphene, *Rev Mod Phys*, **81**(1), pp. 109.
10. Grigorenko, A. N., Polini, M., and Novoselov, K. S. (2012). Graphene plasmonics, *Nat Photonics*, **6**(11), pp. 749–758.

11. Bao, Q. and Loh, K. P. (2012). Graphene photonics, plasmonics, and broadband optoelectronic devices, *ACS Nano*, **6**(5), pp. 3677–3694.

12. Bao, Q., Hoh, H., and Zhang, Y. (2017). *Graphene Photonics, Optoelectronics, and Plasmonics,* 1st Edition, Jenny Stanford Publishing, Singapore.

13. Bonaccorso, F., Sun, Z., Hasan, T., and Ferrari, A. C. (2010). Graphene photonics and optoelectronics, *Nat Photonics*, **4**(9), pp. 611–622.

14. Low, T. and Avouris, P. (2014). Graphene plasmonics for terahertz to mid-infrared applications, *ACS Nano*, **8**(2), pp. 1086–101.

15. Liu, Y., Gong, T., Zheng, Y., Wang, X., Xu, J., Ai, Q., Guo, J., Huang, W., Zhou, S., Liu, Z., Lin, Y., Ren, T.-L., and Yu, B. (2018). Ultra-sensitive and plasmon-tunable graphene photodetectors for micro-spectrometry, *Nanoscale*, **10**(42), pp. 20013–20019.

16. Guo, J., Liu, Y., Lin, Y., Tian, Y., Zhang, J., Gong, T., Cheng, T., Huang, W., and Zhang, X. (2019). Simulation of tuning graphene plasmonic behaviors by ferroelectric domains for self-driven infrared photodetector applications, *Nanoscale*, **11**(43), pp. 20868–20875.

17. Guo, J., Lin, L., Li, S., Chen, J., Wang, S., Wu, W., Cai, J., Zhou, T., Liu, Y., and Huang, W. (2022). Ferroelectric superdomain controlled graphene plasmon for tunable mid-infrared photodetector with dual-band spectral selectivity, *Carbon*, **189**, pp. 596–603.

18. Novoselov, K. S., Fal'ko, V. I., Colombo, L., Gellert, P. R., Schwab, M. G., and Kim, K. (2012). A roadmap for graphene, *Nature*, **490**(7419), pp. 192–200.

19. Ferrari, A. C., Bonaccorso, F., Fal'ko, V., Novoselov, K. S., Roche, S., Bøggild, P., Borini, S., Koppens, F. H., Palermo, V., and Pugno, N. (2015). Science and technology roadmap for graphene, related two-dimensional crystals, and hybrid systems, *Nanoscale*, **7**(11), pp. 4598–4810.

20. Fei, Z., Goldflam, M. D., Wu, J. S., Dai, S., Wagner, M., Mcleod, A. S., Liu, M. K., Post, K. W., Zhu, S., and Janssen, G. C. (2015). Edge and surface plasmons in graphene nanoribbons, *Nano Lett*, **15**(12), pp. 8271.

21. Fei, Z., Rodin, A. S., Andreev, G. O., and Bao, W. (2012). Gate-tuning of graphene plasmons revealed by infrared nano-imaging, *Nature*, **487**(7405), pp. 82.

22. Ni, G. X., Wang, L., Goldflam, M. D., Wagner, M., Fei, Z., Mcleod, A. S., Liu, M. K., Keilmann, F., Özyilmaz, B., and Neto, A. H. C. (2016). Ultrafast optical switching of infrared plasmon polaritons in high-mobility graphene, *Nat Photonics*, **10**(4).

23. Jablan, M., Buljan, H., and Soljacic, M. (2009). Plasmonics in graphene at infrared frequencies, *Phys Rev B*, **80**(24), pp. 245435.

24. Shi, W., Kahn, S., Jiang, L., Wang, S.-Y., Tsai, H.-Z., Wong, D., Taniguchi, T., Watanabe, K., Wang, F., Crommie, M. F., and Zettl, A. (2020). Reversible writing of high-mobility and high-carrier-density doping patterns in two-dimensional van der Waals heterostructures, *Nat Electronics*, **3**(2), pp. 99–105.

25. Spaldin, N. A. and Ramesh, R. (2019). Advances in magnetoelectric multiferroics, *Nat Mater*, **18**(3), pp. 203–212.

26. Xiao, Z., Song, J., Ferry, D. K., Ducharme, S., and Hong, X. (2017). Ferroelectric-domain-patterning-controlled Schottky junction state in monolayer MoS_2, *Phys Rev Lett*, **118**(23), pp. 236801.

27. Baeumer, C., Saldana-Greco, D., Martirez, J. M., Rappe, A. M., Shim, M., and Martin, L. W. (2015). Ferroelectrically driven spatial carrier density modulation in graphene, *Nat Commun*, **6**, pp. 6136.

28. Das, A., Pisana, S., Chakraborty, B., Piscanec, S., Saha, S. K., Waghmare, U. V., Novoselov, K. S., Krishnamurthy, H. R., Geim, A. K., Ferrari, A. C., and Sood, A. K. (2008). Monitoring dopants by Raman scattering in an electrochemically top-gated graphene transistor, *Nat Nanotechnol*, **3**(4), pp. 210–215.

Index